SCHAUM'S OUTLINE OF

THEORY AND PROBLEMS

of

STRENGTH OF MATERIALS

•

BY

WILLIAM A. NASH, PH.D.

Professor of Engineering Mechanics

University of Florida

•

SCHAUM'S OUTLINE SERIES

McGRAW-HILL BOOK COMPANY

New York, St. Louis, San Francisco, Toronto, Sydney

46042

2 3 4 5 6 7 8 9 0 SHSH 8 2 1 0 6 9 8

Preface

Although some of the fundamentals of the statics of rigid bodies were known to the scientists of ancient Greece, no serious thought was given to the problem of the deformations of even the simplest structures until the time of the Renaissance. Then Leonardo da Vinci (1452-1519) and later Galileo (1564-1642) became interested in the statics of deformable bodies and in the mechanical properties of common engineering materials. Galileo's book "Two New Sciences" presented the first written discussion of the properties of structural materials and also the first treatment of the strength of beams. Although some of Galileo's conclusions do not agree with modern ideas, his early work stimulated considerable interest in this new field. In 1678 Robert Hooke (1635-1702) formulated his very famous and exceedingly simple relationship between force and deformation which was perhaps more influential than any other single factor in the development of the theory of Strength of Materials. Hooke's law of proportionality between deformation and force so greatly simplified mathematical analysis that thereafter progress in this field was quite rapid. Jacob Bernoulli (1654-1705) determined the differential equation of a laterally loaded bar, and later Leonard Euler (1707-1783) continued the study of bending action of beams and also investigated the buckling of an axially compressed bar. The first comprehensive discussion of the fiber stresses in a laterally loaded beam was presented in 1776 by Coulomb (1736-1806) and later this same author laid the foundations of the theory for the torsion of bars. Navier (1785-1836) further clarified the problem of bending of beams, and it might perhaps be said that Coulomb and Navier are largely responsible for the presentation of the material that today we call Strength of Materials.

Chronologically, the development of the science of Strength of Materials followed largely after the development of the laws of statics. Statics considered the external effects of a force acting on a body, i.e. the tendency of the forces to change the state of motion of the body. Strength of Materials treats the internal effects of the force, i.e. the state of deformation and stress set up within the boundaries of the body. Briefly, the science of Strength of Materials provides a more comprehensive explanation of the behavior of solids under load than the student has considered previously. Even so, there are many important problems that are beyond the scope of an undergraduate course on this topic and they are reserved for more sophisticated treatments offered in graduate courses under the names of Theory of Elasticity, Theory of Elastic Stability, Theory of Plasticity, Photoelasticity, Theory of a Continuous Media, and a host of other titles. The subject matter of many of these graduate courses is prerequisite to carrying out an ever-increasing number of intricate design problems for industry and is even more essential in research considerations.

This book is designed to supplement standard texts, primarily to assist students in acquiring a more thorough knowledge and proficiency in this basic field. The contents are divided into chapters covering duly-recognized areas of theory and study. Each chapter begins with a summary of the pertinent definitions, principles and theorems, followed by graded sets of solved and supplementary problems. Derivations of formulas and proofs of theorems are included among the solved problems. The problems have been chosen and solutions arranged so that the principles are clearly established. They serve to illustrate and amplify the theory, provide the repetition of basic principles so vital to effective teaching, and bring into sharp focus those fine points without which the student continually feels himself on unsafe ground.

The author is deeply indebted to his wife, Verna B. Nash, for her inspiration and continued assistance in proofreading and in the preparation of the manuscript. He is also indebted to Mr. Roy W. Gregory for painstaking work in the preparation of all drawings and to Mr. Henry Hayden for technical assistance and typographical arrangement. Particular thanks are extended to Prof. Odd Albert of the Polytechnic Institute of Brooklyn for many valuable suggestions and critical review of the entire manuscript.

WILLIAM A. NASH

Gainesville, Florida
September, 1957

Contents

CHAPTER 1

Tension and Compression

INTERNAL EFFECTS OF FORCES

In this book we shall be concerned primarily with what might be called the *internal* effects of forces acting on a body. The bodies themselves will no longer be considered to be perfectly rigid as was assumed in statics; instead, the calculation of the deformations of various shape bodies under a variety of loads will be one of the primary concerns of this study of strength of materials.

AXIALLY LOADED BAR. Perhaps the simplest case to consider at the start will be that of an initially straight metal bar of constant cross-section loaded at its ends by a pair of oppositely directed collinear forces coinciding with the longitudinal axis of the bar and acting through the centroid of each cross-section. For static equilibrium the magnitudes of the forces must be equal. If the forces are directed away from the bar, the bar is said to be in *tension*, if they are directed toward the bar, a state of *compression* exists. These two conditions are illustrated in Figure 1. Under the action of this pair of applied forces, internal resisting forces are set up within the bar and their characteristics may be studied by imagining a plane to be passed through the bar anywhere along its length and oriented perpendicular to the longitudinal axis of the bar. Such a plane is designated as *a-a* in Figure 2*a*. For reasons to be discussed later, this plane should not be "too close" to either end of the bar. If for purposes of

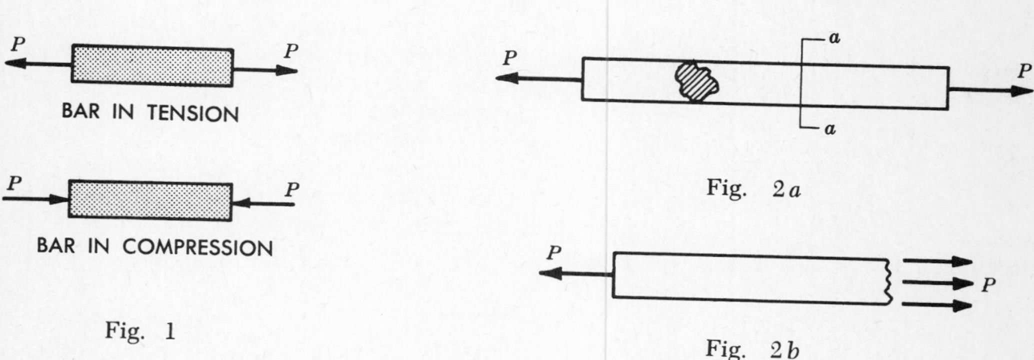

Fig. 1

Fig. 2*a*

Fig. 2*b*

analysis, the portion of the bar to the right of this plane is considered to be removed, as in Figure 2*b*, then it must be replaced by whatever effect it exerts upon the left portion. By this technique of introducing a cutting plane, the originally internal forces now become external with respect to the remaining portion of the body. For equilibrium of the portion to the left this "effect" must be a horizontal force of magnitude P. However, this force P acting normal to the cross-section *a-a* is actually the resultant of distributed forces acting over this cross-section in a direction normal to it.

DISTRIBUTION OF RESISTING FORCES. At this point it is necessary to make some assumption regarding the manner of variation of these distributed forces, and since the applied force *P* acts through the centroid it is commonly assumed that they are uniform across the cross-section. Such a distribution is probably never realized exactly because of the random orientation of the crystalline grains of which the bar is composed. The exact value of the force acting on some very small element of area of the cross-section is a function of the nature and orientation of the crystalline structure at that point. However, over the entire cross-section the variation is described with reasonable engineering accuracy by the assumption of a uniform distribution.

NORMAL STRESS. Instead of speaking of the internal force acting on some small element of area, it is perhaps of more significance and better for comparative purposes, to treat the normal force acting over a *unit* area of the cross-section. The intensity of normal force per unit area is termed the normal *stress* and is expressed in units of force per unit area, e.g. lb/in^2. The phrase *total stress* is sometimes used to denote the resultant axial force in pounds. If the forces applied to the ends of the bar are such that the bar is in tension, then *tensile stresses* are set up in the bar; if the bar is in compression we have *compressive stresses*. It is essential that the line of action of the applied end forces pass through the centroid of each cross-section of the bar.

TEST SPECIMENS. The axial loading shown in Figure 2*a* occurs frequently in structural and machine design problems. To simulate this loading in the laboratory, a test specimen is held in the grips of either an electrically driven gear type testing machine or a hydraulic machine. Both of these machines are commonly used in materials testing laboratories for applying axial tension.

In an effort to standardize materials testing techniques the American Society for Testing Materials, commonly abbreviated A.S.T.M., has issued specifications that are in common use throughout this country. More than a score of different type specimens are prescribed for various metallic and non-metallic materials for both axial tension and axial compression tests. For the present only two of these will be mentioned here, one for metal plates thicker than 3/16 in. and appearing as in Figure 3, the other for metals over 1.5 in. thick and having the appearance shown in Figure 4. The dimensions shown are those specified by the A.S.T.M. but the ends of the test specimens may be of any shape to fit the grips of the testing machine applying the axial load. As may be seen from these figures, the central portion of the specimen is somewhat smaller than the end regions so that failure will not take place in the gripped portion. The rounded fillets shown are provided so that no so-called stress concentrations will arise at the transition between the two lateral dimensions. Ordinarily, a standard gage length in which elongations are measured is marked by punching two very small holes into the surface of the bar either 2 in. or 8 in. apart as shown.

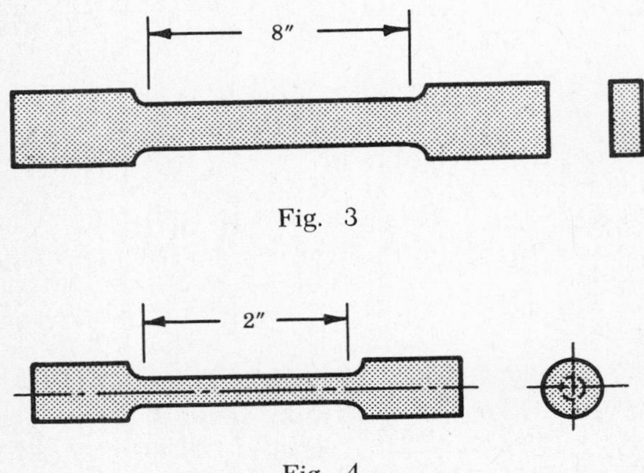

Fig. 3

Fig. 4

NORMAL STRAIN. Let us suppose that one of these tension specimens has been placed in a tension-compression testing machine and tensile forces gradually applied to the ends. The total elongation over the gage length may be measured at any predetermined increments of the axial load by a mechanical strain gage and from these values the elongation per unit length, which is termed *normal strain* and denoted by ϵ, may be found merely by dividing the total elongation Δ by the gage length L, i.e. $\epsilon = \Delta/L$. The strain is usually expressed in units of inches per inch and consequently is dimensionless. The phrase *total strain* is sometimes used to denote the elongation in inches.

STRESS-STRAIN CURVE. As the axial load is gradually increased in increments the total elongation over the gage length is measured at each increment of load and this is continued until fracture of the specimen takes place. Knowing the original cross-sectional area of the test specimen the *normal stress*, denoted by s, may be obtained for any value of the axial load merely by the use of the relation

$$s = \frac{P}{A}$$

where P denotes the axial load in pounds, and A the original cross-sectional area. Having numerous pairs of values of normal stress s and normal strain ϵ, the experimental data obtained may be plotted with these quantities considered as ordinate and abscissa respectively. This is a *stress-strain diagram* of the material for this type of loading. Such a diagram may take many widely different forms, but shown in Figure 5 are several typical plots for common engineering materials. For a metal such as low-carbon structural steel the data will plot approximately as shown in Figure 5*a*, for a so-called brittle material such as cast iron the plot appears as in Figure 5*b*, while for rubber the diagram of 5c is typical.

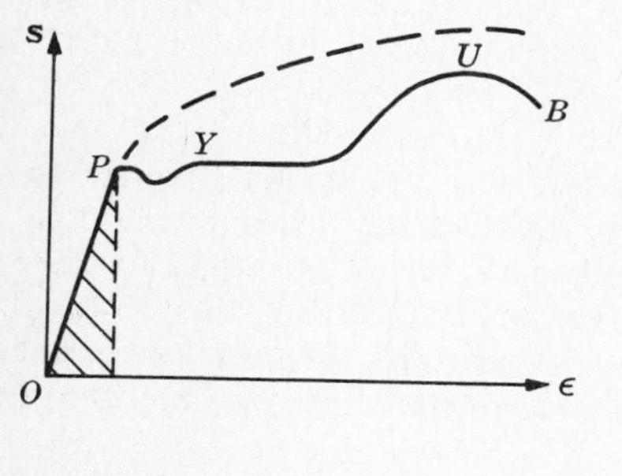

Fig. 5*a*

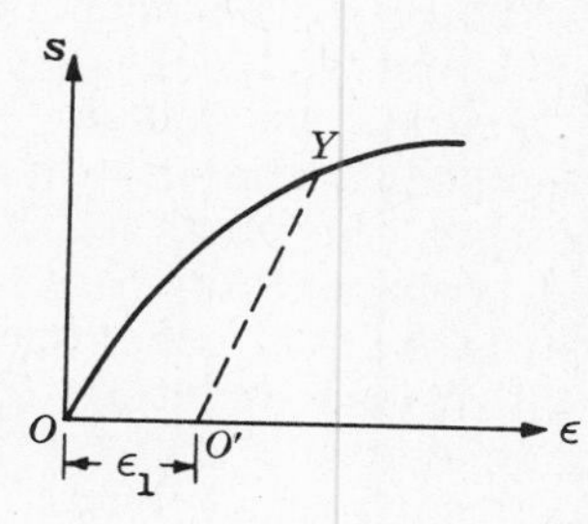

Fig. 5*b*

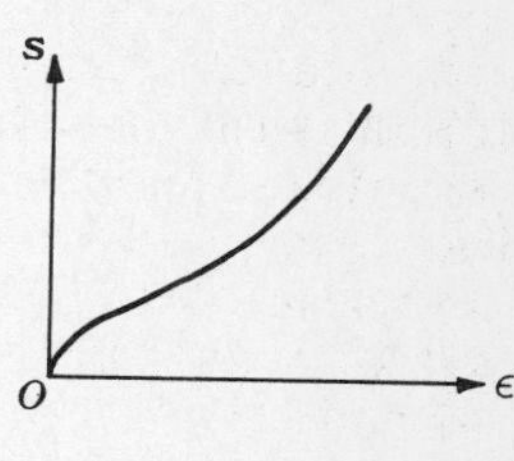

Fig. 5*c*

DUCTILE AND BRITTLE MATERIALS. Metallic engineering materials are commonly classed as either *ductile* or *brittle* materials. A *ductile material* is one having a relatively large tensile strain up to the point of rupture (for example, structural steel or aluminum) whereas a *brittle material* has a relatively small strain up to this same point. An arbitrary strain of 0.05 in/in is frequently taken as the dividing line between these two classes of materials. Cast iron and concrete are examples of brittle materials.

HOOKE'S LAW. For a material whose stress-strain curve is similar to that shown in Figure 5*a*, it is evident that the relation between stress and strain is linear for comparatively small values of strain. This linear relation between elongation and the axial force causing it (since each of these quantities differs only by a constant from the strain or the stress) was first noticed by Sir Robert Hooke in 1678 and bears the name of *Hooke's Law*. To describe this initial linear range of action of the material we may consequently write

$$s = E\epsilon$$

where E denotes the slope of the straight-line portion (OP) of the stress-strain curve in Figure 5*a*.

MODULUS OF ELASTICITY. The quantity E, *i.e.* the ratio of the unit stress to the unit strain, is the *modulus of elasticity* of the material in tension, or, as it is often called, *Young's Modulus*. Values of E for various engineering materials are tabulated in handbooks. Since the unit strain ϵ is a pure number (being a ratio of two lengths) it is evident that E has the same units as does the stress, for example lb/in^2. For many common engineering materials the modulus of elasticity in compression is very nearly equal to that found in tension. It is to be carefully noted that the behavior of materials under load as discussed in this book is restricted (unless otherwise stated) to this linear region of the stress-strain curve.

MECHANICAL PROPERTIES OF MATERIALS

The stress-strain curve shown in Figure 5*a* may be used to characterize several strength characteristics of the material. They are:

PROPORTIONAL LIMIT. The ordinate to the point P is known as the *proportional limit*, i.e. the maximum stress that may be developed during a simple tension test such that the stress is a linear function of strain. For a material having the stress-strain curve shown in Figure 5*b* there is no proportional limit.

ELASTIC LIMIT. The ordinate to a point almost coincident with P is known as the *elastic limit*, i.e., the maximum stress that may be developed during a simple tension test such that there is no permanent or residual deformation when the load is entirely removed. For many materials the numerical values of the elastic limit and the proportional limit are almost identical and the terms are sometimes used synonymously. In those cases where the distinction between the two values is evident the elastic limit is almost always greater than the proportional limit.

ELASTIC RANGE. That region of the stress-strain curve extending from the origin to the proportional limit.

PLASTIC RANGE. That region of the stress-strain curve extending from the proportional limit to the point of rupture.

YIELD POINT. The ordinate to the point Y at which there is an increase in strain with no increase in stress is known as the *yield point* of the material. After loading has progressed to the point Y, yielding is said to take place. Some materials exhibit two points on the stress-strain curve at which there is an increase of strain without an increase of stress. These are called *upper* and *lower yield points*.

ULTIMATE STRENGTH OR TENSILE STRENGTH. The ordinate to the point U, the maximum ordinate to the curve, is known either as the *ultimate strength* or the *tensile strength* of the material.

BREAKING STRENGTH. The ordinate to the point B is called the *breaking strength* of the material.

MODULUS OF RESILIENCE. The work done on a unit volume of material, as a simple tensile force is gradually increased from zero to such a value that the proportional limit of the material is reached, is defined as the *modulus of resilience*. This may be calculated as the area under the stress-strain curve from the origin up to the proportional limit and is represented as the shaded area in Figure 5*a*. The units of this quantity are in-lb/in^3. Thus, resilience of a material is its ability to absorb energy in the elastic range.

MODULUS OF TOUGHNESS. The work done on a unit volume of material as a simple tensile force is gradually increased from zero to the value causing rupture is defined as the *modulus of toughness*. This may be calculated as the entire area under the stress-strain curve from the origin to rupture. Toughness of a material is its ability to absorb energy in the plastic range of the material.

PERCENTAGE REDUCTION IN AREA. The ratio of the decrease in cross-sectional area from the original area upon fracture divided by the *original* area and multiplied by 100 is termed *percentage reduction in area*. It is to be noted that when tensile forces act upon a bar, the cross-sectional area decreases but calculations for the normal stress are usually made upon the basis of the original area. This is the case for the curve shown in Figure 5*a*. As the strains become increasingly larger it is more important to consider the instantaneous values of the cross-sectional area (which are decreasing), and if this is done the *true* stress-strain curve is obtained. Such a curve has the appearance shown by the dotted line in Figure 5*a*.

PERCENTAGE ELONGATION. The ratio of the increase in length (of the gage length) after fracture divided by the initial length, multiplied by 100 is the *percentage elongation*. Both the percentage reduction in area and the percentage elongation are considered to be measures of the *ductility* of a material.

WORKING STRESS. The above-mentioned strength characteristics may be used to select a so-called *working stress*. Throughout this book all working stresses will be within the elastic range of the material. Frequently such a stress is determined merely by dividing either the stress at yield or the ultimate stress by a number termed the *safety factor*. Selection of the safety factor is based upon the designer's judgment and experience. Specific safety factors are sometimes specified in building codes. See Problems 4, 12, 13.

The non-linear stress-strain curve of a brittle material, shown in Figure 5*b*, characterizes several other strength measures that cannot be introduced if the stress-strain curve has a linear region. They are:

YIELD STRENGTH. The ordinate to the stress-strain curve such that the material has a predetermined permanent deformation or "set" when the load is removed is called the *yield strength* of the material. The permanent set is often taken to be either 0.002 or 0.0035 in. per in. These values are of course arbitrary. In Figure 5*b* a set ϵ_1 is denoted on the strain axis and the line $O'Y$ is drawn parallel to the initial tangent to the curve. The ordinate to Y represents the yield strength of the material, sometimes called the *proof stress*.

TANGENT MODULUS. The slope of the tangent to the stress-strain curve at the origin of the plot is known as the *tangent modulus* of the material.

There are other characteristics of a material that are useful in design considerations. They are:

COEFFICIENT OF LINEAR EXPANSION. This is defined as the change of length per unit length of a straight bar subject to a temperature change of one degree. The value of this coefficient is independent of the unit of length but does depend upon the temperature scale used. Usually we will consider the Fahrenheit scale, in which case the coefficient denoted by α is given for steel, for instance, as 6.5×10^{-6} per F°. Temperature changes in a structure give rise to internal stresses just as do applied loads. See Problems 5 and 8.

POISSON'S RATIO. When a bar is subject to a simple tensile loading there is an increase in length of the bar in the direction of the load, but a decrease in the lateral dimensions perpendicular to the load. The ratio of the strain in the lateral direction to that in the axial direction is defined as *Poisson's Ratio*. It is denoted in this book by the Greek letter μ. For most metals it lies in the range 0.25 to 0.35. See Problems 16, 17, 18, 19, 20,

GENERAL FORM OF HOOKE'S LAW. The simple form of Hooke's Law has been given for axial tension when the loading is entirely along one straight line, i.e. uni-axial. Only the deformation in the direction of the load was considered and it was given by

$$\epsilon = \frac{s}{E}$$

In the more general case an element of material is subject to three mutually perpendicular normal stresses s_x, s_y, s_z, which are accompanied by the strains ϵ_x, ϵ_y, ϵ_z respectively. By superposing the strain components arising from lateral contraction due to Poisson's effect upon the direct strains we obtain the general statement of Hooke's Law:

$$\epsilon_x = \frac{1}{E}[s_x - \mu(s_y + s_z)]$$

$$\epsilon_y = \frac{1}{E}[s_y - \mu(s_x + s_z)]$$

$$\epsilon_z = \frac{1}{E}[s_z - \mu(s_x + s_y)]$$

See Problems 17 and 20.

CLASSIFICATION OF MATERIALS

This entire discussion has been based upon the assumptions that two characteristics prevail in the material. They are that we have a:

HOMOGENEOUS MATERIAL, one having the same elastic properties (E, μ) at all points in the body, and an

ISOTROPIC MATERIAL, one having the same elastic properties in all directions at any one point of the body. Not all materials are isotropic. If a material does not possess any kind of elastic symmetry it is called *anisotropic*, or sometimes *aeolotropic*. Instead of having two independent elastic constants (E, μ) as an isotropic material does, such a substance has 21 elastic constants. If the material has three mutually perpendicular planes of elastic symmetry it is said to be *orthotropic*. The number of independent constants is 9 in this case. This book considers only the analysis of isotropic materials.

SOLVED PROBLEMS

1. Determine the total elongation of an initially straight bar of length L, cross-sectional area A, and modulus of elasticity E if a tensile load P acts on the ends of the bar.

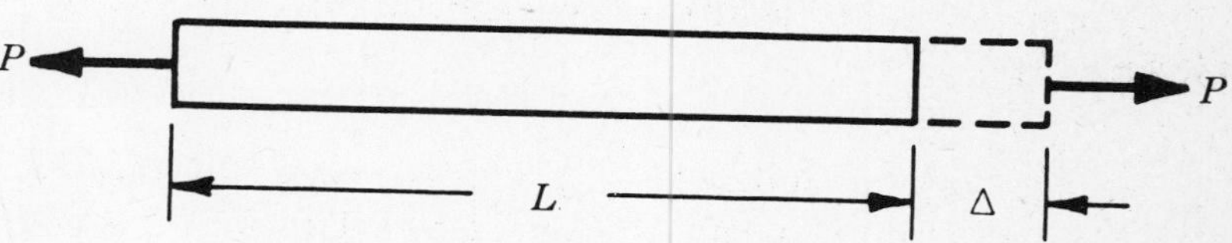

The unit stress in the direction of the force P is merely the load divided by the cross-sectional area, i.e., $s = P/A$. Also the unit strain ϵ is given by the total elongation Δ divided by the original length, i.e. $\epsilon = \Delta/L$. By definition the modulus of elasticity E is the ratio of s to ϵ, i.e.,

$$E = \frac{s}{\epsilon} = \frac{P/A}{\Delta/L} = \frac{PL}{A\Delta} \quad \text{or} \quad \Delta = \frac{PL}{AE}$$

Note that Δ has the units of length, perhaps in. or ft.

2. A surveyors' steel tape 100 ft long has a cross-section of 0.250 in. by 0.03 in. Determine the elongation when the entire tape is stretched and held taut by a force of 12 lb. The modulus of elasticity is $30 \cdot 10^6$ lb/in^2.

$$\text{Elongation } \Delta = \frac{PL}{AE} = \frac{(12)(100 \cdot 12)}{(0.250)(0.03)(30 \cdot 10^6)} = 0.0640 \text{ in.}$$

3. A steel bar of cross-section 1 in^2 is acted upon by the forces shown in Figure (*a*). Determine the total elongation of the bar. For steel, $E = 30 \cdot 10^6$ lb/in^2.

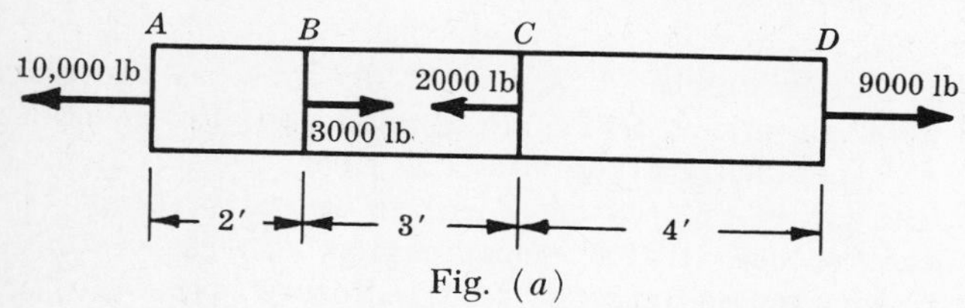

Fig. (*a*)

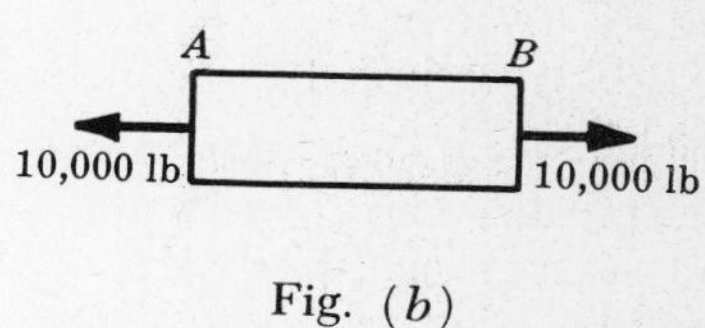

Fig. (*b*)

The entire bar is in equilibrium, hence all portions of it are also. The portion of the bar between A and B has a resultant force of 10,000 lb acting over every cross-section, hence a free-body diagram of this 2 ft length appears as in Figure (*b*) above. The force at the right end of this segment must be 10,000 lb to maintain equilibrium with the applied force at the left end. The elongation of this portion is given by

$$\Delta_1 = \frac{PL}{AE} = \frac{10{,}000(24)}{(1)(30 \cdot 10^6)} = 0.0080 \text{ in.}$$

The force acting in the segment between B and C is found by considering the algebraic sum of the forces to the left of a section between B and C. This indicates that a resultant force of 7000 lb acts to the left, i.e. the section has a tensile force acting upon it. This same result could of course have been obtained by considering the algebraic sum of the forces to the right of this section. Consequently the free-body diagram of the segment BC appears as in Figure (*c*) below.

The elongation of this portion is given by $\Delta_2 = \dfrac{7000(36)}{(1)(30\cdot 10^6)} = 0.0084$ in.

Similarly, the force acting over any cross-section between C and D must be 9000 lb to maintain equilibrium with the applied load at D. The free-body diagram of the segment CD appears as in Fig. (d).

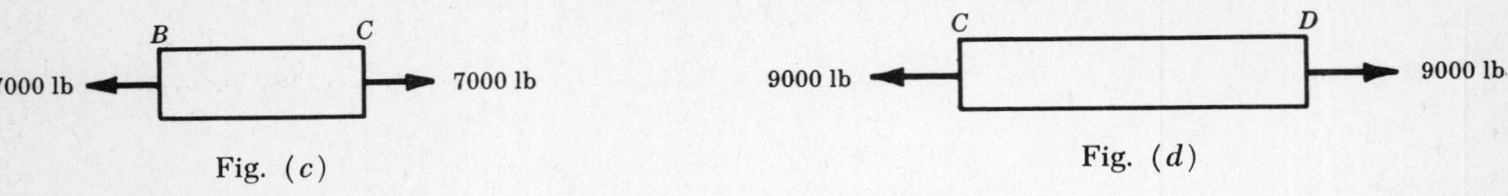

Fig. (c) Fig. (d)

The elongation of this portion is given by $\Delta_3 = \dfrac{(9000)(48)}{(1)(30\cdot 10^6)} = 0.0144$ in.

The total elongation is consequently $\Delta = 0.0080 + 0.0084 + 0.0144 = 0.0308$ in.

4. The Howe truss shown in Fig. (a) supports the single load of 120,000 lb. If the working stress of the material in tension is taken to be 20,000 lb/in², determine the required cross-sectional area of bars DE and AC. Find the elongation of bar DE over its 20 ft length. Assume that the limiting value of the working stress in tension is the only factor to be considered in determining the required area. Take the modulus of elasticity of the bar to be $30\cdot 10^6$ lb/in².

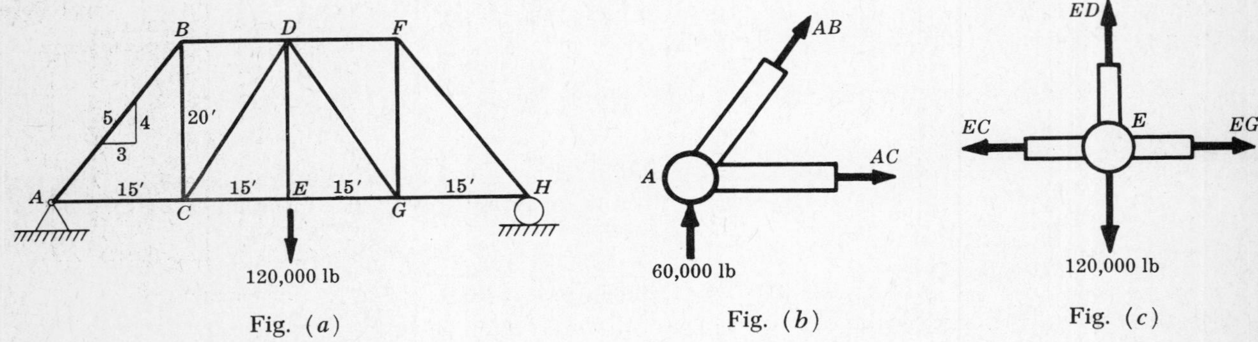

Fig. (a) Fig. (b) Fig. (c)

This truss is statically determinate both externally and internally, i.e., the reactions at the supports may be determined by the equations of static equilibrium and also the axial force in each bar may be found by a simple statics analysis.

It is first necessary to determine the vertical reactions at A and H. By symmetry these are each 60,000 lb. A free-body diagram of the joint at A appears as in Fig. (b). In Fig. (b) the unknown forces in the bars have been denoted as AB and AC, the same designations as the bars themselves, and they have been assumed to be tensile forces. In this manner if they are found to be positive they actually indicate tension. If they are found to be negative they indicate compression, and the signs thus obtained are in agreement with the usual sign convention designating tensile forces as positive and compressive forces as negative. Applying the equations of static equilibrium to the above free-body diagram we have

$$\Sigma F_v = 60{,}000 + \frac{4}{5}(AB) = 0 \qquad \text{or} \qquad AB = -75{,}000 \text{ lb}$$

$$\Sigma F_h = \frac{3}{5}(-75{,}000) + AC = 0 \qquad \text{or} \qquad AC = 45{,}000 \text{ lb}$$

Likewise, a free-body diagram of the point at E appears as in Fig. (c) above. From statics,

$$\Sigma F_v = ED - 120{,}000 = 0 \qquad \text{or} \qquad ED = 120{,}000 \text{ lb}$$

The simple consideration of trusses used here assumes all bars are so-called two-force members, i.e., subject to either axial tension or compression and no other loadings.

For axial loading the stress is given by $s = P/A$, where P is the axial force and A the cross-sectional area of the bar. Here, the stress is given as 20,000 lb/in^2 in each bar and the areas are thus given by

$$A_{DE} = \frac{120,000}{20,000} = 6 \text{ in}^2 \qquad \text{and} \qquad A_{AC} = \frac{45,000}{20,000} = 2.25 \text{ in}^2$$

The elongation of a bar under axial tension is given by $\Delta = \dfrac{PL}{AE}$. For bar DE we have

$$\Delta = \frac{(120,000)(240)}{(6)(30\cdot 10^6)} = 0.160 \text{ in.}$$

5. A series of prismatic bars of rectangular cross-section 2 × 3 in. is used as a clamping device to secure the top on a cylindrical tank containing fluid under pressure. The outside wall of the pressure tank has projecting lugs welded to it and the prismatic bars fit loosely (in the lateral direction) between adjacent lugs. To secure the clamping effect the bar is purposely machined so that it is too short for its flanges (A) to fit over the tank cover resting on top of the lugs. At room temperature, it fails to clear by 0.10 in. The bar (but not the lugs) is then heated so that it can be slipped over the tank top. After it cools it then exerts a force normal to the tank top.

If the total bearing area at one end of the bar (area in contact with the tank top) is 6 in^2, find the unit pressure each bar exerts on the tank top. Also, find the temperature to which the bar must be heated in order that it just clears the top of the tank cover. The bar is steel, for which $\alpha = 6.5\cdot 10^{-6}/^\circ$F.

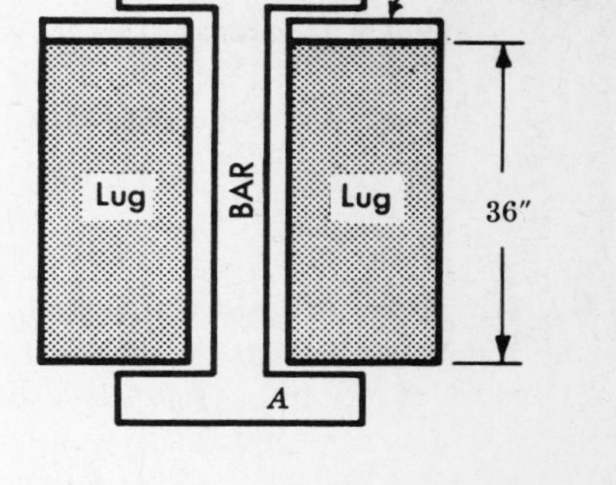

$$0.10 = (6.5\cdot 10^{-6})(36)(\Delta T) \qquad \text{from which} \qquad \Delta T = 426^\circ \text{ F.}$$

The axial force necessary to stretch the bar this same amount is P where

$$0.10 = \frac{P(36)}{(6)(30\cdot 10^6)} \qquad \text{and} \qquad P = 500,000 \text{ lb}$$

The pressure is assumed to be uniformly distributed over the bearing area between the flange and the tank top. Consequently, the pressure is

$$\frac{500,000}{6} = 83,300 \text{ lb/in}^2$$

6. Determine the total increase of length of a bar of constant cross-section hanging vertically and subject to it own weight as the only load. The bar is initally straight.

The normal stress (tensile) over any horizontal cross-section is caused by the weight of the material below that section. The elongation of the element of thickness dy shown is

$$d\Delta = \frac{(Ay\gamma)\,dy}{AE}$$

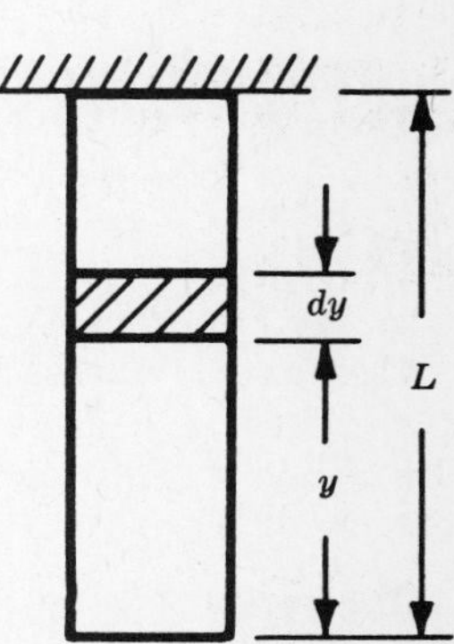

where A denotes the cross-sectional area of the bar and γ its specific weight (weight/unit volume). Integrating, the total elongation of the bar is

$$\Delta = \int_0^L \frac{Ay\gamma\,dy}{AE} = \frac{A\gamma}{AE}\cdot\frac{L^2}{2} = \frac{(A\gamma L)L}{2AE} = \frac{WL}{2AE}$$

where W denotes the total weight of the bar. It is to be noted that the total elongation produced by the weight of the bar is equal to that produced by a load of half its weight applied at the end.

7. A steel wire 1/4 in. in diameter is used for hoisting purposes in building construction. If 500 ft of the wire are hanging vertically, and a load of 300 lb is being lifted at the lower end of the wire, determine the total elongation of the wire. The specific weight of the steel is 0.283 lb/in^3 and $E = 30 \cdot 10^6$ lb/in^2.

The total elongation is caused partially by the applied force of 300 lb and partially by the weight of the wire. The elongation due to the 300 lb load is

$$\Delta_1 = \frac{PL}{AE} = \frac{(300)(500 \cdot 12)}{\frac{\pi}{4}(\frac{1}{4})^2 (30 \cdot 10^6)} = 1.27 \text{ in.}$$

From Problem 6 the elongation due to the weight of the wire is

$$\Delta_2 = \frac{WL}{2AE} = \frac{(\frac{\pi}{4})(\frac{1}{4})^2 (500 \cdot 12)(0.283)(500 \cdot 12)}{2(\frac{\pi}{4})(\frac{1}{4})^2 (30 \cdot 10^6)} = 0.170 \text{ in.}$$

Consequently, the total elongation is $\Delta = 1.27 + 0.17 = 1.44$ in.

8. A straight aluminum wire 100 ft long is subject to a tensile stress of 10,000 lb/in^2. Determine the total elongation of the wire. What temperature change would produce this same elongation? Take $E = 10 \cdot 10^6$ lb/in^2 and α (the coefficient of linear expansion) $= 12.8 \cdot 10^{-6}$/F°.

The total elongation is given by $\Delta = \dfrac{PL}{AE} = \dfrac{(10,000)(100 \cdot 12)}{10 \cdot 10^6} = 1.20$ in.

A rise in temperature of ΔT would cause this same expansion. Then

$$1.20 = (12.8 \cdot 10^{-6})(100 \cdot 12)(\Delta T) \quad \text{and} \quad \Delta T = 78.2° \text{ F.}$$

9. Two prismatic bars are rigidly fastened together and support a vertical load of 10,000 lb as shown. The upper bar is steel having a specific weight of 0.283 lb/in^3, a length of 35 ft, and a cross-sectional area of 10 in^2. The lower bar is brass having a specific weight of 0.300 lb/in^3, a length of 20 ft and a cross-sectional area of 8 in^2. For steel $E = 30 \cdot 10^6$ lb/in^2, for brass $E = 13 \cdot 10^6$ lb/in^2. Determine the maximum stress in each material.

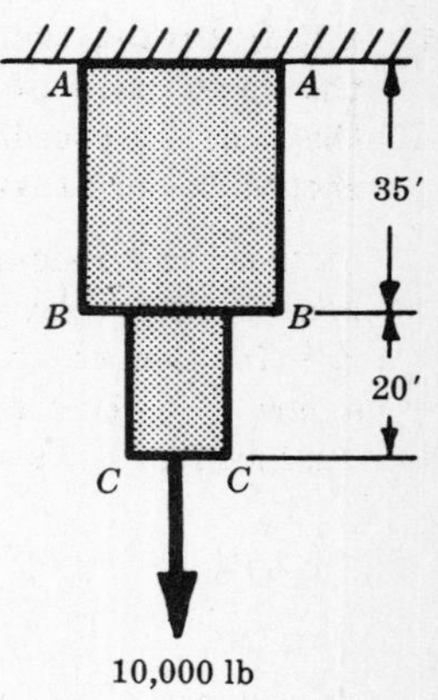

The maximum stress in the brass bar occurs just below the junction at section *B-B*. There, the vertical normal stress is caused by the combined effect of the load of 10,000 lb together with the weight of the entire brass bar below *B-B*.

The weight of the brass bar is

$$W_b = (20 \cdot 12)(8)(0.300) = 576 \text{ lb.}$$

The stress at this section is

$$s = \frac{P}{A} = \frac{10,000 + 576}{8} = 1320 \text{ lb/in}^2.$$

The maximum stress in the steel bar occurs at section *A-A*, the point of suspension, because there the entire weight of the steel and brass bars gives rise to normal stress, whereas at any lower section only a portion of the weight of the steel would be effective in causing stress.

The weight of the steel bar is

$$W_s = (35 \cdot 12)(10)(0.283) = 1185 \text{ lb.}$$

The stress across section *A-A* is $s = \dfrac{P}{A} = \dfrac{10,000 + 576 + 1185}{10} = 1180$ lb/in^2.

10. A solid truncated conical bar of circular cross-section tapers uniformly from a diameter d at its small end to D at the large end. The length of the bar is L. Determine the elongation due to an axial force P applied at each end. See Fig. (a)

The coordinate x describes the distance of a disc-like element of thickness dx from the small end. The radius of this small element is readily found by similar triangles:

$$r = \frac{d}{2} + \frac{x}{L}\left(\frac{D-d}{2}\right)$$

The elongation of this disc-like element may be found by applying the formula for extension due to axial loading, $\Delta = PL/AE$. For the element, this expression becomes

$$d\Delta = \frac{P\,dx}{\pi\left[\frac{d}{2} + \frac{x}{L}\left(\frac{D-d}{2}\right)\right]^2 E}$$

The extension of the entire bar is obtained by summing the elongations of all such elements over the bar. This is of course done by integrating. If Δ denotes the elongation of the entire bar,

$$\Delta = \int_0^L d\Delta = \int_0^L \frac{4P\,dx}{\pi\left[d + \frac{x}{L}(D-d)\right]^2 E} = \frac{4PL}{\pi D d E}$$

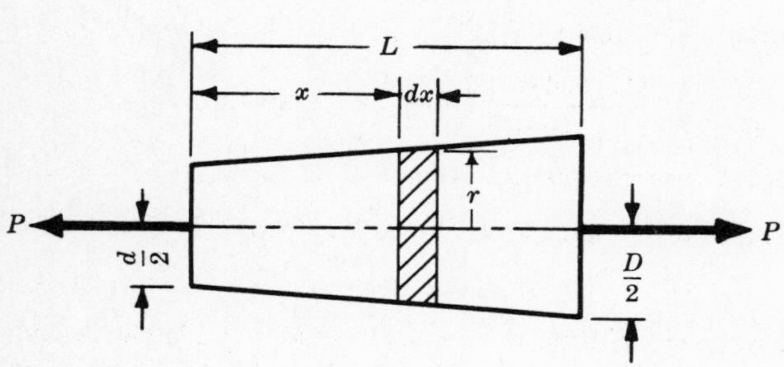

Fig. (a) Prob. 10

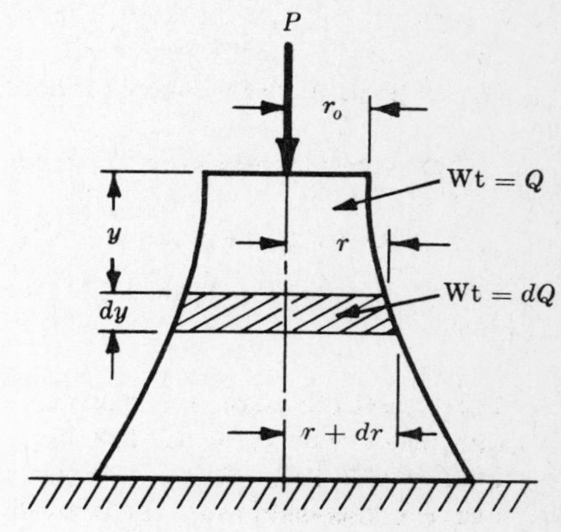

Fig. (b) Prob. 11

11. A body having the form of a solid of revolution supports a load P as shown in Fig. (b). The radius of the upper base of the body is r_0 and the specific weight of the material is γ lb/ft^3. Determine how the radius should vary with the altitude in order that the compressive stress at all cross-sections should be constant. The weight of the solid is not negligible.

Let y be measured from the upper base as shown and let Q denote the weight of that portion of the body of altitude y. Then dQ represents the increment to Q in the increment of altitude dy. Let r and $(r+dr)$ denote the radii of the upper and lower surfaces respectively of this horizontal element and A and $(A+dA)$ the corresponding areas. Considering the normal compressive stresses acting over both surfaces of this element, we have

$$\frac{P+Q}{A} = \frac{P+Q+dQ}{A+dA} = s = \text{constant}$$

from which

$$(1) \qquad \frac{dA}{dQ} = \frac{A}{P+Q} = \frac{1}{s}$$

The increment of area between the upper and lower faces of the element is

$$dA = \pi(r+dr)^2 - \pi r^2 = 2\pi r\,dr$$

The increment of weight is $dQ = \pi r^2 \gamma (dy)$.

Consequently from (1), $\dfrac{2\pi r(dr)}{\pi r^2 \gamma(dy)} = \dfrac{1}{s}$. Integrating, $2 \log r = (\dfrac{\gamma}{s})y + C_1$.

Applying the boundary condition that $r = r_o$ when $y = 0$, we find $C_1 = 2 \log r_o$.

Also from the conditions at the upper base, $s = \dfrac{P}{\pi r_o^2}$. Finally, $r = r_o e^{(\frac{\gamma \pi r_o^2 y}{2P})}$

12. Two identical steel bars are pin connected and support a load of 100,000 lb as shown in Fig. (*a*). Find the required cross-sectional area of the bars so that the normal stress in them is no greater than 30,000 lb/in². Also, find the vertical displacement of the point *B*. Take $E = 30\cdot 10^6$ lb/in².

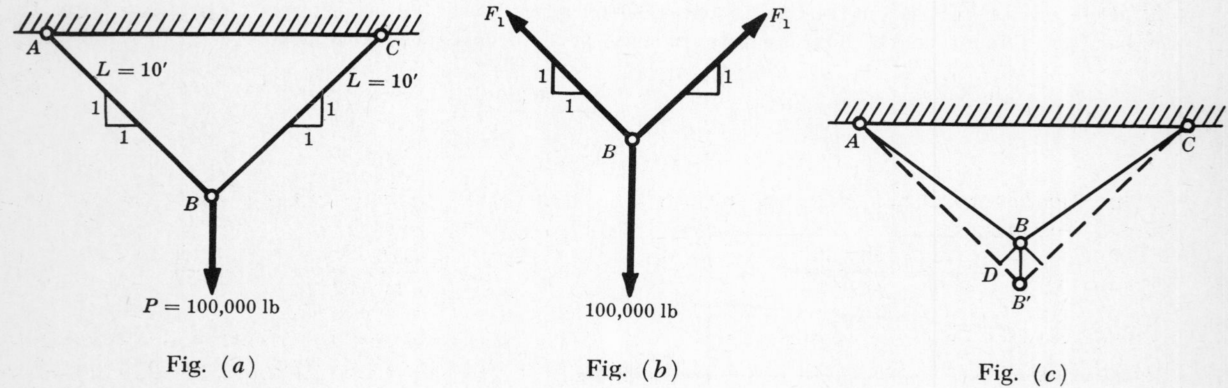

Fig. (*a*) Fig. (*b*) Fig. (*c*)

A free-body diagram of the pin at *B* is shown in Fig. (*b*), where F_1 represents the force (lb) in each bar.

From statics, $\Sigma F_v = 2(\dfrac{1}{\sqrt{2}})F_1 - 100,000 = 0$ or $F_1 = 70,700$ lb.

Hence the required area is $A = \dfrac{70,700}{30,000} = 2.35$ in².

Because our study of strength of materials is restricted to the case of *small* deformations, the basic geometry of the structure is essentially unchanged. Thus, we can denote the position of the deformed bars by the dotted lines shown in Fig. (*c*), and the angle $DB'B$ is very nearly 45°. The elongation of the left bar is represented by DB' and is found from the expression for axial extension (Problem 1) to be

$$DB' = \frac{(70,700)(120)}{(2.35)(30\cdot 10^6)} = 0.120 \text{ in.} \qquad \text{Consequently,} \qquad BB' = \frac{0.120}{\cos 45^\circ} = 0.170 \text{ in.}$$

13. The two steel bars *AB* and *BC* are pinned at each end and support the load of 60,000 lb shown in Fig. (*a*) below. The metal is annealed cast steel, having a yield point of 60,000 lb/in². Safety factors of 2 for tensile members and 3.5 for compressive members are adequate. Determine the required cross-sectional areas of these bars and also the horizontal and vertical components of displacement of point *B*. Take $E = 30\cdot 10^6$ lb/in².

A free-body diagram of the joint at *B* appears as in Fig. (*b*) below if the unknown forces are assumed to be tensile.

From statics: $\Sigma F_v = -60,000 - BC \sin 30^\circ = 0$ or $BC = -120,000$ lb

$\Sigma F_h = -BA - BC \cos 30^\circ = 0$ or $BA = 104,000$ lb

The working stresses are given by $\frac{60,000}{2} = 30,000 \text{ lb/in}^2$ for tension

and $\frac{60,000}{3.5} = 17,100 \text{ lb/in}^2$ for compression.

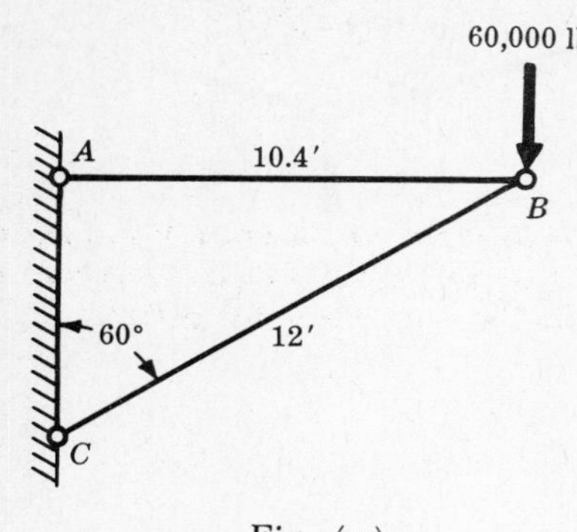

Fig. (a)

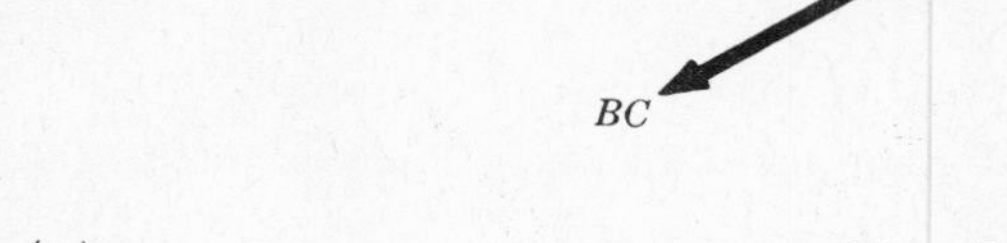

Fig. (b)

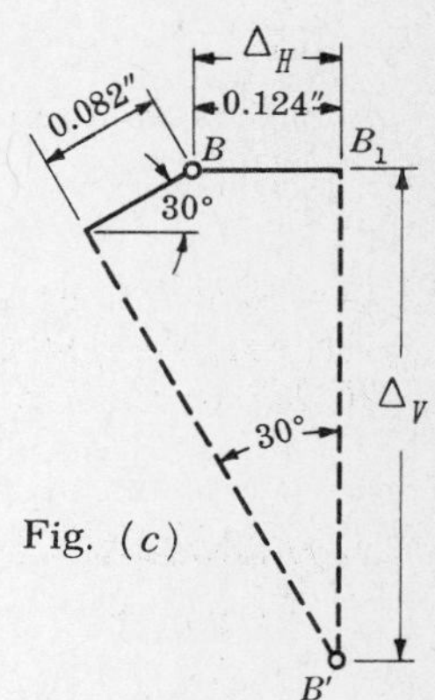

Fig. (c)

The required areas are found by dividing the axial force in each bar by the allowable working stress. Consequently

$$A_{AB} = \frac{104,000}{30,000} = 3.47 \text{ in}^2 \quad \text{and} \quad A_{BC} = \frac{120,000}{17,100} = 7.04 \text{ in}^2.$$

To investigate the displacement of point B it is first necessary to calculate the axial deformation of each of the bars. From the expression derived in Problem 1 we find the extension of AB to be

$$\Delta_{AB} = \frac{(104,000)(10.4 \cdot 12)}{(3.47)(30 \cdot 10^6)} = 0.124 \text{ in.}$$

and the compression of BC to be $\Delta_{BC} = \frac{(120,000)(12 \cdot 12)}{(7.04)(30 \cdot 10^6)} = 0.082$ in.

The location of the point B after deformation has occurred may be determined by realizing that the bar AB actually lengthens 0.124 in. and also rotates as a rigid body about the pin at A. Further, the bar BC shortens 0.082 in. and also rotates about the pin at C.

Fig. (c) above illustrates the movement of point B to its deflected position designated by B'. It is to be observed that the deformations of the structure are small, hence the displacement due to rotation about A of the elongated bar AB may be represented by the straight line B_1B' rather than a circular arc with center at A. The same reasoning applies to the rotation of bar BC. From the geometry of the above sketch we immediately have for the displacement components of point B:

$$\Delta_h = 0.124 \text{ in.}$$

$$\Delta_v = \left[\frac{(0.082 \cos 30^\circ) + 0.124}{\tan 30^\circ}\right] + 0.082 \sin 30^\circ = 0.379 \text{ in.}$$

14. Consider two thin rods or wires as shown in Fig. (a) below, which are pinned at A, B, and C and are initially horizontal and of length L when no load is applied. The weight of each wire is negligible. A force Q is then applied (gradually) at the point B. Determine the magnitude of Q so as to produce a prescribed vertical deflection δ of the point B.

This is an extremely interesting example of a system in which the elongations of all the individual members satisfy Hooke's law and yet for geometric reasons deflection is *not* proportional to force.

Each bar obeys the relation $\Delta = \frac{PL}{AE}$

where P is the axial force in each bar and Δ the axial elongation. Initially each bar is of length L and after the entire load Q has been applied the length is L'. Thus

(1) $$L' - L = \frac{PL}{AE}$$

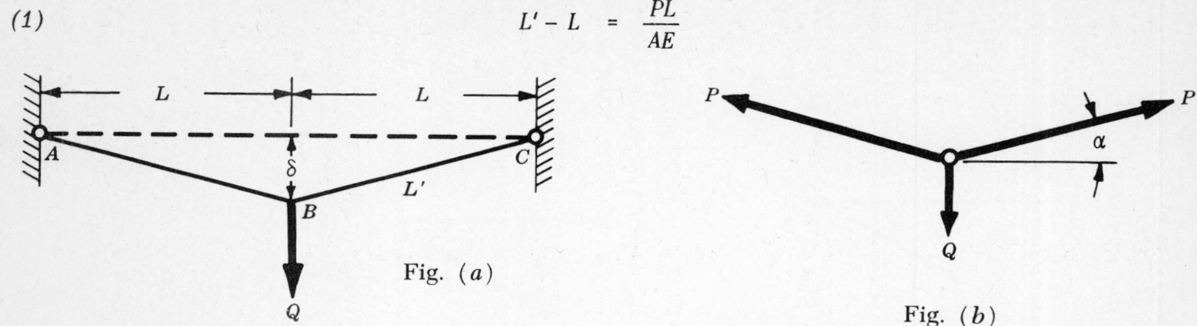

Fig. (a)

Fig. (b)

The free-body diagram of the pin at B is shown in Fig. (b) above. From statics,

$$\Sigma F_v = 2P\sin\alpha - Q = 0 \qquad \text{or} \qquad Q = 2P\left(\frac{\delta}{L'}\right)$$

(2) Using (1), $$Q = 2\left[\frac{(L'-L)AE}{L}\right]\frac{\delta}{L'} = \frac{2\delta AE}{L}\left(1 - \frac{L}{L'}\right)$$

(3) But $$(L')^2 = L^2 + \delta^2$$

(4) Consequently $$Q = \frac{2\delta AE}{L}\left[1 - \frac{L}{\sqrt{L^2+\delta^2}}\right]$$

Also, from the Binomial Theorem we have

(5) $$\sqrt{L^2+\delta^2} = L\sqrt{1+\frac{\delta^2}{L^2}} = L\{1 + \frac{1}{2}\cdot\frac{\delta^2}{L^2} + \cdots\}$$ and thus

(6) $$1 - \frac{L}{L\{1+\frac{1}{2}\cdot\frac{\delta^2}{L^2}\}} \approx 1 - \left(1 - \frac{1}{2}\cdot\frac{\delta^2}{L^2}\right) = \frac{1}{2}\cdot\frac{\delta^2}{L^2}$$

From this we have the approximate relation between force and displacement,

(7) $$Q = \frac{2AE\delta}{L}\left\{\frac{\delta^2}{2L^2}\right\} = \frac{AE\delta^3}{L^3}$$ which corresponds to Equation (4).

Thus, the displacement δ is *not* proportional to the force Q even though Hooke's law holds for each bar individually. It is to be noted that Q becomes more nearly proportional to δ as δ becomes larger, assuming that Hooke's law still holds for the elongations of the bars. In this example superposition does *not* hold. The characteristic of this system is that the action of the external forces is *appreciably* affected by the small deformations which take place. In this event the stresses and displacements are not linear functions of the applied loads and superposition does not apply.

SUMMARY: A material must follow Hooke's law if superposition is to apply. But this requirement alone is not sufficient. We must see whether or not the action of the applied loads is affected by small deformations of the structure. If the effect is substantial, superposition does not hold.

15. For the system discussed in Problem 14 let us consider wires each of initial length 5 ft, cross-sectional area 0.1 in^2 and with $E = 30\cdot 10^6$ lb/in^2. For a load Q of 20 lb determine the central deflection δ by both the exact and the approximate relations given there.

The exact expression relating force and deflection is $Q = \frac{2\delta AE}{L}[1 - \frac{L}{\sqrt{L^2 + \delta^2}}]$.

Substituting the given numerical values, $20 = \frac{2\delta(0.1)(30\cdot10^6)}{(60)}[1 - \frac{60}{\sqrt{(60)^2 + \delta^2}}]$.

Solving by trial and error we find $\delta = 1.131$ in.

The approximate relation between force and deflection is $Q = \frac{AE\delta^3}{L^3}$.

Substituting, $20 = \frac{(0.1)(30\cdot10^6)\delta^3}{(60)^3}$ from which $\delta = 1.129$ in.

16. A square steel bar two inches on a side and 4 ft long is subject to an axial tensile force of 64,000 lb. Determine the decrease in the lateral dimension due to this load. Consider $E = 30\cdot10^6$ lb/in² and $\mu = 0.3$.

The loading is axial, hence the stress in the direction of the load is given by

$$s = \frac{P}{A} = \frac{64,000}{4} = 16,000 \text{ lb/in}^2$$

The simple form of Hooke's law for uniaxial loading states that $E = \frac{s}{\epsilon}$.

The strain ϵ in the direction of the load is thus $\frac{16,000}{30,000,000} = 0.000533$.

The ratio of the lateral strain to the axial strain is denoted as Poisson's ratio, i.e.,

$$\mu = \frac{\text{lateral strain}}{\text{axial strain}}$$

The axial strain has been found to be 0.000533. Consequently, the lateral strain is μ times that value, or

$$(0.3)(0.000533) \doteq 0.000166$$

Since the lateral strain is 0.000166, the change in a two inch length is 0.000332 in. which represents the decrease in the lateral dimension of the bar.

It is to be noted that the definition of Poisson's ratio as the ratio of two strains presumes that only a single uniaxial load acts on the member.

17. Consider a state of stress of an element such that a stress s_x is exerted in one direction, lateral contraction is free to occur in a second (z) direction, but is completely restrained in the third (y) direction. Find the ratio of the stress in the x-direction to the strain in that direction. Also, find the ratio of the strain in the z-direction to that in the x-direction.

Let us examine the general statement of Hooke's law discussed earlier. If in those equations we set $s_z = 0$, $\epsilon_y = 0$ so as to satisfy the conditions of the problem, then Hooke's law becomes

(a) $$\epsilon_x = \frac{1}{E}[s_x - \mu(s_y + 0)]$$

(b) $$\epsilon_y = \frac{1}{E}[s_y - \mu(s_x + 0)] = 0$$

(c) $$\epsilon_z = \frac{1}{E}[0 - \mu(s_x + s_y)]$$

From (b), $$s_y = \mu s_x$$

Consequently, from (a) $$\epsilon_x = \frac{1}{E}[s_x - \mu^2 s_x] = \frac{(1-\mu^2)}{E}s_x$$

Solving this equation for s_x as a function of ϵ_x and substituting in (c) we have

$$\epsilon_z = -\frac{\mu}{E}(s_x + \mu s_x) = -\frac{\mu(1+\mu)}{E}\cdot\frac{\epsilon_x E}{(1-\mu^2)} = -\frac{\mu\epsilon_x}{(1-\mu)}$$

We may now form the ratios $\dfrac{s_x}{\epsilon_x} = \dfrac{E}{1-\mu^2}$ and $-\dfrac{\epsilon_z}{\epsilon_x} = \dfrac{\mu}{1-\mu}$.

The first quantity, $E/(1-\mu^2)$ is usually denoted as the *effective modulus of elasticity* and is useful in the theory of thin plates and shells. The second ratio, $\mu/(1-\mu)$ is called the *effective value of Poisson's ratio.*

18. Consider an elemental block subject to uniaxial tension. Derive approximate expressions for the change of volume per unit volume due to this loading.

The strain in the direction of the forces may be denoted by ϵ_x. The strains in the other two orthogonal directions are then each $(-\mu\epsilon_x)$. Consequently, if the initial dimensions of the element are dx, dy, and dz then the final dimensions are

$$(1+\epsilon_x)dx,\quad (1-\mu\epsilon_x)dy,\quad (1-\mu\epsilon_x)dz$$

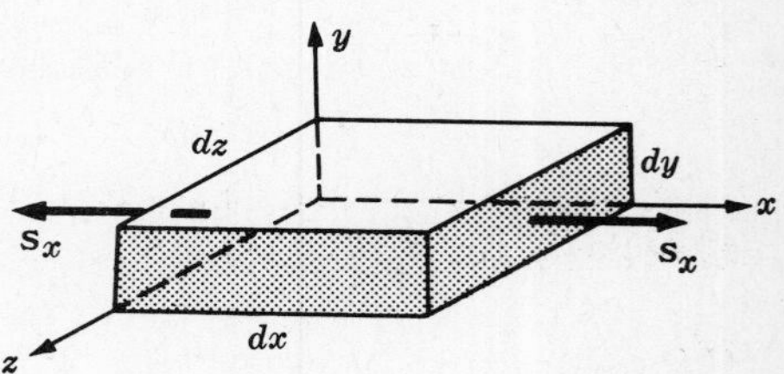

and the volume after deformation is

$$V' = [(1+\epsilon_x)dx][(1-\mu\epsilon_x)dy][(1-\mu\epsilon_x)dz]$$

$$= (1+\epsilon_x)(1-2\mu\epsilon_x)dx\,dy\,dz$$

$$= (1-2\mu\epsilon_x+\epsilon_x)dx\,dy\,dz$$

since the deformations are so small that the *squares* and *products* of strains may be neglected.

Since the initial volume was $dx\,dy\,dz$, the change of volume per unit volume is

$$\frac{\Delta V}{V} = (1-2\mu)\epsilon_x$$

Hence, for a tensile force the volume increases slightly, for a compressive force it decreases.

Also, the cross-sectional area of the element in a plane normal to the direction of the applied force is given approximately by

$$A = (1-\mu\epsilon_x)^2\,dy\,dz = (1-2\mu\epsilon_x)dy\,dz$$

19. A square bar of aluminum 2 in. on a side and 10 in. long is loaded by axial tensile forces at the ends. Experimentally, it is found that the strain in the direction of the load is 0.001 in/in. Determine the volume of the bar when the load is acting. Consider $\mu = 0.33$.

From Problem 18, the change of volume per unit volume is given by

$$\frac{\Delta V}{V} = \epsilon(1-2\mu) = 0.001(1-0.66) = 0.00034$$

Consequently, the change of volume of the entire bar is given by

$$\Delta V = (2)(2)(10)(0.00034) = 0.0136 \text{ in}^3$$

The original volume of the bar in the unstrained state is 40 in^3. Since a tensile force increases the volume, the final volume under load is 40.0136 in^3. It is to be noted that ordinary methods of physical measurement would not lead to accuracy of six significant figures.

20. The general three-dimensional form of Hooke's law in which strain components are expressed as functions of stress components has already been presented. Occasionally it is necessary to express the stress components as functions of the strain components. Derive these expressions.

Given the previous expressions

$$\epsilon_x = \frac{1}{E}[s_x - \mu(s_y + s_z)] \tag{1}$$

$$\epsilon_y = \frac{1}{E}[s_y - \mu(s_x + s_z)] \tag{2}$$

$$\epsilon_z = \frac{1}{E}[s_z - \mu(s_x + s_y)] \tag{3}$$

Let us introduce the notation

$$e = \epsilon_x + \epsilon_y + \epsilon_z \tag{4}$$

$$\theta = s_x + s_y + s_z \tag{5}$$

With this notation, Equations *(1)*, *(2)*, and *(3)* may be readily solved by determinants for the unknowns s_x, s_y, s_z to yield

$$s_x = \frac{\mu E}{(1+\mu)(1-2\mu)} e + \frac{E}{(1+\mu)} \epsilon_x \tag{6}$$

$$s_y = \frac{\mu E}{(1+\mu)(1-2\mu)} e + \frac{E}{(1+\mu)} \epsilon_y \tag{7}$$

$$s_z = \frac{\mu E}{(1+\mu)(1-2\mu)} e + \frac{E}{(1+\mu)} \epsilon_z \tag{8}$$

These are the desired expressions.

Further information may also be obtained from Equations *(1)* through *(5)*. If Equations *(1)*, *(2)*, and *(3)* are added and the symbols e and θ introduced, we have

$$e = \frac{1}{E}(1-2\mu)\theta \tag{9}$$

For the special case of a solid subjected to uniform hydrostatic pressure p, $s_x = s_y = s_z = -p$. Hence

$$e = \frac{-3(1-2\mu)p}{E} \quad \text{or} \quad \frac{p}{e} = -\frac{E}{3(1-2\mu)}$$

The quantity $\frac{E}{3(1-2\mu)}$ is often denoted by K and is called the *bulk modulus* or *modulus of volume expansion* of the material. Physically, the bulk modulus K is a measure of the resistance of a material to change of volume but with no change of shape or form.

We see that the final volume of an element having sides dx, dy, dz prior to loading and subject to strains ϵ_x, ϵ_y, ϵ_z is

$$(1+\epsilon_x)dx\,(1+\epsilon_y)dy\,(1+\epsilon_z)dz = (1+\epsilon_x+\epsilon_y+\epsilon_z)dx\,dy\,dz$$

Thus the ratio of the increase in volume to the original volume is given approximately by

$$e = \epsilon_x + \epsilon_y + \epsilon_z$$

This change of volume per unit volume (e) is defined as the *dilatation*.

SUPPLEMENTARY PROBLEMS

21. A straight bar of uniform cross-section is subject to axial tension. The cross-sectional area of the bar is 1 in^2 and its length is 12 ft. If the total elongation is 0.0910 in. under a load of 19,000 lb, find the modulus of elasticity of the material. *Ans.* $E = 30\cdot10^6$ lb/in^2

22. Compute the height to which a vertical concrete wall may be built if the ultimate compressive strength is 2400 lb/in^2 and a safety factor of 4 is used. The specific weight of concrete is 150 lb/ft^3. *Ans.* h = 576 ft

23. A hollow right-circular cylinder is made of cast iron and has an outside diameter of 3 in. and an inside diameter of 2.5 in. If the cylinder is loaded by an axial compressive force of 10,000 lb, determine the total shortening in a 2 ft length. Also, determine the normal stress under this load. Take the modulus of elasticity to be $15\cdot10^6$ lb/in^2 and neglect any possibility of lateral buckling of the cylinder. *Ans.* Δ = 0.00738 in., s = 4620 lb/in^2

24. A solid circular steel rod 1/4 in. in diameter and 15 in. long is rigidly fastened to the end of a square brass bar one inch on a side and 12 in. long, the geometric axes of the bars lying along the same line. An axial tensile force of 1200 lb is applied at each of the extreme ends. Determine the total elongation of the assembly. For steel, $E = 30\cdot10^6$ lb/in^2 and for brass $E = 13.5\cdot10^6$ lb/in^2. *Ans.* 0.0133 in.

25. The truss shown is pin-connected and supports the single force of 30,000 lb. All bars are made of SAE 1020 steel having a yield point of 35,000 lb/in^2. For tension members a safety factor of 2 is sufficient. Determine the required cross-sectional areas of bars CD and AB. See Fig. (a) below. *Ans.* Area CD = 0.86 in^2, Area AB = 1.07 in^2

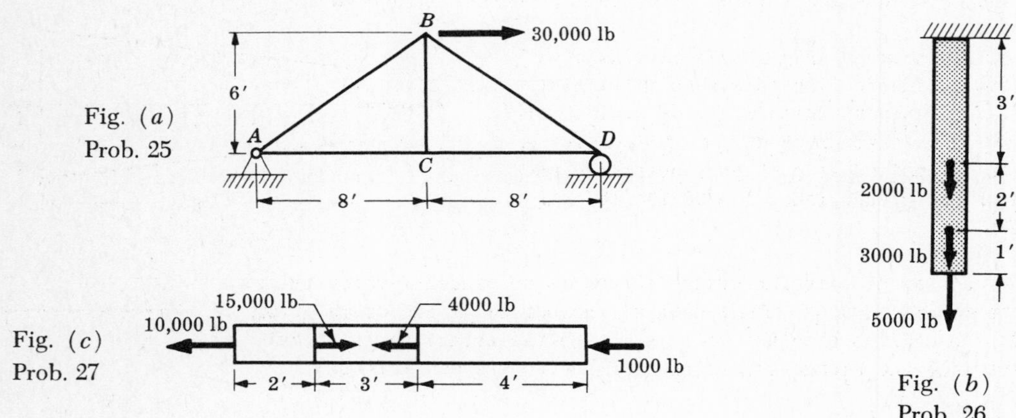

Fig. (a) Prob. 25

Fig. (b) Prob. 26

Fig. (c) Prob. 27

26. A steel bar of uniform cross-section is suspended vertically and carries a load of 5000 lb at its lower extremity, as shown in Fig. (b) above. One foot above this a vertical force of 3000 lb is applied and two feet above this last point a load of 2000 lb is applied. The total length of the bar is 6 ft and its cross-sectional area is 1 in^2. The modulus of elasticity is $30\cdot10^6$ lb/in^2. Determine the total elongation of the bar. *Ans.* 0.0204 in.

27. A brass bar of cross-sectional area 1.5 in^2 is subject to the axial forces shown in Fig. (c) above. Determine the total elongation of the bar. For brass, $E = 13\cdot10^6$ lb/in^2. *Ans.* 0.000616 in.

28. Steel railroad rails are laid with their adjacent ends 1/8 in. apart when the temperature is 60°F. The length of each rail is 39 ft. The material is steel for which $E = 30\cdot10^6$ lb/in^2 and $\alpha = 6.5\cdot10^{-6}$ /°F. (a) Compute the gap between adjacent ends when the temperature is −10°F. (b) At what temperature will adjacent ends just be in contact? (c) Find the compressive stress in the rails when the temperature is 110°F. Neglect any possibility of buckling of the rails. *Ans.* Gap = 0.338 in., T = 101.3°F., s = 1690 lb/in^2

29. The following data were obtained during the tensile test of a circular cold-rolled steel specimen of diameter 0.507 in.

Axial Load (lb)	Elongation in 2" gage length
0	0
1250	0.0004
1850	0.0006
2400	0.0008
3050	0.0010
3640	0.0012
4250	0.0014
4850	0.0016
5450	0.0018
6050	0.0020
6700	0.0022
7250	0.0024
6900	0.0040
6950	0.0080

Axial Load (lb)	Elongation in 2" gage length
6950	0.0120
6950	0.0160
6900	0.0200
6950	0.0240
7000	0.0500
7750	0.1000
9350	0.2000
9900	0.3000
10,100	0.4000
10,100	0.5000
9900	0.6000
9500	0.7000
8900	0.7500

At the time of rupture the final diameter of the bar at the section where failure occurred was 0.295 in. The original 2 in. gage length had increased to 2.750 in.

From the given data determine the proportional limit of the material, the modulus of elasticity, the percentage reduction of area, the percentage elongation, and the breaking strength.

Ans. Proportional limit 36,000 lb/in^2, $E = 30\cdot10^6$ lb/in^2, Percentage reduction of area = 66.3
Percentage elongation = 37.5, Breaking strength = 44,000 lb/in^2

30. A flat steel plate is of trapezoidal form as shown in the adjoining figure. The thickness of the plate is 0.5 in. and it tapers uniformly from a width of 2 in. to 4 in. in a length of 18 in. If an axial force of 10,000 lb is applied at each end, determine the elongation of the plate. Take $E = 30\cdot10^6$ lb/in^2.
Ans. 0.00416 in.

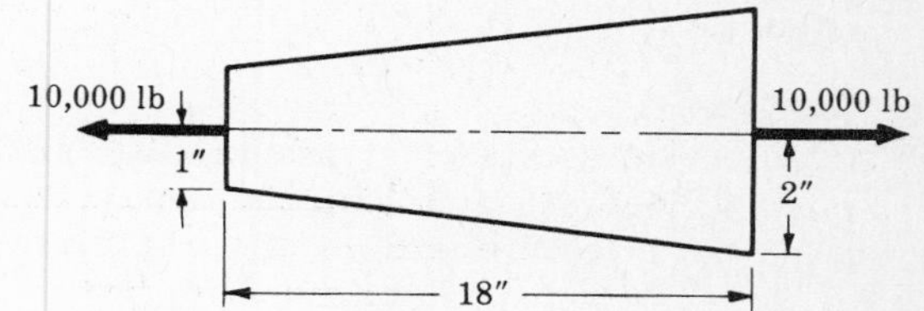

31. A solid conical bar of circular cross-section is suspended vertically as shown in the adjacent figure. The length of the bar is L, the diameter of its base D, the modulus of elasticity is E, and the weight per unit volume γ. Determine the elongation of the bar due to its own weight.

Ans. $\Delta = \dfrac{\gamma L^2}{6E}$

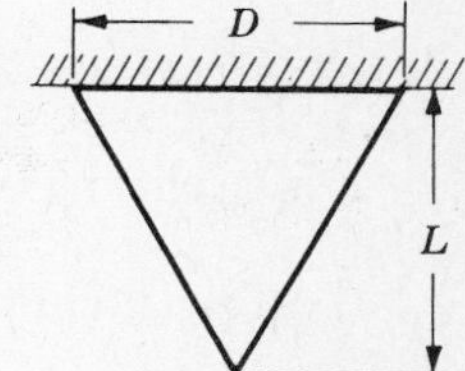

32. The vertical gate AB, shown in the adjacent diagram, may be considered to be absolutely rigid and is pinned at A. It is 10 ft wide and subject to hydrostatic pressure across the entire width. At C a steel tie bar 25 ft long and of cross-sectional area 0.5 in^2 is attached to tie it back to the wall at D. Find the horizontal displacement of point B. Neglect the effect of end restraints at either end of the gate. Take $E = 30\cdot10^6$ lb/in^2.
Ans. Displacement = 1.26 in.

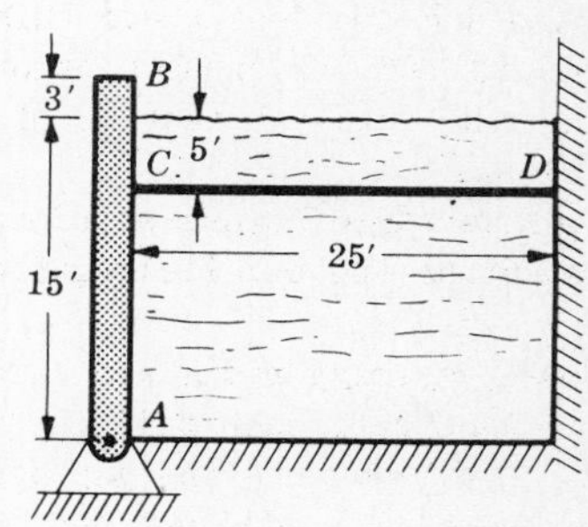

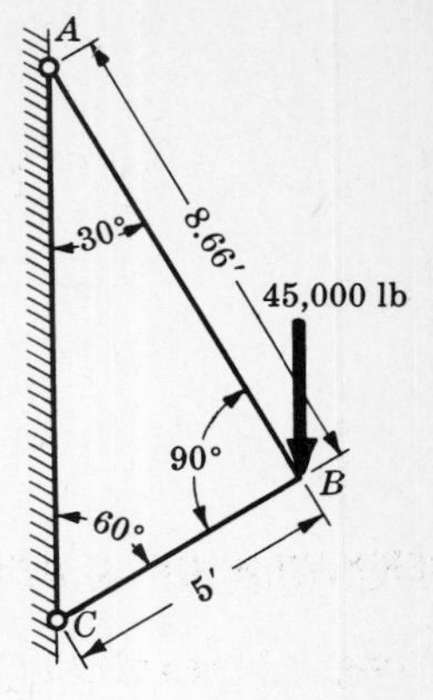

33. The steel bars AB and BC are pinned at each end and support the load of 45,000 lb, as shown in the adjacent diagram. The material is structural steel having a yield point of 35,000 lb/in^2 and safety factors of 2 and 3.5 are satisfactory for tension and compression respectively. Determine the size of each bar and also the horizontal and vertical components of displacement of point B. Take $E = 30 \cdot 10^6$ lb/in^2. Neglect any possibility of lateral buckling of bar BC.

Ans. Area AB = 2.23 in^2
Area BC = 2.25 in^2
Δ_h = 0.0130 in. (to right)
Δ_v = 0.0625 in. (downward)

34. A solid circular brass bar one inch in diameter is subject to an axial tensile force of 10,000 lb. Determine the decrease in the diameter of the bar due to this load. For brass $E = 13.5 \cdot 10^6$ lb/in^2 and $\mu = 0.28$.
Ans. 0.000264 in.

35. A square steel bar is 2 in. on a side and 10 in. long. It is loaded by an axial tensile force of 40,000 lb. If $E = 30 \cdot 10^6$ lb/in^2 and $\mu = 0.3$, determine the change of volume per unit volume.
Ans. 0.000133

36. Consider the square aluminum bar described in Problem 19 but with the axial loading reversed so as to cause compression. The compressive strain is considered to be 0.001 in/in. Determine the volume of the bar when the compressive force has been applied. *Ans.* 39.9864 in^3

37. Consider a state of stress of an element in which a stress s_x is exerted in one direction and lateral contraction is completely restrained in each of the other two directions. Find the effective modulus of elasticity and also the effective value of Poisson's ratio.

Ans. Eff. mod. $= \dfrac{E(1-\mu)}{(1-2\mu)(1+\mu)}$, Eff. Poisson's ratio = 0

38. Consider the state of stress in a bar subject to compression in the axial direction. Lateral expansion is restrained to half the amount it would ordinarily be if the lateral faces were load-free. Find the effective modulus of elasticity.

Ans. Eff. mod. $= \dfrac{E(1-\mu)}{(1-\mu-\mu^2)}$

39. A bar of uniform cross-section is subject to uniaxial tension and develops a strain in the direction of the force of 1/800. Calculate the change of volume per unit volume. Assume $\mu = 1/3$.
Ans. 1/2400 (increase)

40. A straight aluminum rod of diameter 1.25 in. is subjected to an axial tensile force of 10,000 lb. Determine

(*a*) the unit stress
(*b*) the unit strain
(*c*) the elongation in an 8 in. gage length
(*d*) the change of diameter
(*e*) the change of cross-sectional area
(*f*) the change of volume in an 8 in. gage length.

Ans. *a*) 8150 lb/in^2
b) 0.000815 in/in
c) 0.00653 in.
d) −0.000255 in.
e) −0.00050 in^2
f) 0.00400 in^3

Assume $E = 10 \cdot 10^6$ lb/in^2, $\mu = 1/4$.

CHAPTER 2

Statically Indeterminate Force Systems Tension and Compression

DEFINITION OF A DETERMINATE FORCE SYSTEM. If the values of all the external forces which act on a body can be determined by the equations of static equilibrium alone, then the force system is *statically determinate*. The problems in Chapter 1 were all of this type.

EXAMPLES OF DETERMINATE FORCE SYSTEMS. The bar shown in Figure 1 is loaded by the force P. The reactions are R_1, R_2, and R_3. The system is statically determinate because there are three equations of static equilibrium available for the system and these are sufficient to determine the three unknowns.

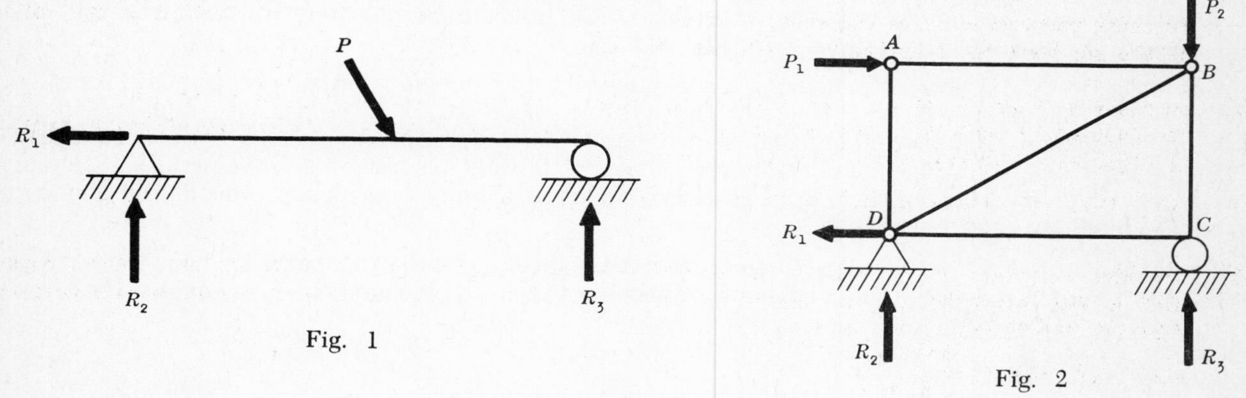

Fig. 1

Fig. 2

The truss *ABCD* shown in Figure 2 is loaded by the forces P_1 and P_2. The reactions are R_1, R_2, and R_3. Again, since there are three equations of static equilibrium available all three unknown reactions may be determined and consequently the external force system is statically determinate.

The above two illustrations refer only to external reactions and the force systems may be defined as statically determinate externally.

DEFINITION OF AN INDETERMINATE FORCE SYSTEM. In many cases the forces acting on a body cannot be determined by the equations of statics alone because there are more unknown forces than there are equations of equilibrium. In such a case the force system is said to be *statically indeterminate*.

EXAMPLES OF INDETERMINATE FORCE SYSTEMS. The bar shown in Figure 3 below is loaded by the force P. The reactions are R_1, R_2, R_3, and R_4. The force system is statically indeterminate because there are four unknown reactions but only three equations of static equilibrium. Such a force system is said to be indeterminate to the first degree.

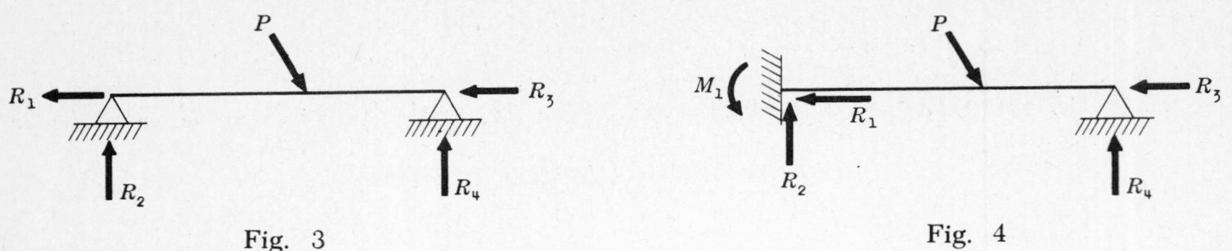

Fig. 3 Fig. 4

The bar shown in Figure 4 above is statically indeterminate to the second degree because there are five unknown reactions R_1, R_2, R_3, R_4, and M_1 but only three equations of static equilibrium. Consequently the values of all reactions cannot be determined by use of statics equations alone.

METHOD OF ANALYSIS. The approach that we will consider here is called the *deformation method* because it considers the deformations in the system. Briefly, the procedure to be followed in analyzing an indeterminate system is to first write all equations of static equilibrium that pertain to the system and then *supplement* these equations with additional equations based upon the deformations of the structure. Enough equations involving deformations must be written so that the total number of equations from both statics and deformations is equal to the number of unknown forces involved. See particularly Problems 1, 2, 3, 4, 7, 12, and 13.

For example, if a system contains five unknown forces yet only three equations of static equilibrium may be written for the system, then it is necessary to supplement the statics equations with two equations based upon the deformations. This system is statically indeterminate to the second degree. The resulting system of five equations is then solved for the five unknowns. Fortunately, in only a very few cases would all unknowns appear in each of the equations.

In this chapter we will treat indeterminate systems involving bars in tension or compression. Other types of indeterminate members will be discussed in subsequent chapters.

SOLVED PROBLEMS

1. The bar shown in Fig. (*a*) is of constant cross-section and is held rigidly between the walls. An axial load P is applied to the bar at a distance L_1 from the left end. Determine the reactions of the walls upon the bar.

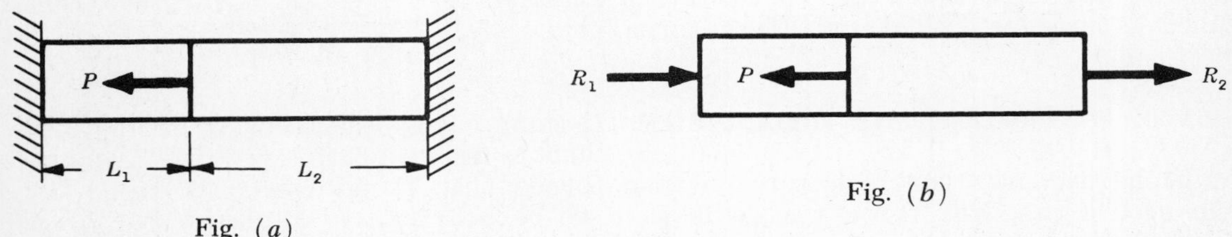

Fig. (*a*) Fig. (*b*)

We must first draw the free-body diagram of the bar, showing the applied load P together with the reactions of the walls upon the bar. These reactions are denoted by R_1 and R_2, as shown in Fig. (*b*).

There is only one equation of static equilibrium, namely

$$\Sigma F_h = R_1 - P + R_2 = 0$$

Since this equation contains two unknowns (R_1 and R_2) the problem is statically indeterminate. Consequently this statics equation must be supplemented by an additional equation based upon the deformations of the bar.

The shortening of the portion of the bar of length L_1 must be equal to the elongation of the region of length L_2. This fact furnishes the basis for the equation concerning deformations. The change of length of a bar due to axial loading was given in Problem 1, Chapter 1. The axial force acting in the left region of the bar is R_1(lb) and in the right region R_2(lb). The equation relating deformations becomes

$$\frac{R_1 L_1}{AE} = \frac{R_2 L_2}{AE}$$

where A denotes the cross-sectional area of the bar and E the modulus of elasticity. From this equation we have $R_1 L_1 = R_2 L_2$, and solving this simultaneously with the statics equation we find

$$R_1 = \frac{PL_2}{L_1 + L_2} \quad \text{and} \quad R_2 = \frac{PL_1}{L_1 + L_2}$$

Knowing these reactions, it is evident that the elongation of the right portion (L_2) of the bar is

$$\Delta_e = \frac{R_2 L_2}{AE} = \frac{PL_1 L_2}{(L_1 + L_2)AE}$$

and the shortening of the left portion (L_1) of the bar is

$$\Delta_c = -\frac{R_1 L_1}{AE} = -\frac{PL_1 L_2}{(L_1 + L_2)AE}$$

Thus

$$\Delta_e = -\Delta_c$$

2. Consider a steel tube surrounding a solid aluminum cylinder, the assembly being compressed between infinitely rigid cover plates by centrally applied forces as shown in Fig. (a). The aluminum cylinder is 3 in. in diameter and the outside diameter of the steel tube is 3.5 in. If $P = 48{,}000$ lb, find the stress in the steel and also in the aluminum. For steel, $E = 30 \cdot 10^6$ lb/in^2 and for aluminum $E = 4 \cdot 10^6$ lb/in^2.

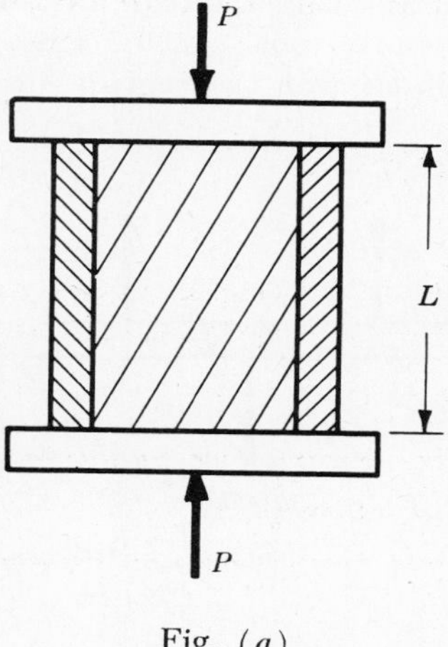

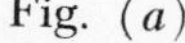

Fig. (a)

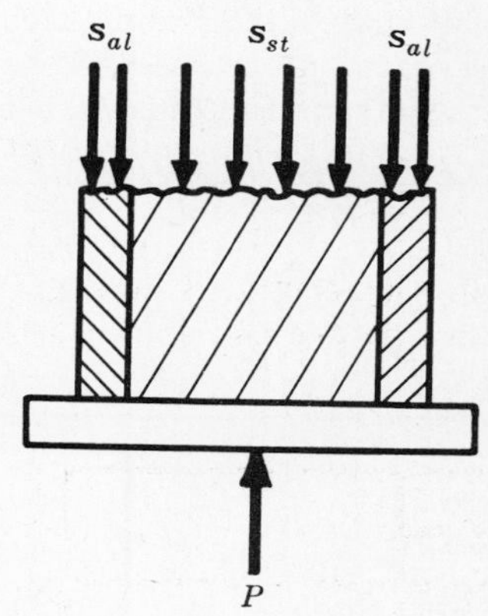

Fig. (b)

Let us pass a horizontal plane through the assembly at any elevation except in the immediate vicinity of the cover plates and then remove one portion or the other, say the upper portion. In that event the portion that we have removed must be replaced by the effect it exerted upon the remaining portion and that effect consists of vertical normal stresses distributed over the two materials. The free-body diagram of the portion of the assembly below this cutting plane is shown in Fig. (b) above where s_{st} and s_{al} denote the normal stresses existing in the steel and aluminum respectively.

Let us denote the resultant force carried by the steel by P_{st} (lb) and that carried by the aluminum by P_{al}. Then

$$P_{st} = A_{st} \cdot s_{st} \quad \text{and} \quad P_{al} = A_{al} \cdot s_{al}$$

where A_{st} and A_{al} denote the cross-sectional areas of the steel tube and the aluminum cylinder respectively. There is only one equation of static equilibrium available for such a force system and it takes the form

$$\Sigma F_v = P - P_{st} - P_{al} = 0$$

Thus, we have one equation in two unknowns, P_{st} and P_{al}, and hence the problem is statically indeterminate. In that event we must supplement the available statics equation by an additional equation derived from the deformations of the structure. Such an equation is readily obtained because the infinitely rigid cover plates force the axial deformations of the two metals to be identical.

The deformation due to axial loading is given by $\Delta = PL/AE$. Equating axial deformations of the steel and the aluminum we have

$$\frac{P_{st} \cdot L}{A_{st} \cdot E_{st}} = \frac{P_{al} \cdot L}{A_{al} \cdot E_{al}}$$

or

$$\frac{P_{st} \cdot L}{\frac{\pi}{4}[(3.5)^2 - (3)^2](30 \cdot 10^6)} = \frac{P_{al} \cdot L}{\frac{\pi}{4}(3)^2(4 \cdot 10^6)} \qquad \text{from which} \qquad P_{st} = 2.71 P_{al}$$

This equation is now solved simultaneously with the statics equation, $P - P_{st} - P_{al} = 0$, and we find $P_{al} = 0.27P$, $P_{st} = 0.73P$.

For a load of $P = 48{,}000$ lb this becomes $P_{al} = 12{,}900$ lb and $P_{st} = 35{,}100$ lb. The desired stresses are found by dividing the resultant force in each material by its cross-sectional area:

$$s_{al} = \frac{12{,}900}{\frac{\pi}{4}(3)^2} = 1820 \text{ lb/in}^2, \qquad s_{st} = \frac{35{,}100}{\frac{\pi}{4}[(3.5)^2 - (3)^2]} = 13{,}700 \text{ lb/in}^2.$$

3. The bar AB is absolutely rigid and is supported by three rods as shown in Fig. (*a*). The two outer rods are steel, and each has a cross-sectional area of 0.50 in^2. The central rod is copper and of area 1.5 in^2. For steel, $E = 30 \cdot 10^6$ lb/in^2 and for copper $E = 17 \cdot 10^6$ lb/in^2. All rods are 7 ft long. The three rods are equally spaced and the applied loads of 12,000 lb are each applied midway between the rods. Neglecting the weight of the bar AB, determine the force in each of the vertical bars. The bar AB remains horizontal after the loads have been applied.

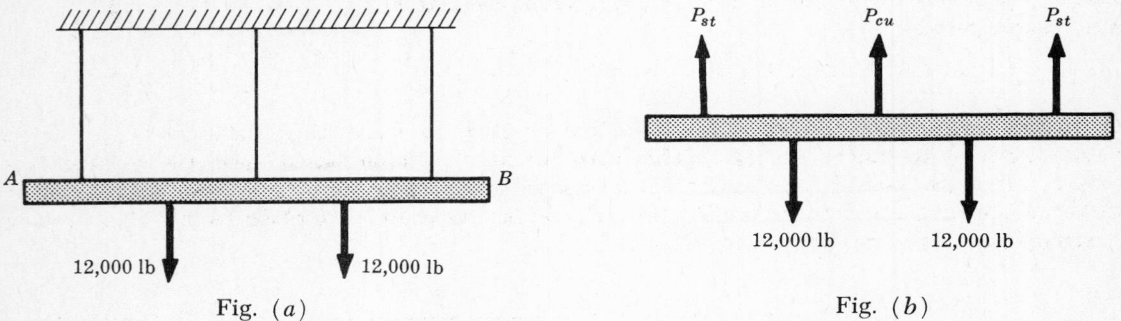

Fig. (*a*) Fig. (*b*)

First draw a free-body diagram of the bar AB showing all the forces acting on it. These forces include the two applied loads and the reactions of the vertical rods upon bar AB. If the force in each of the steel rods is denoted by P_{st} (lb) and that in the copper by P_{cu} (lb), then this diagram appears as in Fig. (*b*) above.

Use has already been made of the condition of symmetry in stating that the forces in the steel rods are equal; hence there remains only one equation of static equilibrium, namely

$$\Sigma F_v = 2P_{st} + P_{cu} - 24{,}000 = 0$$

Thus we have one equation containing two unknowns and the problem is statically indeterminate. This statics equation must be supplemented by an additional equation coming from the deformations of the structure.

Such an equation is readily determined because the elongations of the steel and copper rods are equal. The elongation due to axial loading is $\Delta = PL/AE$, and applying that expression to the steel and copper rods we have

$$\frac{P_{st}(84)}{(0.5)(30\cdot 10^6)} = \frac{P_{cu}(84)}{(1.5)(17\cdot 10^6)} \quad \text{or} \quad P_{st} = 0.588 P_{cu}$$

This equation may now be solved simultaneously with the statics equation to yield

$$2(0.588 P_{cu}) + P_{cu} - 24{,}000 = 0$$

Solving, $P_{cu} = 11{,}000$ lb and $P_{st} = 6500$ lb.

4. Consider a square reinforced concrete pier 12×12 in. in cross-section and 8 ft high. The concrete is strengthened by the addition of eight vertical square steel reinforcing bars, each 3/4 in. on a side and placed symmetrically about the vertical axis of the pier. The pier is loaded by an axial compressive force of 90,000 lb applied to the pier through an absolutely rigid cover plate over the top of the concrete. For steel consider $E = 30\cdot 10^6$ lb/in^2 and for concrete take $E = 2.5\cdot 10^6$ lb/in^2. Determine the stress in the concrete and also in the steel.

Let us cut the pier by a horizontal plane and remove the portion above this plane. The part removed must be replaced by whatever effect it exerted upon the lower portion and that effect consists of vertical forces distributed over the concrete and also over the steel. The free-body diagram of the lower portion has the appearance shown in the adjacent diagram, where P_{st} and P_c denote the resultant forces exerted upon the steel and upon the concrete respectively by the upper portion which has been removed. The force P_c, for example, is actually the resultant of normal stresses that are assumed to be uniformly distributed over the entire cross-sectional area of the concrete. Since the loading is axial it is reasonable to assume a uniform distribution of normal stress, and consequently the resultant P_c lies along the geometric axis of the pier.

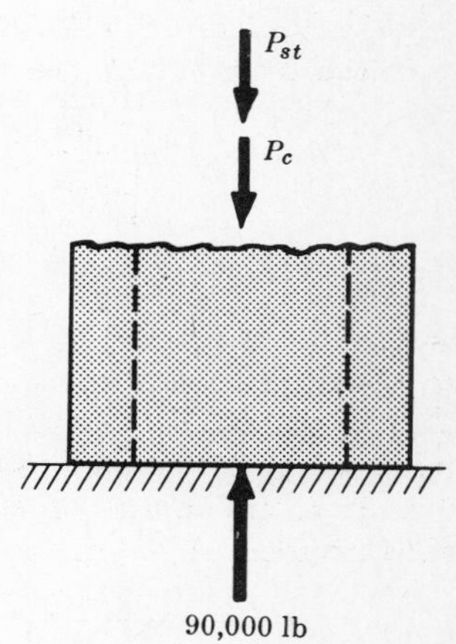

There is only one equation of static equilibrium available for such a system, namely

$$\Sigma F_v = 90{,}000 - P_c - P_{st} = 0$$

This equation contains two unknowns, hence the problem is statically indeterminate. Consequently this equation must be solved in conjunction with another equation based upon the deformations of the structure. Such an equation is easy to obtain since the shortening of the concrete and the steel are equal because of the rigid cover plate. The deformation under axial loading is $\Delta = PL/AE$, and applying this expression to the two materials we have

$$\frac{P_{st}\cdot L}{8(3/4)^2(30\cdot 10^6)} = \frac{P_c\cdot L}{[144 - 8(3/4)^2](2.5\cdot 10^6)}$$

where L denotes the height of the pier. Solving, $P_{st} = 0.388 P_c$. Then

$$90{,}000 - P_c - 0.388 P_c = 0, \qquad P_c = 64{,}900 \text{ lb} \quad \text{and} \quad P_{st} = 25{,}100 \text{ lb}.$$

The stress in the steel is found by dividing the resultant force over all eight bars by their area. Similarly the stress in the concrete is obtained by dividing the resultant force P_c by the cross-sectional area of the concrete. Thus

$$s_{st} = \frac{25{,}100}{8(3/4)^2} = 5600 \text{ lb/in}^2 \quad \text{and} \quad s_c = \frac{64{,}900}{144 - 8(3/4)^2} = 465 \text{ lb/in}^2$$

5. A vertical steel tube has an outside diameter of 36 in, an inside diameter of 35 in, and is filled with concrete. If the yield point of the steel is 45,000 lb/in^2 and a safety factor of 2.25 is adequate and further if the ultimate strength of the concrete is 2500 lb/in^2 and a safety factor of 2.5 is used, what total axial compressive load may be carried? Assume both ends of tube are covered with infinitely rigid plates and neglect the effects of lateral expansion of the two materials. For the steel take $E = 30 \cdot 10^6$ lb/in^2 and for the concrete take $E = 2.5 \cdot 10^6$ lb/in^2. (The ratio of Young's modulus for the steel to the modulus for the concrete is commonly designated by n, i.e., $n = E_{st}/E_c$. Here $n = 12$.)

The cross-sectional area of the concrete is 965 in^2, that of the steel 55.0 in^2. Since the total change in the height of the steel must be equal to that of the concrete, we have

$$\frac{P_c \cdot L}{(965)(2.5 \cdot 10^6)} = \frac{P_{st} \cdot L}{(55)(30 \cdot 10^6)} \quad \text{or} \quad P_c = 1.46 P_{st}$$

where P_c and P_{st} denote the resultant axial forces in the concrete and in the steel respectively. From statics we have only one equation, namely $P = P_c + P_{st}$, where P denotes the total axial load carried.

It is unlikely that the allowable working stress is set up in both materials simultaneously. Perhaps the simplest procedure is to calculate two values of the total axial force, one based upon the assumption that the concrete is stressed to its working stress of 1000 lb/in^2, the other based upon the premise that the working stress of 20,000 lb/in^2 exists in the steel. The smaller of these two loads is naturally the governing value. Thus, if the concrete is stressed to its maximum working stress we have

$$P = 1000(965)[1 + 1/1.46] = 1,625,000 \text{ lb}$$

On the other hand if the steel is stressed to 20,000 lb/in^2 we have

$$P = 20,000(55)[1 + 1.46] = 2,710,000 \text{ lb}$$

Consequently, $P = 1,625,000$ lb is the allowable axial load.

6. The initially straight bar AD has a uniform cross-section and is clamped to the end supports shown. Initially the bar is stress-free. The symmetrically placed loads shown in Fig. (a) are applied to the brackets (whose effect is to be neglected) and it is desired to find the resultant tensile or compressive force acting over any normal cross-section in each of the three regions AB, BC, and CD.

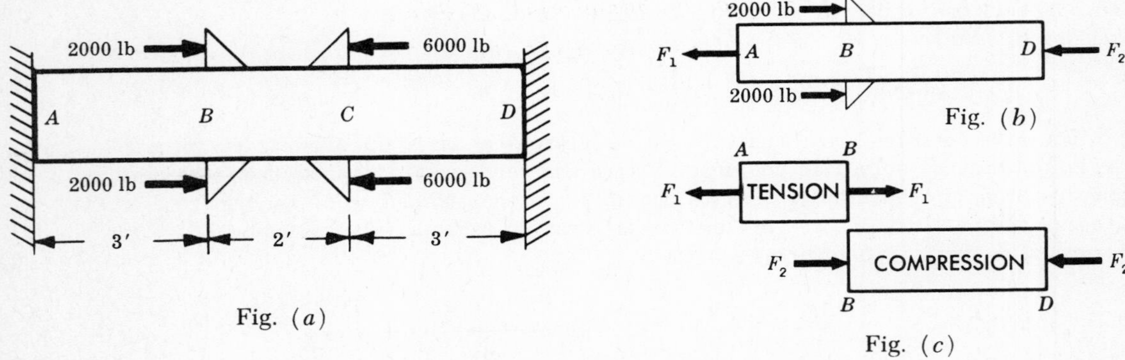

Fig. (a)

Fig. (b)

Fig. (c)

Let us consider first only the total load of 4000 lb applied at B and realize that the bar AD is in equilibrium. Thus there will be two reactions F_1 and F_2 created at the ends of the bar to equilibrate the force of 2(2000) = 4000 lb. Between A and B there will be a tension of F_1 and between B and D there will be a compression as shown in Fig. (b) above. This may also be represented as in Fig. (c) above. Thus F_1 elongates AB and B moves the distance $\Delta_1 = F_1(36)/AE$ to the right. Also, F_2 compresses BD and B moves the distance $\Delta_2 = F_2(60)/AE$.

Evidently $\Delta_1 = \Delta_2$ and we may write $\dfrac{F_1(36)}{AE} = \dfrac{F_2(60)}{AE}$ or $F_1 = (\frac{5}{3}) F_2$.

From statics we have only one equation, $\Sigma F_h = -F_1 - F_2 + 4000 = 0$. Substituting, $(5/3)F_2 + F_2 = 4000$, $F_2 = 1500$ lb (BD is in compression) and $F_1 = 2500$ lb (AB is in tension).

From this the behavior of the distribution of internal axial forces is evident. Due to the load of 2(6000) = 12,000 lb we have

$$(5/8)(12{,}000) = 7500 \text{ lb } (CD \text{ is in tension})$$

$$(3/8)(12{,}000) = 4500 \text{ lb } (AC \text{ is in compression})$$

The resultant axial forces in the various regions of AD are now found by algebraic summation of the above results. The final values of the forces are

$$AB = 2500 - 4500 = -2000 \text{ lb}, \quad BC = -1500 - 4500 = -6000 \text{ lb}, \quad CD = -1500 + 7500 = 6000 \text{ lb}$$

where a positive sign denotes a tensile force and a negative sign a compressive force.

7. The bar AB is considered to be absolutely rigid and is horizontal before the load of 40,000 lb is applied as shown in Fig. (a). The connection at A is a pin, and AB is supported by the steel rod EB and the copper rod CD. The length of CD is 3 ft, of EB is 5 ft. The cross-sectional area of CD is 0.8 in^2, the area of EB is 0.5 in^2. Determine the stress in each of the vertical rods and the elongation of the steel rod. Neglect the weight of AB. For copper $E = 17\cdot 10^6$ lb/in^2, for steel $E = 30\cdot 10^6$ lb/in^2.

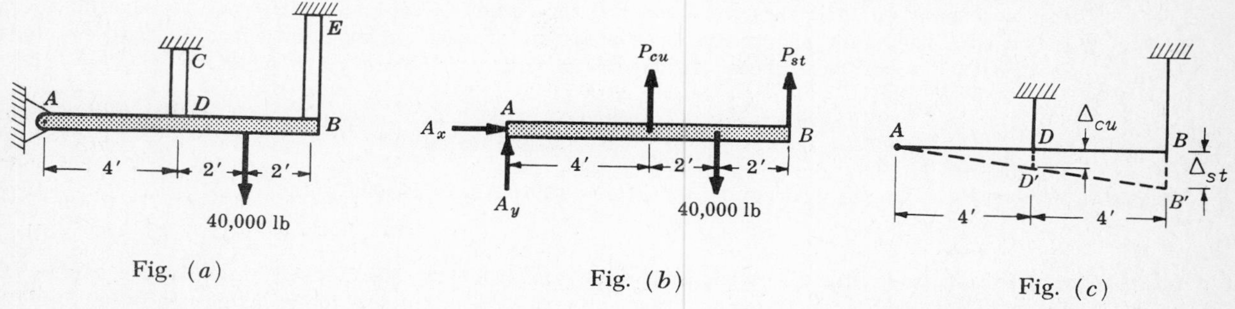

Fig. (a) Fig. (b) Fig. (c)

The first step in the solution is to draw a free-body diagram of the bar AB, showing all forces acting on it. This appears as in Fig. (b) above.

From statics we have

1) $\Sigma F_h = A_x = 0$

2) $\Sigma M_a = 4P_{cu} + 8P_{st} - 40{,}000\,(6) = 0$

3) $\Sigma F_v = A_y + P_{cu} + P_{st} - 40{,}000 = 0$

Since the last two statics equations contain three unknowns the problem is statically indeterminate. Hence we must look to the deformations of the system for another equation. Because the bar AB is rigid the only movement it can undergo is a rigid body rotation about the pin at A as a center. The dotted line in Fig. (c) above indicates the final position of the bar AB after the load of 40,000 lb had been applied. Initially the bar was horizontal, as shown by the solid line.

The lower ends of the rods are originally at D and B, and they move to D' and B' after application of the 40,000 lb load. Because the bar AB is rigid, the similar triangles ADD' and ABB' furnish a simple relation between the deformations of the two vertical bars, namely $\Delta_{cu}/4 = \Delta_{st}/8$, where Δ_{cu} and Δ_{st} denote the elongations of the copper and steel rods respectively. Thus, the additional equation based upon deformations is

$$\Delta_{st} = 2\Delta_{cu}$$

But the elongation under axial loading is given by $\Delta = PL/AE$. Using this equation in the above deformation relation we get

$$\frac{P_{st}\,(60)}{(0.5)(30\cdot 10^6)} = \frac{2P_{cu}\,(36)}{(0.8)(17\cdot 10^6)} \quad \text{or} \quad P_{st} = 1.33\,P_{cu}$$

Solving this equation simultaneously with the statics equation *(2)* we find

$$4P_{cu} + 8(1.33P_{cu}) = 240{,}000, \qquad P_{cu} = 16{,}400 \text{ lb} \quad \text{and} \quad P_{st} = 21{,}800 \text{ lb}$$

The stresses are given by the simple relation $s = P/A$.

In the copper rod, $s_{cu} = 16{,}400/0.8 = 20{,}500 \text{ lb/in}^2$

In the steel rod, $s_{st} = 21{,}800/0.5 = 43{,}600 \text{ lb/in}^2$

8. The copper bar is of uniform cross-section and is rigidly attached to the walls as shown. The rod is 5 ft long and has a cross-sectional area of 2.5 in^2. At a temperature of 80° F the rod is stress-free in the configuration shown. Determine the stress in the rod when the temperature has dropped to 50°F. Assume that the supports do not yield. For copper, $E = 16\cdot10^6$ lb/in^2 and $\alpha = 9.3\cdot10^{-6}$/°F.

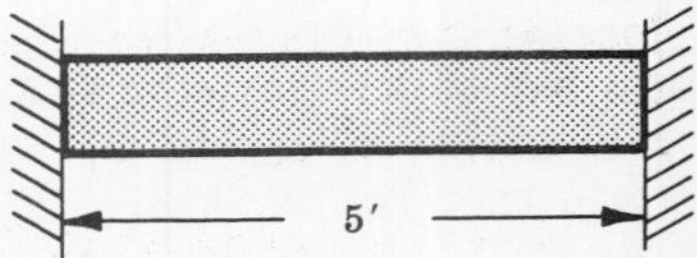

One approach to this problem is to assume that the bar is cut free from the wall at the right end. In that event it is free to contract when the temperature falls and the bar shortens an amount

$$\Delta = (9.3\cdot10^{-6})(60)(30) = 0.0167 \text{ in.}$$

according to the definition of the coefficient of linear expansion (see Chapter 1).

It is next necessary to find the axial force P that must be applied to the bar to stretch it 0.0167 in., i.e. to restore the right end to its true position because we know that actually that end does not move at all when the temperature drops. To determine this force P, use the equation

$$\Delta = \frac{PL}{AE} \quad \text{which gives} \quad 0.0167 = \frac{P(60)}{2.5(16\cdot10^6)} \quad \text{or} \quad P = 11{,}200 \text{ lb}$$

The axial stress set up by this force is $s = P/A = 11{,}200/2.5 = 4500 \text{ lb/in}^2$.

9. The composite bar shown is rigidly attached to the two supports. The left portion of the bar is copper, of uniform cross-sectional area 12 in^2 and length 12 in. The right portion is aluminum, of uniform cross-sectional area 3 in^2 and length 8 in. At a temperature of 80°F the entire assembly is stress-free. The temperature of the structure drops and during this process the right support yields 0.001 in. in the direction of the contracting metal. Determine the minimum temperature to which the assembly may be subjected in order that the stress in the aluminum does not exceed 24,000 lb/in^2. For copper $E = 16\cdot10^6$ lb/in^2, $\alpha = 9.3\cdot10^{-6}$/°F and for aluminum $E = 10\cdot10^6$ lb/in^2, $\alpha = 12.8\cdot10^{-6}$/°F.

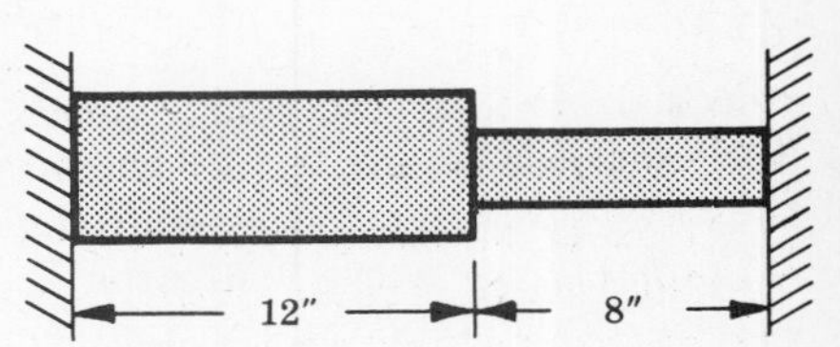

Again, as in the previous problem it is perhaps simplest to consider that the bar is cut just to the left of the supporting wall at the right and is then free to contract due to the temperature drop ΔT. The total shortening of the composite bar is given by

$$(9.3\cdot10^{-6})(12)(\Delta T) + (12.8\cdot10^{-6})(8)(\Delta T)$$

according to the definition of the coefficient of linear expansion. It is to be noted that the shape of the cross-section has no influence upon the change in length of the bar due to a temperature change.

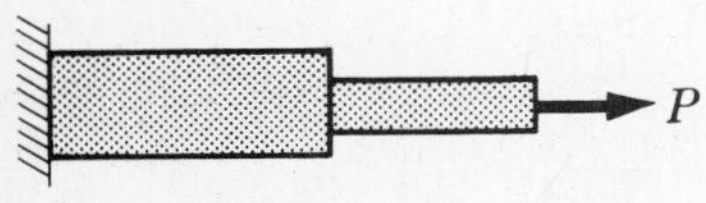

Even though the bar has contracted this amount it is still stress-free. However, this is not the complete analysis because the reaction of the wall at the right has been neglected by cutting the bar there. Consequently, we must represent the action of the wall by an axial force P applied to the bar, as shown in the adjoining diagram. For equilibrium, the resultant force acting over any cross-section of either the copper or the aluminum must be equal to P. The application of the force P stretches the composite bar by an amount $\dfrac{P(12)}{12(16\cdot10^6)} + \dfrac{P(8)}{3(10\cdot10^6)}$.

If the right support were unyielding we would equate the last expression to the expression giving the total shortening due to the temperature drop. Actually the right support yields 0.001 in. and consequently we may write

$$\frac{P(12)}{12(16\cdot10^6)} + \frac{P(8)}{3(10\cdot10^6)} = (9.3\cdot10^{-6})(12)(\Delta T) + (12.8\cdot10^{-6})(8)(\Delta T) - 0.001$$

The stress in the aluminum is not to exceed 24,000 lb/in² and since it is given by the formula $s = P/A$, the maximum force P becomes

$$P = A\cdot s = 3(24,000) = 72,000 \text{ lb}$$

Substituting this value of P in the above equation relating deformations, we find ΔT = 115°F. Therefore the temperature may drop 115°F from the original 80°F. The final temperature would be −35°F.

10. Consider the tapered (conical) steel bar shown which has both ends attached to unyielding supports. The bar is initially stress-free. If the temperature of the entire bar drops 40°F determine the maximum normal stress in the bar. Take $E = 30\cdot10^6$ lb/in² and $\alpha = 6.5\cdot10^{-6}$/°F.

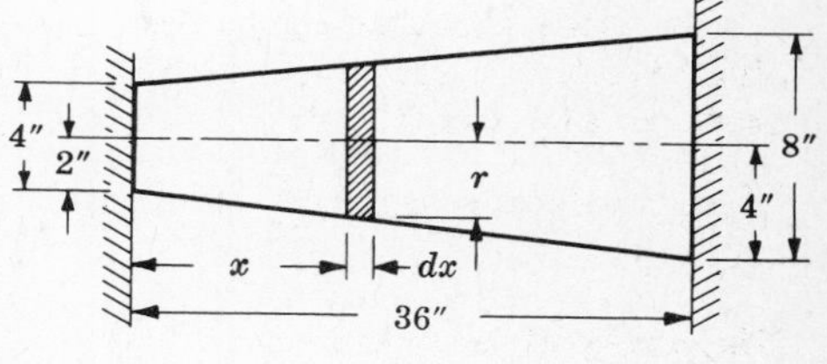

Perhaps the simplest technique for solving this problem is to imagine that one end of the bar, say the right end, is temporarily cut free from its support. In this case the bar contracts an amount

$$(40)(36)(6.5\cdot10^{-6}) = 0.00935 \text{ in.}$$

due to the temperature drop.

Next, let us find the axial force P which must be applied to the "free" right end so as to elongate the bar 0.00935 in., i.e. so that the true boundary condition of complete fixity will be satisfied at the right end. Setting up the coordinate system shown, we have

$$r = 2 + 2x/36 = 2 + x/18$$

Since the angle of taper is comparatively small, the tensile force may be assumed to be uniformly distributed over any one cross-section. Also, since there are no abrupt changes of cross-section we may determine the elongation of the shaded disc-like element of thickness dx by applying $\Delta = PL/AE$, where $L = dx$, to the disc and then integrate over the entire bar:

$$0.00935 = \int_0^{36} \frac{P\,dx}{\pi(2+x/18)^2 E} = \int_0^{36} \frac{324P\,dx}{E\pi(36+x)^2} = \frac{324P}{72E\pi}$$

Solving, P = 196,000 lb, where P is the resultant axial force acting over any cross-section, i.e. the force necessary to restore the bar to its original length.

It is to be noted that the resultant force over every vertical cross-section is P (lb) for equilibrium of any portion of the bar. However, since the cross-sectional area of the bar varies from one end to the other the stress varies from a maximum value at the left end, where the cross-sectional area is a minimum, to a minimum at the right end where the area is a maximum.

The maximum stress at the left end is given by $s_{max} = \dfrac{196,000}{\pi(2)^2} = 15,600$ lb/in².

11. A hollow steel cylinder surrounds a solid copper cylinder and the assembly is subject to an axial loading of 50,000 lb as shown in Fig. (*a*) below. The cross-sectional area of the steel is 3 in², while that of the copper is 10 in². Both cylinders are the same length before the load is applied. Determine the temperature rise of the entire system required to place all of the load on the copper cylinder. The cover plate at the top of the assembly is rigid and for copper $E=16\cdot10^6$ lb/in², $\alpha=9.3\cdot10^{-6}$/°F, while for steel $E = 30\cdot10^6$ lb/in², $\alpha = 6.5\cdot10^{-6}$/°F.

One method of analyzing this problem is to assume that the load as well as the upper cover plate are removed and that the system is allowed to freely expand vertically because of a temperature rise ΔT. In that event the upper ends of the cylinders assume the positions shown by the dotted lines in Fig. (b) below.

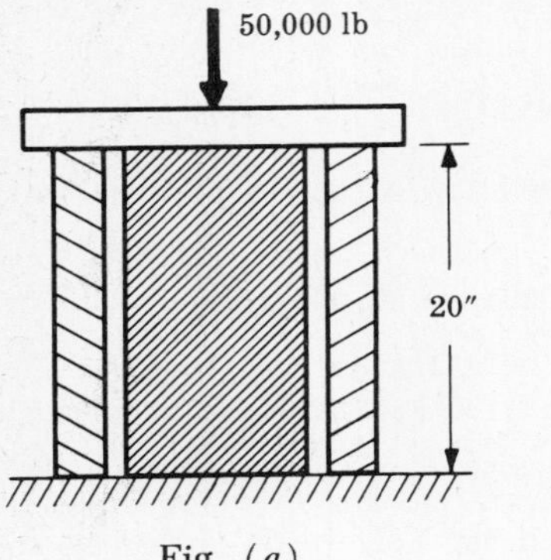

Fig. (a)

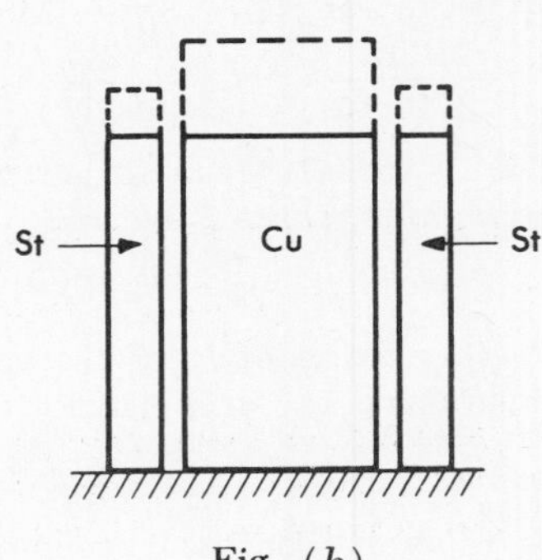

Fig. (b)

The copper cylinder naturally expands upward more than the steel one because the coefficient of linear expansion of copper is greater than that of steel. The upward expansion of the steel is

$$(6.5\cdot10^{-6})(20)(\Delta T),$$

while that of the copper is

$$(9.3\cdot10^{-6})(20)(\Delta T)$$

This is not of course the true situation because the load of 50,000 lb has not as yet been considered. If all of this axial load is carried by the copper then only the copper will be compressed and the compression of the copper is given by

$$\Delta_{cu} = \frac{PL}{AE} = \frac{50,000(20)}{(10)(16\cdot10^6)}$$

The condition of the problem states that the temperature rise ΔT is just sufficient so that all of the load is carried by the copper. Thus, the expanded length of the copper indicated by the dotted lines in the above sketch will be decreased by the action of the force. The net expansion of the copper is the expansion caused by the rise of temperature minus the compression due to the load. The change of length of the steel is due only to the temperature rise. Consequently we may write

$$(9.3\cdot10^{-6})(20)(\Delta T) - \frac{50,000(20)}{(10)(16\cdot10^6)} = (6.5\cdot10^{-6})(20)(\Delta T) \qquad \text{or} \qquad \Delta T = 111^\circ\text{F}$$

12. The rigid bar AD is pinned at A and attached to the bars BC and ED, as shown in Fig. (a). The entire system is initially stress-free and the weights of all bars are negligible. The temperature of bar BC is lowered 50°F and that of the bar ED is raised 50°F. Neglecting any possibility of lateral buckling, find the normal stresses in bars BC and ED. For BC, which is brass, assume $E = 14\cdot10^6$ lb/in², $\alpha = 10.4\cdot10^{-6}$/°F and for ED, which is steel, take $E = 30\cdot10^6$ lb/in², and $\alpha = 6.5\cdot10^{-6}$/°F. The cross-sectional area of BC is 1 in², of ED is 0.5 in².

Let us denote the forces acting on AD by P_{st} and P_{br} acting in the assumed directions shown in the free-body diagram, Fig. (b). Since AD rotates as a rigid body about A (as shown by the dotted line) we have $\Delta_{br}/10 = \Delta_{st}/25$, where Δ_{br} and Δ_{st} denote the axial compression and axial elongation of BC and DE respectively.

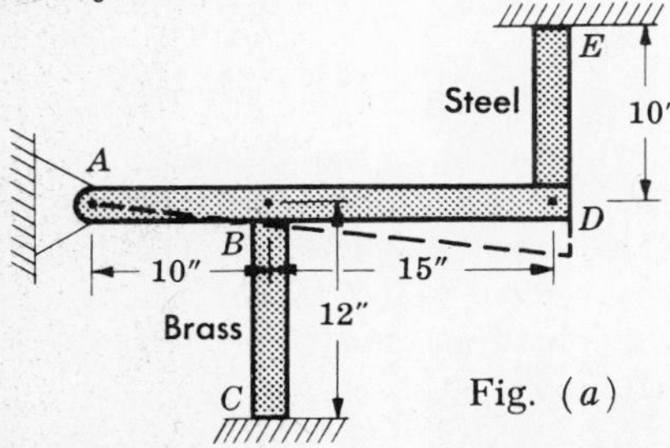

Fig. (a)

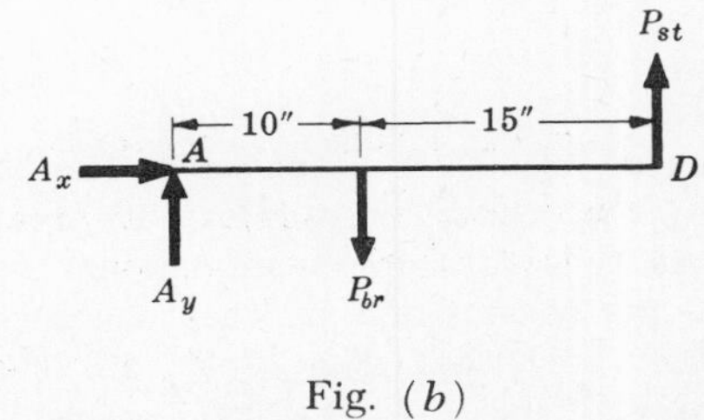

Fig. (b)

The total change of length of BC is composed of a shortening due to the temperature drop as well as a lengthening due to the axial force P_{br}. The total change of length of DE is composed of a lengthening due to the temperature rise as well as a lengthening due to the force P_{st}. Hence we have

$$\frac{2}{5}\left[(6.5\cdot10^{-6})(10)(50) + \frac{P_{st}(10)}{(0.5)(30\cdot10^{6})}\right] = -(10.4\cdot10^{-6})(12)(50) + \frac{P_{br}(12)}{(1)(14\cdot10^{6})}$$

or

$$0.856P_{br} - 0.267P_{st} = 7750$$

From statics,

$$\Sigma M_A = 10P_{br} - 25P_{st} = 0$$

Solving these equations simultaneously, $P_{st} = 4030$ lb and $P_{br} = 10,100$ lb.

Using $s = P/A$ for each bar, we obtain $s_{st} = 8060\ \text{lb/in}^2$ and $s_{br} = 10,100\ \text{lb/in}^2$.

13. Consider the statically indeterminate pin-connected framework shown in Fig. (a). Before the load P is applied the entire system is stress-free. Find the axial force in each bar caused by the vertical load P. The two outer bars are identical and have the cross-sectional area A_i, the middle bar has an area A_v. All bars have the same modulus of elasticity, E.

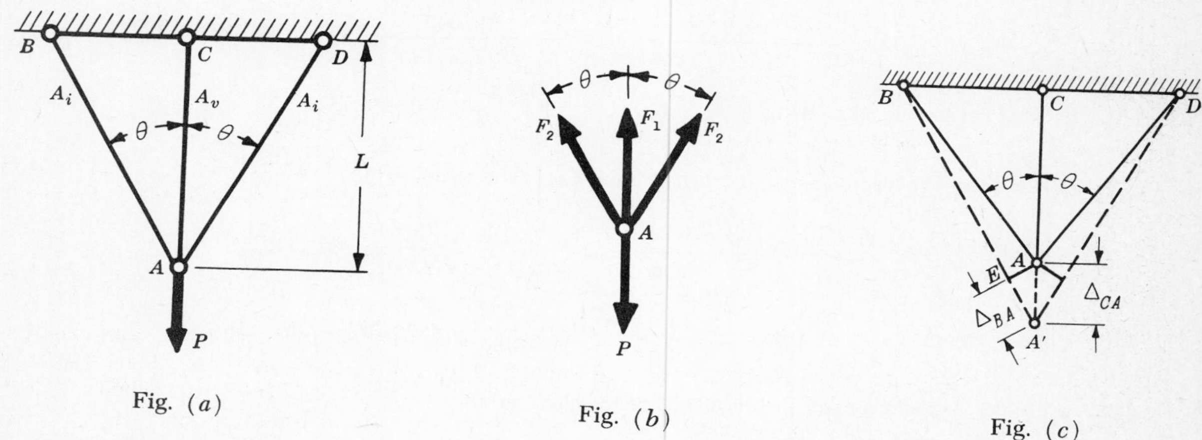

Fig. (a) Fig. (b) Fig. (c)

The free-body diagram of the pin at A appears as in Fig. (b) above, where F_1 and F_2 denote axial forces (lb) in the vertical and inclined bars. From statics we have

$$\Sigma F_v = F_1 + 2F_2\cos\theta - P = 0.$$

This is the only statics equation available since we have made use of symmetry in stating that the forces in the inclined bars are equal. Since it contains two unknowns, F_1 and F_2, the force system is statically indeterminate. Hence we must examine the deformations of the system to obtain another equation. Under the action of the load P the bars assume the positions shown by the dotted lines in Fig. (c) above.

Because the deformations of the system are *small* the basic geometry is essentially unchanged and the angle $BA'A$ may be taken to be θ. AEA' is a right triangle and AE, which is actually an arc having a radius equal in length to the length of the inclined bars, is perpendicular to BA'. The elongation of the vertical bar is thus represented by AA' and that of the inclined bars by EA'. From this small triangle we have the relation

$$\Delta_{BA} = \Delta_{CA}\cos\theta$$

where Δ_{BA} and Δ_{CA} denote elongations of the inclined and vertical bars respectively.

Since these bars are subject to axial loading their elongations may be found by the formula $\Delta = PL/AE$. From that expression we have

$$\Delta_{BA} = \frac{F_2(L/\cos\theta)}{A_i E} \quad \text{and} \quad \Delta_{CA} = \frac{F_1 L}{A_v E}$$

Substituting these in the above equation relating Δ_{BA} and Δ_{CA} we have

$$\frac{F_2 L}{A_i E \cos\theta} = \frac{F_1 L}{A_v E}\cos\theta \quad \text{or} \quad F_2 = F_1\left(\frac{A_i}{A_v}\right)\cos^2\theta$$

Substituting this in the statics equation we find $F_1 + 2F_1(A_i/A_v)\cos^3\theta = P$

or $$F_1 = \frac{P}{1 + 2(A_i/A_v)\cos^3\theta} \quad \text{and} \quad F_2 = \frac{P\cos^2\theta}{(A_v/A_i) + 2\cos^3\theta}$$

14. In the pin-connected framework discussed in Problem 13 the inclined bars each have a cross-sectional area of 1 in^2, the vertical bar has an area of 2 in^2, $BC = CD$ = 12 in., CA = 16 in., and $E = 30\cdot10^6$ lb/in^2. If the applied load P = 12,000 lb, determine the normal stress in each bar as well as the vertical deflection of point A.

Here we have A_i = 1 in^2, A_v = 2 in^2, $\cos\theta = 16/20 = 4/5$, and P = 12,000 lb. From Problem 13, the axial force in the vertical bar CA is

$$F_1 = \frac{P}{1 + 2(A_i/A_v)\cos^3\theta} = \frac{12{,}000}{1 + 2(1/2)(4/5)^3} = 7950 \text{ lb}$$

The normal stress in bar CA is $s_1 = F_1/A_v = 7950/2 = 3980$ lb/in^2.

The axial force in each of the inclined bars is, from Problem 13,

$$F_2 = \frac{P\cos^2\theta}{(A_v/A_i) + 2\cos^3\theta} = \frac{12{,}000(4/5)^2}{(2/1) + 2(4/5)^3} = 2540 \text{ lb}$$

The normal stress in each inclined bar is $s_2 = F_2/A_i = 2540/1 = 2540$ lb/in^2.

The vertical deflection of point A is, from Problem 13,

$$\Delta_{CA} = \frac{F_1 L}{A_v E} = \frac{(7950)(16)}{(2)(30\cdot10^6)} = 0.00212 \text{ in.}$$

SUPPLEMENTARY PROBLEMS

15. A square bar 2 inches on a side is held rigidly between the walls and loaded by an axial force of 40,000 lb as shown. Determine the reactions at the ends of the bar and the extension of the right portion. Take $E = 30\cdot10^6$ lb/in^2.

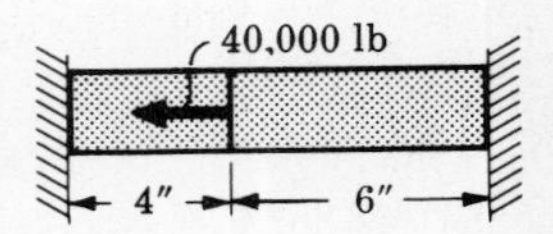

Ans. Left reaction = 24,000 lb, Right reaction = 16,000 lb
Extension = 0.00080 in.

16. A short hollow cast iron shell of square cross-section is filled with concrete. The outside dimension of the cast iron is 18 in. and the wall thickness is 1.5 in. The assembly is compressed by an axial force P of 140,000 lb applied to infinitely rigid cover plates as shown. Determine the stress in each material and the shortening of the member. For concrete take $E = 2.5\cdot10^6$ lb/in^2 and for cast iron consider $E = 15\cdot10^6$ lb/in^2.

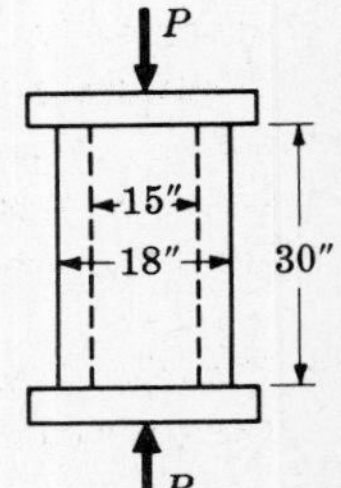

Ans. s_{ci} = 1030 lb/in^2
s_c = 170 lb/in^2
Δ = 0.00205 in.

17. Two initially straight bars are joined together and attached to supports as shown. The left bar is brass for which $E = 14\cdot10^6$ lb/in², $\alpha = 10.4\cdot10^{-6}/°F$ and the right bar is aluminum for which $E = 10\cdot10^6$ lb/in², $\alpha = 12.8\cdot10^{-6}/°F$. The cross-sectional area of the brass bar is 1 in², that of the aluminum bar is 1.5 in². Let us suppose that the system is initially stress-free and that the temperature then drops 40°F.

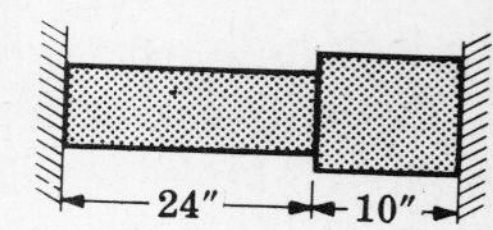

(*a*) If the supports are unyielding, find the normal stress in each bar.

(*b*) If the right support yields 0.005 in., find the normal stress in each bar. The weight of the bars is negligible.

Ans. *a*) s_{br} = 6350 lb/in², s_{al} = 4230 lb/in² *b*) s_{br} = 4250 lb/in², s_{al} = 2840 lb/in²

18. A steel tube 2 in. outside diameter and 1.75 in. inside diameter surrounds a solid brass cylinder 1.5 in. in diameter. Both are joined to a rigid cover plate at each end. The assembly is stress-free at a temperature of 80°F. If the temperature is then raised to 250°F., determine the stresses in each material. For brass $E = 14\cdot10^6$ lb/in², $\alpha = 10.4\cdot10^{-6}/°F$; for steel $E = 30\cdot10^6$ lb/in², $\alpha = 6.5\cdot10^{-6}/°F$.
Ans. s_{st} = 10,500 lb/in², s_{br} = −4400 lb/in²

19. A short reinforced concrete pier is subjected to an axial compressive load. Both ends are covered with infinitely rigid plates so that the overall deformations of the steel and the concrete are identical. If the stress set up in the concrete is 900 lb/in², find the stress in the steel. For the steel take $E = 30\cdot10^6$ lb/in², and consider $n = 12$ ($n = E_{st}/E_c$). Neglect the effects of lateral expansion of both steel and concrete under this load. *Ans.* s_{st} = 10,800 lb/in²

20. A compound bar is composed of a strip of copper between two cold-rolled steel plates. The ends of the assembly are covered with infinitely rigid cover plates and an axial tensile load P is applied to the bar by means of a force acting on each rigid plate as shown in Fig. (*a*) below. The width of all bars is 4 in., the steel plates are each 1/4 in. thick and the copper is 3/4 in. thick. Determine the maximum load P that may be applied. The ultimate strength of the steel is 80,000 lb/in² and that of the copper is 30,000 lb/in². A safety factor of 3 based upon the ultimate strength of each material is satisfactory. For steel $E = 30\cdot10^6$ lb/in² and for copper $E = 13\cdot10^6$ lb/in².
Ans. P = 76,200 lb

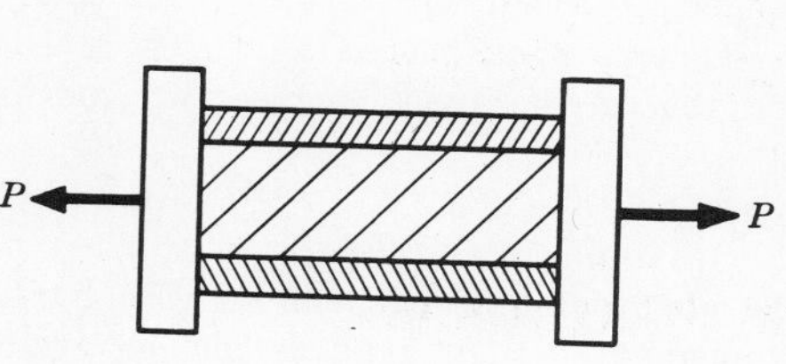

Fig. (*a*) Prob. 20

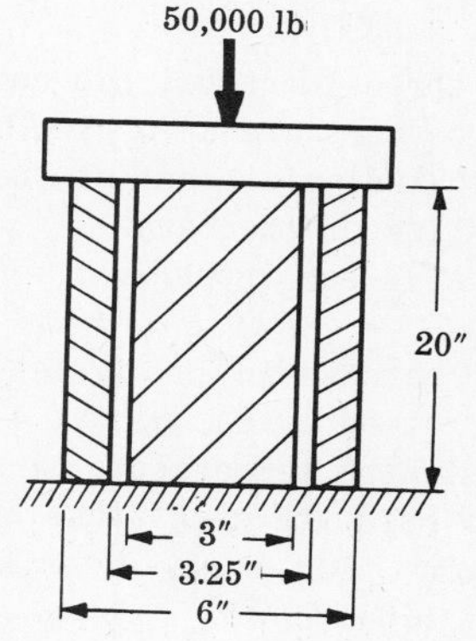

Fig. (*b*) Prob. 21

21. An aluminum right-circular cylinder surrounds a steel cylinder as shown in Fig. (*b*) above. The axial compressive load of 50,000 lb is applied through the infinitely rigid cover plate shown. If the aluminum cylinder is originally 0.010 in. longer than the steel before any load is applied, find the normal stress in each when the temperature has dropped 50°F and the entire load is acting. For steel take $E = 30\cdot10^6$ lb/in², $\alpha = 6.5\cdot10^{-6}/°F$, and for aluminum assume $E = 10\cdot10^6$ lb/in², $\alpha = 12.8\cdot10^{-6}/°F$.
Ans. s_{st} = 945 lb/in², s_{al} = 2170 lb/in²

22. The rigid horizontal bar AB is supported by three vertical wires as shown in Fig. (*a*) below and carries a load of 24,000 lb. The weight of AB is negligible and the system is stress-free before the

24,000 lb load is applied. After the load is applied, the temperature of all three wires is raised 25°F. Find the stress in each wire as well as the location of the applied load in order that AB remain horizontal. For the steel wire take $E = 30 \cdot 10^6$ lb/in², $\alpha = 6.5 \cdot 10^{-6}$/°F, for the brass wire $E = 14 \cdot 10^6$ lb/in², $\alpha = 10.4 \cdot 10^{-6}$/°F, and for the copper $E = 17 \cdot 10^6$ lb/in², $\alpha = 9.3 \cdot 10^{-6}$/°F. Neglect any possibility of lateral buckling of any of the wires.
Ans. $s_{st} = 32{,}300$ lb/in², $s_{br} = 22{,}400$ lb/in², $s_{cu} = 21{,}400$ lb/in², $x = 0.273$ ft

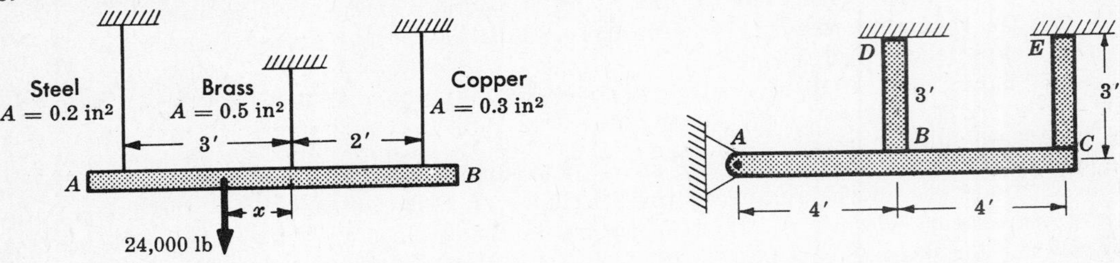

Fig. (*a*) Prob. 22 Fig. (*b*) Prob. 23

23. The bar AC is absolutely rigid and is pinned at A and attached to bars DB and CE as shown in Fig. (*b*) above. The weight of AC is 10,000 lb and the weights of the other two bars are negligible. Consider the temperature of both bars DB and CE to be raised 70°F. Find the resulting normal stresses in these two bars. DB is copper for which $E = 15 \cdot 10^6$ lb/in², $\alpha = 9.3 \cdot 10^{-6}$/°F and the cross-sectional area is 2 in², while CE is steel for which $E = 30 \cdot 10^6$ lb/in², $\alpha = 6.5 \cdot 10^{-6}$/°F and the cross-section is 1 in². Neglect any possibility of lateral buckling of the bars.
Ans. $s_{st} = 9100$ lb/in², $s_{cu} = -4100$ lb/in²

24. Consider the rigid bar BD which is supported by the two wires shown in Fig. (*c*) below. The wires are initially stress-free and the weights of all members are to be neglected. Find the tension in each wire after the load P has been applied to the extreme end of the bar. The two wires have the same modulus of elasticity.
Ans. Force in $AD = \dfrac{2P}{A_1 L_2^2 H / 2A_2 L_1^3 + 2H/L_2}$, Force in $AC = \dfrac{2P}{4HA_2 L_1^2 / A_1 L_2^3 + H/L_1}$

25. Consider three identical pin connected bars arranged as shown in Fig. (*d*) below and supporting the load P. The bars are at angles of 120° with one another. Find the axial force in each bar and also the vertical displacement of the point of application of the load. Neglect any possibility of lateral buckling of the bars.
Ans. Force in each upper bar = $P/3$, Force in lower bar = $-2P/3$, $\Delta_v = 2PL/3AE$

26. The three bars shown in Fig. (*e*) below support the vertical load of 5000 lb. The bars are all stress-free and joined by the pin at A before the load is applied. The load is put on gradually and simultaneously the temperature of all three bars decreases 15°F. Calculate the stress in each bar. The outer bars are each brass and of cross-sectional area 0.4 in². The central bar is steel and of area 0.3 in². For brass $E = 13 \cdot 10^6$ lb/in² and $\alpha = 10.4 \cdot 10^{-6}$/°F and for steel $E = 30 \cdot 10^6$ lb/in² and $\alpha = 6.3 \cdot 10^{-6}$/°F. *Ans.* $s_{br} = 3550$ lb/in², $s_{st} = 10{,}000$ lb/in²

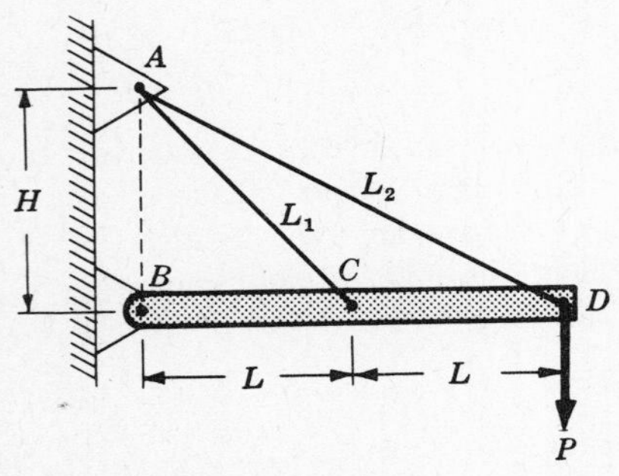

Fig. (*c*) Prob. 24

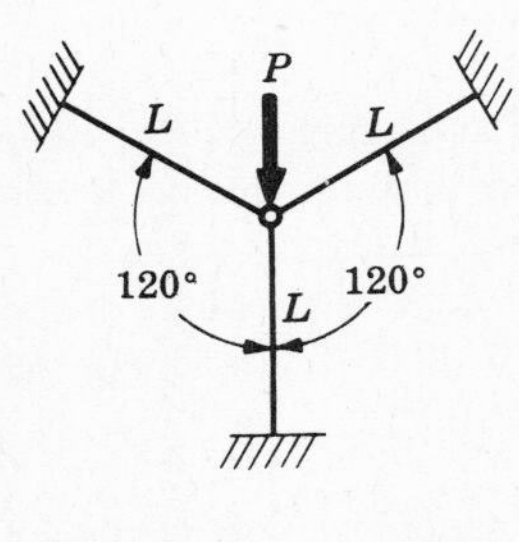

Fig. (*d*) Prob. 25

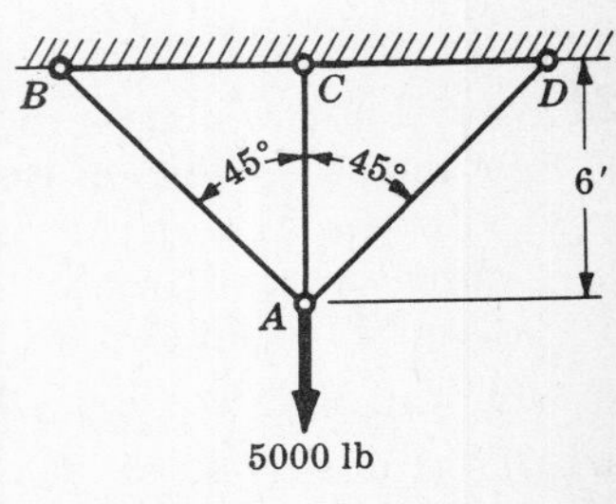

Fig. (*e*) Prob. 26

CHAPTER 3

Thin-Walled Cylinders and Spheres

In Chapters 1 and 2 we examined various cases involving uniform normal stresses acting in bars. Another application of uniformly distributed normal stresses occurs in the approximate analysis of thin-walled cylinders and spheres subject to internal gas or liquid pressure.

NATURE OF STRESSES. If the cylinder shown in the adjoining diagram is subject to a uniform internal pressure, normal stresses arise in two directions in the cylinder wall. The stresses acting in the direction of the geometric axis of the cylinder are called axial or longitudinal stresses and those acting in a perpendicular direction are circumferential or tangential stresses. These stresses are assumed to act on an element as shown and they act in the plane of the cylinder wall.

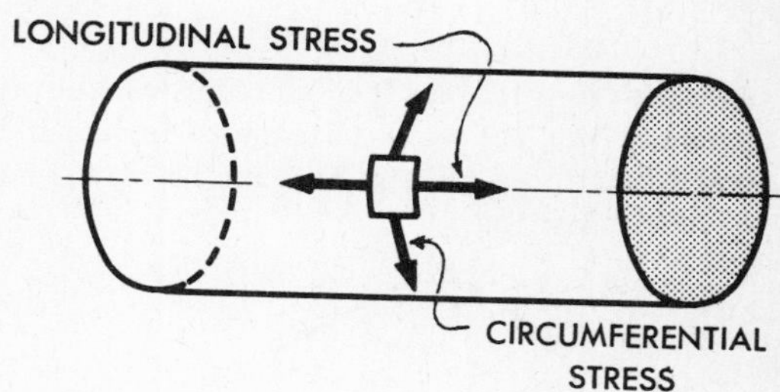

ASSUMPTIONS. It is assumed that the tensile or compressive stresses existing in the walls of the cylinder or sphere may be considered to be *uniformly* distributed over the thickness of the wall. Also the loads, stresses, and deformations in cylindrical shells are assumed to be symmetric about the axis of the cylinder. See Problems 1,2,3, 4,5,9,10. The stresses and deformations in spherical shells are considered to be symmetric about the center of the sphere. See Problem 7.

LIMITATIONS. The ratio of the wall thickness to the radius of curvature should not exceed approximately 0.10. Also there must be no discontinuities in the structure. This simplified treatment presented here does not permit consideration of reinforcing rings on a cylindrical shell as shown in the figure below, nor does it give an accurate indication of the stresses and deformations in the vicinity of end closure plates on cylindrical pressure vessels. Even so the treatment is satisfactory in many design problems.

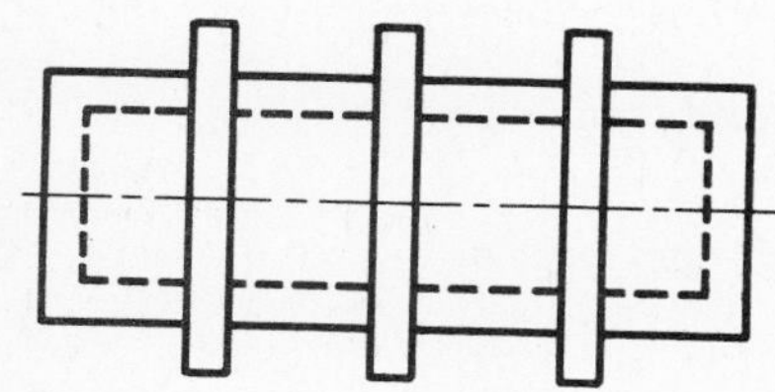

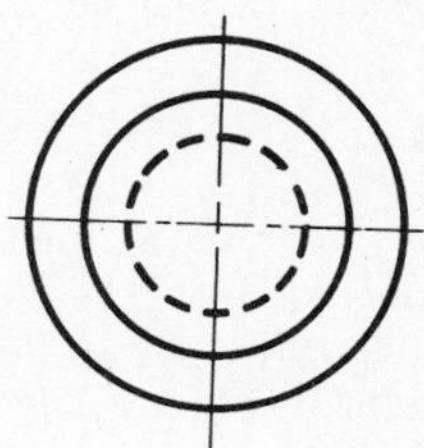

The problems presented here are concerned with stresses arising from a uniform internal pressure acting on a cylinder or sphere. The formulas for the various stresses will be correct if the sense of the pressure is reversed, i.e. if external pressure acts on the container. However, it is to be noted that an additional consideration, beyond the scope of this book, must then be taken into account. Not only must the stress distribution be investigated but another study of an entirely different nature must also be carried out to determine the load at which the shell will *buckle* due to the compression. A buckling or instability failure may take place even though the peak stress is far below the maximum allowable working stress of the material.

APPLICATIONS. Liquid storage tanks and containers, water pipes, boilers, submarine hulls, and certain airplane components are common examples of thin-walled cylinders and spheres.

SOLVED PROBLEMS

1. Consider a thin-walled cylinder closed at both ends by cover plates and subject to a uniform internal pressure p. The wall thickness is h and the inner radius r. Neglecting the restraining effects of the end plates, calculate the longitudinal and circumferential normal stresses existing in the walls due to this loading.

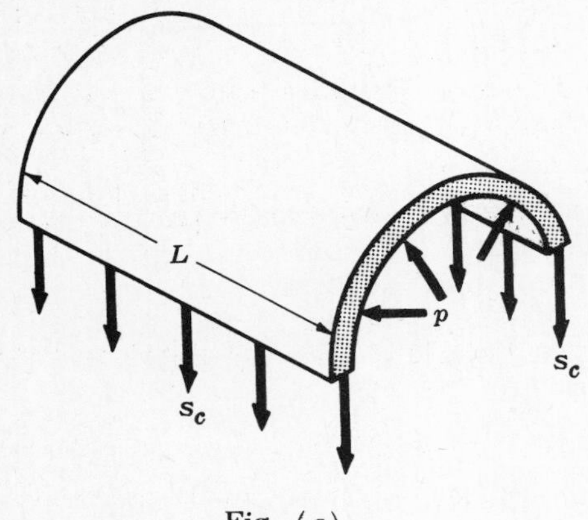

Fig. (a)

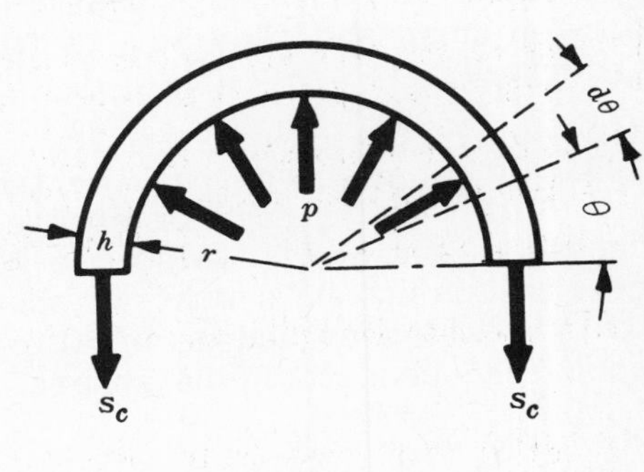

Fig. (b)

To determine the circumferential stress s_c let us consider a section of the cylinder of length L to be removed from the vessel. The free-body diagram of half of this section appears as in Figure (a) above. Note that the body has been cut in such a way that the originally *internal* effect (s_c) now appears as an *external* force to this free-body. Figure (b) shows the forces acting on a cross-section.

The horizontal components of the radial pressures cancel one another by virtue of symmetry about the vertical centerline. In the vertical direction we have the equilibrium equation

$$\Sigma F_v \;=\; -2s_c hL \;+\; \int_0^{\pi} pr(d\theta)(\sin\theta)L \;=\; 0$$

Integrating,
$$2s_c hL \;=\; -\,p\,r\,L\,[\cos\theta]_0^{\pi} \qquad \text{or} \qquad s_c \;=\; \frac{pr}{h}$$

Note that the resultant vertical force due to the pressure p could have been obtained by multiplying the pressure by the horizontal *projected area* upon which the pressure acts.

To determine the longitudinal stress s_L consider a section to be passed through the cylinder normal to its geometric axis. The free-body diagram of the remaining portion of the cylinder is shown in the adjoining figure. For equilibrium,

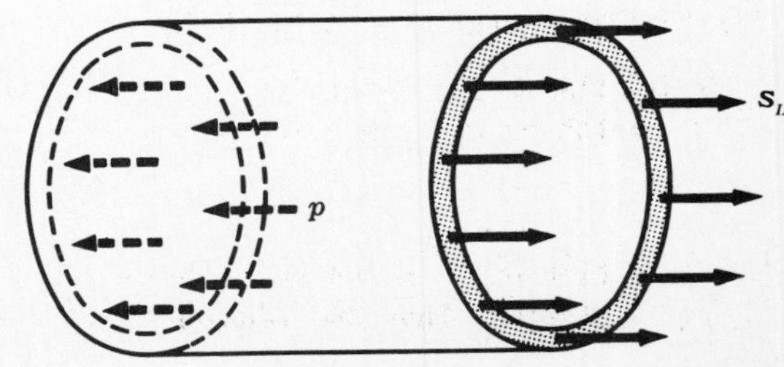

$$\Sigma F_h = -p\pi r^2 + 2\pi r h s_L = 0 \qquad \text{or} \qquad s_L = \frac{pr}{2h}$$

Consequently, the circumferential stress is twice the longitudinal stress. Thus if the water in a closed pipe freezes, the pipe will rupture along a line running longitudinally along the cylinder. These rather simple expressions for stresses are not accurate in the immediate vicinity of the end closure plates.

2. A cast iron water pipe 8 in. in inside diameter is to be subjected to an internal pressure of 200 lb/in^2. What should be the minimum thickness of the pipe in order that the stress should not exceed a working stress of 3500 lb/in^2?

From Problem 1 we know that the stress in the circumferential direction is always the one to be used in design considerations. Then

$$s_c = \frac{pr}{h}, \qquad 3500 \text{ lb/in}^2 = \frac{(200 \text{ lb/in}^2)(4 \text{ in.})}{h} \qquad \text{and} \qquad h = 0.228 \text{ in.}$$

3. The tank of an air compressor consists of a cylinder closed by hemispherical ends. The cylinder is 24 in. in inside diameter and is subjected to an internal pressure of 500 lb/in^2. If the material is a steel whose yield point is 36,000 lb/in^2 and a safety factor of 3.5 is used, calculate the required wall thickness of the cylinder. Neglect localized effects at the juncture of cylinder and hemisphere.

The ends of the tank are closed. Hence according to Problem 1 there exists a circumferential stress in the cylinder wall given by $s_c = pr/h$ and a longitudinal stress in the cylinder given by $s_L = pr/2h$.

Since the circumferential stress is twice the longitudinal stress, it is the critical one for design purposes and it must not exceed the allowable working stress of 36,000/3.5 lb/in^2. For the circumferential stress we thus have $\dfrac{36,000}{3.5} = \dfrac{500(12)}{h}$ or $h = 0.585$ in.

A more complete analysis would include a study of the stresses in the hemispherical ends.

4. A steam boiler is to be 5 ft in inside diameter. It is subject to an internal pressure of 125 lb/in^2. What will be the tension in the shell per linear inch of longitudinal seam? Per linear inch of circumferential seam?

From Problem 1 we know that internal pressure acting on a hollow cylinder gives rise to a circumferential stress $s_c = pr/h$ and also a longitudinal stress $s_L = pr/2h$. For the given data of $p = 125$ lb/in^2, $r = 30$ in., we may write

$$(1) \quad s_c = \frac{125(30)}{h} = \frac{3750}{h} \qquad \text{and} \qquad (2) \quad s_L = \frac{125(30)}{2h} = \frac{1875}{h}$$

Although the thickness h is unknown, we may still calculate the force per unit length of either seam. Let us denote the force per unit length of the longitudinal seam by T_c. Inspection of the first sketch in Problem 1 indicates that T_c is given by the simple relation

$$(3) \quad T_c = s_c \cdot h$$

The direction of T_c, of course, agrees with the direction of s_c shown. Comparison of equations (1) and (3) indicates that $T_c = 3750$ lb/in. Thus the tension per linear inch of longitudinal seam is 3750 lb.

Similarly we may denote the force per unit length of circumferential seam by T_L. Inspection of the third sketch of Problem 1 indicates that

$$(4) \quad T_L = s_L \cdot h$$

the direction of T_L being the same as that of s_L. Comparison of equations (2) and (4) indicates that $T_L = 1875$ lb/in. Thus the tension per inch of circumferential seam is 1875 lb.

5. A vertical steel stand-pipe, i.e. a cylindrical tank open at the top and having a vertical axis, is 8 ft in inside diameter and 80 ft high. The tank is filled with water having a specific weight of 62.4 lb/ft^3. The material is structural steel having a yield point of 35,000 lb/in^2 and a safety factor of 2 is to be used. What is the required thickness of steel plate necessary at the bottom of the tank if the welded longitudinal seam is presumed to be as strong as the solid metal? What thickness is required if the seam is only 85 percent as effective as the solid metal?

The pressure p (in any direction) at the base of the stand-pipe is given by the formula $p = wh$, where w represents the weight of the liquid per unit volume and h denotes the height of the column of water above the base. This formula is immediately evident if we consider that the pressure on a square foot of the base numerically equals the weight of a column of water one square foot in cross-section and h feet high. Hence the pressure at the base is

$$p = 62.4(80) = 5000 \text{ lb/ft}^2 \qquad \text{or} \qquad p = 34.6 \text{ lb/in}^2$$

Since this pressure is hydrostatic it is acting in all directions with the same magnitude, and in particular it is acting radially against the inside wall of the stand-pipe as shown in the adjacent figure. As may be seen from the equation $p = wh$, the radial pressure decreases toward the top of the tank as shown in the sketch, but the maximum value occurs at the base and consequently that is the region that must be considered for design purposes.

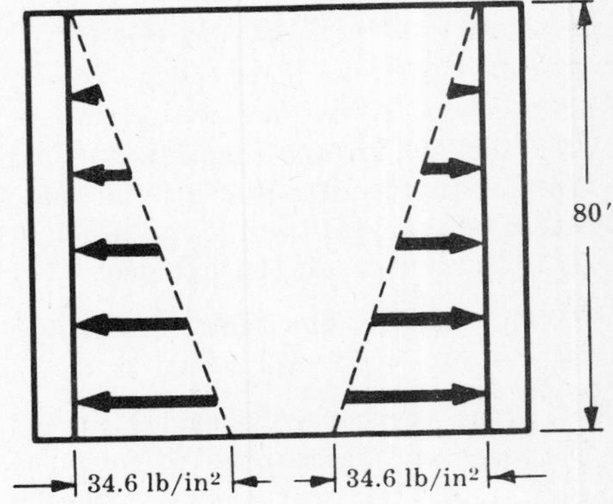

Since the top of the tank is open, there is no longitudinal stress and from Problem 1 it is known that the circumferential stress at any point in the stand-pipe is given by $s_c = pr/h$. Considering the region at the base of the tank, this equation becomes

$$\frac{35{,}000}{2} = \frac{34.6(48)}{h} \qquad \text{or} \qquad h = 0.095 \text{ in.}$$

This assumes that the longitudinal seams are as strong as the solid metal. Actually this thickness should be increased slightly to allow for the effects of corrosion.

If the longitudinal seams are only 85% as strong as the solid metal, the required thickness is $h = 0.095/0.85 = 0.112$ in.

6. Calculate the increase in the radius of the cylinder considered in Prob. 1 due to the internal pressure p.

Let us consider the longitudinal and circumferential loadings separately. Due to radial pressure p *only* the circumferential stress is given by $s_c = pr/h$, and because $s = E\epsilon$ the circumferential strain is given by $\epsilon_c = pr/Eh$.

It is to be noted that ϵ_c is a unit strain. The length over which it acts is merely the circumference of the cylinder which is $2\pi r$. Hence the total elongation of the circumference is

$$\Delta = \epsilon_c(2\pi r) = 2\pi pr^2/Eh$$

The final length of the circumference is thus: $2\pi r + 2\pi pr^2/Eh$. Dividing this circumference by 2π we find the new radius of the deformed cylinder to be: $r + pr^2/Eh$. The increase in radius is: pr^2/Eh.

Due to the axial pressure p *only*, longitudinal stresses $s_L = pr/2h$ are set up. These longitudinal stresses give rise to longitudinal strains $\epsilon_L = pr/2Eh$. As in Chap. 1 an extension in the direction of loading, which is the longitudinal direction here, is accompanied by a decrease in the dimension perpendicular to the load. Thus here the circumferential dimension decreases. The ratio of the strain in the lateral direction to that in the direction of loading was defined in Chapter 1 to be Poisson's ratio, denoted by μ. Consequently the above strain ϵ_L induces a circumferential strain equal to $-\mu\epsilon_L$

and if this strain is denoted by ϵ_c' we have $\epsilon_c' = -\mu pr/2Eh$ which tends to decrease the radius of the cylinder as shown by the negative sign.

In a manner exactly analogous to the treatment of the increase of radius due to radial loading only, the decrease of radius corresponding to the strain ϵ_c' is given by $\frac{\mu}{2}\cdot\frac{pr^2}{Eh}$. The resultant increase of radius due to the internal pressure p is thus

$$\Delta r = \frac{pr^2}{Eh} - \frac{\mu}{2}\cdot\frac{pr^2}{Eh} = \frac{pr^2}{Eh}(1 - \frac{\mu}{2})$$

7. Consider a closed thin-walled spherical shell subject to a uniform internal pressure p. The inside radius of the shell is r and its wall thickness is h. Derive an expression for the tensile stress existing in the wall.

For a free-body diagram, let us consider exactly half of the entire sphere. This body is acted upon by the applied internal pressure p as well as the forces that the other half of the sphere, which has been removed, exerts upon the half under consideration. Because of the symmetry of loading and deformation, these forces may be represented by circumferential tensile stresses s_c as shown in the adjoining free-body diagram.

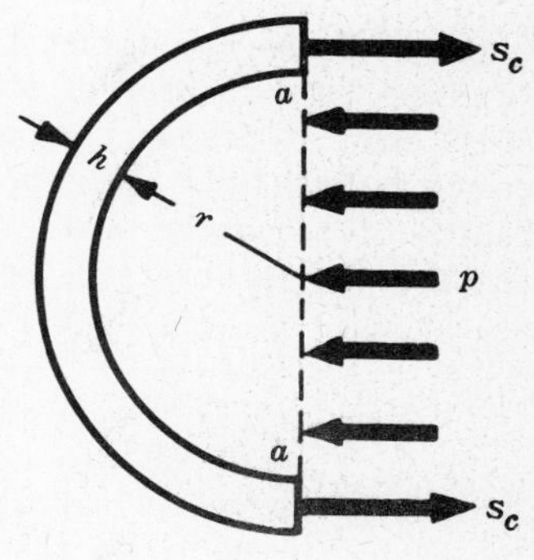

This free-body diagram represents the forces acting on the hemisphere, the diagram showing only a projection of the hemisphere on a vertical plane. Actually the pressure p acts over the entire inside surface of the hemisphere and in a direction perpendicular to the surface at every point. However, as mentioned in Problem 1, it is permissible to consider the force exerted by this same pressure p upon the *projection* of this area which in this case is the vertical circular area denoted by a-a. This is possible because the hemisphere is symmetric about the horizontal axis and the vertical components of the pressure annul one another. Only those components of pressure parallel to the horizontal axis produce the tensile stress s_c. For equilibrium we have

$$\Sigma F_h = s_c \cdot 2\pi rh - p\pi r^2 = 0 \qquad \text{or} \qquad s_c = \frac{pr}{2h}$$

From symmetry this circumferential stress is the same in all directions at any point in the wall of the sphere.

8. A 60 ft diameter spherical tank is to be used to store gas. The shell plating is 0.5 in. thick and the working stress of the material is 18,000 lb/in^2. What is the maximum permissible gas pressure p?

From Problem 7, the tensile stress is uniform in all directions and is given by

$$s_c = \frac{pr}{2h}. \qquad \text{Substituting,} \quad 18{,}000 = \frac{p(30\times 12)}{2(0.5)} \quad \text{and} \quad p = 50 \text{ lb/in}^2.$$

9. Consider a laminated pressure vessel composed of two thin co-axial cylinders as shown below. In the state prior to assembly there is a slight "interference" between these shells, i.e. the inner one is too large to slide into the outer one. The outer cylinder is heated, placed on the inner and allowed to cool, thus providing a "shrink fit". If both cylinders are steel and the mean diameter of the assembly is 4 in., find the tangential stresses in each shell arising from the "shrinking" if the initial interference (of diameters) is 0.010 in. The thickness of the inner shell is 0.10 in., that of the outer shell 0.08 in.

There is evidently an interfacial pressure p acting between the adjacent faces of the two shells as shown below.

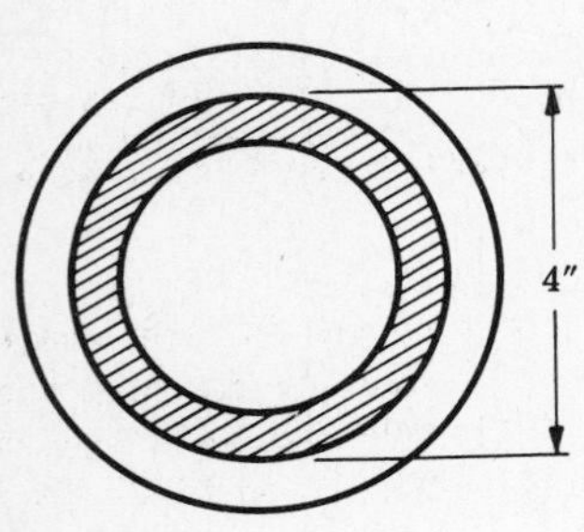

LAMINATED PRESSURE VESSEL

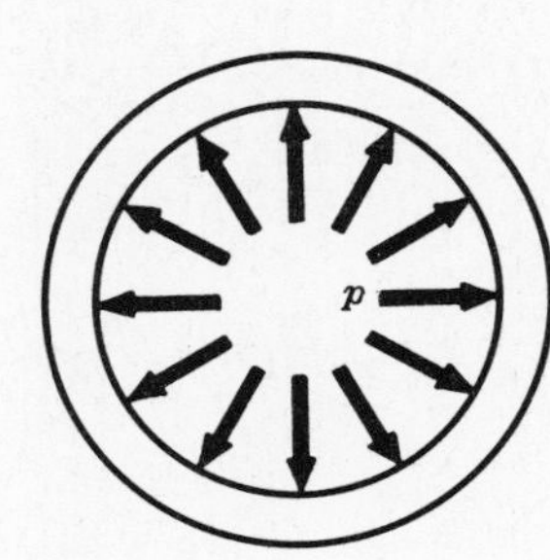

OUTER CYLINDER

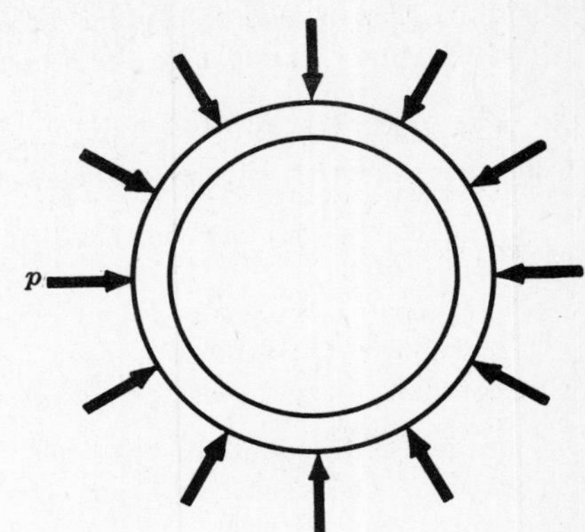

INNER CYLINDER

It is to be noted that there are no external applied loads. The pressure p may be considered to increase the diameter of the outer shell and decrease the diameter of the inner so that the inner shell may fit inside the outer. The radial expansion of a cylinder due to a radial pressure p was found in Problem 6 to be pr^2/Eh. No longitudinal forces are acting in this problem. The increase in radius of the outer shell due to p, plus the decrease in radius of the inner one due to p, must equal the initial interference between radii, or 0.010/2 in. Thus we have

$$\frac{p(2)^2}{(30 \cdot 10^6)(0.10)} + \frac{p(2)^2}{(30 \cdot 10^6)(0.08)} = 0.005 \qquad \text{or} \qquad p = 1670 \text{ lb/in}^2$$

This pressure, illustrated in the above figures, acts between the cylinders after the outer one has been "shrunk" onto the inner one. In the inner cylinder this pressure p gives rise to a stress

$$s_c = \frac{pr}{h} = -\frac{1670(2)}{0.10} = -33,400 \text{ lb/in}^2$$

In the outer cylinder the circumferential stress due to the pressure p is

$$s_c' = \frac{pr}{h} = \frac{1670(2)}{0.08} = 41,700 \text{ lb/in}^2$$

If, for example, the laminated shell is subject to a uniform internal pressure, these "shrink fit" stresses would merely be added algebraically to the stresses found by the use of the simple formulas given in Problem 1.

10. The thin steel cylinder just fits over the inner copper cylinder as shown. Find the tangential stresses in each shell due to a temperature rise of 60°F. Do not consider the effects introduced by the accompanying longitudinal expansion. This arrangement is sometimes used for storing corrosive fluids. Take

$$E_{st} = 30 \times 10^6 \text{ lb/in}^2, \quad \alpha_{st} = 6.5 \times 10^{-6}/^\circ\text{F}$$

$$E_{cu} = 13 \times 10^6 \text{ lb/in}^2, \quad \alpha_{cu} = 9.3 \times 10^{-6}/^\circ\text{F}.$$

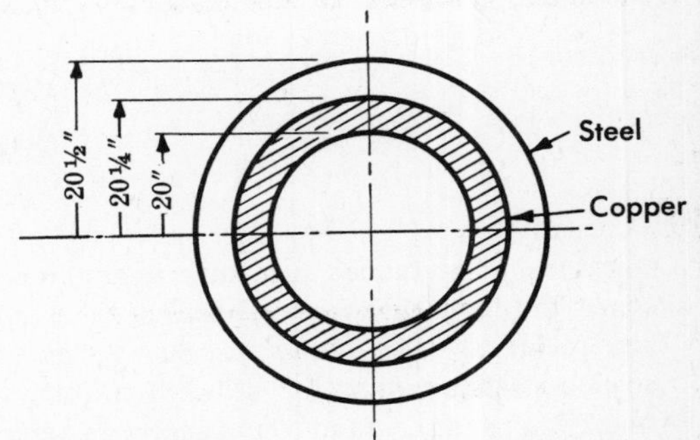

The simplest approach is to first consider the two shells to be separated from one another so that they are no longer in contact.

Due to the temperature rise of 60°F the circumference of the steel shell increases an amount

$$2\pi(20.375)(60)(6.5 \times 10^{-6}) = 0.0498 \text{ in.}$$

Also, the circumference of the copper shell increases an amount

$$2\pi(20.125)(60)(9.3 \times 10^{-6}) = 0.0705 \text{ in.}$$

Thus the "interference" between the radii, i.e. the difference in radii, of the two shells (due to the heating) is $\dfrac{0.0705 - 0.0498}{2\pi} = 0.00345$ in. Again, there are no external loads acting on either cylinder.

However, from the statement of the problem the adjacent surfaces of the two shells are obviously in contact after the temperature rise. Hence there must be an inter-facial pressure p between the two surfaces, i.e. a pressure tending to increase the radius of the steel shell and decrease the radius of the copper shell so that the copper shell may fit inside the steel one. Such a pressure is shown in the following free-body diagrams.

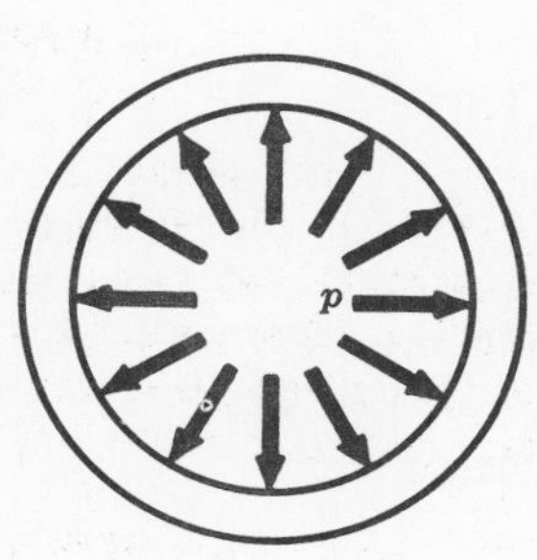

STEEL CYLINDER

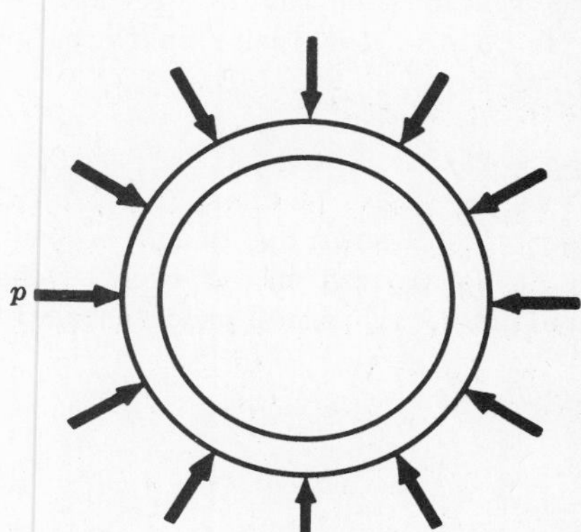

COPPER CYLINDER

In Problem 6 the change of radius of a cylinder due to a uniform radial pressure p (with no longitudinal forces acting) was found to be pr^2/Eh. Consequently the increase of radius of the steel shell due to p added to the decrease of radius of the copper one, due to p, must equal the "interference"; thus

$$\frac{p(20.375)^2}{(30 \times 10^6)(0.25)} + \frac{p(20.125)^2}{(13 \times 10^6)(0.25)} = 0.00345 \qquad \text{or} \qquad p = 19.2 \text{ lb/in}^2$$

This interfacial pressure creates the required continuity at the common surface of the two shells when they are in contact. Using the formula for the tangential stress, $s_c = pr/h$, we find the tangential stresses in the steel and copper shells to be respectively

$$s_{st} = \frac{19.2(20.375)}{0.25} = 1560 \text{ lb/in}^2 \qquad \text{and} \qquad s_{cu} = -\frac{19.2(20.125)}{0.25} = -1550 \text{ lb/in}^2$$

11. Consider a thin-walled cylinder of wall thickness 0.16 in. and shell diameter 5 in. The shell is reinforced by a single layer of closely wound steel wire of diameter 0.04 in., the wire being wound on to the shell with an initial tension of 10,000 lb/in^2 before any internal pressure is applied to the cylinder. Determine the stress in the wire and also in the shell after application of a uniform internal radial pressure of 500 lb/in^2.

Since the internal pressure is entirely radial there is no stress in the direction of the axis of the cylinder. The pressure of the wire on the shell prior to application of the 500 lb/in^2 load is equivalent to a uniform external radial pressure p acting on the shell. It is convenient to consider a one inch length of the shell; and since the wire is 0.04 in. in diameter, then 25 adjacent turns of wire reinforce this one inch length of shell. The free-body diagram of the 25 turns of wire in contact with a one inch length of the cylinder is shown in the adjoining figure. Only half of each turn of wire is shown, the other half having been removed and its effect represented by the initial tension of 10,000 lb/in^2 in the wire. Obviously it would not be useful to diagram the entire turn, because that would not lead to any relation between p and the initial tension.

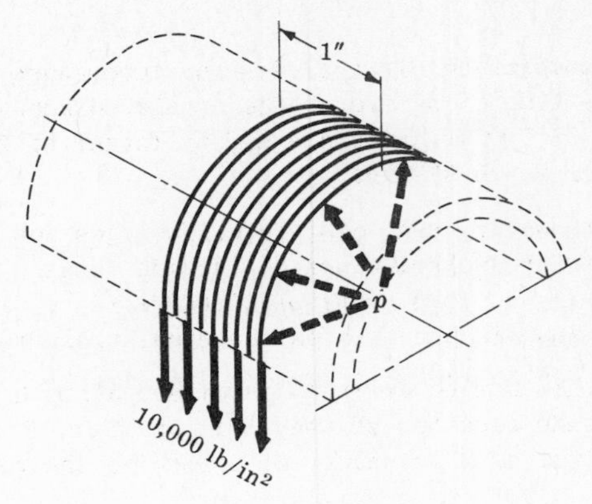

Summing forces vertically in the free-body diagram and remembering that there are 25 turns of wire in the one inch length being considered, we have

$$\Sigma F_v = 5p - 25(2)(10,000)\frac{\pi}{4}(0.04)^2 = 0$$

with the factor of 2 entering in the second term because the wire is cut at both ends of a diameter. Note that p is considered to act over the horizontal projected area. Solving, p = 126 lb/in^2. This pressure applied to the shell produces an initial circumferential compression in the cylinder given by

$$s_c = \frac{pr}{h} = \frac{126(2.5)}{0.16} = 1970 \text{ lb/in}^2$$

If now the load of 500 lb/in^2 is applied to the interior of the shell, the resulting circumferential stress s is resisted by both the shell and the winding. This stress s is identical in both the cylinder and the wire. We may then draw the adjoining free-body diagram of the upper half of a one inch length of shell reinforced by 25 turns of wire. For equilibrium in the vertical direction we have

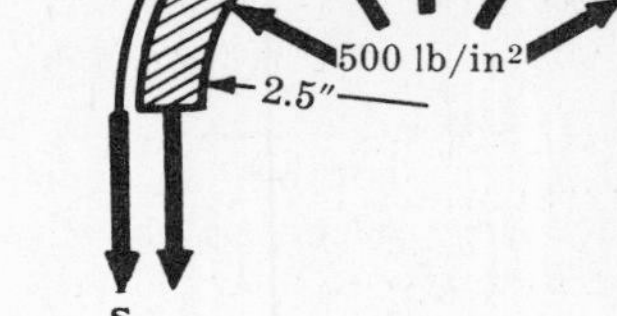

$$\Sigma F_v = 500(5)(1) - s[2(0.16) + 2(25)\frac{\pi}{4}(0.04)^2] = 0$$

or s = 6530 lb/in^2, where s represents the unit stress in either the shell or the wire winding.

The stress in the circumferential direction of the shell is consequently

$$s_1 = 6530 - 1970 = 3560 \text{ lb/in}^2$$

and the final stress in the wire is

$$s_2 = 10,000 + 6530 = 16,530 \text{ lb/in}^2$$

With no wire winding the circumferential stress in the shell is readily found to be 7820 lb/in^2. This illustrates the reinforcing effect of such a winding.

SUPPLEMENTARY PROBLEMS

12. A compressed air cylinder for laboratory use ordinarily carries approximately 2300 lb/in^2 pressure at the time of delivery. The outside diameter of such a cylinder is 10 in. If the steel has a yield point of 33,000 lb/in^2 and a safety factor of 2.5 is adequate, calculate the required wall thickness. *Ans.* h = 0.875 in.

13. For use in rural districts, fuel gas for home use is frequently stored in cylinders closed by either hemispherical or ellipsoidal ends. Consider such a tank 36 in. in diameter and made of steel having a yield point of 30,000 lb/in^2. The shell thickness is 0.5 in. Based upon a safety factor of 3, what is the greatest internal pressure such a tank can withstand? *Ans.* p = 278 lb/in^2

14. A thin-walled cylinder is closed at both ends and contains oil under a pressure of 120 lb/in^2. The inside diameter of the shell is 16 in. If the yield point of the material is 38,000 lb/in^2 and a safety factor of 3 is selected, determine the required wall thickness. *Ans.* h = 0.076 in.

15. A vertical cylindrical gasoline storage tank is 85 ft in diameter and is filled to a depth of 40 ft with gasoline whose specific gravity is 0.74. If the yield point of the shell plating is 35,000 lb/in^2 and a safety factor of 2.5 is adequate, calculate the required wall thickness at the bottom of the tank neglecting any localized bending effects there. *Ans.* h = 0.466 in.

16. A spherical tank for storing gas under pressure is 80 ft in diameter and is made of structural steel 5/8 in. thick. The yield point of the material is 35,000 lb/in^2 and a safety factor of 2.5 is adequate. Determine the maximum permissible internal pressure, assuming the welded seams between the various plates are as strong as the solid metal. Also, determine the permissible pressure if the seams are 75 percent as strong as the solid metal. *Ans.* p = 36.5 lb/in^2, p = 27.4 lb/in^2

17. To assist motorists who have tire trouble, many service stations bring a small tank filled with compressed air to the scene of any difficulty. A typical tank is 12 in. in diameter and when filled carries a pressure of 175 lb/in^2. The tank is cylindrical and is closed by hemispherical ends. Neglect the bending effect in the vicinity of the juncture of these two elements. Calculate the required wall thickness of the cylinder and also of the sphere based upon a safety factor of 4. Assume a yield point of 30,000 lb/in^2 for the steel plating. *Ans.* Cylinder h = 0.140 in., Sphere h = 0.07 in.

18. Calculate the increase in the radius of the spherical shell mentioned in Problem 7 due to the internal pressure.
Ans. $\Delta r = \dfrac{pr^2}{2Eh}(1-\mu)$

19. Derive an expression for the increase of volume per unit volume of a thin-walled circular cylinder subjected to a uniform internal pressure p. The ends of the cylinder are closed by circular plates. Assume that the radial expansion is constant along the length.
Ans. $\dfrac{\Delta V}{V} = \dfrac{pr}{Eh}(\dfrac{5}{2}-2\mu)$

20. Calculate the increase of volume per unit volume of a thin-walled steel circular cylinder closed at both ends and subjected to a uniform internal pressure of 80 lb/in^2. The wall thickness is 1/16 in., the radius 13.5 in. and μ = 1/3. Consider $E = 30\cdot 10^6$ lb/in^2. *Ans.* $\Delta V/V = 1.06\cdot 10^{-3}$

21. Consider a laminated cylinder consisting of a thin steel shell "shrunk" on an aluminum one. The thickness of each is 0.10 in. and the mean diameter of the assembly is 4 in. The initial "interference" of the shells prior to assembly is 0.004 in. measured on a diameter. Find the tangential stresses in each shell caused by this "shrink fit". For aluminum $E = 10\cdot 10^6$ lb/in^2 and for steel $E = 30\cdot 10^6$ lb/in^2.
Ans. s_{st} = 7500 lb/in^2, s_{al} = −7500 lb/in^2

CHAPTER 4

Direct Shear Stresses

DEFINITION OF SHEAR FORCE. If a plane is passed through a body, a force acting along this plane is called a shear force. It will be denoted by F_s.

DEFINITION OF SHEAR STRESS. The shear force, divided by the area over which it acts, is called the shear stress. It is denoted in this book by s_s. Thus

$$s_s = \frac{F_s}{A}$$

COMPARISON OF SHEAR AND NORMAL STRESSES. Let us consider a bar cut by a plane *a-a* perpendicular to its axis, as shown in the adjacent figure. A normal stress *s* is perpendicular to this plane. This is the type of stress considered in Chapters 1, 2, and 3.

A shear stress is one acting *along* the plane, as shown by the stress s_s. Hence the distinction between normal stresses and shear stresses is one of *direction*.

ASSUMPTIONS. It is necessary to make some assumption regarding the manner of distribution of shear stresses, and for lack of any more precise knowledge it will be taken to be uniform in all problems discussed in this chapter. Thus the expression $s_s = F_s/A$ indicates an average shear stress over the area.

APPLICATIONS. Riveted joints (Problem 7), wood test specimens (Problem 5), and keys used to lock pulleys to shafts (Problem 8) are common examples of systems involving shear stresses.

DEFORMATIONS DUE TO SHEAR STRESSES. Let us consider the deformation of a plane rectangular element cut from a solid where the forces acting on the element are known to be the shearing stresses s_s in the directions shown in Figure (*a*) below.

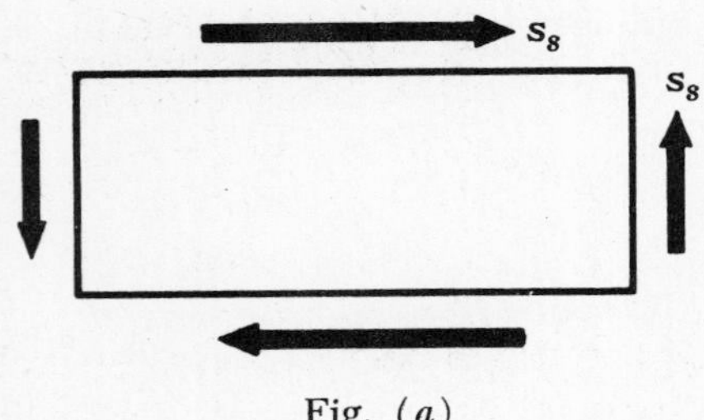

Fig. (*a*)

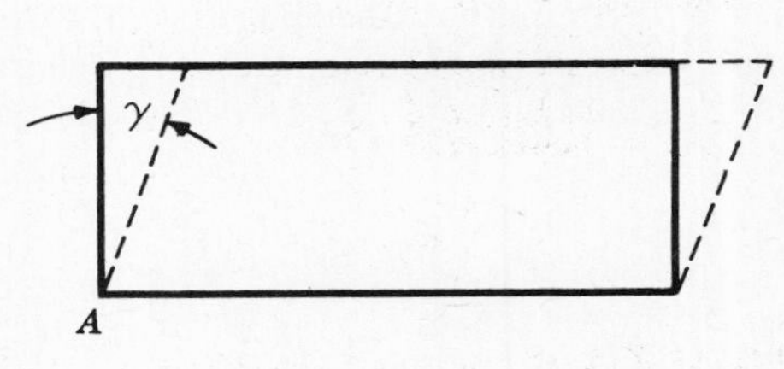

Fig. (*b*)

The faces of the element parallel to the plane of the paper are assumed to be load-free. Since there are no normal stresses acting on the element, the lengths of the sides of the originally rectangular element will not change when the shearing stresses assume the value s_s. However, there will be a distortion of the originally right *angles* of the element, and after this distortion due to the shearing stresses the element assumes the configuration shown by the dotted lines in Fig. (*b*) above.

SHEAR STRAIN. The change of angle at the corner *A* of the element is defined as the shear strain. It must be expressed in radian measure and is usually denoted by γ.

MODULUS OF ELASTICITY IN SHEAR. The ratio of the shear stress s_s to the shear strain γ is called the modulus of elasticity in shear and is usually denoted by G. Thus

$$G = \frac{s_s}{\gamma}$$

G is also known as the *modulus of rigidity*.

The units of G are the same as those of the shear stress, e.g. lb/in², since the shear strain is dimensionless. The experimental determination of G and the region of linear action of s_s and γ will be discussed in Chapter 5. Stress-strain diagrams for various materials may be drawn for shearing loads, just as they were drawn for normal loads in Chapter 1. They have the same general appearance as those sketched in Chapter 1 but the numerical values associated with the plots are of course different.

SOLVED PROBLEMS

1. Consider the bolted joint shown in Fig. (*a*) below. The force *P* is 7500 lb and the diameter of the bolt is 0.5 in. Determine the average value of the shearing stress existing across either of the planes *a-a* or *b-b*.

Lacking any more precise information we can only assume that the force *P* is equally divided between the sections *a-a* and *b-b*. Consequently a force of 7500/2 = 3750 lb acts across either of these planes over a cross-sectional area of $\frac{1}{4}\pi(0.5)^2 = 0.197$ in².

Thus the average shearing stress across either plane is $s_s = \dfrac{P/2}{A} = \dfrac{3750}{0.197} = 19{,}000$ lb/in².

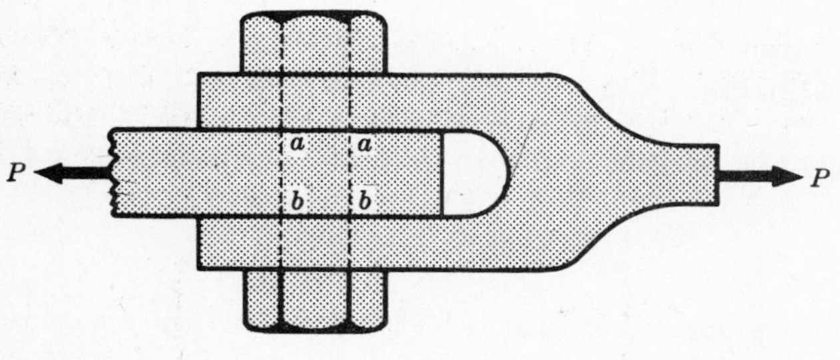

Fig. (*a*)

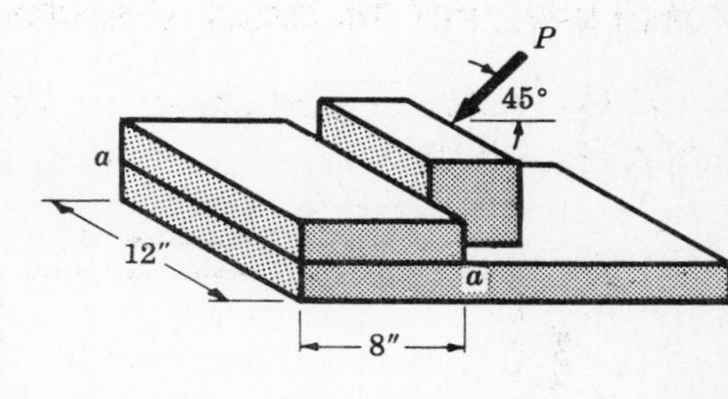

Fig. (*b*)

2. Referring to Fig. (*b*) above, the force *P* tends to shear off the stop along the plane *a-a*. If *P* = 8000 lb, determine the average shearing stress on the plane *a-a*.

Only the horizontal component of *P* is effective in producing this shearing action. It is given by 8000 cos 45° = 5650 lb.

The average shearing stress across plane a-a is therefore $s_s = \frac{P \cos 45^\circ}{A} = \frac{5650}{12(8)} = 59 \text{ lb/in}^2$.

3. Low carbon structural steel has a shearing ultimate strength of approximately 45,000 lb/in^2. Determine the force P necessary to punch a 1 in. diameter hole through a plate of this steel 3/8 in. thick. If the modulus of elasticity in shear for this material is 12×10^6 lb/in^2, find the shear strain at the edge of this hole when the shear stress is 21,000 lb/in^2.

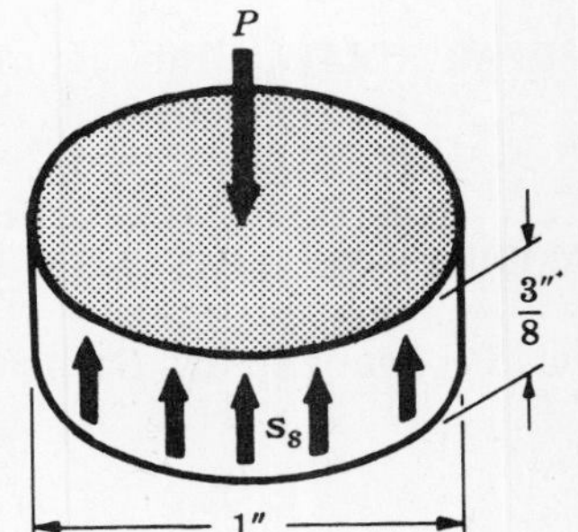

Let us assume uniform shearing on a cylindrical surface 1 in. in diameter and 3/8 in. thick as shown in the adjoining sketch. For equilibrium we find the force P to be

$$P = s_s A = \pi(1)(3/8)(45{,}000) = 53{,}100 \text{ lb}$$

To determine the shear strain γ when the shear stress s_s is 21,000 lb/in^2, we employ the definition $G = s_s/\gamma$ to obtain

$$\gamma = \frac{s_s}{G} = \frac{21{,}000}{12{,}000{,}000} = 0.00175 \text{ radian}$$

4. Consider the rectangular specimen shown, 1 in. × 2 in. in cross-section, that is sometimes used for determining the tensile strength of wood. For white oak, having a shearing ultimate strength parallel to the grain of 900 lb/in^2, determine the minimum length of specimen necessary in the grips, a, in order that a shear failure in the grips will not occur before the specimen fails in tension. The tensile fracture occurs at a load P of 6600 lb.

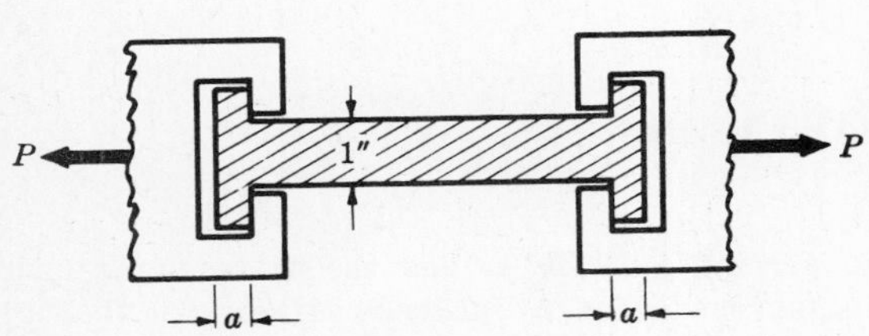

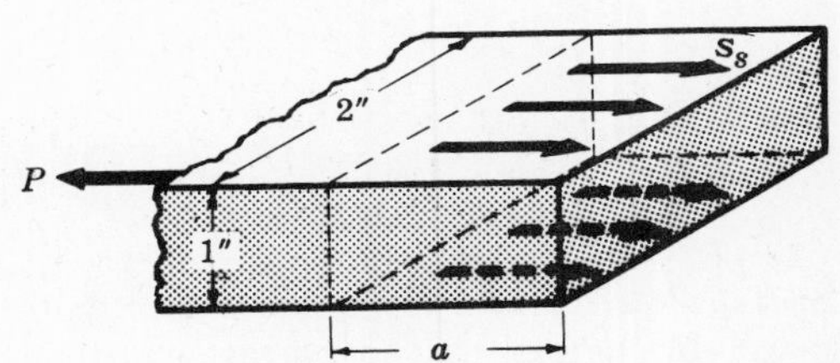

The shear stresses act as shown over the area at the right end and also on a corresponding area at the left end of the specimen.

Assuming a uniform distribution of shearing stress, we have

$$s_s = \frac{P}{A}, \qquad 900 = \frac{6600}{2(2)(a)} \qquad \text{and} \qquad a = 1.84 \text{ in.}$$

Naturally, the length used in the grips should be somewhat greater than 1.84 in. to be certain that the tensile fracture occurs first.

5. In the wood industries, inclined blocks of wood are sometimes used to determine the *compression-shear* strength of glued joints. Consider the pair of glued blocks A and B which are 1.5 in. deep in a direction perpendicular to the plane of the paper. Determine the shearing ultimate strength of the glue if a vertical force of 9000 lb is required to cause rupture of the joint. It is to be noted that a good glue causes a large proportion of the failure to occur in the wood.

Let us consider the equilibrium of the lower block, A. The reactions of the upper block B upon the lower one consist of both normal and shearing forces appearing as in the perspective and orthogonal views shown.

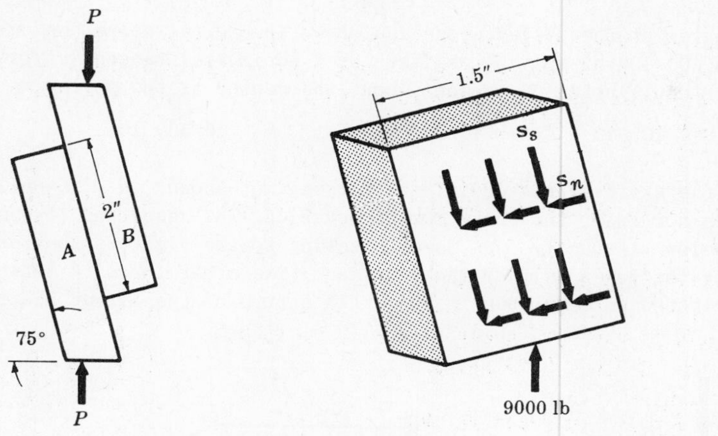

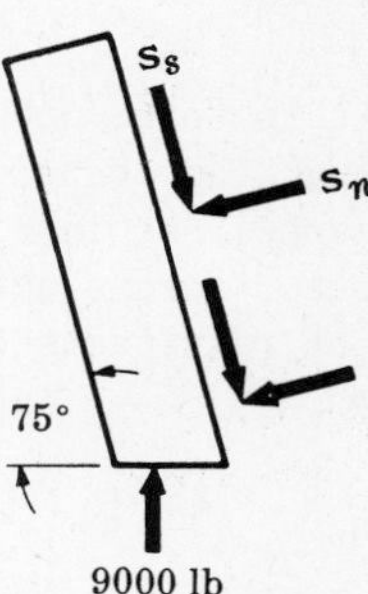

Referring to the sketch on the right we see that for equilibrium in the horizontal direction

$$\Sigma F_h = s_s(2)(1.5)\cos 75^\circ - s_n(2)(1.5)\cos 15^\circ = 0 \qquad \text{or} \qquad s_n = 0.269\,s_s$$

For equilibrium in the vertical direction we have

$$\Sigma F_v = 9000 - s_s(2)(1.5)\sin 75^\circ - s_n(2)(1.5)\sin 15^\circ = 0$$

Substituting $s_n = 0.269\,s_s$ and solving, we find $s_s = 2900 \text{ lb/in}^2$.

6. The shearing stress in a piece of structural steel is 15,000 lb/in². If the modulus of rigidity G is 12,000,000 lb/in², find the shearing strain γ.

By definition, $G = \dfrac{s_s}{\gamma}$. Then the shearing strain $\gamma = \dfrac{s_s}{G} = \dfrac{15,000}{12,000,000} = 0.00125$ radian.

7. A single rivet is used to join two plates as shown in the accompanying diagram. If the diameter of the rivet is 3/4 in. and the load P is 7000 lb, what is the average shearing stress developed in the rivet?

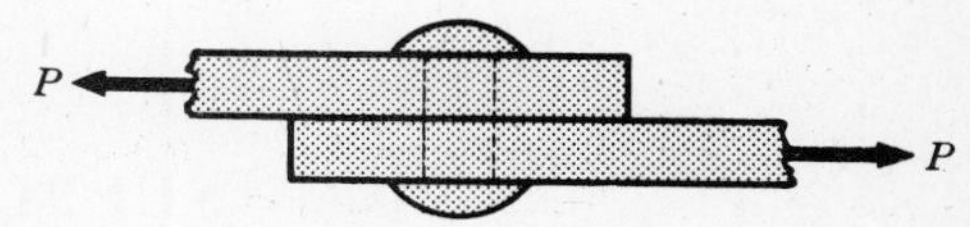

Here the average shear stress in the rivet is P/A where A is the cross-sectional area of the rivet.

Consequently the average shear is given by $s_s = \dfrac{7000}{\frac{\pi}{4}(0.75)^2} = 15,900 \text{ lb/in}^2$.

8. Shafts and pulleys are usually fastened together by means of a key, as shown in Fig. (a) below. Con-

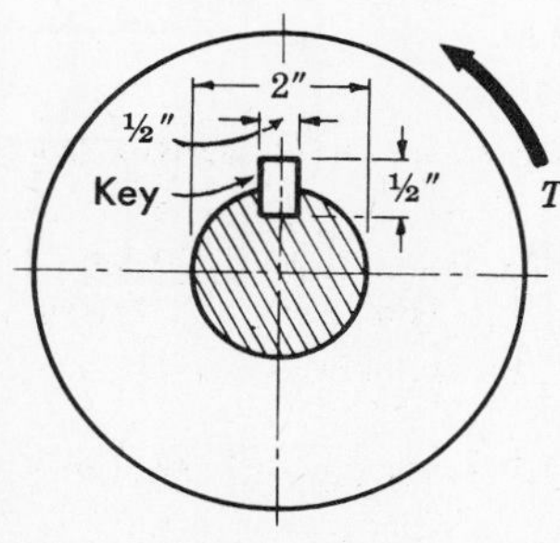

Fig. (a)

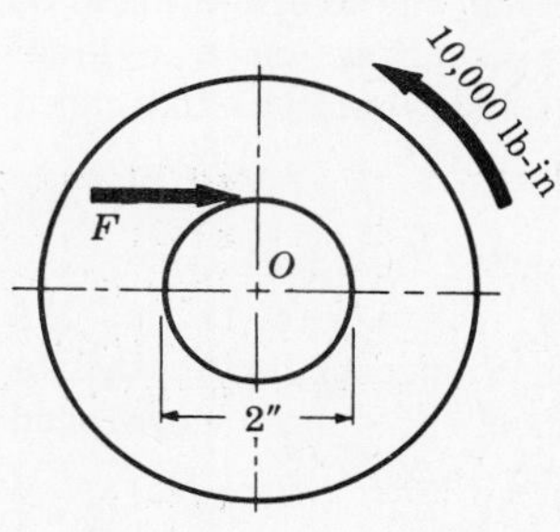

Fig. (b)

sider a pulley subject to a turning moment T of 10,000 lb-in. keyed by a $\frac{1}{2}\times\frac{1}{2}\times 3$ in. key to the shaft. The shaft is 2 in. in diameter. Determine the shear stress on a horizontal plane through the key.

Drawing a free-body diagram of the pulley alone, as shown in Fig. (b) above, we see that the applied turning moment of 10,000 lb-in. must be resisted by a horizontal tangential force F exerted on the pulley by the key. For equilibrium of moments about the center of the pulley we have

$$\Sigma M_O = 10{,}000 - F(1) = 0 \qquad \text{or} \qquad F = 10{,}000 \text{ lb}$$

It is to be noted that the shaft exerts additional forces, not shown, on the pulley. These act through the center O and do not enter the above moment equation. The resultant forces acting on the key appear as in Fig. (c) below. Actually the force F acting to the right is the resultant of distributed forces acting over the lower half of the left face. The other forces F shown likewise represent resultants of distributed force systems. The exact nature of the force distribution is not known.

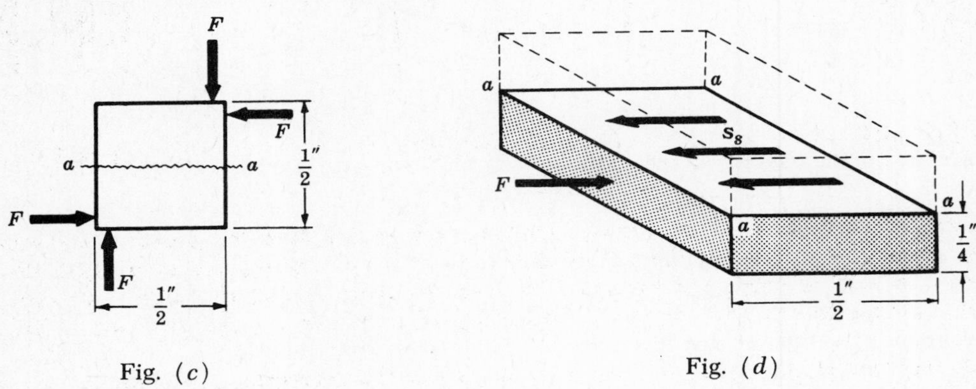

Fig. (c) Fig. (d)

The free-body diagram of the portion of the key below a horizontal plane a-a through its mid-section is shown in Fig. (d) above. For equilibrium in the horizontal direction we have

$$\Sigma F_h = 10{,}000 - s_s(\tfrac{1}{2})(3) = 0 \qquad \text{or} \qquad s_s = 6670 \text{ lb/in}^2$$

This is the horizontal shear stress in the key.

SUPPLEMENTARY PROBLEMS

9. In Problem 1, if the maximum allowable working stress in shear is 14,000 lb/in^2, determine the required diameter of the bolt in order that this value is not exceeded. *Ans.* d = 0.585 in.

10. Consider a steel bolt 3/8 in. in diameter and subject to an axial tensile load of 2000 lb as shown in the adjoining diagram. Determine the average shearing stress in the bolt head, assuming shearing on a cylindrical surface of the same diameter as the bolt, as indicated by the dotted lines. *Ans.* s_s = 5420 lb/in^2

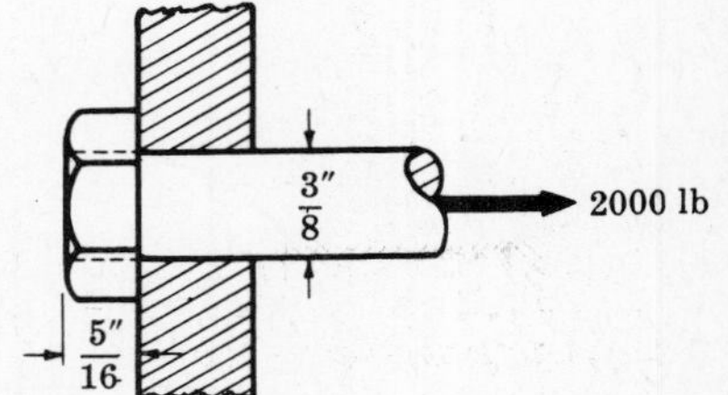

11. A circular punch $\frac{3}{4}$ in. in diameter is used to punch a hole through a steel plate $\frac{1}{2}$ in. thick. If the force necessary to drive the punch through the metal is 61,000 lb, determine the maximum shearing stress developed in the material. *Ans.* s_s = 51,800 lb/in^2

12. In structural practice, steel clip angles are commonly used to transfer loads from horizontal girders to vertical columns. If the reaction of the girder upon the angle is a downward force of 10,000 lb as shown in Fig. (*a*) below and if two 7/8 in. diameter rivets resist this force, find the average shearing stress in each of the rivets. *Ans.* s_s = 8300 lb/in²

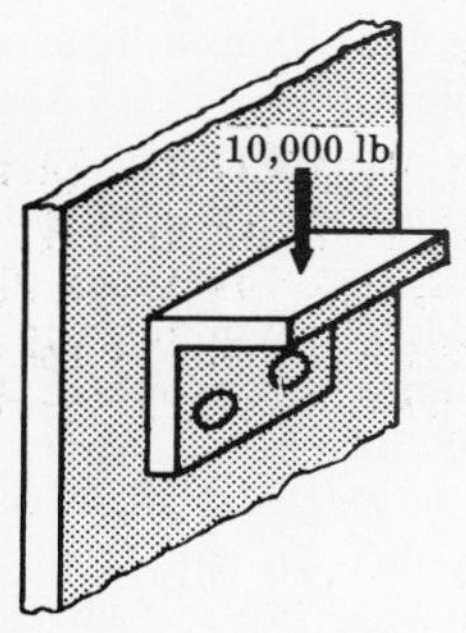

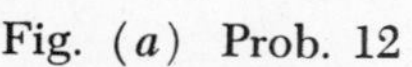

Fig. (*a*) Prob. 12

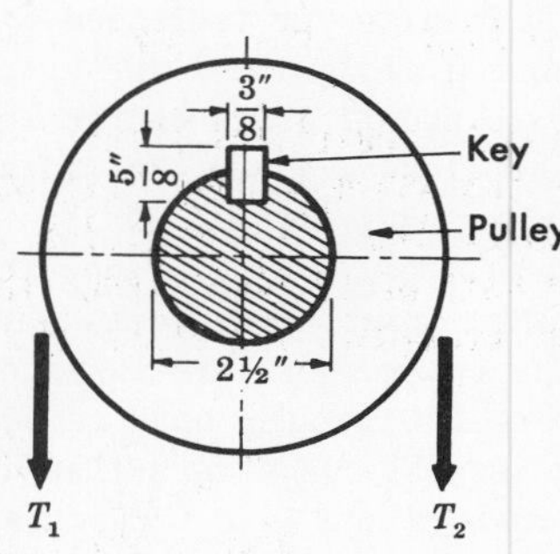

Fig. (*b*) Prob. 13

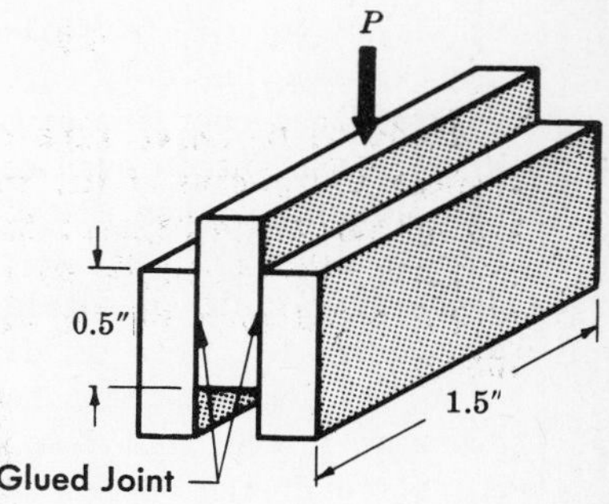

Fig. (*c*) Prob. 14

13. A pulley is keyed (to prevent relative motion) to a 2½ in. diameter shaft. The unequal belt pulls, T_1 and T_2, on the two sides of the pulley give rise to a net turning moment of 1200 lb-in. The key is 3/8 in. by 5/8 in. in cross-section and 3 in. long, as shown in Fig. (*b*) above. Determine the average shearing stress acting on a horizontal plane through the key. *Ans.* s_s = 855 lb/in²

14. The arrangement shown in Fig. (*c*) above is often used to determine the shearing strength of a glued joint. If the load *P* at rupture is 2500 lb, what is the average shearing stress in the joint at this instant? *Ans.* s_s = 1670 lb/in²

15. Fig. (*d*) below shows another type of apparatus for determining shearing strengths of cylindrical specimens. The specimen is clamped between blocks A_1, A_2 and B_1, B_2 and a downward force *P* is applied on block *C*. What force *P* must be applied in order to fracture a round hot rolled steel bar ¾ in. in diameter and having a shearing ultimate strength of 105,000 lb/in²? *Ans.* *P* = 93,000 lb

16. Consider the balcony-type structure shown in Fig. (*e*) below. The horizontal balcony is loaded by a total load of 20,000 lb distributed in a radially symmetric fashion. The central support is a shaft 20 in. in diameter and the balcony is welded at both the upper and lower surfaces to this shaft by welds 3/8 in. on a side (or leg) as shown in the enlarged view at the right. Determine the average shearing stress existing between the shaft and the weld. *Ans.* 424 lb/in²

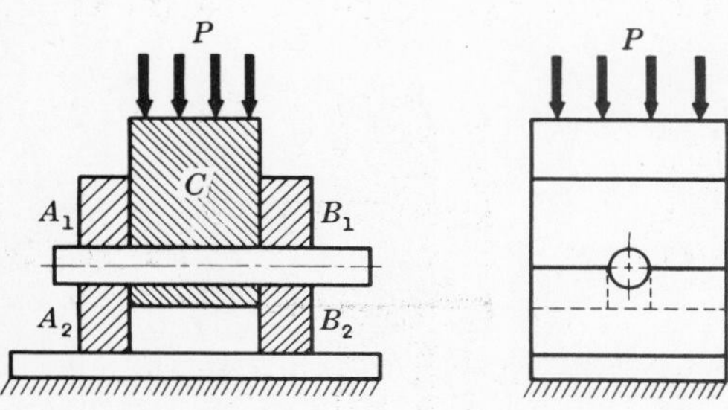

Fig. (*d*) Prob. 15

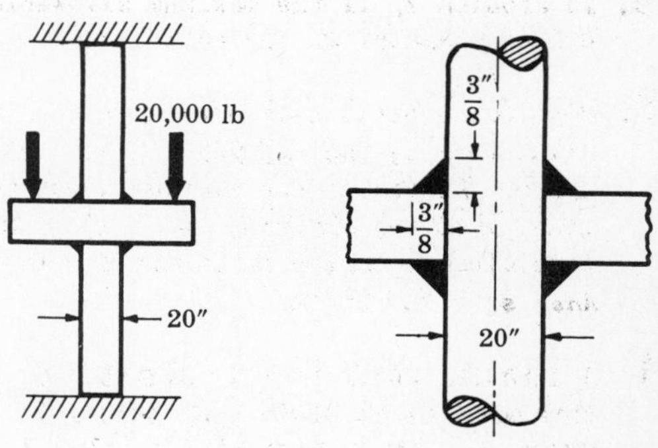

Fig. (*e*) Prob. 16

17. In certain bridge and roof trusses the diagonals, vertical posts, and horizontal chords are joined together on a pin. Consider the arrangement of parallel eye-bars shown in Fig. (*a*) below connected by a 6 in. diameter pin of structural steel. If the tension in each eye-bar is 220,000 lb, determine the average shearing stress in the pin. Also, calculate the shearing strain corresponding to this shearing stress if $G = 12\cdot 10^6$ lb/in². *Ans.* s_s = 7750 lb/in², γ = 0.000648

18. A thin vertical cylindrical shell 125 ft in diameter is loaded by a uniformly distributed load along its upper edge but only partially supported along its lower extremity as shown in Fig. (*b*) below. If the total load over the top is 950,000 lb and if 80 ft of the lower edge are unsupported, find the average shear stress over each of the sections *a-a* and *b-b* if the shell is concrete, 8 in. thick and 22 ft high. *Ans.* s_s = 46 lb/in²

19. A copper tube 2 in. in outside diameter and of wall thickness $\frac{1}{4}$ in. fits loosely over a solid steel circular bar 1-7/16 in. in diameter. The two members are fastened together by two metal pins each 5/16 in. in diameter and passing transversely through both members, one pin being near each end of the assembly. At room temperature the assembly is stress-free when the pins are in position. The temperature of the entire assembly is then raised 75°F. Calculate the average shear stress in the pins. For copper $E = 13\cdot 10^6$ lb/in², $\alpha = 9.3\cdot 10^{-6}$/°F; for steel $E = 30\cdot 10^6$ lb/in², $\alpha = 6.5\cdot 10^{-6}$/°F. *Ans.* s_s = 17,900 lb/in²

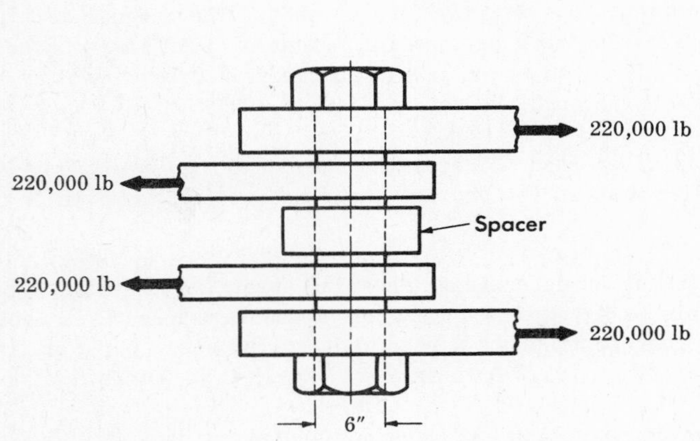

Fig. (*a*) Prob. 17

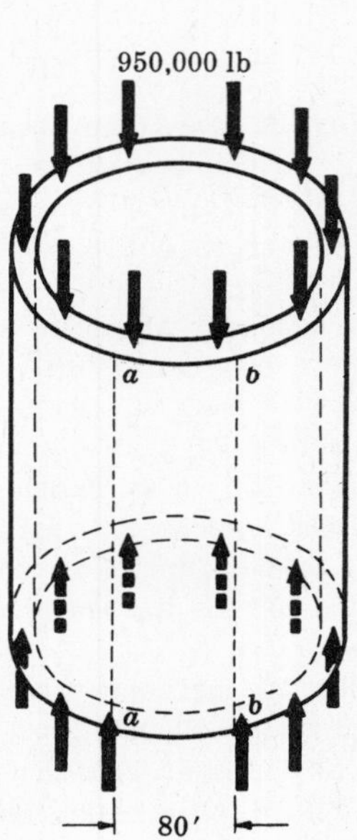

Fig. (*b*) Prob. 18

CHAPTER 5

Torsion

DEFINITION OF TORSION. Consider a bar rigidly clamped at one end and twisted at the other end by a torque $T(=Fd)$ applied in a plane perpendicular to the axis of the bar as shown. Such a bar is said to be in torsion.

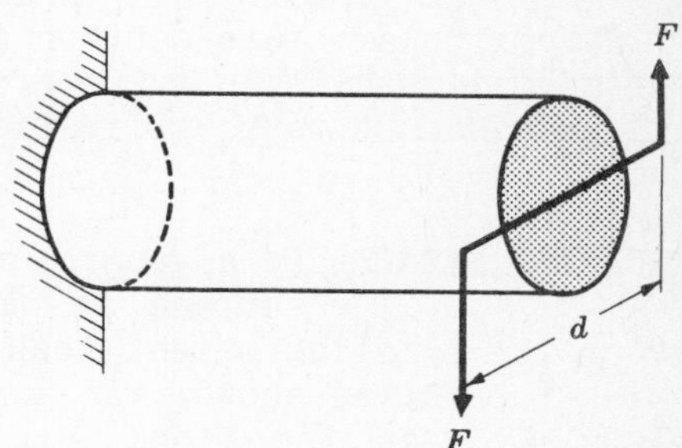

EFFECTS OF TORSION. The effects of a torsional load applied to a bar are (1) to impart an angular displacement of one end cross-section with respect to the other end, and (2) to set up shearing stresses on any cross-section of the bar perpendicular to its axis.

TWISTING MOMENT. Occasionally a number of couples act along the length of a shaft. In that case it is convenient to introduce a new quantity, the *twisting moment* which for any section along the bar is defined to be the algebraic sum of the moments of the applied couples that lie to one side of the section in question. The choice of side in any case is of course arbitrary.

POLAR MOMENT OF INERTIA. For a hollow circular shaft of outer diameter D_o with a concentric circular hole of diameter D_i the polar moment of inertia of the cross-sectional area, usually denoted by J, is given by

$$J = \frac{\pi}{32}(D_o^4 - D_i^4)$$

The polar moment of inertia for a solid shaft is obtained by setting $D_i = 0$. See Prob. 1.

This quantity J is simply a mathematical property of the geometry of the cross-section. It has no physical significance but it does occur in the study of the stresses set up in a circular shaft subject to torsion.

Occasionally it is convenient to rewrite the above equation in the form

$$J = \frac{\pi}{32}(D_o^2 + D_i^2)(D_o^2 - D_i^2) = \frac{\pi}{32}(D_o^2 + D_i^2)(D_o + D_i)(D_o - D_i)$$

This last form is useful in numerical evaluation of J in those cases where the difference $(D_o - D_i)$ is small. See Prob. 9.

TORSIONAL SHEARING STRESS. For either a solid or a hollow circular shaft subject to a twisting moment T the torsional shearing stress s_s at a distance ρ from the center of the shaft is given by

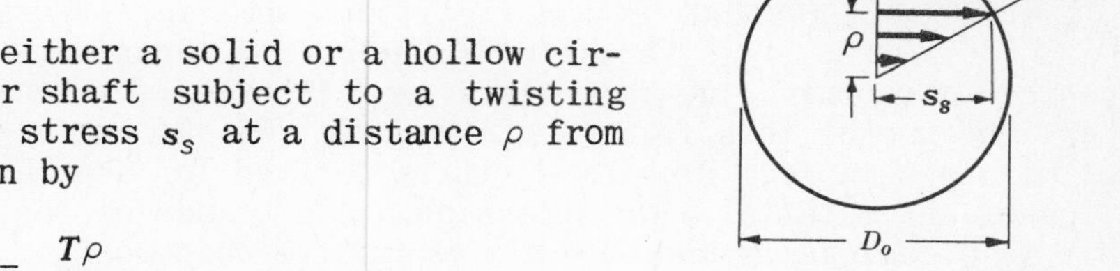

$$s_s = \frac{T\rho}{J}$$

The derivation of this equation is discussed in detail in Prob. 2. For applications, see Prob. 5, 7, 8, 11, 13, 14, 16, 18. This stress distribution varies from zero at the center of the shaft (if it is solid) to a maximum at the outer fibers, as shown in the above figure. Only the torsion of shafts of solid or hollow circular cross-sections will be discussed here.

ASSUMPTIONS. In the derivation of the formula $s_s = T\rho/J$ it is assumed that a plane section of the shaft normal to its axis prior to loading remains plane after the torques have been applied and further that a diameter in the section before deformation remains a diameter, or straight line, after deformation. Because of the polar symmetry of a circular shaft these assumptions seem to be reasonable. However, for a shaft of non-circular cross-section these assumptions are no longer true; it is known from experimental results that the cross-sections of non-circular shafts warp out of their original planes during application of external loads.

SHEARING STRAIN. If a generator a-b is marked on the surface of the unloaded bar, then after the twisting moment T has been applied this line moves to a-b', as shown. The angle γ, measured in radians, between the final and original positions of the generator is defined as the shearing strain at the surface of the bar. The same definition would hold at any interior point of the bar.

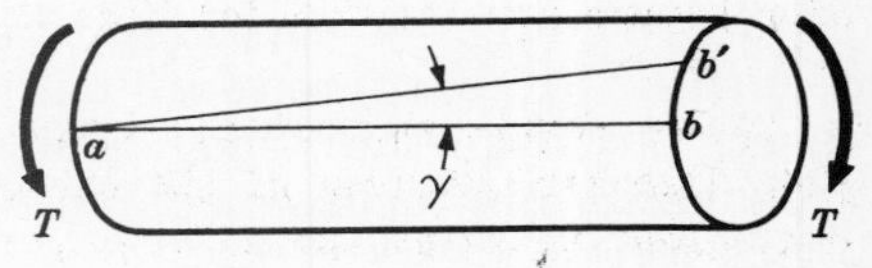

MODULUS OF ELASTICITY IN SHEAR. The ratio of the shear stress s_s to the shear strain γ is called the modulus of elasticity in shear and, as in Chapter 4, is given by

$$G = \frac{s_s}{\gamma}$$

Again the units of G are the same as those of shear stress, since the shear strain is dimensionless.

ANGLE OF TWIST. If a shaft of length L is subject to a constant twisting moment T along its length, then the angle θ through which one end of the bar will twist relative to the other is

$$\theta = \frac{TL}{GJ}$$

where J denotes the polar moment of inertia of the cross-section. This equation is derived in Problem 3. For applications see Problems 8, 11-17.

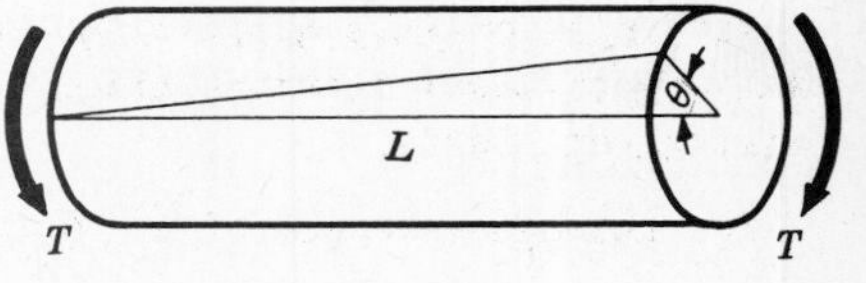

THE MODULUS OF RUPTURE is the fictitious shearing stress obtained by substituting in the equation $s_s = T\rho/J$ the maximum torque T carried by a shaft when tested to rupture. The variable ρ in this case is taken to be the outer radius of the bar. The use of this formula at the point of rupture is not of course justified because, as may be seen from Problem 2, it is derived for use only within the linear range of action of the material. The stress obtained by use of the formula is then not a true stress but is occasionally useful for comparative purposes.

STATICALLY INDETERMINATE PROBLEMS frequently arise in the case of torsional loadings. One example of this is a shaft composed of two materials, a tube of one material surrounding a tube or solid bar of another material with the entire assembly being subjected to a twisting moment. As usual, the applicable statics equations must be supplemented by additional equations based upon the deformations of the structure in order to furnish a number of equations equal to the number of unknowns. The unknowns in this case would be the twisting moments carried by each material. The equation based upon deformations for this case would state that the angles of twist of the various materials were equal. (See Problems 15-17.)

SOLVED PROBLEMS

1. Derive an expression for the polar moment of inertia of the cross-sectional area of a hollow circular shaft. What does this expression become for the special case of a solid circular shaft?

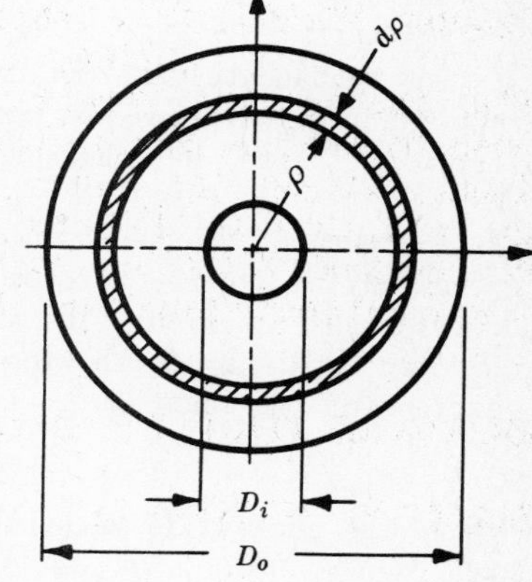

Let D_o denote the outside diameter of the shaft and D_i the inside diameter. Because of the circular symmetry involved, it is most convenient to adopt the polar coordinate system shown.

By definition, the polar moment of inertia is given by the integral

$$J = \int_A \rho^2 \, da$$

where A indicates that the integral is to be evaluated over the entire cross-sectional area.

To evaluate this integral it is desirable to select some element of area, da, so that ρ is constant for all points in the element. The thin ring-shaped element of radius ρ and radial thickness $d\rho$ shown constitutes a convenient element. It is assumed that the thickness $d\rho$ of the ring is small compared to ρ. The area of the ring-shaped element is given by $da = 2\pi\rho(d\rho)$. Hence the polar moment of inertia is given by

$$J = \int_{\frac{1}{2}D_i}^{\frac{1}{2}D_o} \rho^2 (2\pi\rho)\,d\rho = 2\pi\left[\frac{\rho^4}{4}\right]_{\frac{1}{2}D_i}^{\frac{1}{2}D_o} = \frac{\pi}{32}(D_o^4 - D_i^4)$$

The units of this quantity are evidently $(\text{length})^4$, e.g. in^4. It is not necessary to attempt to attach any physical interpretation to the quantity J. It will be found useful in problems involving the twisting of shafts.

For the special case of a solid circular shaft, the above expression becomes

$$J = \frac{\pi}{32}D^4$$

where D denotes the diameter of the shaft.

2. Derive an expression for the relationship between the applied twisting moment acting on a shaft of circular cross-section and the shearing stress at any point in the shaft.

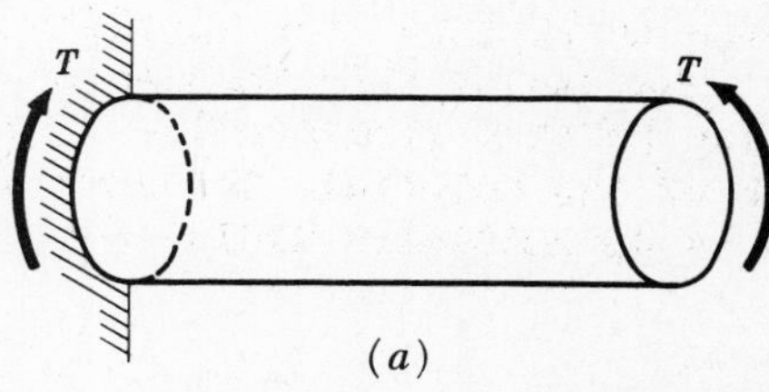

(a)

The shaft is shown loaded by the two torques T and consequently is in static equilibrium. To determine the distribution of shearing stress in the shaft, let us cut the shaft by a plane passing through it in a direction perpendicular to the geometric axis of the bar. Further, let us specify that this cutting plane is not "too close" to either end of the body, where the loads T are applied. The use of such a cutting plane is in accordance with the usual procedure followed in Strength of Materials, i.e. to cut the body in such a manner that the forces under investigation become external to the new body formed. These forces (or stresses) were of course internal effects with regard to the original, uncut body.

The free-body diagram of the portion of the shaft to the left of this plane now appears as in the adjoining figure. Obviously a torque T must act over the cross-section cut by the plane. This is true since the entire shaft is in equilibrium, hence any portion of it also is. The torque T acting on the cut section represents the effect of the right portion of the shaft on the left portion. Since the right portion has been removed, it must be replaced by its effect on the left portion. This effect is represented by the torque T. This torque is of course a resultant of shearing stresses distributed over the cross-section. It is now necessary to make certain assumptions in order to determine the nature of the variation of shear stress intensity over the cross-section.

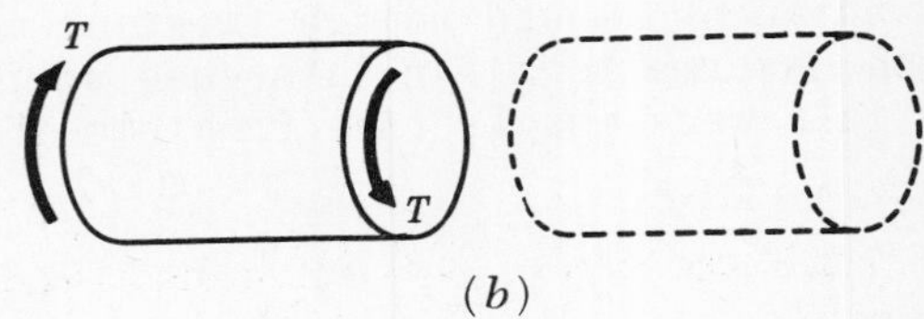

(b)

One fundamental assumption is that a plane section of the shaft normal to its axis before loads are applied remains plane and normal to the axis after loading. This may be verified experimentally for circular shafts, but this assumption is not valid for shafts of non-circular cross-section.

A generator on the surface of the shaft, denoted by O_1A in the accompanying figure, deforms into the configuration O_1B after torsion has occurred. The angle between these configurations is denoted by α. By definition, the shearing unit strain γ on the surface of the shaft is

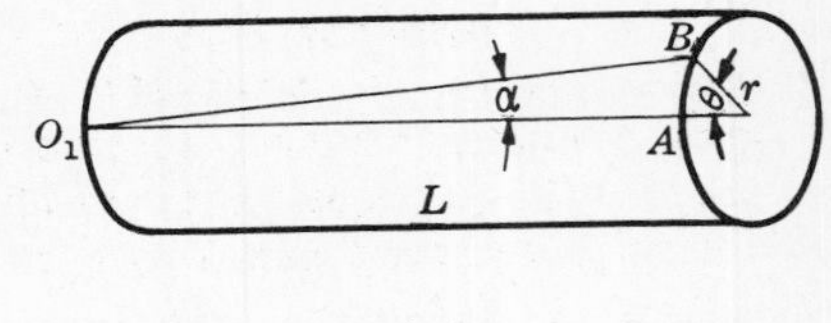

(c)

$$\gamma = \tan\alpha \approx \alpha$$

where the angle α is measured in radians. From the geometry of the figure,

$$\alpha = \frac{AB}{L} = \frac{r\theta}{L}; \quad \text{hence } \gamma = \frac{r\theta}{L}$$

But since a diameter of the shaft prior to loading is assumed to remain a diameter after torsion has occurred, the shearing unit strain at a general distance ρ from the center of the shaft may likewise be written

$$\gamma_\rho = \frac{\rho\theta}{L}$$

Consequently the shearing strains of the longitudinal fibers vary linearly as the distances from the center of the shaft.

If we assume that we are concerned only with the linear range of action of the material where the shearing stress is proportional to shearing strain, then it is evident that the shearing stresses of the longitudinal fibers vary linearly as the distances from the center of the shaft. Obviously the distribution of shearing stresses is symmetric around the geometric axis of the shaft. They have the appearance shown in the adjoining figure. For equilibrium, the sum of the moments of these distributed shearing forces over the entire circular cross-section is equal to the applied twisting moment. Also, the sum of the moments of these forces is exactly equal to the torque T shown in Figure (b) above.

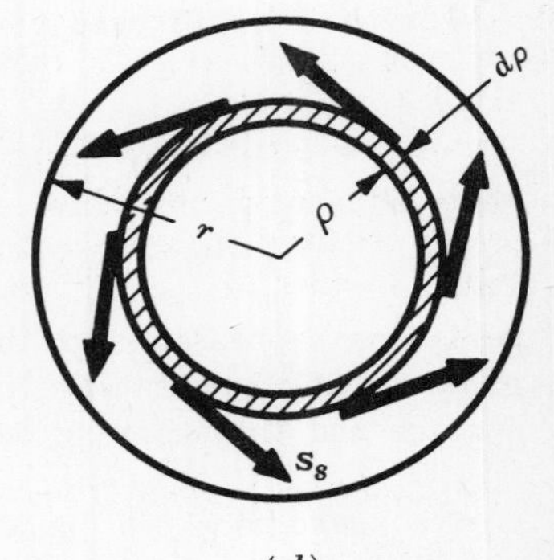

(d)

Thus we have $$T = \int_0^r s_s\,\rho\cdot da$$

where da represents the area of the shaded ring-shaped element shown in Fig. (d). However, the shearing stresses vary as the distances from the geometric axis; hence

$$(s_s)_\rho/\rho = (s_s)_r/r = \text{constant}$$

where the subscripts on the shearing stress denote the distances of the element from the axis of the shaft. Consequently we may write

$$T = \int_0^r \frac{(s_s)_\rho}{\rho}(\rho^2)\,da = \frac{(s_s)_\rho}{\rho}\int_0^r \rho^2\,da$$

since the ratio $\dfrac{(s_s)_\rho}{\rho}$ is a constant. However, the expression $\int_0^r \rho^2\,da$ is by definition (see Problem 1) the polar moment of inertia of the cross-sectional area. Values of this for solid and hollow circular shafts are derived in Problem 1. Hence the desired relationship is

$$T = \frac{(s_s)_\rho J}{\rho} \qquad \text{or} \qquad (s_s)_\rho = \frac{T\rho}{J}$$

3. Derive an expression for the angle of twist of a circular shaft as a function of the applied twisting moment.

Let L denote the length of the shaft, J the polar moment of inertia of the cross section, T the applied twisting moment (assumed constant along the length of the bar), and G the modulus of elasticity in shear. The angle of twist in a length L is represented by θ in the adjoining diagram.

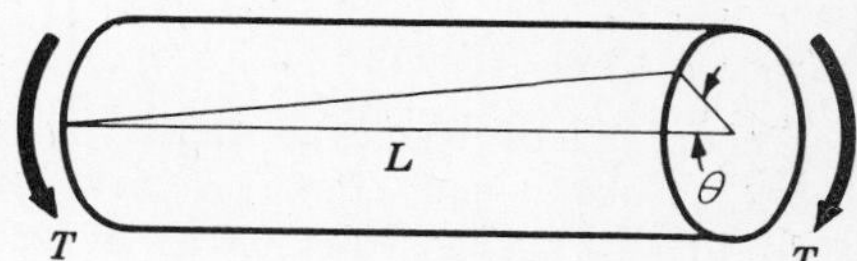

From Problem 2 we have at the outer fibers where $\rho = r$:

$$\gamma_r = \frac{r\theta}{L} \qquad \text{and also} \qquad (s_s)_r = \frac{Tr}{J}$$

By definition, the shearing modulus is given by $G = \dfrac{s_s}{\gamma} = \dfrac{Tr/J}{r\theta/L} = \dfrac{TL}{J\theta}$ from which $\theta = \dfrac{TL}{GJ}$.

Note that θ is expressed in radians, i.e. it is dimensionless. One consistent set of units would be to take T in lb-in, L in in., G in lb/in^2, and J in in^4.

Occasionally the angle of twist in a unit length is useful. It is often denoted by ϕ and is given by

$$\phi = \frac{\theta}{L} = \frac{T}{GJ}$$

4. Derive a relationship between the twisting moment acting on a rotating shaft, the horsepower transmitted by it, and its angular velocity, which is assumed to be constant.

Let us denote the twisting moment acting on the shaft by T, the angular velocity in rev/min by n, and the horsepower by hp. Also, let us consider a one-minute time interval. During this interval the twisting moment does an amount of work given by the product of the moment times the angular displacement in radians, or $T \times 2\pi n$. If T is measured in lb-in, then the work has these same units. By definition if work is being done at the rate of 33,000 ft-lb/min = 12(33,000) in-lb/min, it is equivalent to one horsepower. The horsepower transmitted by the shaft is consequently

$$\text{hp} = \frac{T \times 2\pi n}{12(33{,}000)} \qquad \text{from which} \qquad T = \frac{63{,}000 \times \text{hp}}{n}$$

where n is in rev/min and T is in lb-in.

5. If a twisting moment of 10,000 lb-in is impressed upon a $1\frac{3}{4}$ in. diameter shaft, what is the maximum shearing stress developed? Also, what is the angle of twist in a 4 ft length of the shaft? The material is steel for which $G = 12\times10^6$ lb/in^2.

From Problem 1 the polar moment of inertia of the cross-sectional area is

$$J = \frac{\pi}{32}(D_o)^4 = \frac{\pi}{32}(\frac{7}{4})^4 = 0.92 \text{ in}^4$$

The torsional shearing stress s_s at any distance ρ from the center of the shaft was shown in Problem 2 to be: $(s_s)_\rho = T\rho/J$. The maximum shear stress is developed at the outer fibers and there at $\rho = 7/8$ in. we have

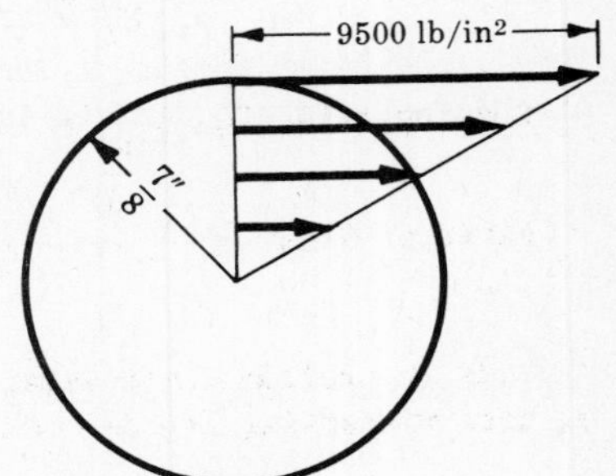

$$(s_s)_{max} = \frac{10,000(7/8)}{0.92} = 9500 \text{ lb/in}^2$$

Hence the shear stress varies linearly from zero at the center of the shaft to 9500 lb/in^2 at the outer fibers as shown.

The angle of twist θ in a 4 ft length of the shaft is

$$\theta = \frac{TL}{GJ} = \frac{10,000(48 \text{ in.})}{12\times10^6(0.92)} = 0.0435 \text{ radian}$$

6. A coupling of the type shown is frequently used for connecting the ends of two shafts. This is a so-called solid-coupling, the two portions being bolted together by six 3/4 in. diameter bolts. If the solid shaft transmits 65 hp at 250 rpm, determine the mean shear in the bolts.

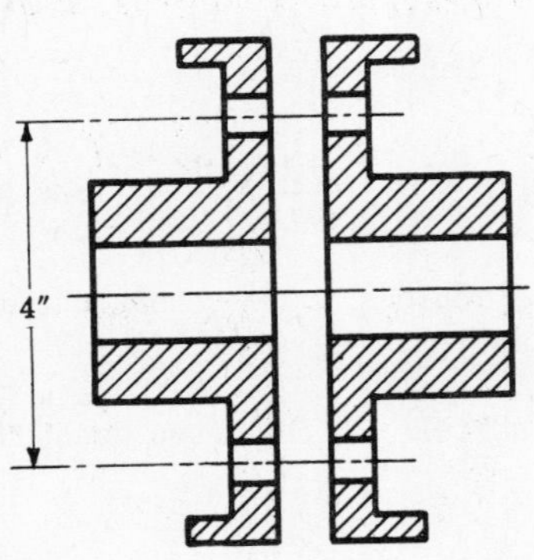

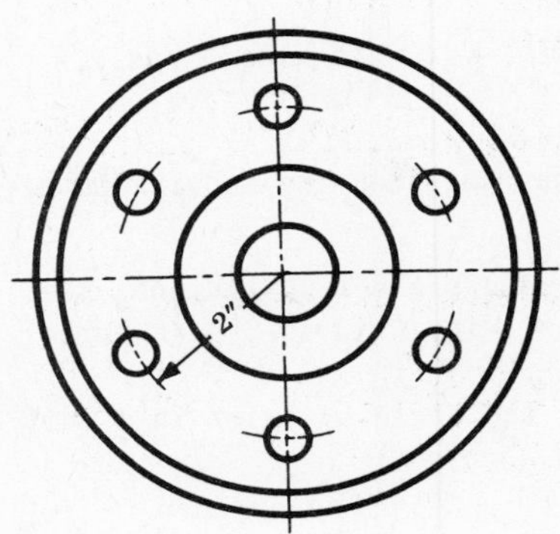

The torque transmitted is $T = \frac{63,000 \times hp}{n} = \frac{63,000(65)}{250} = 16,400$ lb-in. The tangential force acting 2 in. from the center of the shaft to give rise to this torque is $16,400/2 = 8200$ lb.

The mean shear stress in each of the bolts is then $s_s = \frac{8200}{6\times\frac{\pi}{4}(3/4)^2} = 3100$ lb/in^2. It has been assumed that the radius of the bolts is small compared to the radius of the bolt circle.

Note that this is the mean shear stress that appeared in Chapter 4 and that the torsional shear stress does not enter into the problem.

7. Consider a solid circular shaft and also a hollow circular shaft whose inside diameter is 3/4 of the outside. Compare the weights of equal lengths of these two shafts required to transmit a given torsional load if the maximum shear stresses developed in the two shafts are equal.

For the solid shaft of diameter d the shearing stress is given by $(s_s)_\rho = T\rho/J$. The maximum shearing stress occurs at the outer fibers where $\rho = d/2$. Hence

$$(s_s)_{max} = \frac{T(d/2)}{(\pi/32)d^4} = \frac{16T}{\pi d^3} \quad \text{or} \quad \frac{T}{(s_s)_{max}} = \frac{\pi d^3}{16}$$

For the hollow shaft of outer diameter D the maximum shearing stress still occurs at the outer fibers where $\rho = D/2$. Then

$$(s_s)_{max} = \frac{T(D/2)}{\frac{\pi}{32}\{D^4 - (\frac{3}{4}D)^4\}} = \frac{16\,T}{\pi(0.684)D^3} \quad \text{or} \quad \frac{T}{(s_s)_{max}} = \frac{\pi(0.684)D^3}{16}$$

But the ratio $T/(s_s)_{max}$ is a constant for both shafts; hence $0.684\,D^3 = d^3$ or $D = 1.135\,d$.

$$\text{Ratio of weights} = \frac{D^2 - (3D/4)^2}{d^2} = \frac{0.4375\,D^2}{d^2} = \frac{0.4375(1.135\,d)^2}{d^2} = 0.563$$

Thus the hollow shaft weighs only 56.3% as much as the solid one, which illustrates the efficiency of hollow shafting as compared to solid.

8. A hollow steel shaft 10 ft long must transmit a torque of 250,000 lb-in. The total angle of twist in this length is not to exceed 2.5° and the allowable shearing stress is 12,000 lb/in^2. Determine the inside and outside diameters of the shaft if $G = 12 \times 10^6$ lb/in^2.

Let d_o and d_i designate the outside and inside diameters of the shaft respectively. From Problem 3 the angle of twist θ is given by $\theta = TL/GJ$, where θ is expressed in radians. Consequently in the 10 ft length we have

$$2.5 \text{ deg} \times \frac{1 \text{ rad}}{57.3 \text{ deg}} = \frac{250{,}000(120 \text{ in.})}{12 \times 10^6 \times \frac{\pi}{32}(d_o^4 - d_i^4)} \quad \text{or} \quad d_o^4 - d_i^4 = 583.65$$

The maximum shearing stress occurs at the outer fibers where $\rho = d_o/2$. Then

$$(s_s)_{max} = \frac{T(d_o/2)}{\frac{\pi}{32}(d_o^4 - d_i^4)}, \qquad 12{,}000 = \frac{250{,}000(d_o/2)}{\frac{\pi}{32}(d_o^4 - d_i^4)} \quad \text{and} \quad d_o^4 - d_i^4 = 106.10\,d_o$$

Thus $106.10\,d_o = 583.65$ or $d_o = 5.50$ in. Substituting, $d_i = 4.27$ in.

9. Let us consider a thin-walled tube subject to torsion. Derive an approximate expression for the allowable twisting moment if the working stress in shear is a given constant s_w. Also, derive an approximate expression for the strength-weight ratio of such a tube. It is assumed the tube does not buckle.

The polar moment of inertia of a hollow circular shaft of outer diameter D_o and inner diameter D_i is $J = \frac{\pi}{32}(D_o^4 - D_i^4)$. If R denotes the outer radius of the tube, then $D_o = 2R$ and further, if t denotes the wall thickness of the tube, then $D_i = 2R - 2t$.

The polar moment of inertia J may be written in the alternate form

$$J = \frac{\pi}{32}\{(2R)^4 - (2R-2t)^4\} = \frac{\pi}{2}\{R^4 - (R-t)^4\} = \frac{\pi}{2}\{4R^3t - 6R^2t^2 + 4Rt^3 - t^4\}$$

$$= \frac{\pi}{2}R^4\{4(t/R) - 6(t/R)^2 + 4(t/R)^3 - (t/R)^4\}$$

Neglecting squares and higher powers of the ratio t/R, since we are considering a thin-walled tube, this becomes, approximately,

$$J = 2\pi R^3 t$$

The ordinary torsion formula is $T = s_w J/R$. For a thin-walled tube this becomes, for the allowable twisting moment,

$$T = 2\pi R^2 t\, s_w$$

The weight W of the tube is $W = \gamma L A$ where γ is the specific weight of the material, L the length of the tube and A the cross-sectional area of the tube. The area is given by

$$A = \pi\{R^2 - (R-t)^2\} = \pi\{2Rt - t^2\} = \pi R^2\{2t/R - (t/R)^2\}$$

Again neglecting the square of the ratio t/R for a thin tube, this becomes

$$A = 2\pi R t$$

The strength-weight ratio is defined to be T/W. This is given by

$$\frac{T}{W} = \frac{2\pi R^2 t\, s_w}{2\pi R t L\gamma} = \frac{R\, s_w}{L\gamma}$$

This ratio is of considerable importance in aircraft design.

10. A solid circular shaft has a slight taper extending uniformly from one end to the other. Denote the radius at the small end by a, that at the large end by b. Determine the error committed if the angle of twist for a given length is calculated using the mean radius of the shaft. The radius at the larger end is 1.2 times that at the smaller end.

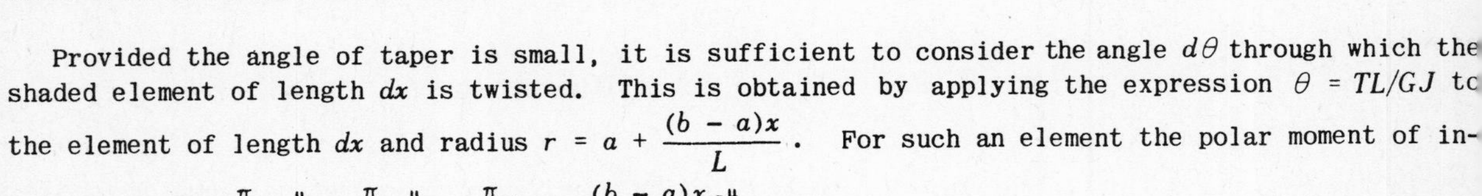

Let us set up a coordinate system with the variable x denoting the distance from the small end of the shaft. The radius at a section at the distance x from the small end is

$$r = a + \frac{(b-a)x}{L}$$

where L is the length of the bar.

Provided the angle of taper is small, it is sufficient to consider the angle $d\theta$ through which the shaded element of length dx is twisted. This is obtained by applying the expression $\theta = TL/GJ$ to the element of length dx and radius $r = a + \frac{(b-a)x}{L}$. For such an element the polar moment of inertia is $J = \frac{\pi}{32}D^4 = \frac{\pi}{2}r^4 = \frac{\pi}{2}\left[a + \frac{(b-a)x}{L}\right]^4$. Thus

$$d\theta = \frac{T\,dx}{G\frac{\pi}{2}\left[a + \frac{(b-a)x}{L}\right]^4}$$

The angle of twist in the length L is found by integrating the last equation. Thus

$$\theta = \frac{2T}{G\pi}\int_0^L \frac{dx}{\left[a + \frac{(b-a)x}{L}\right]^4} = \frac{2T}{G\pi}\left(-\frac{1}{3}\right)\left(\frac{L}{b-a}\right)\left[\frac{1}{\left[a + \frac{(b-a)x}{L}\right]^3}\right]_0^L = \frac{2TL}{3G\pi(b-a)}\left(-\frac{1}{b^3} + \frac{1}{a^3}\right)$$

If $b = 1.2a$, this becomes $\theta = \dfrac{1.40433\,TL}{G\pi a^4}$.

For a solid shaft of uniform radius $1.1a$, $\theta_1 = \dfrac{TL}{G\frac{\pi}{2}(1.1a)^4} = \dfrac{1.36602\,TL}{G\pi a^4}$.

Percent error $= \dfrac{0.03831}{1.40433} \times 100 = 2.73\%$

11. A solid circular shaft has a uniform diameter of 2 in. and is 10 ft long. At its midpoint 65 hp are delivered to the shaft by means of a belt passing over a pulley. This power is used to drive two machines, one at the left end of the shaft consuming 25 hp and one at the right end consuming the remaining 40 hp. Determine the maximum shearing stress in the shaft and also the relative angle of twist between the two extreme ends of the shaft. The shaft turns at 200 rpm and the material is steel for which $G = 12 \times 10^6$ lb/in^2.

In the left half of the shaft we have 25 hp which corresponds to a torque T_1 given by

$$T_1 = \frac{63{,}000 \times \text{hp}}{n} = \frac{63{,}000(25)}{200} = 7880 \text{ lb-in}$$

Similarly, in the right half we have 40 hp corresponding to a torque T_2 given by

$$T_2 = \frac{63{,}000(40)}{200} = 12{,}600 \text{ lb-in}$$

The maximum shearing stress consequently occurs in the outer fibers in the right half and is given by the ordinary torsion formula:

$$(s_s)_\rho = \frac{T\rho}{J} \quad \text{or} \quad s_s = \frac{12{,}600(1)}{\frac{\pi}{32}(2)^4} = 8000 \text{ lb/in}^2$$

The angle of twist of the left end relative to the center is $\theta_1 = \dfrac{7880(60)}{12 \cdot 10^6\, \frac{\pi}{32}(2)^4} = 0.0250$ rad.

The angle of twist of the right end relative to the center is $\theta_2 = \dfrac{12{,}600(60)}{12 \cdot 10^6\, \frac{\pi}{32}(2)^4} = 0.0401$ rad, in the same direction as θ_1.

The relative angle of twist between the two ends of the shaft is $\theta = \theta_2 - \theta_1 = 0.015$ radian.

12. Consider the two solid circular shafts connected by the 2 in. and 10 in. pitch diameter gears. The shafts are assumed to be supported by the bearings in such a manner that they undergo no bending. Find the angular rotation of D, the right end of one shaft, with respect to A, the left end of the other, caused by the torque of 2500 lb-in applied at D. The left shaft is steel for which $G = 12 \cdot 10^6$ lb/in^2 and the right is brass for which $G = 5 \cdot 10^6$ lb/in^2.

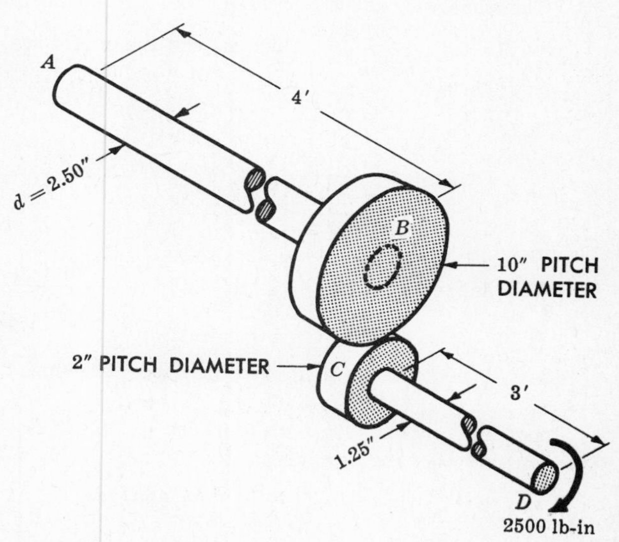

A free-body diagram of the right shaft reveals that a tangential force F must act on the smaller gear as shown. For equilibrium, $F = 2500$ lb.

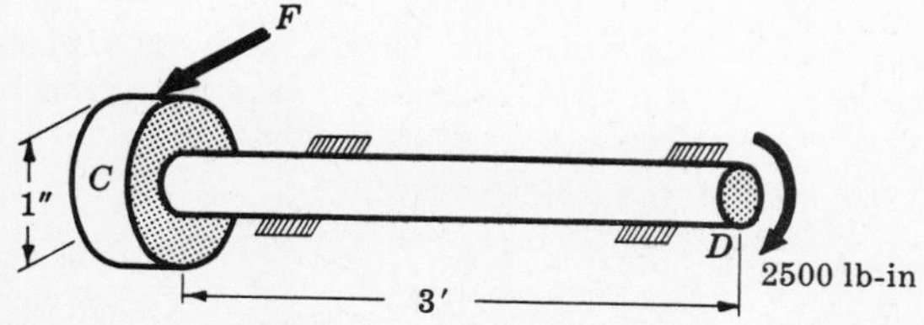

The angle of twist of the right shaft is given by $\theta_1 = \frac{TL}{GJ} = \frac{2500(36)}{5\cdot 10^6 \frac{\pi}{32}(1.25)^4} = 0.0750$ rad.

A free-body diagram of the left shaft is shown in the adjoining figure. The force F is equal and opposite to that acting on the small gear C. This force F acts 5 in. from the center line of the left shaft, hence it imparts a torque of 5(2500) = 12,500 lb-in to the shaft AB. Because of this torque there is a rotation of end B with respect to end A given by the angle θ_2, where

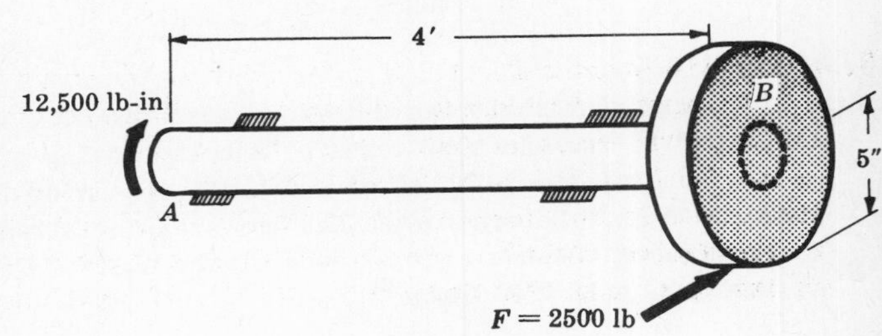

$$\theta_2 = \frac{12{,}500(48)}{12\cdot 10^6 \frac{\pi}{32}(2.5)^4} = 0.0130 \text{ rad.}$$

It is to be carefully noted that this angle of rotation θ_2 induces a *rigid body* rotation of the entire shaft CD because of the gears. In fact, the rotation of CD will be in the same ratio to that of AB as the ratio of the pitch diameters, or 5:1. Thus a rigid body rotation of 5(0.0130) rad is imparted to shaft CD. Superposed on this rigid body movement of CD is the angular displacement of D with respect to C previously denoted by θ_1.

Hence the resultant angle of twist of D with respect to A is $\theta = 5(0.0130) + 0.075 = 0.140$ rad.

13. The compound shaft shown is of steel for which $G = 12\cdot 10^6$ lb/in^2. The stress concentration arising because of the abrupt change of cross-section is to be disregarded.

At the lower extremity the shaft is subject to a torque of 50,000 lb-in in the direction shown, at the junction it is subject to a torque of 80,000 lb-in in the direction opposite to that of the first torque. Determine the maximum shearing stress in each portion of the shaft and also the angles of twist at B and at C.

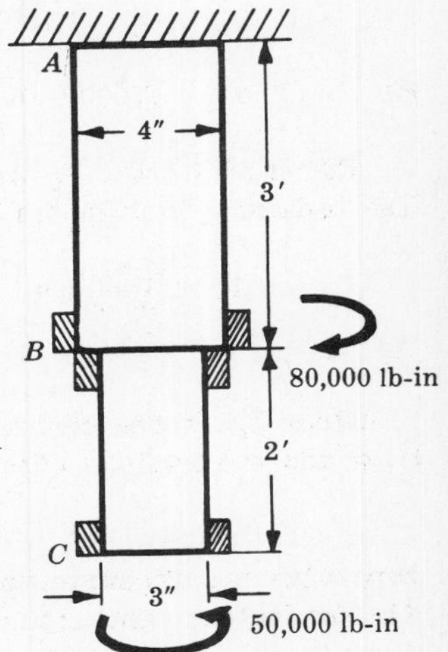

The torque acting in the region BC is obviously 50,000 lb-in. In the region AB it is 50,000 − 80,000 = −30,000 lb-in, i.e. opposite in direction to that in BC.

The shearing stress in each of these regions is given by the formula $(s_s)_\rho = \frac{T\rho}{J}$. Thus at the outer fibers of each of these shafts we have

$$(s_s)_{AB} = \frac{30{,}000(2)}{\frac{\pi}{32}(4)^4} = 2380 \text{ lb/in}^2 \quad \text{and} \quad (s_s)_{BC} = \frac{50{,}000(1.5)}{\frac{\pi}{32}(3)^4} = 9400 \text{ lb/in}^2$$

The angle of twist at B is $\theta_1 = \frac{30{,}000(36)}{12\cdot 10^6 \frac{\pi}{32}(4)^4} = 0.00357$ radian (clockwise looking downward).

This is the true or absolute value of the angle of twist at B.

Let us temporarily consider the junction B to be fixed in space in its unstrained position and calculate the angle of rotation of section C with respect to B. This angle of twist is given by

$$\theta_2 = \frac{50{,}000(24)}{12\cdot 10^6 \frac{\pi}{32}(3)^4} = 0.01260 \text{ radian (counterclockwise)}$$

However, this is not the true angle of twist at C because the section B is not fixed in space but instead rotates through 0.00357 radian in the opposite direction. Hence the true angle of rotation of C relative to its original unstrained position is

$$\theta_3 = 0.0126 - 0.00357 = 0.00903 \text{ radian (counterclockwise)}$$

14. A compound shaft consists of a 2 ft length of brass rod securely joined to a 2 ft length of aluminum bar. Each has a diameter 2.5 in. The proportional limit of the brass in shear is 15,000 lb/in^2 and that of the aluminum alloy is 22,000 lb/in^2. A safety factor of 2 is to be applied to each of these strengths. If the angle of twist at the lower end is not to exceed 1°, what is the maximum twisting moment the shaft can carry? It is rigidly clamped at its upper end. For brass $G = 5\cdot10^6$ lb/in^2; for aluminum $G = 4\cdot10^6$ lb/in^2.

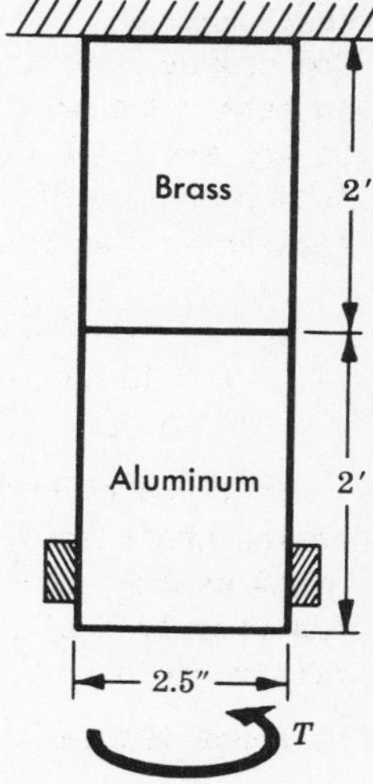

Perhaps the simplest method of solving the problem is to determine three values of the twisting moment. The first is that torque that is just sufficient to set up the working stress in shear in the brass, the second torque sets up the working stress in shear in the aluminum, and the third creates a 1° twist of the entire shaft. The allowable torque is then the minimum of these three values.

The first and second torques, T_1 and T_2, are found from the simple torsion formula:

$$\frac{15,000}{2} = \frac{T_1(1.25)}{\frac{\pi}{32}(2.5)^4} \qquad \text{and} \qquad \frac{22,000}{2} = \frac{T_2(1.25)}{\frac{\pi}{32}(2.5)^4}$$

or $\qquad T_1 = 23,000$ lb-in $\qquad$ and $\qquad T_2 = 33,800$ lb-in

The third torque T_3 gives rise to a 1° angle of twist of the entire shaft. It may be found from the ordinary formula for torsional deformations to be:

$$1° \times \frac{1 \text{ rad}}{57.3°} = \frac{T_3(24)}{5\cdot10^6\,\frac{\pi}{32}(2.5)^4} + \frac{T_3(24)}{4\cdot10^6\,\frac{\pi}{32}(2.5)^4} \qquad \text{from which} \qquad T_3 = 6200 \text{ lb-in}$$

Since T_3 is the minimum of these three values, the angle of twist is the controlling factor in design and the maximum torque that may be applied is 6200 lb-in.

15. Determine the reactive torques at the fixed ends of the circular shaft loaded by the couples shown in Fig. (*a*) below. The cross-section of the bar is constant along the length.

Let us assume that the reactive torques T_L and T_R are positive in the directions shown in Fig. (*b*). From statics we have

$$(1) \qquad T_L - T_1 + T_2 - T_R = 0$$

This is the only equation of static equilibrium and it contains two unknowns. Hence this problem is

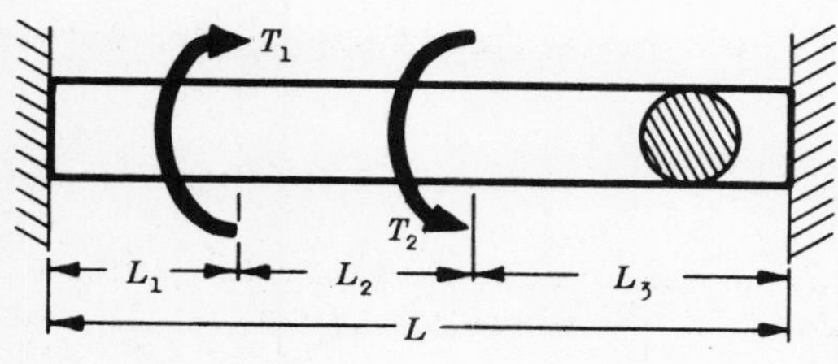

Fig. (*a*)

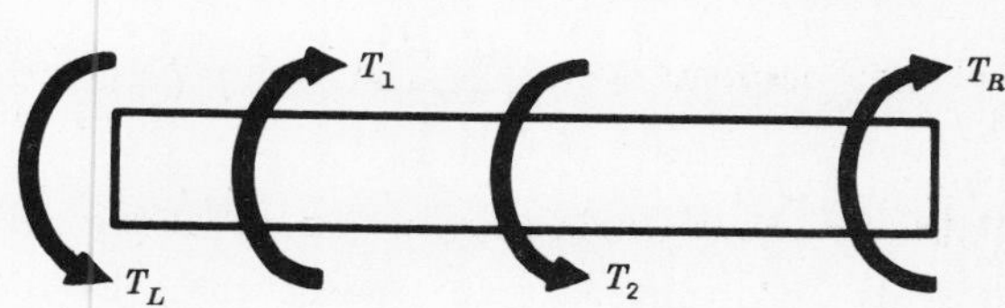

Fig. (*b*)

statically indeterminate and it is necessary to augment this equation with another equation based on the deformations of the system.

The variation of torque with length along the bar may be represented by the following plot:

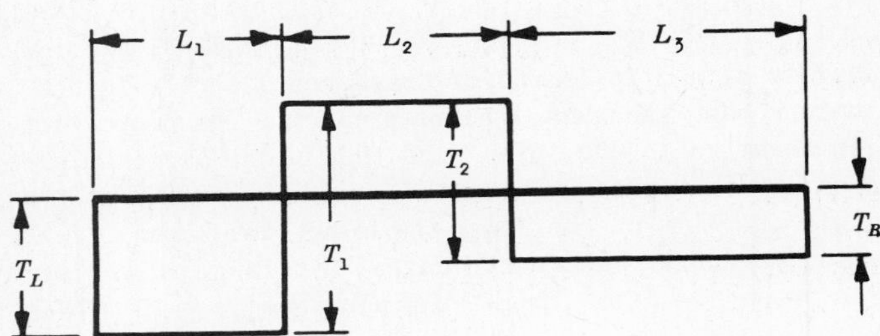

The free-body diagram of the left region of length L_1 appears as in Fig. (*a*) below.

Working from left to right along the shaft, the twisting moment in the central region of length L_2 is given by the algebraic sum of the torques to the left of this section, i.e. $(T_1 - T_L)$. The free-body diagram of this region appears as in Fig. (*b*) below.

Finally, the free-body diagram of the right region of length L_3 appears as in Fig. (*c*) below.

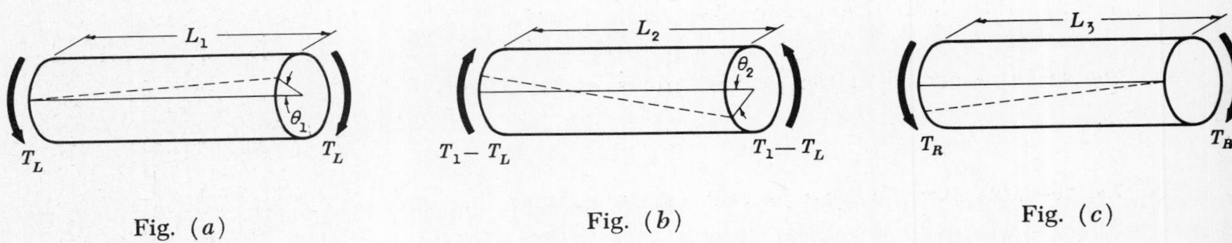

Fig. (*a*) Fig. (*b*) Fig. (*c*)

Let θ_1 denote the angle of twist at the point of application of T_1, and θ_2 the angle at T_2. Then from a consideration of the regions of lengths L_1 and L_3 we immediately have

$$(2) \quad \theta_1 = \frac{T_L L_1}{GJ} \qquad \text{and} \qquad (3) \quad \theta_2 = \frac{T_R L_3}{GJ}$$

The original position of a generator on the surface of the shaft is shown by the solid line in each of the above diagrams and the deformed position by dotted lines. Consideration of the central region of length L_2 reveals that the angle of twist of its right end with respect to its left end is $(\theta_1 + \theta_2)$. Hence, since the torque causing this deformation is $(T_1 - T_L)$, we have

$$(4) \quad \theta_1 + \theta_2 = \frac{(T_1 - T_L)L_2}{GJ}$$

Solving equations (1) through (4) simultaneously, we find

$$T_L = T_1\left(\frac{L_2 + L_3}{L}\right) - T_2\left(\frac{L_3}{L}\right) \qquad \text{and} \qquad T_R = -T_1\left(\frac{L_1}{L}\right) + T_2\left(\frac{L_1 + L_2}{L}\right)$$

It is of interest to examine the behavior of a generator on the surface of the shaft. Originally it was, of course, straight over the entire length L, but after application of T_1 and T_2 it has the appearance shown by the broken line in the adjoining figure.

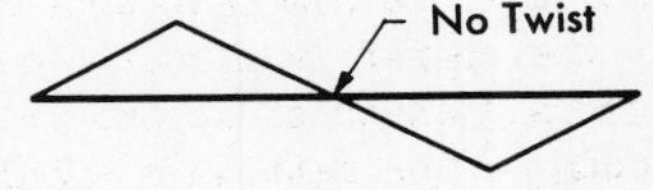

16. Consider a composite shaft fabricated from a 2 in. diameter solid aluminum alloy, $G = 4\cdot10^6$ lb/in^2, surrounded by a hollow steel circular shaft of outside diameter 2.5 in. and inside diameter 2 in., $G = 12\cdot10^6$ lb/in^2. The two metals are rigidly connected at their juncture. If the composite shaft is loaded by a twisting moment of 14,000 lb-in, calculate the shearing stress at the outer fibers of the steel and also at the extreme fibers of the aluminum.

Let T_1 = torque carried by the aluminum shaft and T_2 = torque carried by the steel. For static equilibrium of moments about the geometric axis we have

$$T_1 + T_2 = T = 14{,}000$$

when T = external applied twisting moment. This is the only equation from statics available in this problem and hence, since it contains the two unknowns T_1 and T_2 it is necessary to supplement it with an additional equation coming from the deformations of the shaft. The structure is thus statically indeterminate.

Such an equation is easily found, since the two materials are rigidly joined; hence their angles of twist must be equal. In a length L of the shaft we have, using the formula $\theta = TL/GJ$,

$$\frac{T_1 L}{4\cdot10^6 \frac{\pi}{32}(2)^4} = \frac{T_2 L}{12\cdot10^6 \frac{\pi}{32}[(2.5)^4 - (2)^4]} \qquad \text{or} \qquad T_1 = 0.231\,T_2$$

This equation, together with the statics equation, may be solved simultaneously to yield

T_1 = 2600 lb-in (carried by aluminum) and T_2 = 11,400 lb-in (carried by steel)

The shearing stress at the extreme fibers of the steel tube is $s_2 = \dfrac{11{,}400(1.25)}{\frac{\pi}{32}[(2.5)^4-(2)^4]} = 6300$ psi.

The shearing stress at the extreme fibers of the aluminum is $s_1 = \dfrac{2600(1)}{\frac{\pi}{32}(2)^4} = 1650$ psi.

17. A solid steel circular shaft is surrounded by a thick, hollow copper shell which is rigidly attached to the steel. The entire assembly is subjected to a twisting moment. If the copper carries 1.5 times as much torque as does the steel, find the ratio of external to internal diameter of the copper tube. For copper $G = 6\cdot10^6$ lb/in^2, for steel $G = 12\cdot10^6$ lb/in^2.

Since the two metals are rigidly attached, the angles of twist of the two are equal. This angle of twist is given by $\theta = TL/GJ$. Hence if T is the torque carried by the steel, we have

$$\frac{TL}{12\cdot10^6 \frac{\pi}{32} d_i^4} = \frac{(1.5\,T)L}{6\cdot10^6 \frac{\pi}{32}(d_o^4 - d_i^4)} \qquad \text{from which} \qquad \frac{d_o}{d_i} = \sqrt{2} = 1.414$$

where d_o and d_i are the outer and inner diameters of the copper tube.

18. If the maximum allowable shearing stress in the copper tube of Problem 17 is 8000 lb/in^2 and if the maximum allowable shearing stress in the steel is 12,000 lb/in^2, determine the limiting torque that the compound shaft can carry. The diameter of the steel shaft is 2.5 in. and, as in Problem 17, the copper carries 1.5 times as much torque as does the steel.

Perhaps the simplest approach is to determine two values of torque, one based upon the assumption that the copper is stressed to its maximum allowable shear stress, the other assuming that the steel develops a shear of 12,000 lb/in^2. It is unlikely that the same torque develops the critical stresses in each of the materials simultaneously. The smaller of these two torques is then the limiting value that the compound shaft can carry.

Let us assume that a shear stress of 8000 lb/in^2 is developed in the outer fibers of the copper tube. This tube has an outside diameter of $2.5\sqrt{2}$ = 3.54 in. and an inside diameter of 2.5 in. To find the torque T_c carried by the copper tube:

$$8000 = \frac{T_c(3.54/2)}{\frac{\pi}{32}[(3.54)^4 - (2.5)^4]} \qquad \text{or} \qquad T_c = 52,500 \text{ lb-in}$$

The torque carried by the steel in this case is $T_s = \frac{52,500}{1.5} = 35,000$ lb-in.

The torque carried by the compound shaft is the sum of these torques or 87,500 lb-in.

Next, let us assume that a shear stress of 12,000 lb/in^2 is developed in the extreme fibers of the steel. The torque T_s' carried by the steel in this case is:

$$12,000 = \frac{T_s'(2.5/2)}{\frac{\pi}{32}(2.5)^4} \qquad \text{or} \qquad T_s' = 36,900 \text{ lb-in}$$

Hence the torque carried by the copper is $T_c' = 1.5(36,900) = 55,400$ lb-in. The total torque carried by the compound shaft according to this assumption is consequently 92,300 lb-in.

Thus the first of these two values is the controlling one and the limiting torque that the assembly can carry is 87,500 lb-in if neither working stress is to be exceeded.

SUPPLEMENTARY PROBLEMS

19. If a solid circular shaft of 1.25 in. diameter is subject to a torque T of 2,500 lb-in causing an angle of twist of 3.12 degrees in a 5 ft length, determine the shear modulus of the material.
Ans. $G = 11.5 \cdot 10^6$ lb/in^2

20. Let us consider a hollow circular shaft of outside diameter 5 in. and inside diameter 3 in. By experiment it is determined that the shearing stress at the inside fibers is 7000 lb/in^2. What is the shearing stress at the outside fibers? *Ans.* 11,700 lb/in^2

21. Determine the maximum shearing stress in a 4 in. diameter solid shaft carrying a torque of 228,000 lb-in. What is the angle of twist per unit length if the material is steel for which $G = 12 \cdot 10^6$ lb/in^2? *Ans.* 18,100 lb/in^2, 0.000755 rad/in

22. Determine the maximum horsepower a solid steel shaft 2.25 in. in diameter can transmit at 250 rpm if the working stress in shear is 11,000 lb/in^2. *Ans.* 98 hp

23. A hollow steel shaft 18 ft long has an outside diameter of 5 in. and an inside diameter of 2.5 in. The shaft is attached to an engine delivering 250 hp at a speed of 150 rpm. Calculate the maximum shearing stress in the shaft and the twist in the 18 ft length. Take $G = 12 \cdot 10^6$ lb/in^2.
Ans. 4570 lb/in^2, 0.0327 rad

24. A propeller shaft in a ship is 14 in. in diameter. The allowable working stress in shear is 7500 lb/in^2 and the allowable angle of twist is 1 degree in 15 diameters of length. If $G = 12 \cdot 10^6$ lb/in^2, determine the maximum torque the shaft can transmit. *Ans.* 3,780,000 lb-in

25. Consider the same shaft described in Problem 24 but with a 7 in. axial hole bored throughout its length. The conditions on working stress and angle of twist remain as before. By what percentage is the torsional load carrying capacity reduced? By what percentage is the weight of the shaft reduced?
Ans. 6.25%, 25%

26. Compare the torques that may be carried by two shafts of the same cross-sectional area, one a hollow circular shaft whose radial thickness is 1.25 in. and the other a solid circular shaft 5 in. in diameter. The maximum shearing stress is the same for both shafts. *Ans.* Ratio of torques = 1.69

27. A hollow steel shaft is to transmit 7500 hp at 120 rpm. If the allowable shearing stress is 12,000 lb/in^2 and the ratio of the outer diameter to the inner is 2, determine the outer diameter. Also, find the angle of twist in a 40 ft length. $G = 12 \cdot 10^6$ lb/in^2. *Ans.* 12.1 in., 4.59°

28. Determine the diameter of a solid steel shaft that will transmit 200 hp at a speed of 250 rpm if the allowable shearing stress is 12,000 lb/in^2. Also, determine the dimensions of a hollow steel shaft whose inside diameter is three-fourths of its outside diameter for these same conditions. What is the ratio of the angles of twist per unit length for these two shafts?
Ans. Diameter = 2.77 in., Outer diameter = 3.15 in., Ratio = 0.88

29. Consider a solid circular shaft transmitting 1800 hp at 350 rpm. Determine the necessary diameter of the shaft so that (*a*) it does not twist through an angle greater than 1 degree in a length of 20 diameters, and also (*b*) the shear stress does not exceed 9000 lb/in^2. The shaft is steel for which $G = 12 \cdot 10^6$ lb/in^2. *Ans.* 6.80 in.

30. A compound shaft is composed of a 24 in. length of solid copper 4 in. in diameter joined to a 32 in. length of solid steel 4.5 in. in diameter. A torque of 120,000 lb-in is applied to each end of the shaft. Find the maximum shear stress in each material and the total angle of twist of the entire shaft. For copper $G = 6 \cdot 10^6$ lb/in^2, for steel $G = 12 \cdot 10^6$ lb/in^2.
Ans. In the copper, 9520 lb/in^2; in the steel, 6700 lb/in^2; $\theta = 0.027$ rad

31. The vertical shaft and pulleys keyed to it may be considered to be weightless. The shaft rotates with a uniform angular velocity. The known belt pulls are indicated and the three pulleys are rigidly keyed to the shaft as shown in Fig. (*a*) below. If the working stress in shear is 7500 lb/in^2 determine the necessary diameter of a solid circular shaft. Neglect bending of the shaft because of the proximity of the bearings to the pulleys. *Ans.* 1.21 in.

32. Determine the number of bolts required in a coupling joining two shafts each 2.5 in. in diameter and carrying a torque of 110,000 lb-in. The allowable shearing stress in the bolts is 12,000 lb/in^2, the diameter of the bolt circle is 7 in. and the bolts are $\frac{3}{4}$ in. in diameter. *Ans.* 6 bolts

33. Consider the compound steel shaft shown in Fig. (*b*) below consisting of two solid circular bars. The stress concentration at the junction of the two is to be neglected. The maximum allowable shearing stress is 11,000 lb/in^2 and the maximum allowable angle of twist over the entire 5 ft length is 1 degree. What is the torque-carrying capacity of this shaft? For this material, $G = 12 \cdot 10^6$ lb/in^2.
Ans. 47,200 lb-in

34. Determine the reactive torques at the fixed ends of the circular shaft loaded by the three couples shown in Fig. (*c*) below. The cross-section of the bar is constant along the length.
Ans. $T_L = 3600$ lb-in, $T_R = 13,600$ lb-in

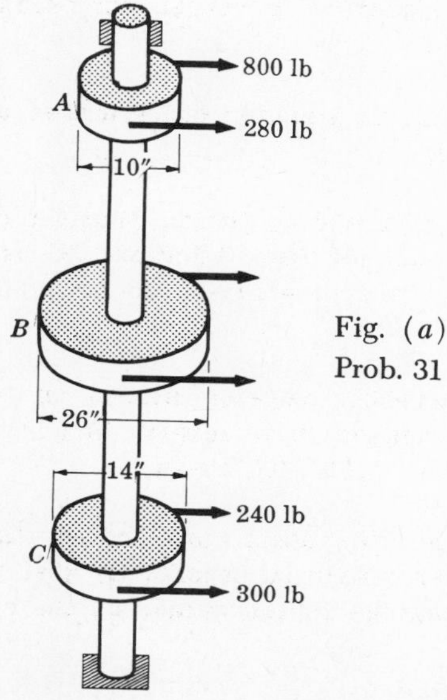

Fig. (*a*) Prob. 31

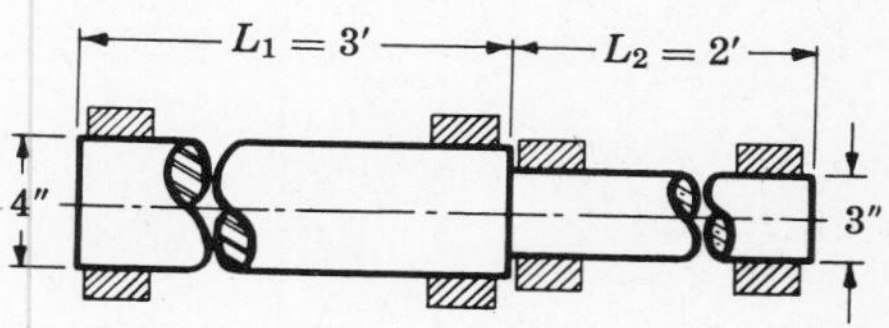

Fig. (*b*) Prob. 33

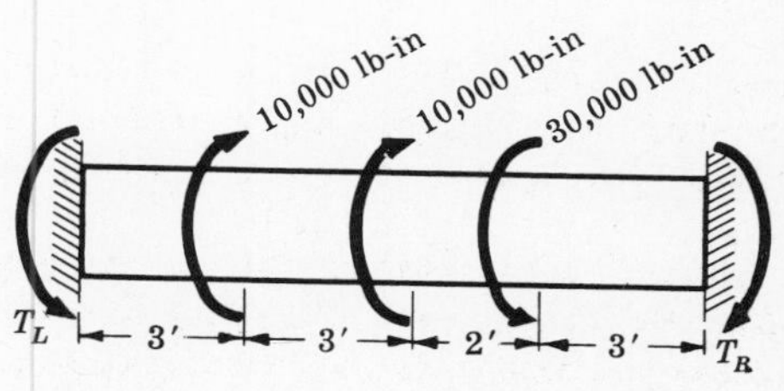

Fig. (*c*) Prob. 34

35. A compound shaft is formed by surrounding a 2.5 in. diameter solid brass shaft by a steel tube having a wall thickness of 0.25 in. The metals are securely attached at their juncture. Determine the increase in torque-carrying capacity of the composite shaft over that of the brass shaft alone. For brass $G = 5\cdot 10^6$ lb/in^2, for steel $G = 12\cdot 10^6$ lb/in^2. The working stress in shear for brass is 7500 lb/in^2, for steel it is 12,000 lb/in^2. *Ans.* 100%

36. Consider a hollow steel shaft of inside diameter 2 in. and outside diameter 3 in. surrounded by an aluminum tube of 0.25 in. wall thickness. Such compound shafts are sometimes used in the presence of corrosive elements. The two metals are bonded together securely at their juncture. If a twisting moment of 65,000 lb-in is applied to the assembly, find the maximum shearing stress in the steel and also in the aluminum. For steel $G = 12\cdot 10^6$ lb/in^2, for aluminum $G = 4\cdot 10^6$ lb/in^2.
Ans. In the steel $s_s = 11{,}300$ lb/in^2, in the aluminum $s_s = 4400$ lb/in^2

CHAPTER 6

Shearing Force and Bending Moment

DEFINITION OF A BEAM. A bar subject to forces or couples that lie in a plane containing the longitudinal axis of the bar is called a beam. The forces are understood to act perpendicular to the longitudinal axis.

CANTILEVER BEAMS. If a beam is supported at only one end and in such a manner that the axis of the beam cannot rotate at that point, it is called a cantilever beam. This type of beam is illustrated in the adjacent figure. The left end of the bar is free to deflect but the right end is rigidly clamped. The right end is usually said to be "restrained". The reaction of the supporting wall at the right upon the beam consists of a vertical force together with a couple acting in the plane of the applied loads shown.

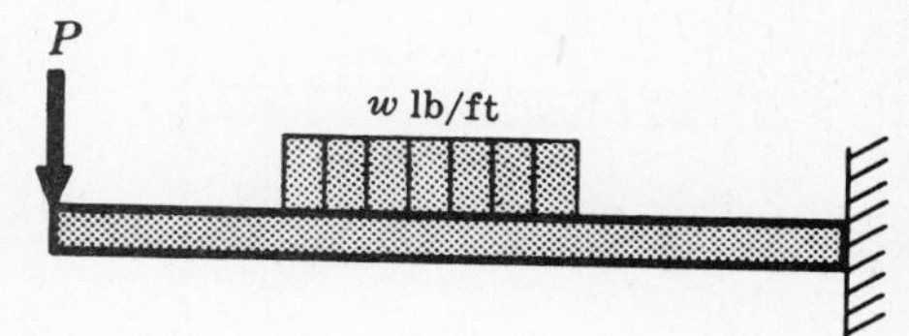

SIMPLE BEAMS. A beam that is freely supported at both ends is called a simple beam. The term "freely supported" implies that the end supports are capable of exerting only forces upon the bar and are not capable of exerting any moments. Thus there is no restraint offered to the angular rotation of the ends of the bar at the supports as the bar deflects under the loads. Two simple beams are sketched below.

It is to be observed that at least one of the supports must be capable of undergoing horizontal movement so that no force will exist in the direction of the axis of the beam. If neither end were free to move horizontally, then some axial force would arise in the beam as it deforms under load. Problems of this nature are not considered in this book.

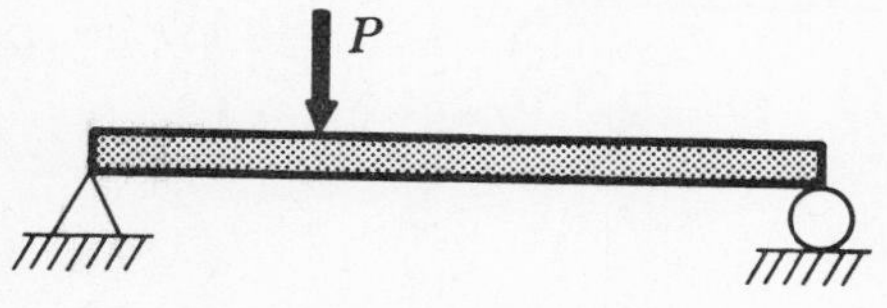

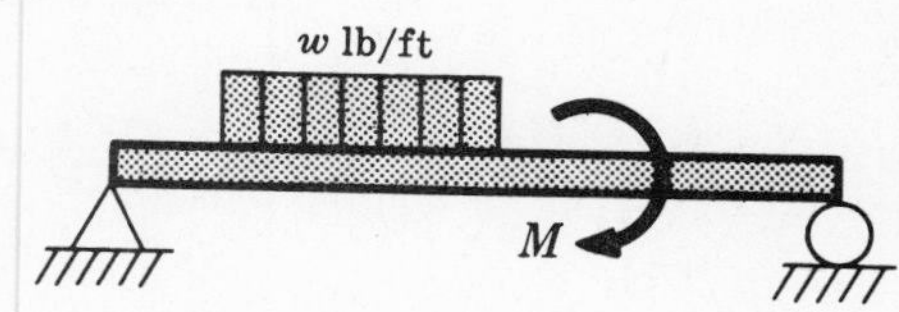

The first beam above is said to be subject to a concentrated force, the second is loaded by a uniformly distributed load as well as a couple.

OVERHANGING BEAMS. A beam freely supported at two points and having one or both ends extending beyond these supports is termed an overhanging beam. Two examples of this appear below.

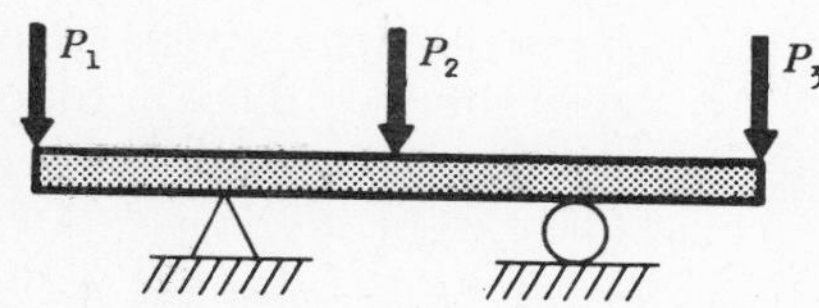

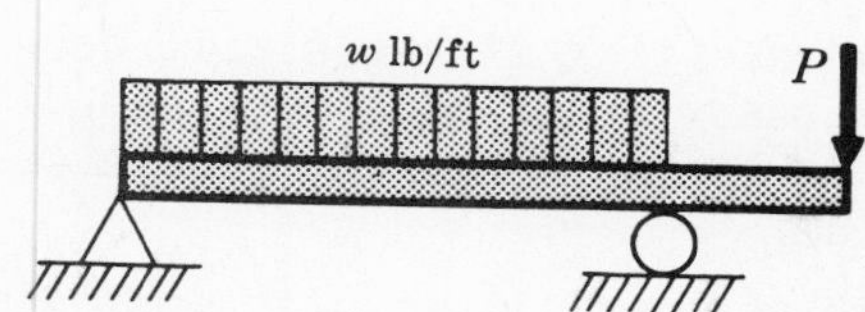

STATICALLY DETERMINATE BEAMS. All of the beams considered above, the cantilevers, simple beams, and overhanging beams, are ones in which the reactions of the supports may be determined by use of the equations of static equilibrium. The values of these reactions are independent of the deformations of the beam. Such beams are said to be statically determinate.

STATICALLY INDETERMINATE BEAMS. If the number of reactions exerted upon the beam exceeds the number of equations of static equilibrium, then the statics equations must be supplemented by equations based upon the deformations of the beam. In this case the beam is said to be statically indeterminate.

A cantilever-type beam that is supported at the extreme end, a beam rigidly clamped at both ends, and a beam extending over three or more supports are all examples of indeterminate beams. These have the following appearance.

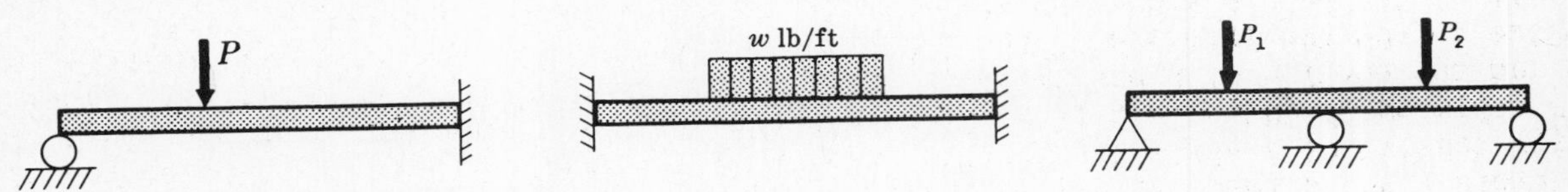

This type of beam will be discussed in Chapter 11.

TYPES OF LOADING. Loads commonly applied to a beam may consist of concentrated forces (applied at a point), uniformly distributed loads, in which case the magnitude is expressed as a certain number of pounds per foot of length of the beam, or uniformly varying loads, as shown below.

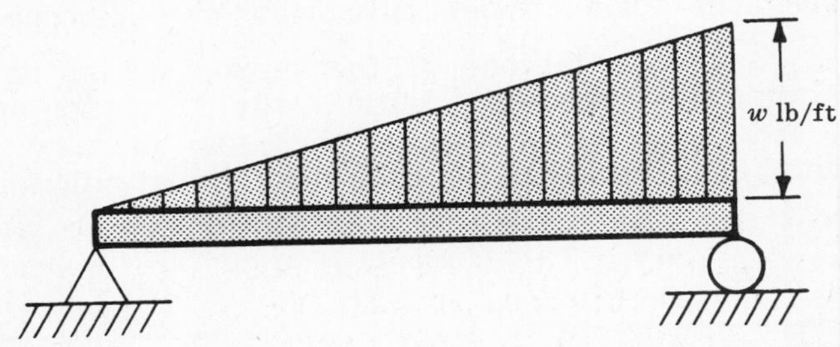

A beam may also be loaded by an applied couple. The magnitude of the couple is usually expressed in lb-ft or lb-in.

In this book, only gradually applied loads are considered. Impact, or dynamic, loads on a beam require a considerably more difficult type of analysis.

INTERNAL FORCES AND MOMENTS IN BEAMS. When a beam is loaded by forces and couples, internal stresses arise in the bar. In general, both normal and shearing stresses will occur. In order to determine the magnitude of these stresses at any section of the beam, it is necessary to know the resultant force and moment acting at that section. These may be found by applying the equations of static equilibrium.

This is perhaps illustrated most simply by considering as an example a particular loading, say several concentrated forces acting on a simple beam as shown in Fig. 1 below.

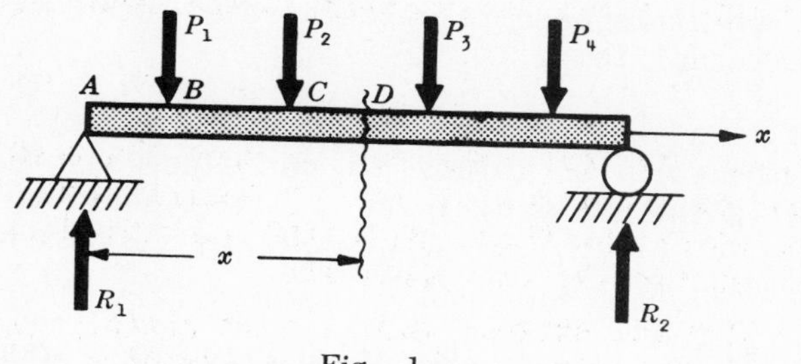

Fig. 1

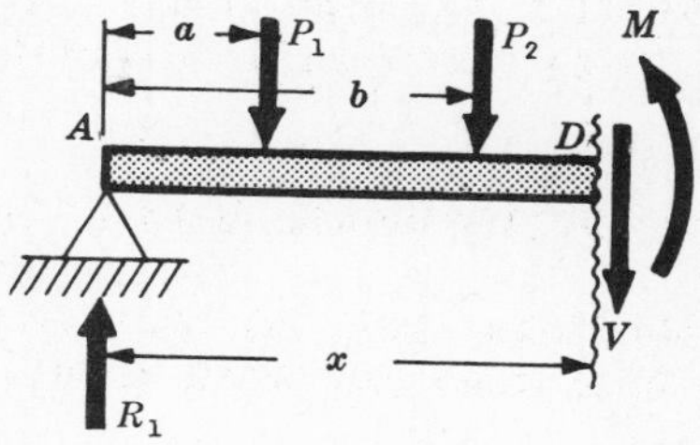

Fig. 2

It is desired to study the internal stresses across the section at D, located a distance x from the left end of the beam. To do this let us consider the beam to be cut at D and the portion of the beam to the right of D removed. The portion removed must then be replaced by the effect it exerted upon the portion to the left of D and this effect will consist of a vertical shearing force together with a couple, as represented by the vectors V and M respectively in the free-body diagram of the left portion of the beam shown in Fig. 2 above.

The force V and the couple M hold the left portion of the bar in equilibrium under the action of the forces R_1, P_1, P_2. The quantities V and M are taken to be positive if they have the senses indicated above.

RESISTING MOMENT. The couple M shown in Fig. 2 above is called the resisting moment at section D. The magnitude of M may be found by use of a statics equation which states that the sum of the moments of all forces about an axis through D and perpendicular to the plane of the page is zero. Thus

$$\Sigma M_O \;=\; M - R_1 x + P_1(x-a) + P_2(x-b) \;=\; 0$$

or

$$M \;=\; R_1 x - P_1(x-a) - P_2(x-b)$$

Thus the resisting moment M is the moment at point D created by the moments of the reaction at A and the applied forces P_1 and P_2. The resisting moment M is the resultant couple due to stresses that are distributed over the vertical section at D. These stresses act in a horizontal direction and are tensile in certain portions of the cross-section and compressive in others. Their nature will be discussed in detail in Chapter 8.

RESISTING SHEAR. The vertical force V shown in Fig. 2 above is called the resisting shear at section D. For equilibrium of forces in the vertical direction,

$$\Sigma F_v \;=\; R_1 - P_1 - P_2 - V \;=\; 0$$

or

$$V \;=\; R_1 - P_1 - P_2$$

This force V is actually the resultant of shearing stresses distributed over the vertical section at D. The nature of these stresses will be studied in Chapter 8.

BENDING MOMENT. The algebraic sum of the moments of the external forces to one side of the section D about an axis through D is called the bending moment at D. This is represented by

$$R_1 x - P_1(x-a) - P_2(x-b)$$

for the loading considered above. This quantity is considered in each of the Problems 1 through 15 inclusive. Thus the bending moment is opposite in direction to the resisting moment but is of the same magnitude. It is usually denoted by M also. Ordinarily the bending moment rather than the resisting moment is used in calculations because it can be represented directly in terms of the external loads.

SHEARING FORCE. The algebraic sum of all the vertical forces to one side, say the left side, of section D is called the shearing force at that section. This is represented by

$$R_1 - P_1 - P_2$$

for the above loading. The shearing force is opposite in direction to the resisting shear but of the same magnitude. Usually it is denoted by V. It is ordinarily used in calculations, rather than the resisting shear. This quantity is considered in each of the Problems 1 through 15 inclusive.

SIGN CONVENTIONS. The customary sign conventions for shearing force and bending moment are represented by the following diagrams.

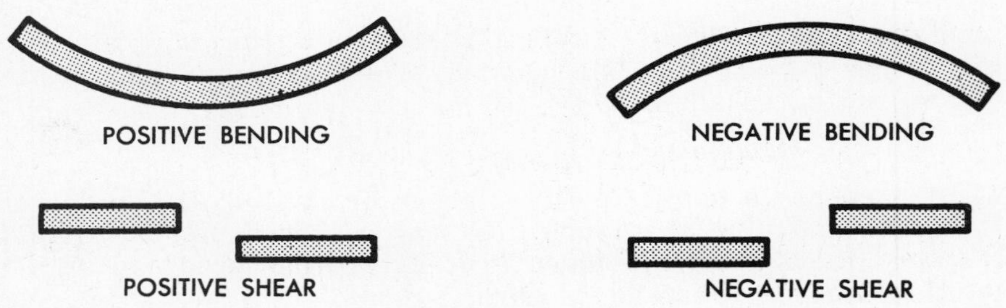

Thus a force that tends to bend the beam so that it is concave upward, as shown in the upper left-hand diagram, is said to produce a positive bending moment. A force that tends to shear the left portion of the beam upward with respect to the right portion, as shown in the lower left-hand diagram, is said to produce a positive shearing force.

An easier method for determining the algebraic sign of the bending moment at any section is to say that upward external forces produce positive bending moments, downward forces yield negative bending moments.

SHEAR AND MOMENT EQUATIONS. Usually it is convenient to introduce a coordinate system along the beam, with the origin at one end of the beam. It will be desirable to know the shearing force and bending moment at all sections along the beam and for this purpose two equations are written, one specifying the shearing force V as a function of the distance, say x from one end of the beam, the other giving the bending moment M as a function of x.

SHEARING FORCE AND BENDING MOMENT DIAGRAMS. The plots of these equations for V and M are known as shearing force and bending moment diagrams respectively. In these plots the abscissas (horizontals) indicate the position of the section along the beam and the ordinates (verticals) to the two diagrams represent the values of the shearing force and bending moment respectively. Thus these diagrams represent graphically the variation of shearing force and bending moment at any section along the length of the bar. From these plots it is quite easy to determine the maximum value of each of these quantities.

RELATION BETWEEN SHEARING FORCE AND BENDING MOMENT. A simple beam with several applied loads is sketched below. The coordinate system with origin at the left end A is established and distances to various sections in the beam are denoted by the variable x.

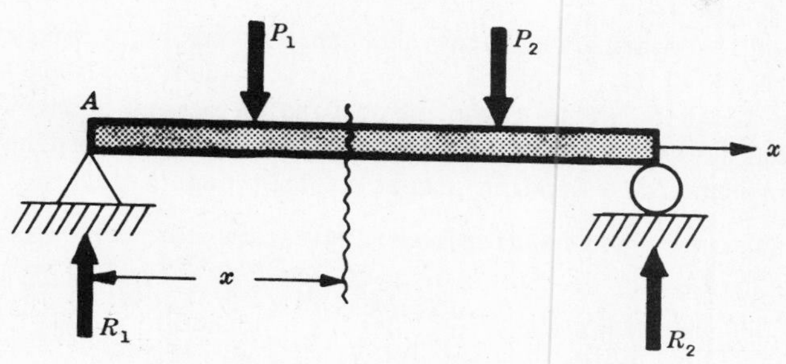

For any value of x the shearing force V and the bending moment M are related by the equation

$$V = \frac{dM}{dx}$$

This relation is derived in Problem 7. For applications see Problems 8, 10, 12, 13.

SOLVED PROBLEMS

1. For the cantilever beam shown in Fig. (a) below, write equations for the shearing force and bending moment at any point along the length of the bar. Also, draw shearing force and bending moment diagrams approximately to scale.

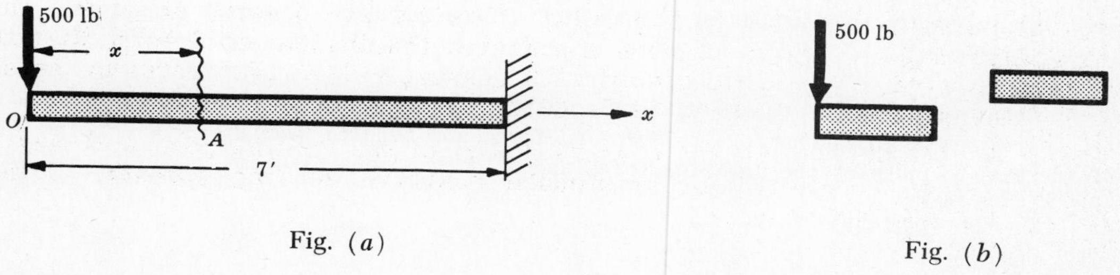

Fig. (a) Fig. (b)

In this particular problem it is not necessary to first determine the reactions of the support. Let us choose the axis of the beam as the x-axis of a coordinate system with origin O at the left end of the bar.

Let us consider any vertical section through the beam at a general distance x from the left end. The 500 lb force tends to shear the portion of the beam to the left of the section at x downward with respect to the portion to the right of the section. That is, if the bar were cut at the section the two portions would move to the relative positions shown in Fig. (b) above. This, according to our sign convention, is negative shear. Hence the shearing force V at any section x is simply the algebraic sum of all the forces to the left of that section, which is in this case 500 lb. Thus

$$V = -500 \text{ lb}$$

Also, our sign convention states that downward forces produce negative bending moments. Hence the bending moment M at the section x due to the 500 lb force is the moment of that force about an axis perpendicular to the plane of the page through the point A. Thus the equation for the bending moment M is

$$M = -500x \text{ lb-ft}$$

From the above equation for shearing force it is evident that this quantity is constant along the length of the bar and thus should be plotted as a horizontal straight line, the ordinate to this line at any point representing the shearing force of −500 lb at this same point. Since the shearing force is negative, this horizontal line should be plotted below the axis, as shown in the second of the figures below.

The equation for the bending moment indicates that this quantity is zero at the left end of the beam and at the right end, where $x = 7$ ft, takes on the value of $-500(7) = -3500$ lb-ft. Since the moment equation is a first degree function of x, the plot of bending moment along the beam is a straight line connecting O at the left and −3500 at the right end of the beam. The ordinate at any point to this inclined straight line represents the bending moment at that same point.

Consequently the shearing force and bending moment diagrams have the appearance shown below.

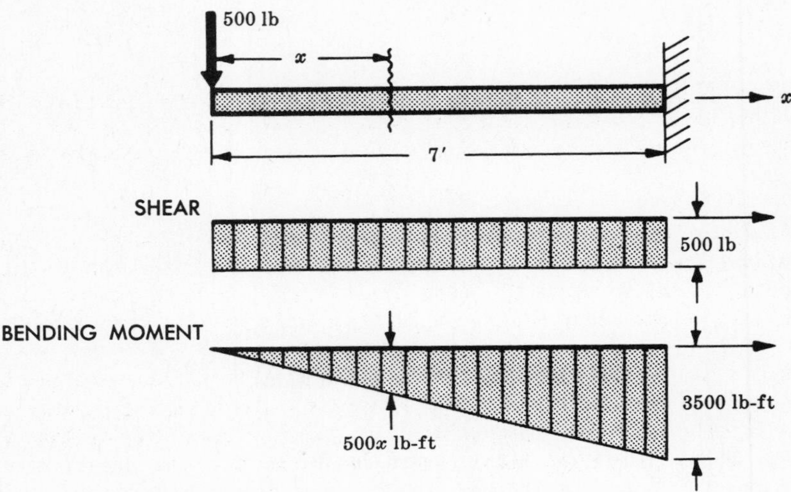

The last two plots are thus graphs of the shearing force and bending moment equations.

2. For the cantilever beam subject to the uniformly distributed load of w lb per ft of length, as shown in Fig. (a) below, write equations for the shearing force and bending moment at any point along the length of the bar. Also plot the shearing force and bending moment diagrams approximately to scale.

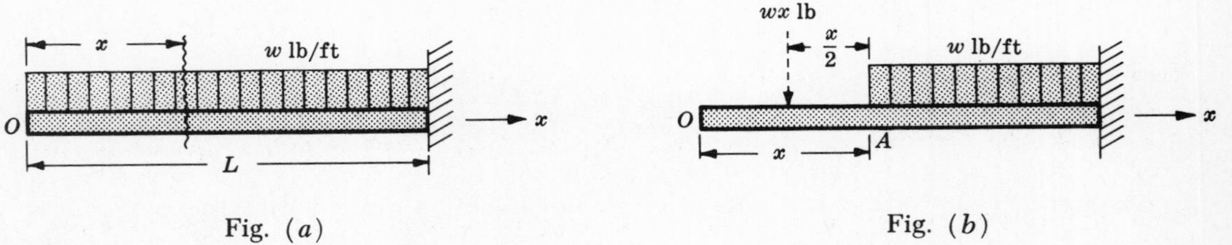

Fig. (a) Fig. (b)

Again, it is not necessary to determine the reactions at the supporting wall. We shall choose the axis of the beam as the x-axis of a coordinate system with origin O at the left end of the bar. To determine the shearing force and bending moment at any section of the beam a distance x from the free end, we may replace the portion of the distributed load to the left of this section by its resultant. As shown by the dotted vector in Fig. (b) above, the resultant is a downward force of wx lb acting midway between O and the section x. Note that none of the load to the right of the section is included in calculating this resultant. Such a resultant force tends to shear the portion of the bar to the left of the section downward with respect to the portion to the right. By our sign convention this constitutes negative shear.

The shearing force at this section x is defined to be the sum of the forces to the left of the section. In this case, the sum is wx lb acting downward; hence

$$V = -wx \text{ lb}$$

This equation indicates that the shear is zero at $x = 0$ and when $x = L$ it is $-wL$. Since V is a first-degree function of x, the shearing force plots as a straight line connecting these values at the ends of the beam. It has the appearance shown in Fig. (c) below. The ordinate to this inclined line at any point represents the shearing force at that same point.

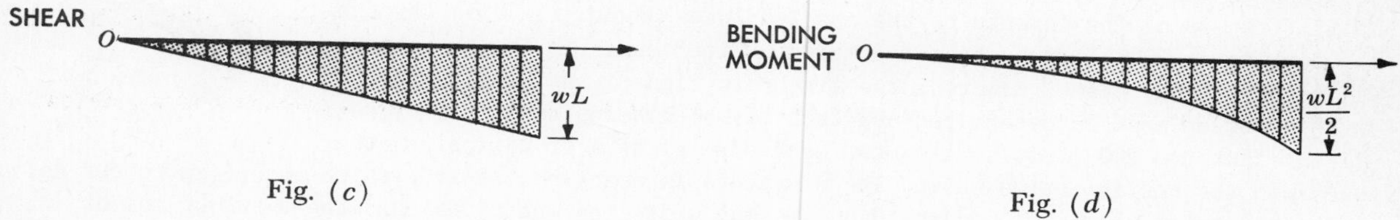

Fig. (c) Fig. (d)

The bending moment at this same section x is defined to be the sum of the moments of the forces to the left of this section about an axis through point A and perpendicular to the plane of the page. This sum of the moments is given by the moment of the resultant, wx lb about an axis through A. It is

$$M = -wx(x/2) \text{ lb-ft}$$

The minus sign is necessary because downward loads indicate negative bending moments. By this equation the bending moment is zero at the left end of the bar and $-wL^2/2$ at the clamped end when $x = L$. The variation of bending moment is parabolic along the bar and may be plotted as in Fig. (d) above. The ordinate to this parabola at any point represents the bending moment at that same point.

It is to be noted that a downward uniform load as considered here leads to a bending moment diagram that is concave downward. This could be established by taking the second derivative of M with respect to x, the derivative in this particular case being $-w$. Since the second derivative is negative, the rules of calculus tell us that the curve must be concave downward.

3. Consider a cantilever beam loaded only by the couple of 200 lb-ft applied as shown in Fig. (a) below. Write equations for the shearing force and bending moment at any point along the length of the bar. Plot shearing force and bending moment diagrams for this loading.

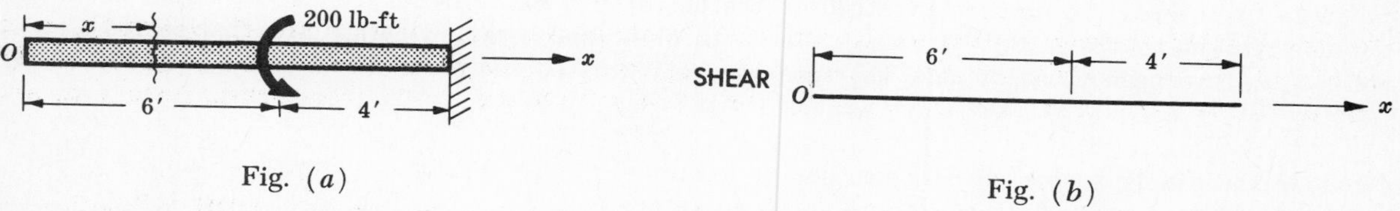

Fig. (a) Fig. (b)

Once again it is not necessary to determine the reaction at the wall although it is obvious that this reaction must consist only of a moment of magnitude 200 lb-ft in a clockwise direction. We shall choose the axis of the beam as the x-axis of the coordinate system with origin O at the left end of the bar.

This problem presents certain features that were not present in the earlier problems because the couple is not applied at the end of the bar. The shearing force at any section along the length of the bar is the algebraic sum of the applied vertical forces to the left of the section chosen. Since the couple has no force effect in any direction, there are no vertical forces applied to the bar and consequently the shearing force V for all values of x is zero. This may be represented graphically in the shearing force diagram by a horizontal straight line coinciding with the axis of the plot as shown in Fig. (b) above.

In determining an equation for the bending moment it is apparent that two distinct regions must be considered along the length of the beam. One is the 6 ft length to the left of the applied couple, the other the 4 ft length between the couple and the wall. Usually it is convenient to denote values of x

in the former region by the designation $0 < x < 6$ ft

and those lying in the latter region by $6 < x < 10$ ft.

If we first consider any section at a distance x from the left end where $0 < x < 6$, then the bending moment, which is defined to be the sum of the moments of the forces to the left of this section about an axis through this section and perpendicular to the plane of the page, is evidently zero since that region is not subject to any applied loads. For any section through the beam to the right of the couple of 200 lb-ft, the sum of the moments of the applied loads about the axis through the section is 200 lb-ft because the moment of the couple is the same about all points in the plane. However, the algebraic sign of the bending moment in this region must be determined and that may be done quite simply by realizing that the 200 lb-ft couple must bend the right 4 ft region of the beam into the configuration shown. The curvature is concave downward, which according to our definition constitutes negative bending. Hence we may write the equations for the bending moment at any section x in the form

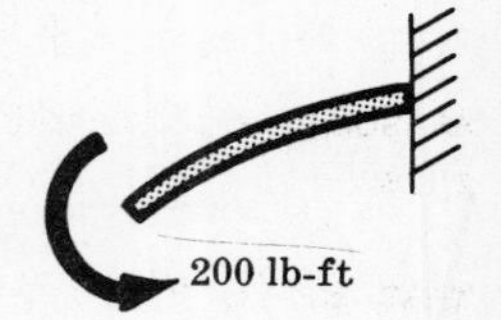

$$M = 0 \quad \text{for} \quad 0 < x < 6 \text{ ft}, \qquad M = -200 \text{ lb-ft} \quad \text{for} \quad 6 < x < 10 \text{ ft}$$

The loaded beam together with plots of the shear and moment equations is shown below.

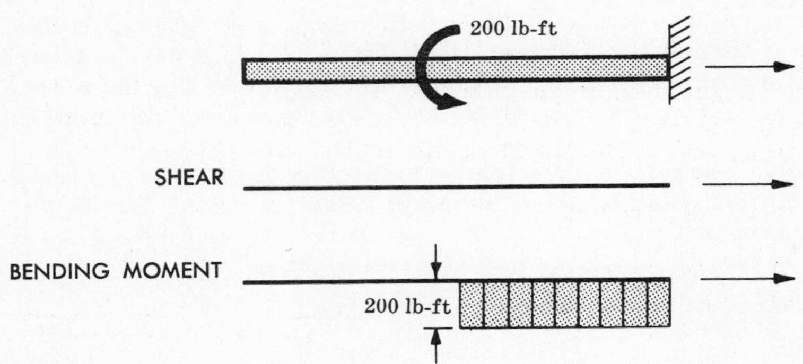

It is of interest to consider instead of the above beam 10 ft in length, a 4 ft long beam loaded at its free end by the 200 lb-ft couple. Then the shear and moment diagrams for that entire 4 ft beam would also have been found to have the appearances indicated above for the 4 ft region at the right.

4. Consider the simply supported beam subject to a single concentrated load of 4000 lb as shown. Write equations for the shearing force and bending moment at any position in the beam and plot the shear and moment diagrams.

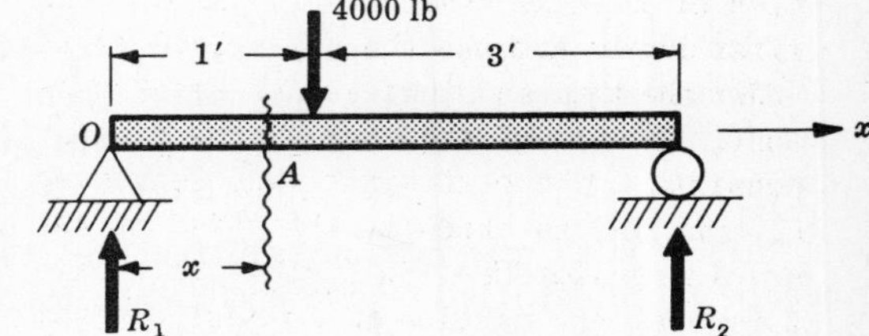

It is necessary to first determine the external reactions R_1 and R_2. Taking moments about point O,

$$\Sigma M_O = 4R_2 - 4000(1) = 0, \qquad R_2 = 1000 \text{ lb}$$

For equilibrium in the vertical direction,

$$\Sigma F_v = R_1 + 1000 - 4000 = 0, \qquad R_1 = 3000 \text{ lb}$$

Again, we shall introduce an x-axis coinciding with the axis of the beam and having an origin O at the left end of the beam.

Let us consider the vertical shearing force at any section a distance x from the left end. If we first restrict ourselves to the region to the left of the 4000 lb load, the shearing force consists entirely of the reaction R_1 = 3000 lb because that is the only force to the left of this section. This force tends to shear the portion of the beam to the left of x upward with respect to the remainder of the bar, i.e. to the relative positions shown in the adjoining figure. According to our sign convention this is positive shear. Hence in this region to the left of the load,

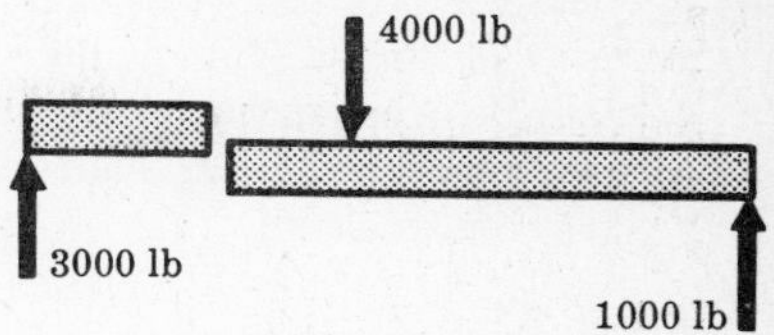

$$V = 3000 \text{ lb} \quad \text{for} \quad 0 < x < 1 \text{ ft}$$

As soon as x exceeds 1 ft the shearing force which is the algebraic sum of the forces to the left of x is

$$V = 3000 - 4000 = -1000 \text{ lb} \quad \text{for} \quad 1 < x < 4 \text{ ft}$$

That is, in the region to the right of the 4000 lb load the left reaction contributes a positive shearing force, the 4000 lb load a negative shearing force, and the resultant is downward or negative. These two equations are necessary to define the shearing force along the length of the bar.

In the region to the left of the load the bending moment at the section x is the moment of the 3000 lb reaction about an axis perpendicular to the plane of the page through A. This moment is

$$M = 3000x \text{ lb-ft} \quad \text{for} \quad 0 < x < 1 \text{ ft}$$

and it is positive since upward forces cause positive bending moments. As soon as we consider a section to the right of the 4000 lb load, the bending moment is due partially to the 3000 lb reaction and partly to the 4000 lb load. This moment is

$$M = 3000x - 4000(x-1) \text{ lb-ft} \quad \text{for} \quad 1 < x < 4 \text{ ft}$$

Again, two equations are required to define this quantity along the length of the bar. It is to be noted that the first expression for M is true only if x is less than 1 ft. There is no way to combine these two into one equation that will hold along the entire beam.

Plots of these equations for shearing force and bending moment are quite simple. To the left of the load the shearing force is constant (3000 lb) and hence is represented by a horizontal line BC as shown in the second of the accompanying sketches. To the right it is again constant (−1000 lb) and consequently is represented by another horizontal line DE. The bending moment in the left region increases linearly from zero at the left support to a maximum value of 3000 lb-ft under the load. In the right portion of the bar it must again be a linear function of x since the equation is of the first degree and has the value 3000 lb-ft under the load and zero at the right support. Thus the bending moment diagram consists of two straight line segments, FG and GH, as shown in the third of the adjoining sketches. In fact it is always true that the portion of a moment diagram between points of application of two concentrated forces is a straight line.

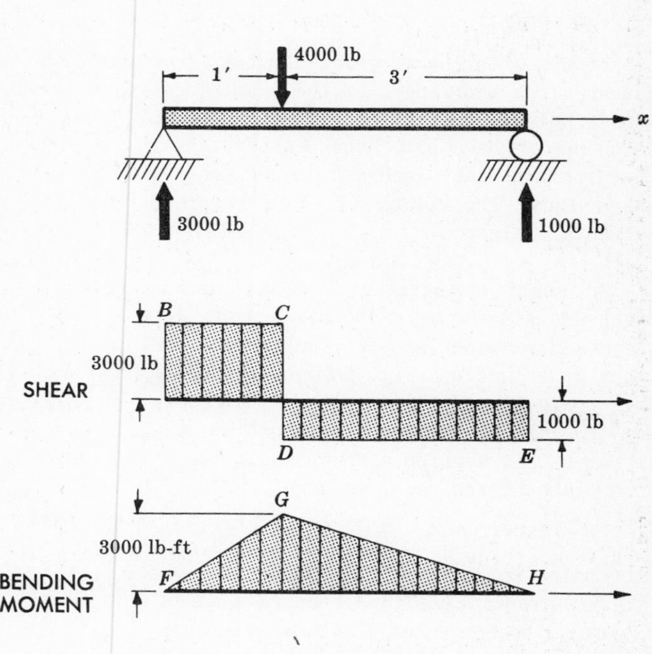

It is to be observed that the magnitude of the discontinuity or *jump* in the shear diagram at x = 1 ft is equal to the magnitude of the concentrated force applied at that same point. This is always true under the point of application of a concentrated force.

5. Write equations for the shearing force and bending moment at any point along the beam and draw the shearing force and bending moment diagrams for the simply supported beam subject to the three concentrated loads shown.

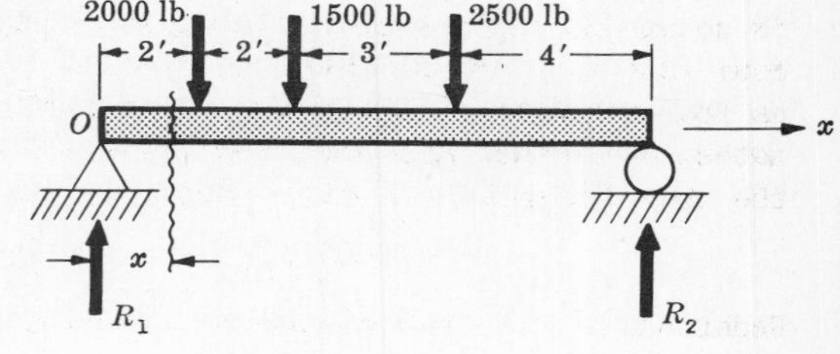

It is first necessary to find the reactions R_1 and R_2 by static equilibrium. Thus:

$$\Sigma M_O = 11R_2 - 2000(2) - 1500(4) - 2500(7) = 0$$

$$\Sigma F_v = R_1 + 2500 - 2000 - 1500 - 2500 = 0$$

from which $R_2 = 2500$ lb and $R_1 = 3500$ lb.

The x-axis coincides with the axis of the beam and has its origin at the left end as shown.

Evidently four equations will be needed to define the shearing force and four more for the bending moment since there are that many regions between concentrated forces.

Let us first examine the shearing force. Working from left to right along the beam and remembering that the shearing force at any section a distance x from the left end is given by the algebraic sum of the forces to the left of that section, we have

$$V = 3500 \text{ lb} \quad \text{for} \quad 0 < x < 2 \text{ ft}$$

According to our sign convention the reaction R_1 produces positive shear.

$$V = 3500 - 2000 = 1500 \text{ lb} \quad \text{for} \quad 2 < x < 4 \text{ ft}$$

According to our sign convention the 2000 lb force produces negative shear.

$$V = 3500 - 2000 - 1500 = 0 \text{ lb} \quad \text{for} \quad 4 < x < 7 \text{ ft}$$

$$V = 3500 - 2000 - 1500 - 2500 = -2500 \text{ lb} \quad \text{for} \quad 7 < x < 11 \text{ ft}$$

Since the shearing force is constant in each of these regions, it may be plotted as four horizontal lines along the length of the beam. The ordinates to these lines have the four values found above. In any problem involving concentrated forces such as we have here, the negative shear between the last load and R_2 should always equal the reaction R_2 with reversed algebraic sign.

We shall now examine the bending moment. Working from left to right along the beam and remembering that the bending moment at a section x is defined to be the algebraic sum of the moments of the forces to the left of this section about an axis through this section and perpendicular to the plane of the page, we have

$$M = 3500x \text{ lb-ft} \quad \text{for} \quad 0 < x < 2 \text{ ft}$$

According to our sign convention upward forces produce positive bending moments, downward forces negative bending moments.

$$M = 3500x - 2000(x-2) \text{ lb-ft} \quad \text{for} \quad 2 < x < 4 \text{ ft}$$

$$M = 3500x - 2000(x-2) - 1500(x-4) \text{ lb-ft} \quad \text{for} \quad 4 < x < 7 \text{ ft}$$

$$M = 3500x - 2000(x-2) - 1500(x-4) - 2500(x-7) \text{ lb-ft} \quad \text{for} \quad 7 < x < 11 \text{ ft}$$

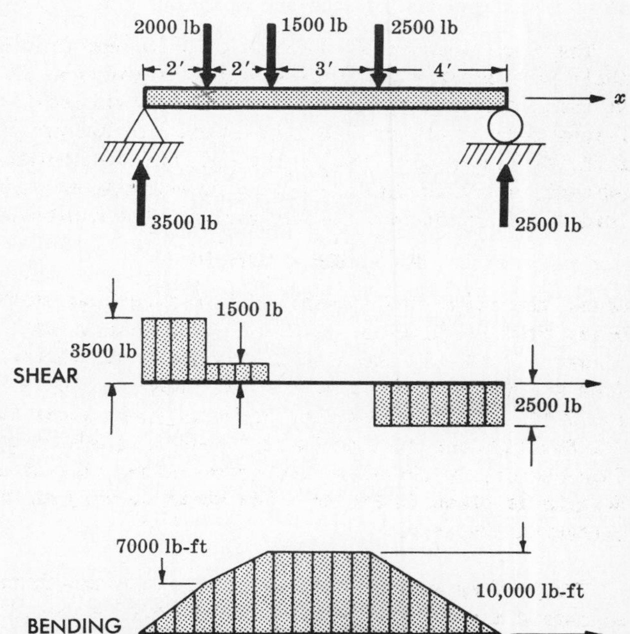

These four equations completely define the bending moment along the length of the beam and there is no way to replace them by a single equation that will be exactly equivalent to the four. Since all four equations are first-degree functions of x, it is evident that the bending moment may be plotted as four straight line segments. To determine this plot it is only necessary to establish three ordinates, one under each load, and these are readily given by the above equations. For example, under the 2000 lb load the bending moment is

$$M_{x=2} = 3500(2) = 7000 \text{ lb-ft}$$

Under the second load we may use the equation for $2 < x < 4$ with x set equal to 4 ft. Thus we find

$$M_{x=4} = 3500(4) - 2000(4-2) = 10{,}000 \text{ lb-ft}$$

Under the 2500 lb load we use the equation for $4 < x < 7$ with x set equal to 7 ft. Thus

$$M_{x=7} = 3500(7) - 2000(7-2) - 1500(7-4) = 10{,}000 \text{ lb-ft}$$

Since the two ends of the bar are simply supported, the bending moment at each of those points is zero.

The shearing force and the bending moment diagrams together with the sketch of the beam are plotted above. These last two plots illustrate the variation of shearing force and bending moment at any point in the beam.

6. Consider the simply supported beam 10 ft long and subject to a uniformly distributed vertical load of 120 lb per ft of length, as shown in Fig. (a) below. Draw shearing force and bending moment diagrams.

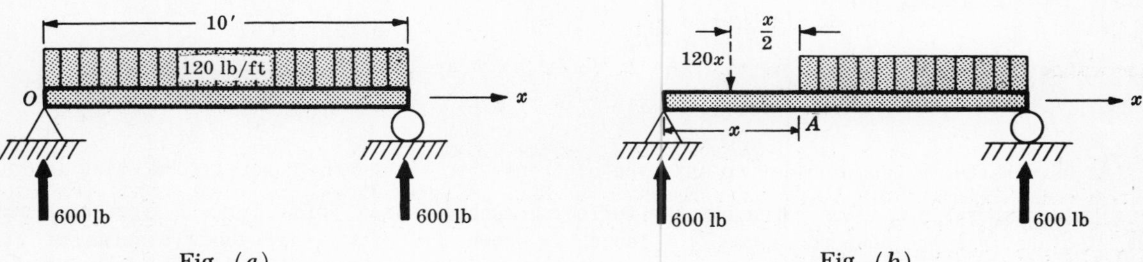

Fig. (a) Fig. (b)

The total load on the beam is 1200 lb, and from symmetry each of the end reactions is 600 lb. We shall now consider any cross-section of the beam at a distance x from the left end. The shearing force at this section is given by the algebraic sum of the forces to the left of this section and these forces consist of the 600 lb reaction and the distributed load of 120 lb/ft extending over a length x ft. We may replace the portion of the distributed load to the left of the section at x by its resultant, which is $120x$ lb acting downward as shown by the dotted vector in Fig. (b) above. None of the load to the right of x is included in this resultant. The shearing force at x is then given by

$$V = 600 - 120x \text{ lb}$$

Since there are no concentrated loads acting on the beam, this equation is valid at all points along its length. Evidently the shearing force varies linearly from V = 600 lb at $x = 0$ to $V = 600 - 1200 = -600$ lb at $x = 10$ ft. The variation of shearing force along the length of the bar may then be represented by a straight line connecting these two end-point values. The shear diagram is shown in Fig. (c). The shear is zero at the center of the beam.

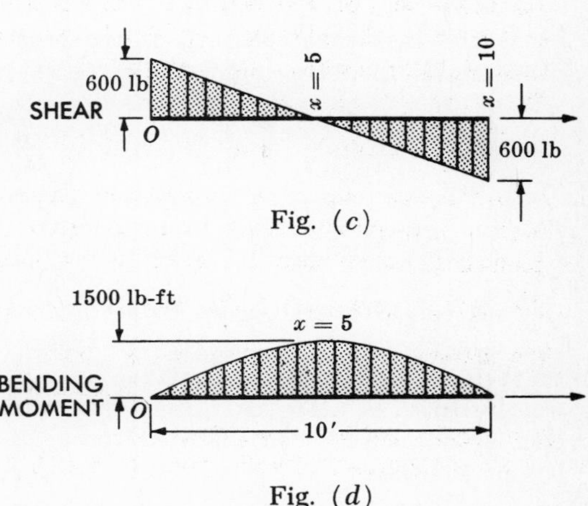

Fig. (c)

Fig. (d)

The bending moment at the section x is given by the algebraic sum of the moments of the 600 lb reaction and the distributed load of $120x$ lb about an axis through A perpendicular to the plane of the paper. Remembering that upward forces give positive bending moments, we have

$$M = 600x - 120x(x/2) \text{ lb-ft}$$

Again, this equation holds along the entire length of the beam. It is to be noted that since the load is uniformly distributed the resultant indicated by the dotted vector acts at a distance $x/2$ from A, i.e. at the midpoint of the uniform load to the left of the section x where the bending moment is being calculated. From the above equation it is evident that the bending moment is represented by a parabola along the length of the beam. Since the bar is simply supported the moment is zero at either end and, because of the symmetry of loading, the bending moment must be a maximum at the center of the beam where $x = 5$ ft. The bending moment at that point is

$$M_{x=5} = 600(5) - 60(5)^2 = 1500 \text{ lb-ft}$$

The parabolic variation of bending moment along the length of the bar may thus be represented by the ordinates to the bending moment diagram shown in Fig. (d) above.

7. Derive a relationship between the shearing force and bending moment at any point in a beam.

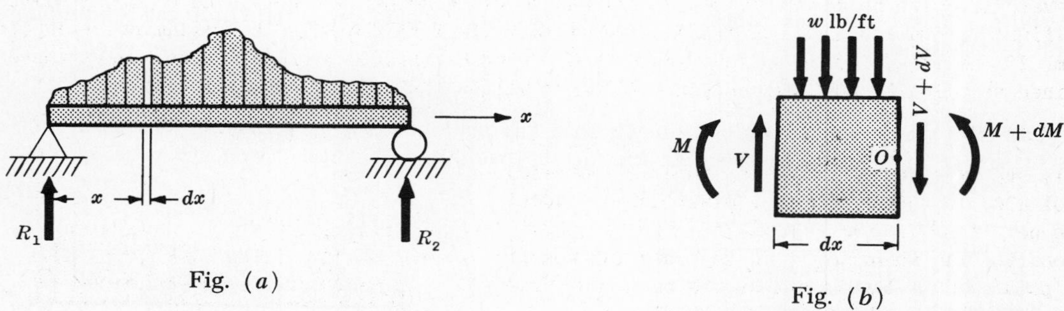

Fig. (a) Fig. (b)

Let us consider a beam subject to any type of transverse load of the general form shown in Fig. (a).

Simple supports are illustrated but the following consideration holds for all types of beams. We will isolate from the beam the element of length dx shown and draw a free-body diagram of it. The shearing force V acts on the left side of the element, and in passing through the distance dx the shearing force will in general change slightly to an amount $(V+dV)$. The bending moment M acts on the left side of the element and it changes to $(M+dM)$ on the right side. Since dx is extremely small, the applied load may be taken as uniform over the top of the beam and equal to w lb/ft. The free-body diagram of this element thus appears as in Fig. (b) above. For equilibrium of moments about O, we have

$$\Sigma M_O = M - (M + dM) + V\,dx - w(dx)\left(\frac{dx}{2}\right) = 0$$

or

$$dM = V\,dx + \frac{w}{2}(dx)^2$$

Since the last term consists of the product of two differentials, it is negligible compared with the other terms involving only one differential. Hence

$$dM = V\,dx$$

or

$$V = \frac{dM}{dx}$$

Thus the shearing force is equal to the rate of change of the bending moment with respect to x.

This equation will prove to be of considerable value in drawing shearing force and bending moment diagrams for the more complicated types of loading. For example, from this equation it is evident that if the shearing force is positive at a certain section of the beam then the slope of the bending moment diagram is also positive at that point. Also, it demonstrates that an abrupt change in shear, corresponding to a concentrated force, is accompanied by an abrupt change in the slope of the bending moment diagram.

Further, at those points where the shear is zero, the slope of the bending moment diagram is zero. At these points where the tangent to the moment diagram is horizontal, the moment may have a maximum or minimum value. This follows from the ordinary calculus technique of obtaining maximum or minimum values of a function by equating the first derivative of the function to zero. Thus in the accompanying sketch if the curves shown represent portions of a bending moment diagram then critical values may occur at points A and B.

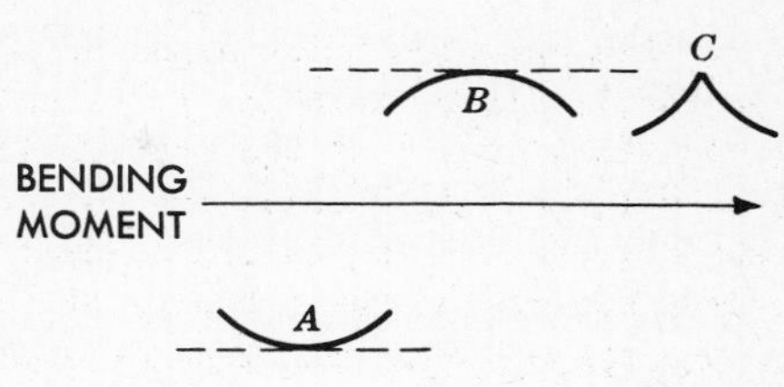

To establish the direction of concavity at a point such as A or B, we may form the second derivative of M with respect to x, i.e. d^2M/dx^2. If the value of this second derivative is positive, then the moment diagram is concave upward, as at A, and the moment assumes a minimum value. If the second derivative is negative the moment diagram is concave downward, as at B, and the moment assumes a maximum value.

However, it is to be carefully noted that the calculus method of obtaining critical values by use of the first derivative does not indicate possible maximum values at a cusp-like point in the moment diagram, if one occurs, such as that shown at C. If such a point is present, the moment there must be determined numerically and then compared to other values that are possibly critical.

3. A simply supported beam is subject to a concentrated force of 4000 lb together with a distributed load of 1600 lb per ft of length applied as shown. Write equations for the shearing force and bending moment at any point along the length of the beam and draw the shearing force and bending moment diagrams.

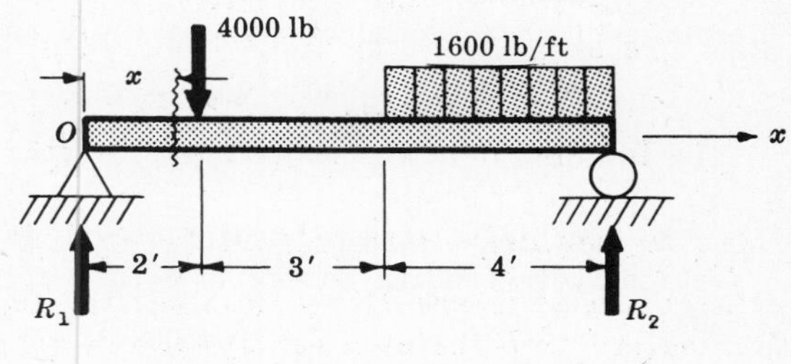

It is first necessary to determine the reactions R_1 and R_2. From statics we may write

$$\Sigma M_O = 9R_2 - 4000(2) - 6400(7) = 0, \quad R_2 = 5870 \text{ lb}$$

$$\Sigma F_v = R_1 - 4000 - 6400 + 5870 = 0, \quad R_1 = 4530 \text{ lb}$$

It is to be noted that for the purpose of determining external reactions it is always permissible to replace the entire distributed load, which in this case is 1600 lb/ft, by its resultant. Thus the resultant of 6400 lb may be used in the first of the above equations, and since the 1600 lb/ft is a uniformly distributed load the resultant acts through the midpoint of the length of 4 ft over which the distributed load is acting.

Introducing the x-axis shown with origin at the left end of the beam, it is evident that in the extreme left region where $0 < x < 2$ ft, the shearing force is due entirely to the reaction R_1, which tends to shear the left portion of the beam upward with respect to the right portion. This constitutes positive shear, hence

$$V = 4530 \text{ lb} \quad \text{for} \quad 0 < x < 2 \text{ ft}$$

Thus the shear plots as a horizontal straight line in this region.

The shearing force to the right of the 4000 lb load is influenced by both R_1 and the 4000 lb force. This load produces negative shear, and we have

$$V = 4530 - 4000 = 530 \text{ lb} \quad \text{for} \quad 2 < x < 5 \text{ ft}$$

Again the shear plots as a horizontal straight line.

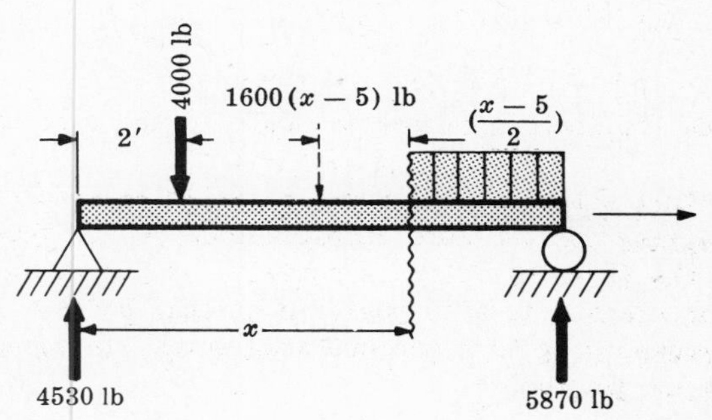

For values of x greater than 5 ft, the distributed load of 1600 lb/ft enters the equation for the shearing force. In contrast to the earlier replacement of the entire distributed load by its resultant for the purpose of determining reactions, it is now to be carefully noted that this is no longer possible. Since we are working from left to right along th beam, we may replace only the portion of the distributed load lying to the *left* of the section x b its resultant. This is illustrated above for any value of x in excess of 5 ft where the resultant per tains to the load lying between x = 5 ft and the section x shown. Since the load is 1600 lb/ft and i acts over a length of $(x-5)$ ft the resultant, indicated by the dotted vector, is $1600(x-5)$ lb. I acts at the midpoint of the load lying to the *left* of the section x.

The shearing force at the section x is again the algebraic sum of the forces to the left of x:

$$V = 4530 - 4000 - 1600(x-5) \text{ lb} \quad \text{for} \quad 5 < x < 9 \text{ ft}$$

Thus in this region the shearing force is a first-degree function of x and the values at the end-point of this region are readily found by substitution in this equation. Substituting x = 5 ft,

$$V_{x=5} = 530 \text{ lb}$$

Substituting x = 9 ft,

$$V_{x=9} = -5870 \text{ lb}$$

The shearing force diagram is now readily plotted. In the left and central portions of the bar it is represented by two horizontal straight lines having ordinates of 4530 lb and 530 lb respectively. In the right region it is represented by an inclined straight line joining the ordinates of 530 at x = 5 and −5870 at x = 9. It is shown in the adjoining sketch.

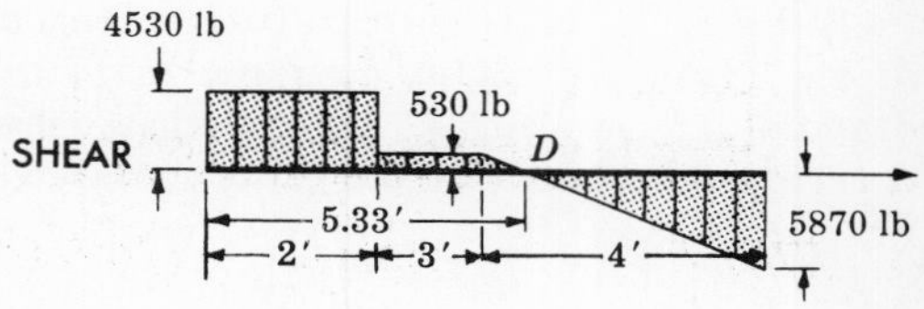

The point where the shear is zero under the distributed load is found by setting V = 0 in the shear equation for that region. Doing this we find

$$4530 - 4000 - 1600(x-5) = 0 \qquad \text{from which} \qquad x = 5.33 \text{ ft}$$

This is point D in the above shear diagram.

The equations for the bending moment in the left and central regions are readily written. At an section x the bending moment in either of these regions is

$$M = 4530x \text{ lb-ft} \quad \text{for} \quad 0 < x < 2 \text{ ft}$$

and

$$M = 4530x - 4000(x-2) \text{ lb-ft} \quad \text{for} \quad 2 < x < 5 \text{ ft}$$

From the first of these equations, the moment is zero at x = 0 and is 9060 lb-ft at x = 2 ft. From the second equation, the moment at x = 5 is obtained by substituting x = 5:

$$M_{x=5} = 4530(5) - 4000(5-2) = 10,650 \text{ lb-ft}$$

From the above two equations the moment diagram is evidently a straight line in each of these regions. These two straight lines merely join the end-point values of 0, 9060 and 10,650 at x = 0, x = 2, and x = 5 ft respectively.

For values of x greater than 5 ft the distributed load enters the moment equation. In calculating the bending moment at a section x in this region it is again convenient to replace the portion of the distributed load lying to the left of x by its resultant, as shown in the previous sketch. Using this resultant to find the bending moment at x due to the uniform load, we find

$$M = 4530x - 4000(x-2) - 1600(x-5)\left(\frac{x-5}{2}\right) \text{ lb-ft} \quad \text{for} \quad 5 < x < 9 \text{ ft}$$

Thus in this region the moment plots as a parabola. In fact, this is always true under a uniformly distributed load. If x = 5 ft is substituted in this equation, the moment is found to be 10,650 lb-ft, just as was found by using the equation for the central portion of the beam. The right end of the bar is simply supported, hence the moment is zero at x = 9 ft. It is of considerable interest to calculate the moment at x = 5.33 ft, since that is where the shear is zero. From Problem 7, since $V = dM/dx$ the slope of the moment diagram must be zero at this point. Substituting x = 5.33 ft, we find

$$M_{x=5.33} = 4530(5.33) - 4000(3.33) - 1650(0.33)\left(\frac{0.33}{2}\right)$$

$$= 10,810 \text{ lb-ft}$$

The moment diagram may now be plotted. It consists of two straight lines in the left and central regions and a parabola in the right portion. This parabola has a horizontal tangent at $x = 5.33$ ft and evidently that is the point of maximum moment. The moment diagram appears at the right.

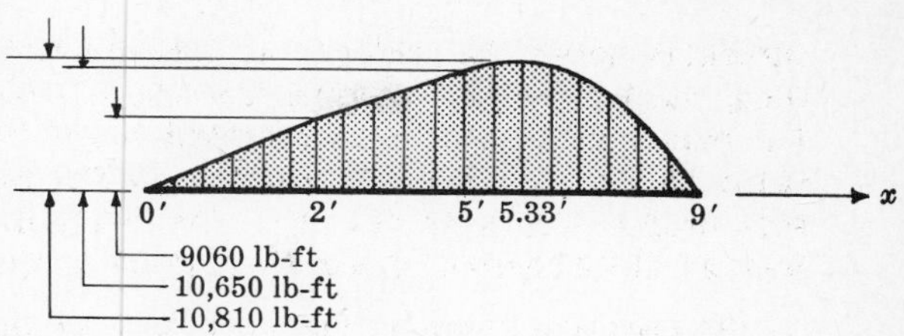

It may be noted from the shear diagram that there is only a gradual change of shear at $x = 5$ ft. Since $V = dM/dx$ at all points along the bar, then there is only a gradual change in the slope to the moment diagram at this point. Hence the straight line and the parabola in the moment diagram have a common tangent at $x = 5$ ft.

9. A simply supported beam is loaded by the couple of 1000 lb-ft, as shown in Fig. (a) below. Draw the shearing force and bending moment diagrams due to this loading.

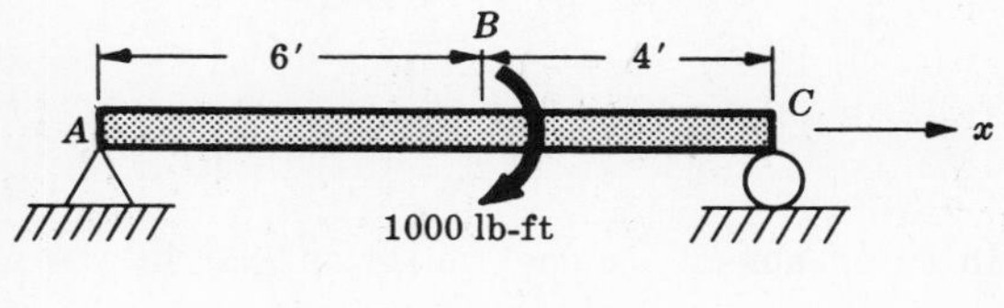

Fig. (a)

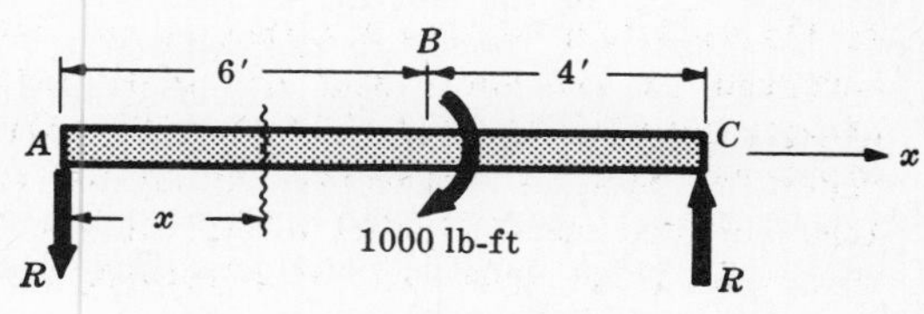

Fig. (b)

The beam is loaded by one couple, and the only possible manner in which equilibrium may be created is for the reactions at the supports A and C to constitute another couple. Thus these reactions appear as in Fig. (b) above. For equilibrium,

$$\Sigma M_A = 10R - 1000 = 0 \qquad \text{from which} \qquad R = 100 \text{ lb}$$

Thus the two forces R shown constitute the reactions necessary for equilibrium.

Introducing the x-axis shown with origin at the left end of the bar, it is evident that two regions of the beam must be considered in the analysis. In the region to the left of the 1000 lb-ft couple the shear is due to the left reaction R and we have

$$V = -100 \text{ lb} \qquad \text{for} \qquad 0 < x < 6 \text{ ft}$$

The bending moment in this region is negative since R is downward and it is given by

$$M = -100x \text{ lb-ft} \qquad \text{for} \qquad 0 < x < 6 \text{ ft}$$

As soon as we pass to the right of the applied couple of 1000 lb-ft we must consider the shear and moment again. Since a couple has no force effect in any direction, the couple being composed of two equal and opposite parallel forces, the shearing force is the same as in the left region, namely

$$V = -100 \text{ lb} \qquad \text{for} \qquad 6 < x < 10 \text{ ft}$$

The bending moment consists of the moment of the left reaction about an axis through the section at x together with the moment of the 1000 ft-lb couple. The algebraic sign of the bending moment due to this couple may be determined by assuming that it alone acts on the region BC of the beam, in which case it evidently causes bending of the nature shown in the adjoining diagram. According to our sign convention, this is positive bending. Hence the bending moment due to the 1000 lb-ft couple is positive and we have

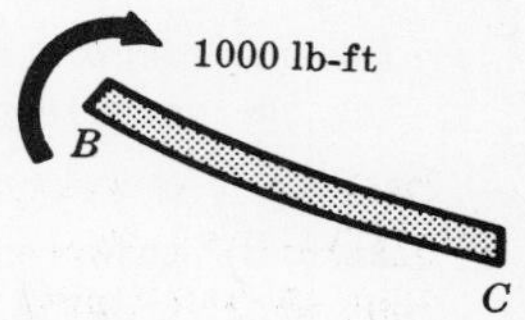

$$M = -100x + 1000 \qquad \text{for} \qquad 6 < x < 10 \text{ ft}$$

By this equation the bending moment at $x = 6$ ft is 400 lb-ft. Actually this is the moment slightly to the right of the point of application of the couple. Also, at $x = 10$ ft, $M = 0$ from this equation.

According to the previous equation for $0 < x < 6$ ft, the bending moment at $x = 6$ ft is −600 lb-ft. This is really the moment just to the left of the applied couple. The moment at $x = 0$ is zero.

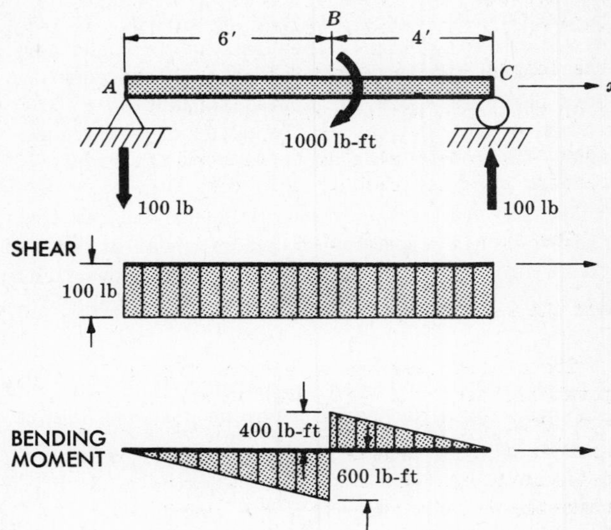

Thus the shear is constant, −100 lb everywhere across the beam, and the bending moment diagram is a straight line in each of the two regions. It is zero at the ends and has the values −600 lb-ft and 400 lb-ft on the left and right sides of B respectively. The shearing force and bending moment diagrams appear at the right.

Since $V = dM/dx$ and the shear has the same value at all points along the beam, then the slope of the bending moment diagram must be constant. Therefore the two inclined straight lines in the bending moment diagram are parallel.

From this it may be seen that whenever a couple acts on a bar the bending moment diagram exhibits an abrupt discontinuity or *jump* at the point where the couple is applied.

10. The simply supported beam, shown in Fig. (a) below, carries a vertical load that increases uniformly from zero at the left end to a maximum value of 600 lb/ft of length at the right end. Draw the shearing force and bending moment diagrams.

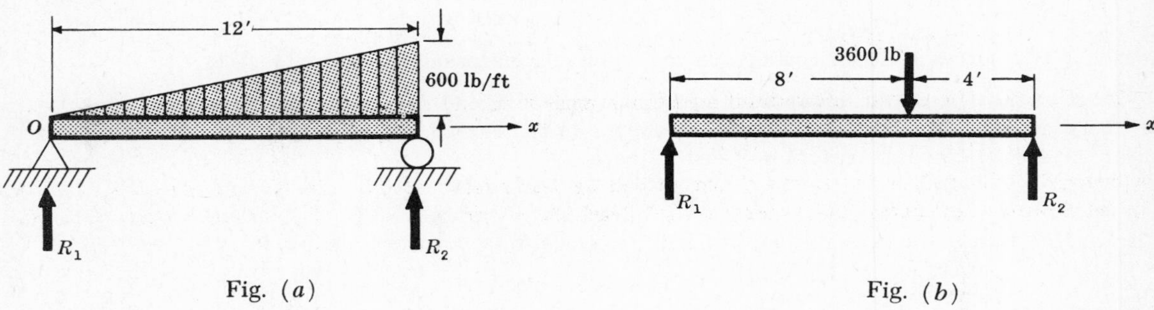

Fig. (a) Fig. (b)

For the purpose of determining the reactions R_1 and R_2 the entire distributed load may be replaced by its resultant which will act through the centroid of the triangular loading diagram. Since the load varies from 0 at the left end to 600 lb/ft at the right end, the average intensity is 300 lb/ft acting over a length of 12 ft. Hence the total load is 3600 lb applied 8 ft to the right of the left support. The free-body diagram to be used in determining the reactions is shown in Fig. (b) above. Applying the equations of static equilibrium to this bar, we find R_1 = 1200 lb and R_2 = 2400 lb.

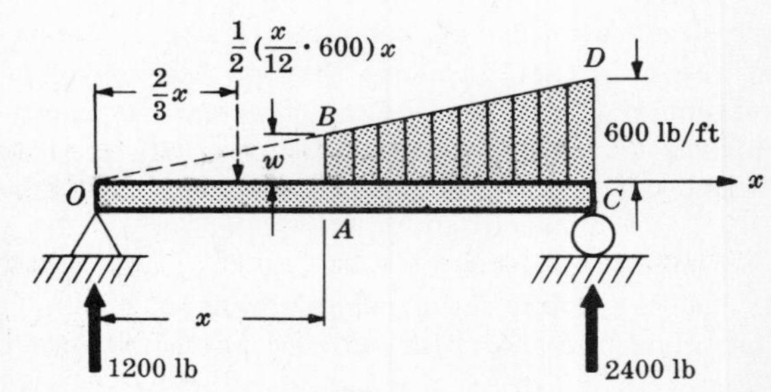

Fig. (c)

However, this resultant cannot be used for the purpose of drawing shear and moment diagrams. We must consider the distributed load and determine the shear and moment at a section a distance x from the left end as shown in the adjoining Fig. (c). At this section x the load intensity w may be found from the similar triangles OAB and OCD as follows:

$$w/x = 600/12 \qquad \text{or} \qquad w = (\frac{x}{12})600 \text{ lb/ft}$$

The average load intensity over the length x is $\frac{1}{2}(x/12)600$ lb/ft because the load is zero at the left end. The total load acting over the length x is the average intensity of loading multiplied by the length, or $\frac{1}{2}(\frac{x}{12}600)x$ lb. This acts through the centroid of the triangular region OAB shown, i.e. through a point located a distance $\frac{2}{3}x$ from O. The resultant of this portion of the distributed load is indicated by the dotted vector in Fig. (c) above. No portion of the load to the right of the section x is included in this resultant force.

The shearing force at A is now readily found to be $V = 1200 - \frac{1}{2}(\frac{x}{12}600)x = 1200 - 25x^2$

and the bending moment at A is given by $M = 1200x - \frac{1}{2}(\frac{x}{12}600)x(\frac{x}{3}) = 1200x - \frac{25}{3}x^3$.

These equations are true along the entire length of the beam. The shearing force thus plots as a parabola, having a value 1200 lb when $x = 0$ and -2400 lb when $x = 12$ ft. The bending moment is a third degree polynomial. It vanishes at the ends and assumes a maximum value where the shear is zero. This is true because $V = dM/dx$, hence the point of zero shear must be the point where the tangent to the moment diagram is horizontal. This point of zero shear may be found by setting $V = 0$:

$$0 = 1200 - 25x^2 \quad \text{or} \quad x = 6.94 \text{ ft}$$

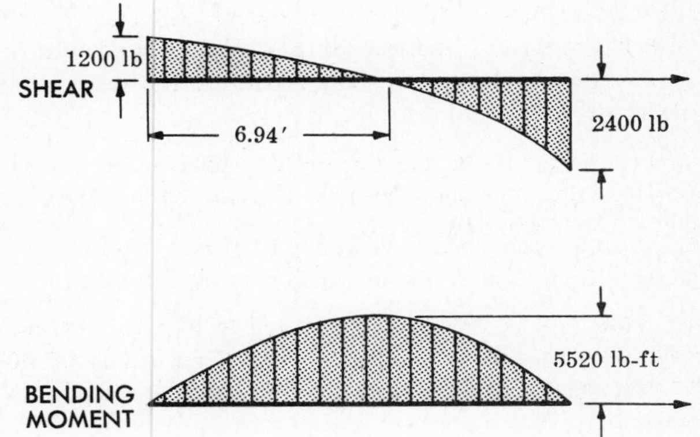

The bending moment at this point is found by substitution in the general expression given above:

$$M_{x=6.94} = 1200(6.94) - \frac{25}{3}(6.94)^3$$

$$= 5520 \text{ lb-ft}$$

The plots of the shear and moment equations appear in the two diagrams above.

1. The beam AC is supported at B and C and loaded by the couple of 1600 lb-ft applied at A, as shown in Fig. (a) below. Determine the reactions and draw the shearing force and bending moment diagrams.

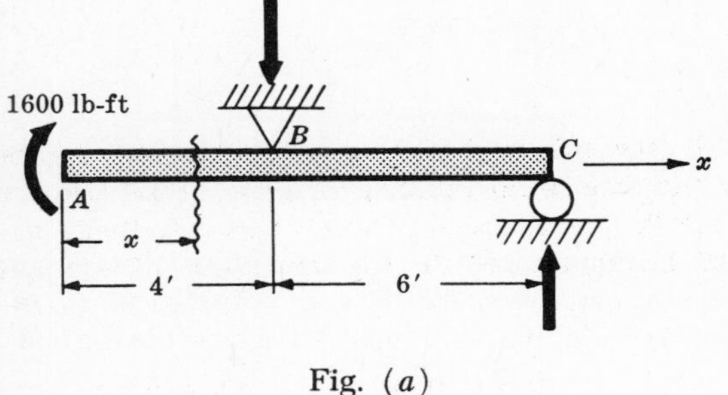

Fig. (a)

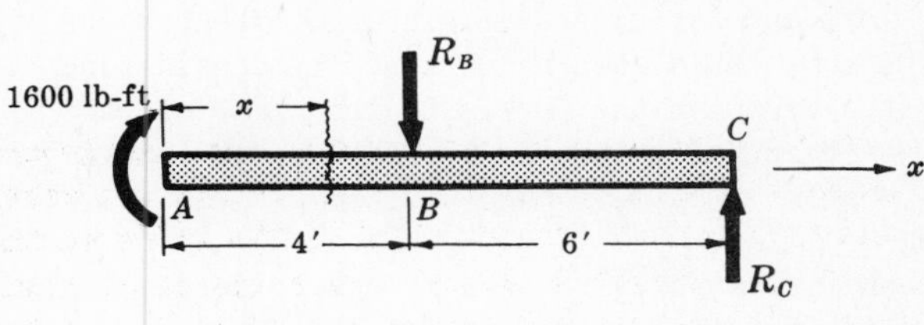

Fig. (b)

The free-body diagram of the bar may be drawn as in Fig. (b) above, with the assumed positive directions of the reactions at B and C as indicated.

For static equilibrium, we have: $\Sigma M_B = 1600 - 6R_C = 0$ or $R_C = 267$ lb

$\Sigma F_v = -R_B + 267 = 0$ or $R_B = 267$ lb

Since both results are positive, the assumed directions of R_B and R_C were correct.

The x-axis is introduced as usual with origin at A. Working from left to right along the beam it is evident that there are no vertical shearing forces acting in the region AB; hence we may write

$$V = 0 \quad \text{for} \quad 0 < x < 4 \text{ ft}$$

At any section a distance x from the left end in this region, the bending moment is due entirely to the applied couple of 1600 lb-ft. It is necessary to determine the algebraic sign of this moment and this may be done quite easily by realizing that such a couple produces curvature of AB which is concave upward. By our sign convention this constitutes positive bending, hence we have

$$M = 1600 \text{ lb-ft} \quad \text{for} \quad 0 < x < 4 \text{ ft}$$

Of course, the moment of a couple is the same about all points in the plane.

For values of x greater than 4 ft, the reaction R_B enters the equations for shear and moment. The shearing force due to R_B is negative since it tends to shear the region to the left of any section x downward with respect to the right region. Consequently we have

$$V = -267 \text{ lb} \quad \text{for} \quad 4 < x < 10 \text{ ft}$$

In BC the bending moment at a distance x from A is due partially to the couple of 1600 lb-ft and partially to the moment of the reaction R_B about an axis through the section x and perpendicular to the plane of the page, and we have

$$M = 1600 - 267(x-4) \text{ lb-ft} \quad \text{for} \quad 4 < x < 10 \text{ ft}$$

Substituting $x = 4$ ft in this equation, we find $M_{x=4} = 1600$ lb-ft. At $x = 10$ ft, the equation yields $M_{x=10} = 0$. The above equation is of the first degree in x, hence the bending moment diagram in the region BC plots as a straight line with the values of 1600 lb-ft at B and 0 at C.

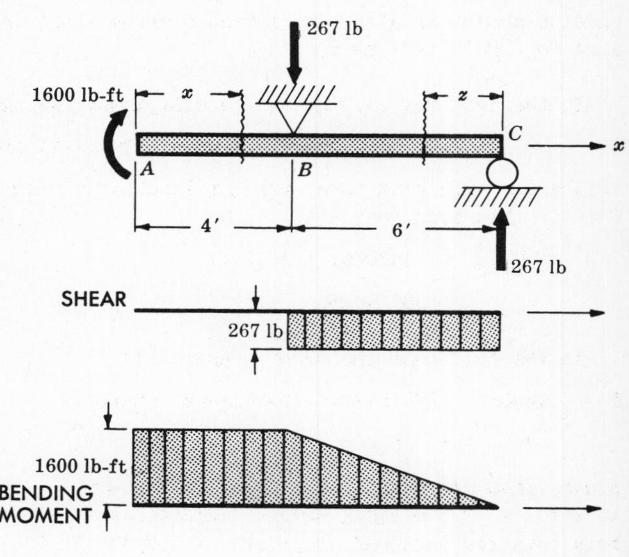

The loading, shearing force, and bending moment diagrams appear at the right. According to the equations derived above, the shear is zero in AB and -267 lb in BC; hence the plot consists of two horizontal lines. The bending moment is constant (1600 lb-ft) in AB and decreases linearly to 0 between B and C.

It is to be observed that the bending moment diagram in the region BC could have been obtained somewhat more simply by introducing a new coordinate z measured positive to the left and having its origin at C. The bending moment could then be obtained by considering the moment of the forces to the *right* of this section designated by z. This moment is evidently

$$M = 267z \quad \text{for} \quad 0 < z < 6 \text{ ft}$$

and it is positive because upward forces imply positive bending moments. This obviously plots as a straight line in the region BC. The concept of introducing a new coordinate z running positive to the left and then considering forces to the right of such a section is frequently very convenient.

12. Consider the beam with overhanging ends loaded by the three concentrated forces shown. Find equations for the shearing force and bending moment at any point along the bar and plot the corresponding diagrams.

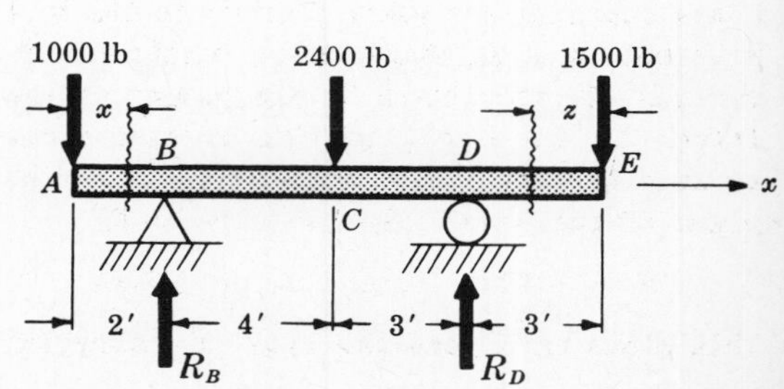

The reactions are readily determined from statics:

$$\Sigma M_B = 1000(2) + R_D(7) - 2400(4) - 1500(10) = 0$$

$$\Sigma F_v = R_B + 3230 - 1000 - 2400 - 1500 = 0$$

from which $R_D = 3230$ lb and $R_B = 1670$ lb.

The x-axis coincides with the axis of the beam and has its origin at A. It is most convenient to keep the origin at A throughout the problem, rather than shifting it successively to B, C, etc., in considering the various regions along the beam.

For any value of x the shearing force is simply given by the algebraic sum of the forces to the left of the section. There are four regions between concentrated forces, hence four equations are necessary to describe the shearing force:

In AB, $V = -1000$ lb for $0 < x < 2$ ft

In BC, $V = -1000 + 1670 = 670$ lb for $2 < x < 6$ ft

In CD, $V = -1000 + 1670 - 2400 = -1730$ lb for $6 < x < 9$ ft

In DE, $V = -1000 + 1670 - 2400 + 3230 = 1500$ lb for $9 < x < 12$ ft

Thus in each of these four regions the shearing force is a constant; hence it plots as a horizontal straight line in each region, as shown in the shearing diagram below. Notice that at the point of application of each concentrated force, including the reactions, the *jump* in the ordinate to the shear diagram is equal in magnitude to the concentrated force at that point.

In AB the bending moment is given by the moment of the 1000 lb force about an axis perpendicular to the plane of the page through the section x. Downward forces indicate negative bending moments, hence

$$M = -1000x \quad \text{for} \quad 0 < x < 2 \text{ ft}$$

The fact that the beam is overhanging between A and B does not complicate the determination of the bending moment at all. The moment diagram in AB is consequently a straight line, varying from zero at A to -2000 lb-ft at B.

In the next region, BC, the bending moment is given by

$$M = -1000x + 1670(x-2) \text{ lb-ft} \quad \text{for} \quad 2 < x < 6 \text{ ft}$$

This again will plot as a straight line. The bending moment at $x = 6$ ft is found by substituting $x = 6$ ft in this equation:

$$M_{x=6} = -1000(6) + 1670(4) = 680 \text{ lb-ft}$$

In the region CD the bending moment is

$$M = -1000x + 1670(x-2) - 2400(x-6) \text{ lb-ft} \quad \text{for} \quad 6 < x < 9 \text{ ft}$$

Again, this plots as a straight line in CD. At $x = 9$ ft the bending moment indicated by this equation is found by putting $x = 9$ ft:

$$M_{x=9} = -1000(9) + 1670(7) - 2400(3) = -4500 \text{ lb-ft}$$

The equation for the bending moment in DE is perhaps found most simply by introducing a new coordinate z taken positive to the left with origin at E. The bending moment at this section z is then given by the moment of the forces to the *right* of this section about an axis through z and perpendicular to the plane of the page. Then we have in DE,

$$M = -1500z \quad \text{for} \quad 0 < z < 3 \text{ ft}$$

This plots as a straight line in the region DE.

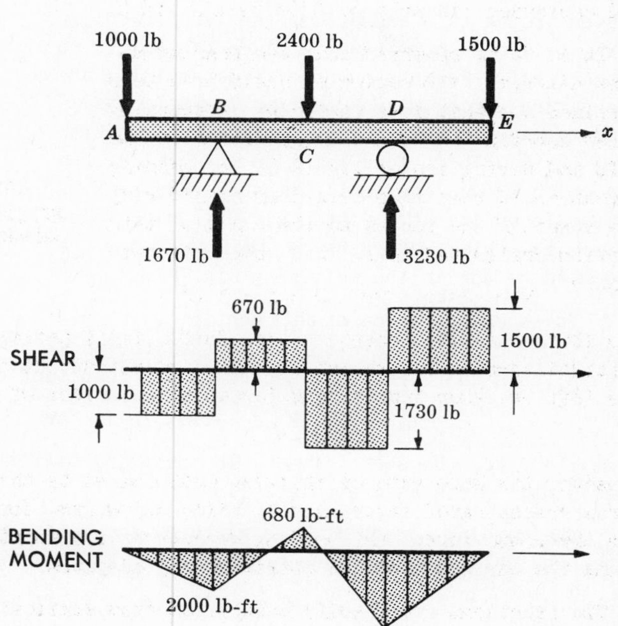

Thus the bending moment diagram plots as a series of straight lines as shown above.

It is to be observed that in those regions such as BC and DE where the shearing force V is positive, the slope of the bending moment diagram is also positive. This is to be expected from the relation $V = dM/dx$. Similarly, in AB and CD both the shearing force and the slope of the bending moment diagram are negative.

13. The beam ABC is simply supported at B and C, overhangs in the region AB and carries a uniformly distributed load of 120 lb per ft of length of the beam, as shown in Fig. (a) below. Draw the shearing force and bending moment diagrams.

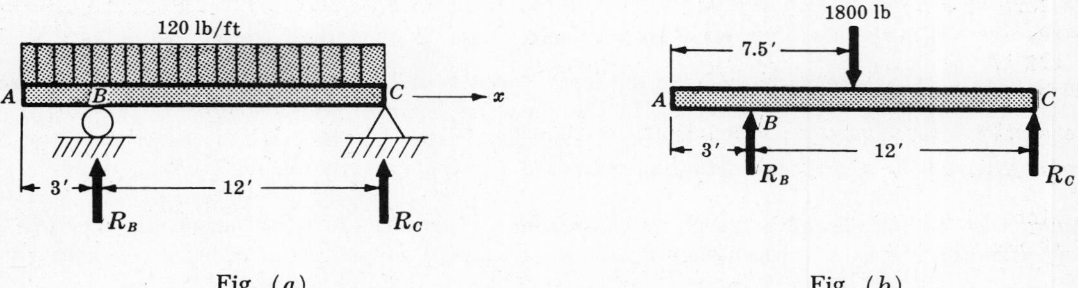

Fig. (a) Fig. (b)

For the purpose of determining the reactions R_B and R_C the entire distributed load may be replaced by its resultant. The resultant is a force = 120 lb/ft × 15 ft = 1800 lb acting through the midpoint of the load, i.e. 7.5 ft from either end. The resultant is indicated by the 1800 lb vector in the free-body diagram, Fig. (b) above. For static equilibrium,

$$\Sigma M_B = R_C(12) - 1800(4.5) = 0 \quad \text{or} \quad R_C = 675 \text{ lb}$$

$$\Sigma F_v = R_B - 1800 + 675 = 0 \quad \text{or} \quad R_B = 1125 \text{ lb}$$

We will introduce an x-axis coinciding with the horizontal beam and having its origin at A. Even though this end is free (not supported) it is still most convenient to place the origin at that point. For the purpose of determining the shearing force at any section in AB located a distance x from A, we may replace that portion of the 120 lb/ft load lying to the *left* of the section by its resultant. Thus the resultant is $120x$ lb acting at a distance $x/2$ from the section x. This resultant is indicated by the dotted vector in the adjoining Fig. (c).

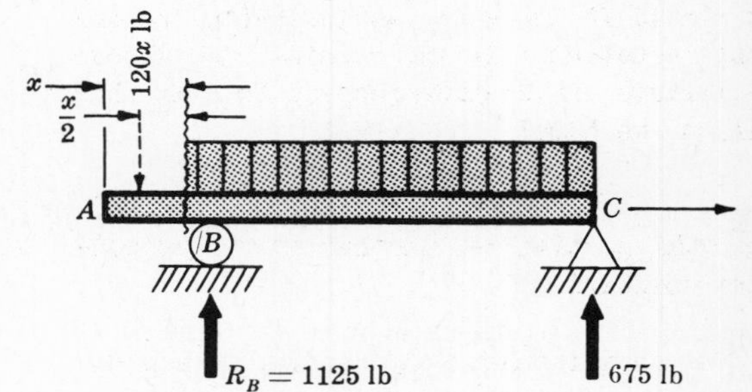

Fig. (c)

The shearing force at this section x is then the sum of the forces to the left of the section, this sum being represented by the resultant of $120x$ lb. Thus we may write

$$V = -120x \text{ lb} \quad \text{for} \quad 0 < x < 3 \text{ ft}$$

At $x = 3$ ft, the shear is -360 lb according to this equation. The shearing force in this region consequently plots as a straight line.

The bending moment at x is given by the moment of this resultant force about an axis through x and perpendicular to the plane of the page. It is given by

$$M = -120x(x/2) \text{ lb-ft} \quad \text{for} \quad 0 < x < 3 \text{ ft}$$

Evidently the plot of the bending moment along the length of the bar is parabolic in this region, varying from zero at A to -540 lb-ft at B as may be seen by substituting $x = 3$ in the above equation.

To establish the direction of concavity of the bending moment diagram we may form the second derivative of M with respect to x. This gives

$$d^2M/dx^2 = -120 \quad \text{for} \quad 0 < x < 3 \text{ ft}$$

The fact that the second derivative is negative everywhere in this region indicates that the curve is concave downward.

As soon as we pass to the right of the reaction at point B, this concentrated force of 1125 lb must be included in the equations for shearing force and bending moment. The portion of the distributed load to the left of any section x may still be replaced by its resultant of $120x$ lb acting downward at a distance $x/2$ to the left of the section x, as shown in the adjoining Fig. (d).

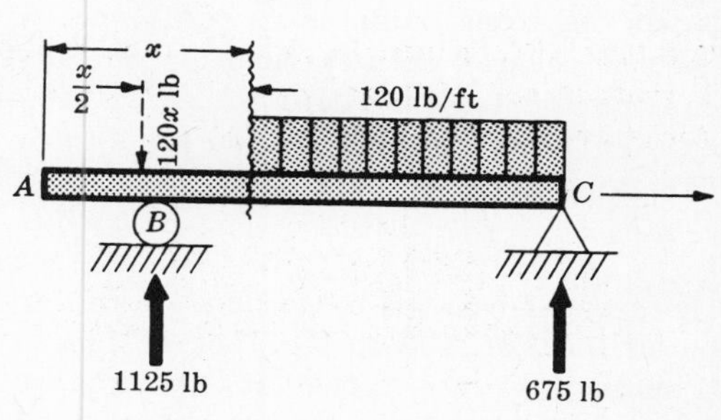

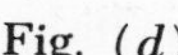
Fig. (d)

The shearing force at x is given by

$$V = -120x + 1125 \text{ lb} \quad \text{for} \quad 3 < x < 15 \text{ ft}$$

which plots as a straight line in the region BC. At x = 3 ft the shear is

$$V_{x=3} = -120(3) + 1125 = 765 \text{ lb}$$

at x = 15 ft the shear is

$$V_{x=15} = -120(15) + 1125 = -675 \text{ lb}$$

The bending moment at x is given by

$$M = -120x(x/2) + 1125(x-3) \text{ lb-ft} \quad \text{for} \quad 3 < x < 15 \text{ ft}$$

which plots as a parabola. Substituting x = 3 ft, this equation yields

$$M_{x=3} = -120(3)(1.5) = -540 \text{ lb-ft}$$

which of course agrees with the value of the bending moment at this point obtained by using the equation for the region AB. The bending moment at x = 15 ft is zero, as indicated by the above equation. Again, forming the second derivative of M with respect to x in this region, we have

$$d^2M/dx^2 = -120$$

so this portion of the curve is also concave downward.

The shearing force and bending moment diagrams may then be plotted as shown.

In plotting the bending moment diagram in BC, it is helpful to first determine the location of D, the point where the shearing force is zero. This may be found by setting V = 0 in the equation for shearing force in BC:

$$0 = -120x + 1125 \quad \text{or} \quad x = 9.38 \text{ ft}$$

This locates the point D. Since $V = dM/dx$, the tangent to the bending moment diagram is horizontal at D, the point of zero shear. This is a critical value of moment that should be investigated. It is to be remembered that the calculus method of determining

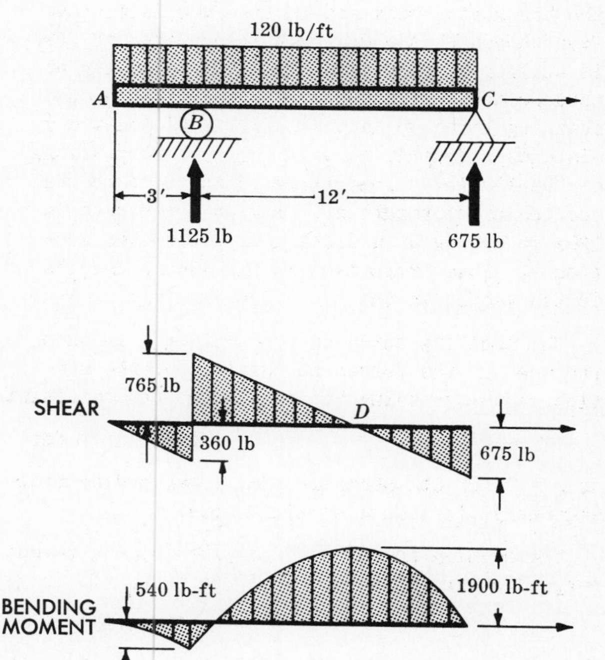

maximum values does detect peak values such as at D but fails to indicate peaks at a cusp-like point such as occurs in the moment diagram at B. Hence all points of each type must be investigated in determining the maximum bending moment in a beam. The moment at point D is found by substitution to be

$$M_{x=9.38} = -60(9.38)^2 + 1125(9.38-3) = 1900 \text{ lb-ft}$$

The equation $V = dM/dx$ shows that in those regions such as AB and DC where the shear is negative, the slope of the moment diagram is also negative. Likewise in BD where the shear is positive, the slope of the moment diagram is positive. Also, since the shear changes abruptly at B, the slope of the moment diagram changes abruptly in passing from the curve to the left of B to that to the right. Hence a common tangent could not exist between the two parabolas constituting the bending moment diagram. Full use of the relation $V = dM/dx$ usually enables a moment diagram to be plotted with the actual calculation of only a few values.

14. The horizontal beam AD is loaded by a uniformly distributed load of 400 lb per foot of length and is also subject to the concentrated force of 3000 lb applied as shown. Draw the shearing force diagram and also draw the bending moment diagram in parts.

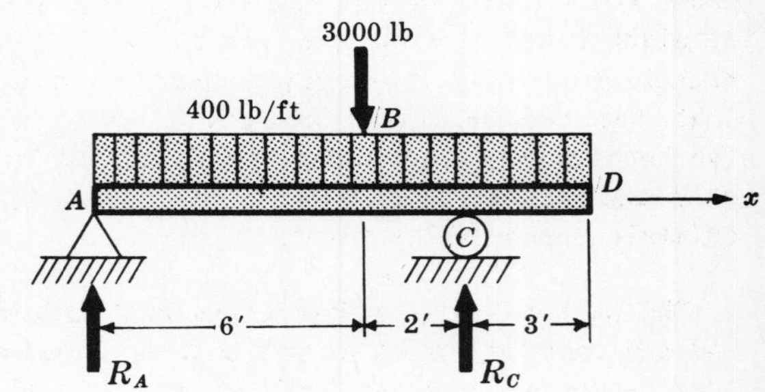

From statics the following equilibrium equations may be written:

$$\Sigma M_A = 8R_C - 3000(6) - 400(11)(5.5) = 0$$

$$\Sigma F_v = R_A + 5275 - 3000 - 400(11) = 0$$

from which $R_C = 5275$ lb and $R_A = 2125$ lb.

The usual x-axis is introduced with the origin at point A. Three regions must be considered in writing equations for the shearing force, namely AB, BC, CD. In a manner exactly analogous to that outlined in Problem 13, the equations for shearing force may be written as

$$(1) \quad V = 2125 - 400x \text{ lb} \quad \text{for} \quad 0 < x < 6 \text{ ft}$$

$$(2) \quad V = 2125 - 400x - 3000 \text{ lb} \quad \text{for} \quad 6 < x < 8 \text{ ft}$$

$$(3) \quad V = 2125 - 400x - 3000 + 5275 \text{ lb} \quad \text{for} \quad 8 < x < 11 \text{ ft}$$

From (1), the shear at $x = 0$ is 2125 lb. Also, the shear just to the left of the 3000 lb load is found by substituting $x = 6$ ft in (1); the result is −275 lb. The shear immediately to the right of the 3000 lb load is found by substituting $x = 6$ ft in equation (2); this yields

$$V_{x=6} = 2125 - 400(6) - 3000 = -3275 \text{ lb}$$

The shear just to the left of point C is found by substituting $x = 8$ ft in equation (2); this gives

$$V_{x=8} = 2125 - 400(8) - 3000 = -4075 \text{ lb}$$

The shear immediately to the right of point C is found by substituting $x = 8$ ft in (3); the result is

$$V_{x=8} = 2125 - 400(8) - 3000 + 5275 = 1200 \text{ lb}$$

From equations (1), (2) and (3) it is evident that the shearing force diagram plots as a straight line in each of the three regions. The values of the shearing force at the end points of these intervals have just been established, so these values may be plotted and then connected by straight lines to give the shearing force diagram as shown at the right.

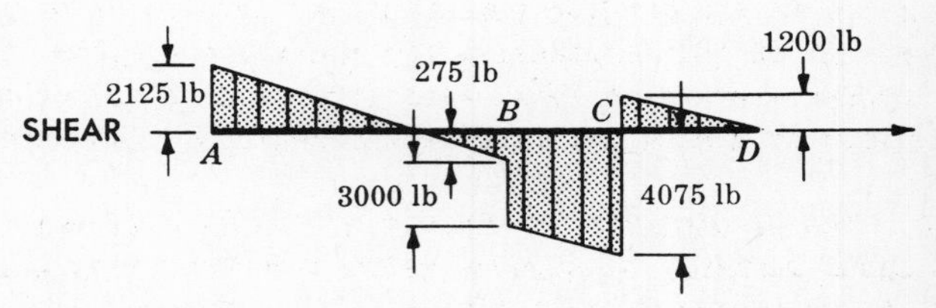

The bending moment diagram will be plotted in a different manner than previously. The technique will be to consider each loading on the bar separately and plot the bending moment due to it alone as if no other forces were acting on the structure. The moment diagram is then said to be plotted in *parts*. As will be seen in a later chapter dealing with deflections of beams this method is often very convenient, although the choice as to whether to use it or the conventional type of plot presented in earlier problems depends upon the purpose for which the diagram is being drawn. More will be said about this later.

Let us work from left to right along the beam. The moment diagram may be considered to consist of four parts, one due to the reaction R_A, another due to the uniformly distributed load, a third due to the 3000 lb force, and the last due to the reaction R_C. At any section a distance x from the point A the bending moment due to R_A only is simply $2125x$ lb-ft. This value is positive because R_A acts upward. This same expression holds for all values of x along the bar. This is a first-degree function of x, hence the bending moment due to R_A only plots as a straight line. At $x = 0$, the moment is zero, and substituting $x = 11$ ft in the above expression it is seen that the bending moment at point D is 23,375 lb-ft. The bending moment at any section x due to this force only may thus be represented by the ordinates to the triangle shown at the right.

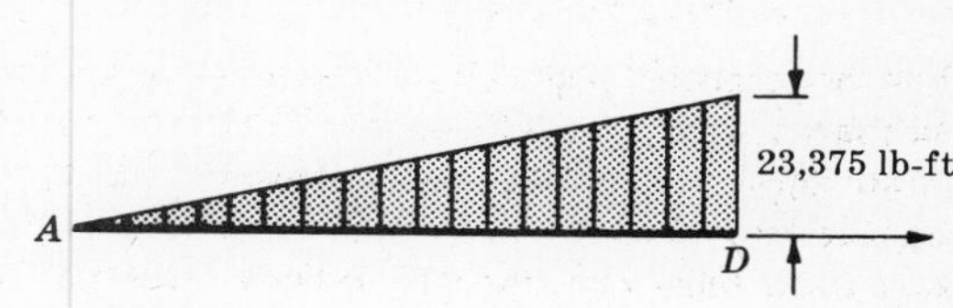

The uniformly distributed load will be treated next. All other loads are temporarily disregarded and the bending moment at any section x due to the uniform load is calculated. This proceeds in the same manner as discussed before, i.e. the portion of the load to the left of the section x is replaced by its resultant, indicated by the dotted vector in Fig. (a) below.

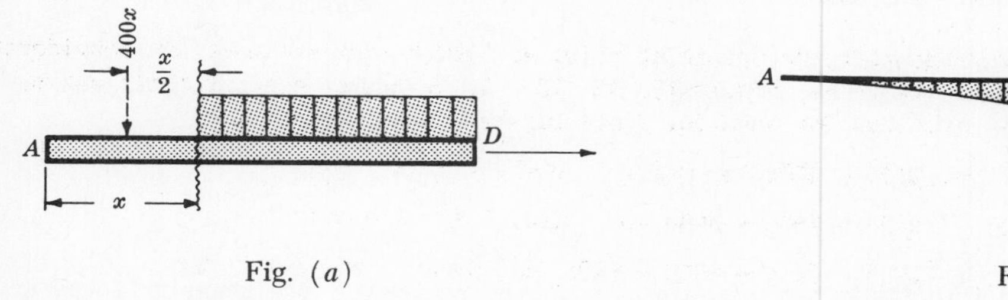

Fig. (a)

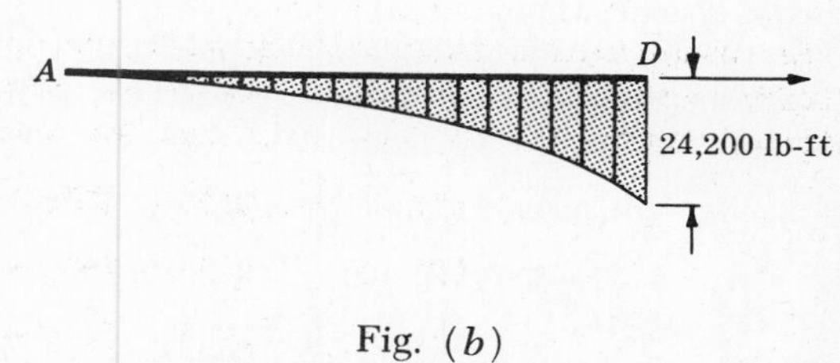

Fig. (b)

Due to the distributed load only, the bending moment at any section x anywhere along the beam is given by

$$-400x(x/2) \text{ lb-ft}$$

When $x = 0$ this expression vanishes, and when $x = 11$ ft it is equal to $-24,200$ lb-ft. It plots as a parabola since the expression is of the second degree, as shown in Fig. (b) above.

As we work from left to right along the beam, the influence of the 3000 lb load is not apparent until we pass to the right of point B. After that, at any section x the bending moment due to this force alone, temporarily disregarding all other forces, is given by

$$-3000(x-6) \text{ lb-ft} \qquad \text{for} \qquad 6 < x < 11 \text{ ft}$$

It is to be noted that x is always measured from the point A. When $x = 6$ ft the bending moment due to this force only vanishes, and when $x = 11$ ft this expression has the value $-15,000$ lb-ft. It is a first-degree expression in x, hence the bending moment due to this force alone plots as a straight line in the region BD, as shown at the right.

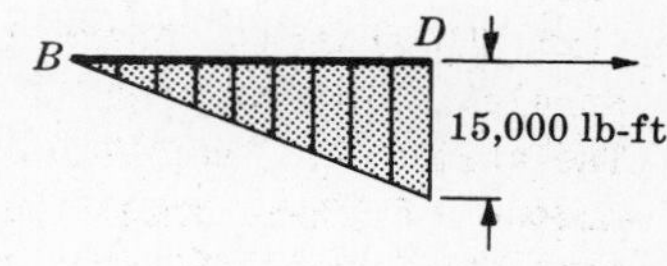

The bending moment diagram due to R_C only may be formed in an analogous manner. As soon as we consider sections anywhere in the region CD, the force R_C will give rise to a bending moment. Due to this force only, there is a bending moment of $5275(x-8)$ lb-ft for $8 < x < 11$ ft. When $x = 8$ ft this value is zero, and when $x = 11$ ft the bending moment due to R_C only is $5275(11-8) = 15,825$ lb-ft. This is a first-degree

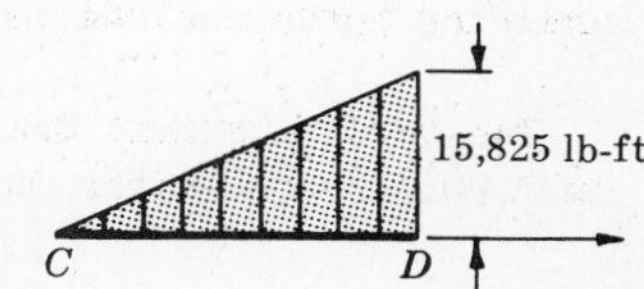

expression in x and hence the moment diagram due to R_C only also appears as a triangle as shown above.

The moment diagrams due to the various loadings have each been obtained as if there were only that one load acting on the beam. Actually of course all loads act simultaneously, so the true value of the moment at any point is the algebraic sum of the values indicated by the above four plots. It is customary to plot all of these individual diagrams together, as shown at the right.

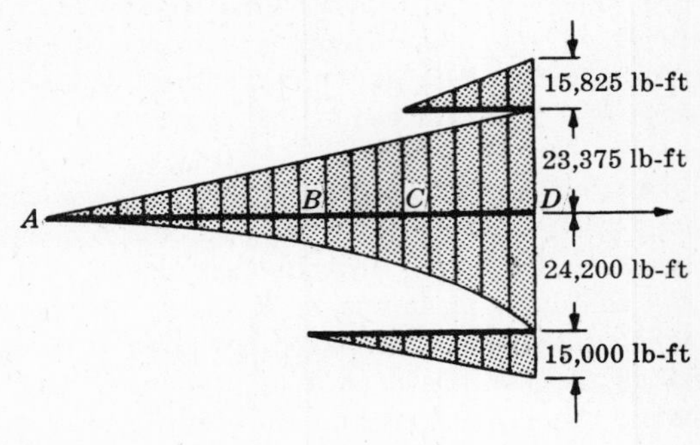

Notice that the horizontal bases of the two small triangular diagrams are shifted so that there is no overlapping of the various figures. This is not mandatory but makes for ease of interpretation. The algebraic sum of the four ordinates at D equals zero, which is necessary because that is a free end. The composite type of diagram discussed in previous problems can now be obtained by summing ordinates at every point of the above diagram. Between A and B, only two quantities would be included in the sum; between B and C, three quantities; and between C and D, four.

15. The beam AE is simply supported at B and D and overhangs both ends. It is subject to a uniformly distributed load of 600 lb per ft of length as well as a couple of magnitude 10,000 lb-ft applied at C. Draw the shearing force diagram and also draw the bending moment diagram in parts.

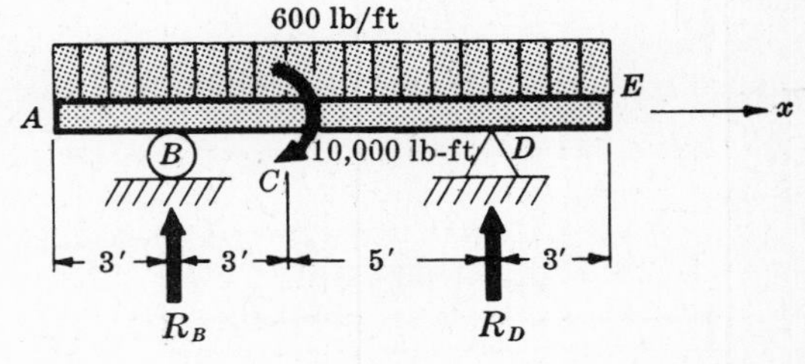

The reactions may be determined by the following equations of static equilibrium:

$$\Sigma M_B = 8R_D - 10{,}000 - 600(14)(4) = 0, \qquad R_D = 5450 \text{ lb}$$

$$\Sigma F_v = R_B + 5450 - 600(14) = 0, \qquad R_B = 2950 \text{ lb}$$

The x-axis is introduced with its origin at point A. In the region AB, the shearing force at any section a distance x from point A is given by the resultant of the distributed load to the left of this section. This resultant is evidently a force of $600x$ lb acting downward. Thus we have

$$(1) \qquad V = -600x \qquad \text{for} \qquad 0 < x < 3 \text{ ft}$$

Substituting $x = 3$ ft, this equation yields a shearing force at that point of -1800 lb. The shear at $x = 0$ is, of course, zero.

As soon as we pass to the right of B the reaction R_B appears in the shearing force equation. For any section a distance x from A the shearing force in the region BD is obtained by summing the applied forces to the left of this section. This sum is given by

$$(2) \qquad V = -600x + 2950 \text{ lb} \qquad \text{for} \qquad 3 < x < 11 \text{ ft}$$

Note that the applied couple at C does not enter the equations for shearing force because the couple does not have any force effect in any direction. It does, however, enter the equation indirectly since it influences the values of the reactions R_B and R_D. Substituting $x = 3$ ft and $x = 11$ ft in (2),

$$V_{x=3} = 1150 \text{ lb} \quad \text{and} \quad V_{x=11} = -3650 \text{ lb}$$

In considering values of x greater than 11 ft, the reaction R_D must be included in the equation for shearing force. Summing forces to the left of a section x in the region DE, we find

$$(3) \qquad V = -600x + 2950 + 5450 \text{ lb} \qquad \text{for} \qquad 11 < x < 14 \text{ ft}$$

Substituting $x = 11$ ft and $x = 14$ ft in this equation (3), we find

$$V_{x=11} = 1800 \text{ lb} \quad \text{and} \quad V_{x=14} = 0 \text{ lb}$$

The shearing force at any point along the bar is defined by one of the three equations (1), (2), or (3) depending upon the region in which the point x lies. Since V is a first-degree function of x in each of these regions, the shearing force diagram plots as a straight line in each of these three regions. The values of the ordinates at the end points of each of these regions have already been obtained by substitution. In AB these end-point values were 0 and -1800 lb. In BD they were 1150 lb and -3650 lb. Finally, in DE they were found to be 1800 lb and 0. These values may be plotted at the corresponding points along the beam and the ordinates representing these values connected by a straight line in each region. In this manner the above shearing force diagram is obtained.

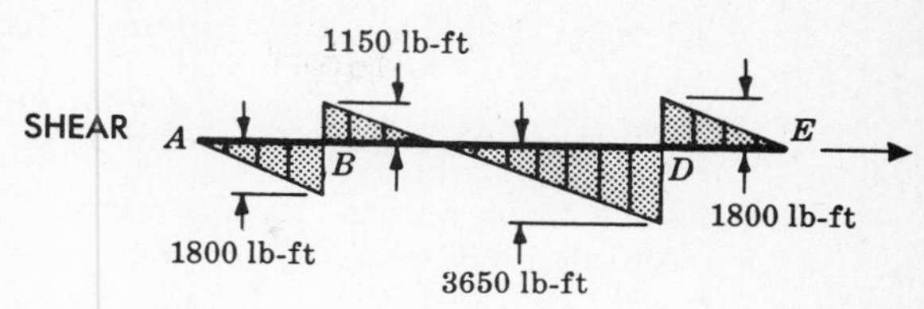

The magnitude of the vertical *jump* at each of the points B and D is of course equal to the value of the concentrated reactions R_B and R_D at each of these points.

In plotting the bending moment diagram in parts, each of the loads, including the reactions, are considered individually as if there were no other loads acting on the beam. Beginning with the uniform load of 600 lb/ft, a section at a distance x from the left end A is considered and the bending moment at this section due to the distributed load only is calculated. The resultant of the distributed forces lying to the left of this section is indicated by the dotted vector in the figure above. The moment of this resultant about an axis through the section at x and perpendicular to the plane of the page is

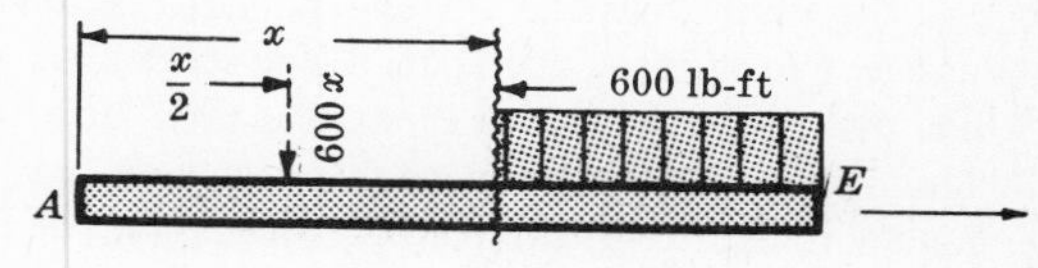

$$M = -600x(x/2) = -300x^2 \text{ lb-ft} \quad \text{for} \quad 0 < x < 14 \text{ ft}$$

Thus the bending moment diagram representing the distributed load only is parabolic. At $x = 0$ the moment is 0 and at the right end, $x = 14$ ft, the above equation yields the following value:

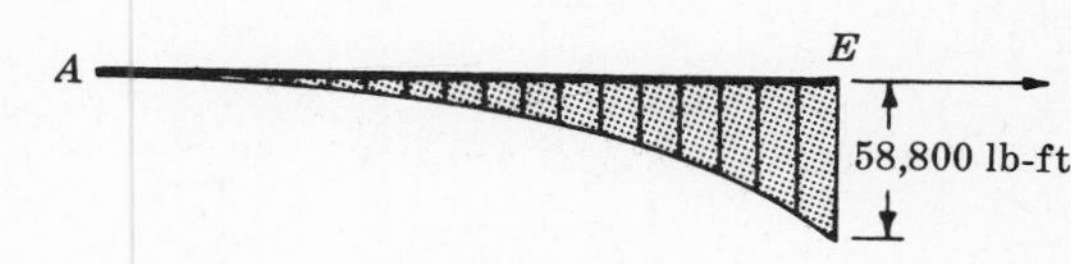

$$M_{x=14} = -300(14)^2 = -58,800 \text{ lb-ft}$$

This "part" of the bending moment diagram thus has the appearance indicated in the figure above.

Since we are working from left to right along the beam, the moment due to the reaction R_B does not come into consideration until we consider values of x greater than 3 ft. Then, due to this load only, the moment of this 2950 lb force about an axis through the section x is given by

$$M = 2950(x-3) \text{ lb-ft} \quad \text{for} \quad 3 < x < 14 \text{ ft}$$

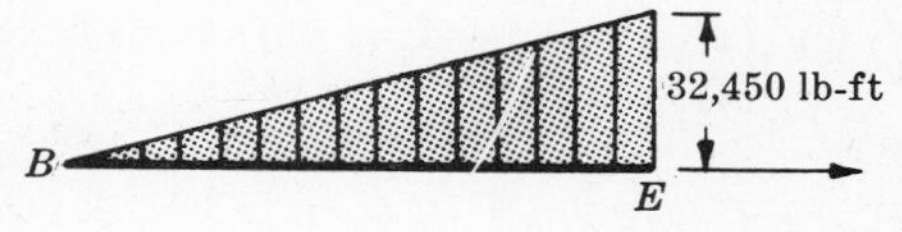

Since this is a first-degree function of x the bending moment due to R_B only plots as a straight line. According to this equation, the bending moment is zero at $x = 3$ ft and is $2950(14-3) = 32,450$ lb-ft at the point E. These two end-point values may now be connected by a straight line to yield the above moment diagram due to R_B only.

Progressing to the right along the beam, we next consider the applied couple of 10,000 lb-ft at the point C. For sections located at a distance x from point A, where x lies to the right of point C, this applied couple appears in the bending moment diagram. While it is true that the moment of this couple is the same about all points in the plane, it does not appear in the bending moment diagram until we consider values of x greater than 6 ft because the bending moment takes into account only the moments of those forces and couples to the left of the section x being considered. This applied couple produces curvature as shown in the adjoining figure. According to our sign conven-

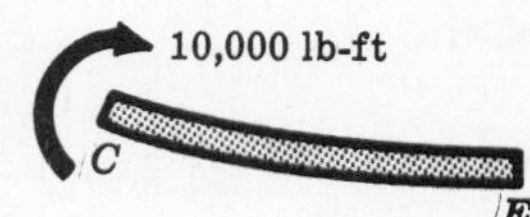

tion this constitutes positive bending. Hence we have for the applied couple only

$$M = 10{,}000 \text{ lb-ft} \quad \text{for} \quad 6 < x < 14 \text{ ft}$$

This constant value plots as a horizontal straight line, as shown in the adjoining figure.

Finally, the reaction R_D appears in the calculation of the bending moment at sections lying to the right of point D. For such a section at the distance x to the right of point A, the bending moment due to R_D only is

$$M = 5450(x-11) \text{ lb-ft} \quad \text{for} \quad 11 < x < 14 \text{ ft}$$

This too plots as a straight line. At point D this moment is zero, and substituting $x = 14$ ft we find $M = 16{,}350$ lb-ft at point E. Connecting these two end-point values by a straight line yields the adjoining moment diagram due to R_D only.

The four "parts" of the bending moment diagram are finally plotted together, as shown at the right. The horizontal bases of each of the "parts" have been displaced vertically so as to avoid overlapping of the various diagrams.

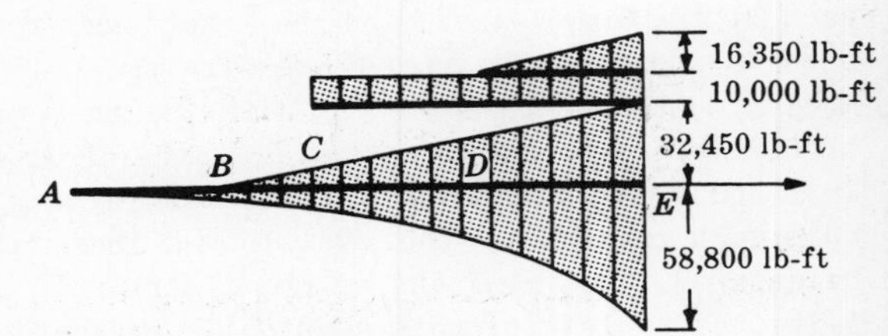

The analysis could just as well have been carried out by working from right to left. The resulting diagram by parts would have had an entirely different appearance than that shown above.

SUPPLEMENTARY PROBLEMS

For the following three cantilever beams of Problems 16, 17, 18 loaded as shown, write equations for the shearing force and bending moment at any point along the length of the beam. Also, draw the shearing force and bending moment diagrams.

16.

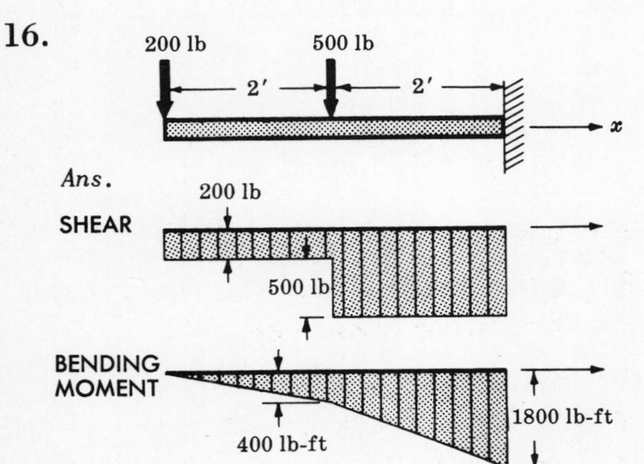

$V = -200$ lb for $0 < x < 2$ ft

$V = -700$ lb for $2 < x < 4$ ft

$M = -200x$ lb-ft for $0 < x < 2$ ft

$M = -200x - 500(x-2)$ lb-ft for $2 < x < 4$ ft

17.

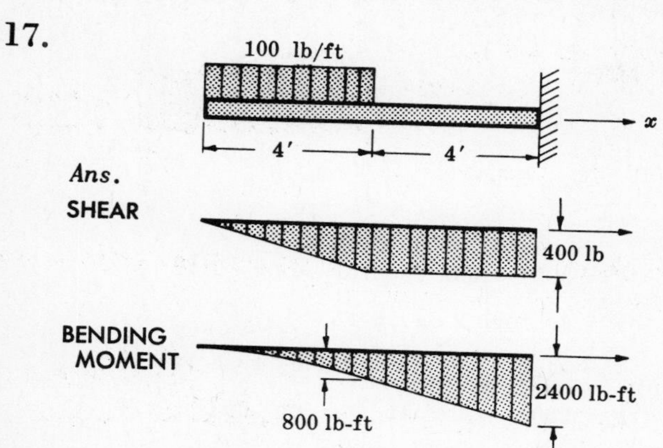

$V = -100x$ lb for $0 < x < 4$ ft

$V = -400$ lb for $4 < x < 8$ ft

$M = -50x^2$ lb-ft for $0 < x < 4$ ft

$M = -400(x-2)$ lb-ft for $4 < x < 8$ ft

18.

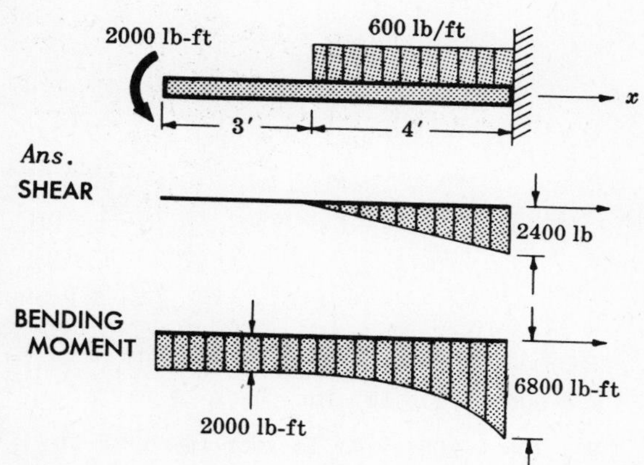

$V = 0$ for $0 < x < 3$ ft

$V = -600(x-3)$ lb for $3 < x < 7$ ft

$M = -2000$ lb-ft for $0 < x < 3$ ft

$M = -2000 - 300(x-3)^2$ lb-ft for $3 < x < 7$ ft

For the following nine beams of Problems 19-27 simply supported at the ends and loaded as shown write equations for the shearing force and bending moment at any point along the length of the beam. Also, draw the shearing force and bending moment diagrams.

19.

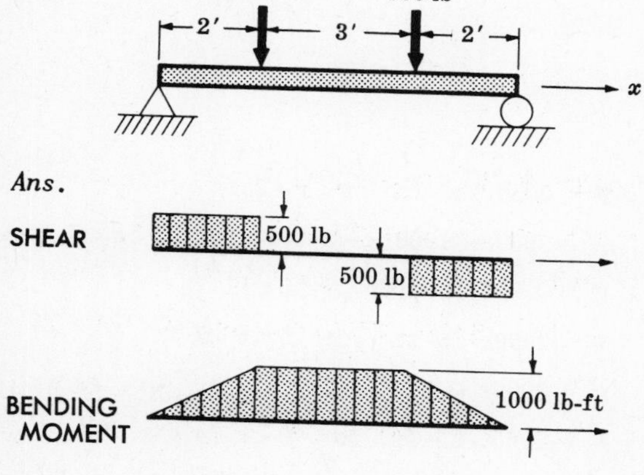

$V = 500$ lb for $0 < x < 2$ ft

$V = 0$ for $2 < x < 5$ ft

$V = -500$ lb for $5 < x < 7$ ft

$M = 500x$ lb-ft for $0 < x < 2$ ft

$M = 1000$ lb-ft for $2 < x < 5$ ft

$M = 1000 - 500(x-5)$ lb-ft for $5 < x < 7$ ft

20.

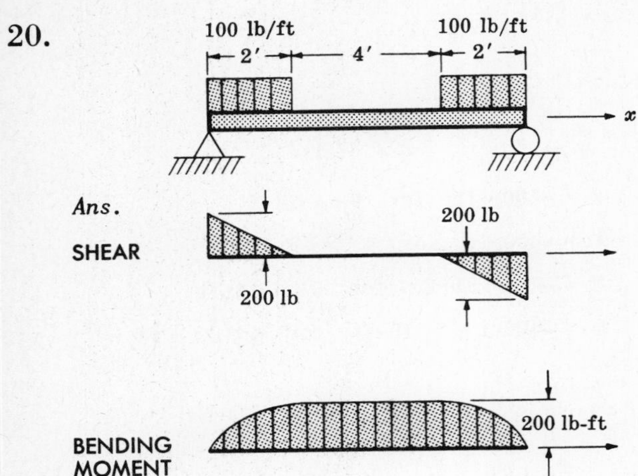

$V = 200 - 100x$ lb for $0 < x < 2$ ft

$V = 0$ for $2 < x < 6$ ft

$V = -100(x-6)$ lb for $6 < x < 8$ ft

$M = 200x - 50x^2$ lb-ft for $0 < x < 2$ ft

$M = 200$ lb-ft for $2 < x < 6$ ft

$M = 200 - 50(x-6)^2$ lb-ft for $6 < x < 8$ ft

21.

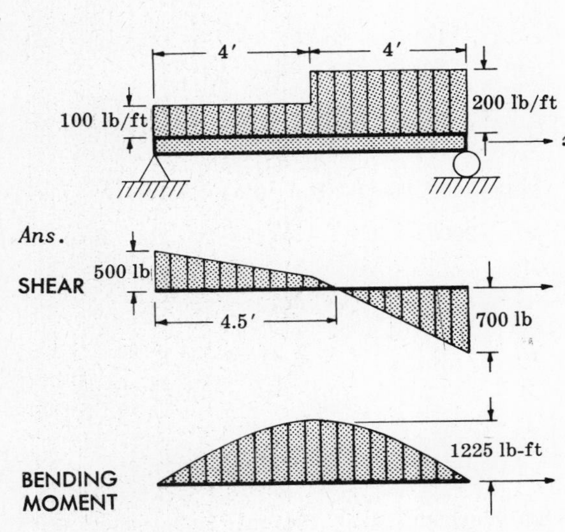

$V = 500 - 100x$ lb for $0 < x < 4$ ft

$V = 100 - 200(x-4)$ lb for $4 < x < 8$ ft

$M = 500x - 50x^2$ lb-ft for $0 < x < 4$ ft

$M = 500x - 400(x-2) - 100(x-4)^2$ lb-ft for $4 < x < 8$ ft

22.

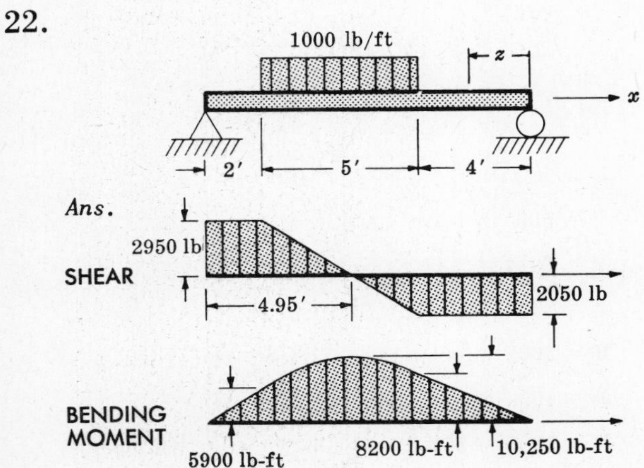

$V = 2950$ lb for $0 < x < 2$ ft

$V = 2950 - 1000(x-2)$ lb for $2 < x < 7$ ft

$V = -2050$ lb for $7 < x < 11$ ft

$M = 2950x$ lb-ft for $0 < x < 2$ ft

$M = 2950x - 500(x-2)^2$ lb-ft for $2 < x < 7$ ft

$M = 2050z$ lb-ft for $0 < z < 4$ ft

23.

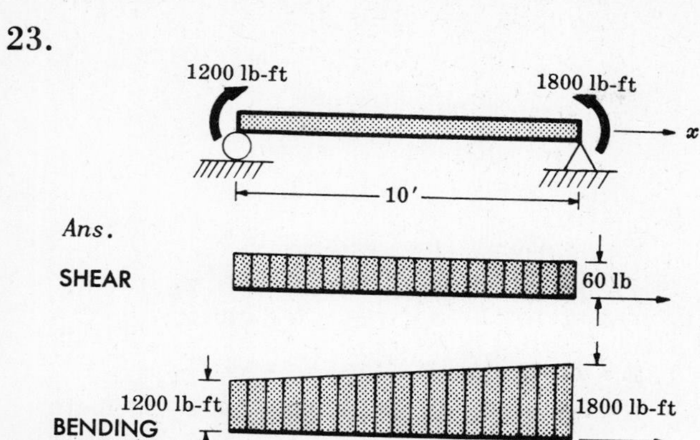

$V = 60$ lb for $0 < x < 10$ ft

$M = 1200 + 60x$ lb-ft for $0 < x < 10$ ft

24.

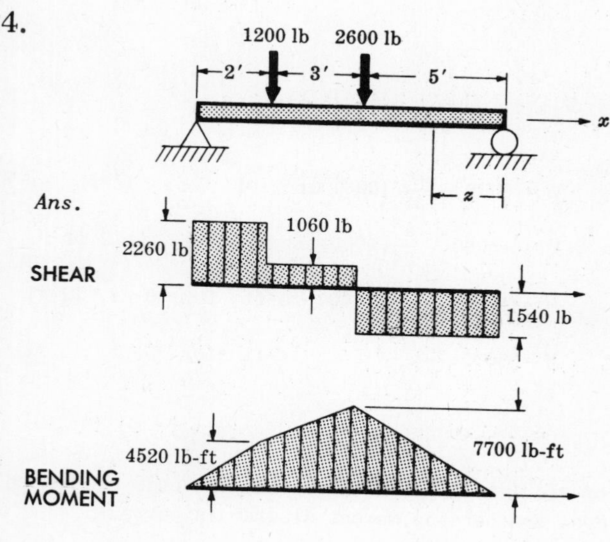

$V = 2260$ lb for $0 < x < 2$ ft

$V = 1060$ lb for $2 < x < 5$ ft

$V = -1540$ lb for $5 < x < 10$ ft

$M = 2260x$ lb-ft for $0 < x < 2$ ft

$M = 2260x - 1200(x-2)$ lb-ft for $2 < x < 5$ ft

$M = 1540z$ lb-ft for $0 < z < 5$ ft

25.

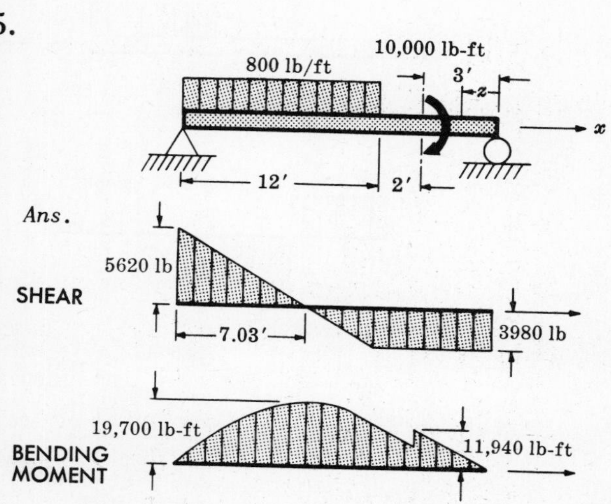

$V = 5620 - 800x$ lb for $0 < x < 12$ ft

$V = -3980$ lb for $12 < x < 17$ ft

$M = 5620x - 400x^2$ lb-ft for $0 < x < 12$ ft

$M = 5620x - 9600(x-6)$ lb-ft for $12 < x < 14$ ft

$M = 3980z$ for $0 < z < 3$ ft

26.

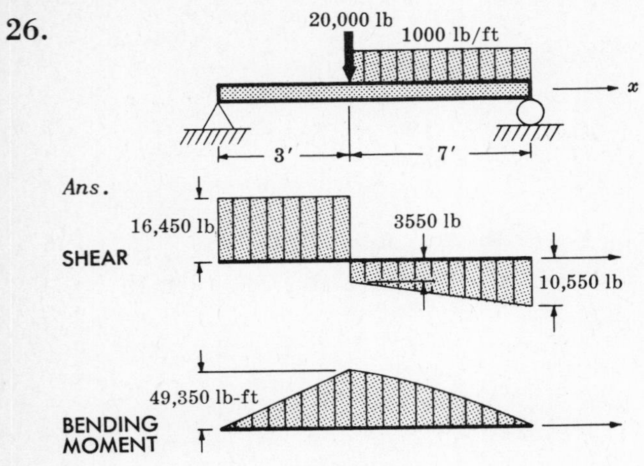

$V = 16{,}450$ lb for $0 < x < 3$ ft

$V = 16{,}450 - 20{,}000 - 1000(x-3)$ lb for $3 < x < 10$ ft

$M = 16{,}450x$ lb-ft for $0 < x < 3$ ft

$M = 16{,}450x - 20{,}000(x-3) - 500(x-3)^2$ lb-ft for $3 < x < 10$ ft

27.

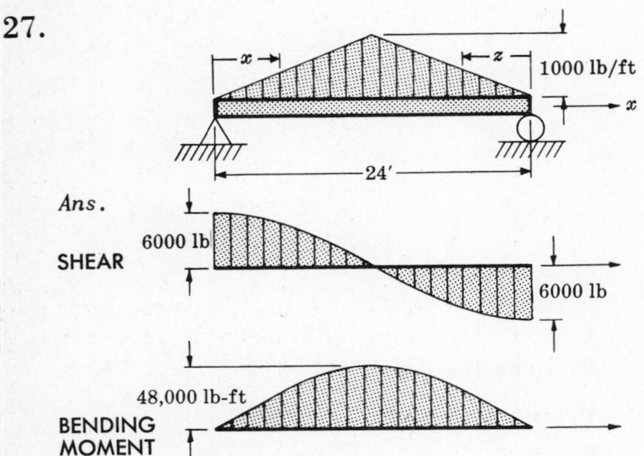

$V = 6000 - \frac{x^2}{24}(1000)$ lb for $0 < x < 12$ ft

$V = -6000 + \frac{z^2}{24}(1000)$ lb for $0 < z < 12$ ft

$M = 6000x - \frac{x^3}{72}(1000)$ lb-ft for $0 < x < 12$ ft

$M = 6000z - \frac{z^3}{72}(1000)$ lb-ft for $0 < z < 12$ ft

For the following two simply supported beams of Problems 28-29 with overhanging ends and loaded as shown, draw the shearing force diagram and also draw the bending moment diagram in parts.

28. 29.

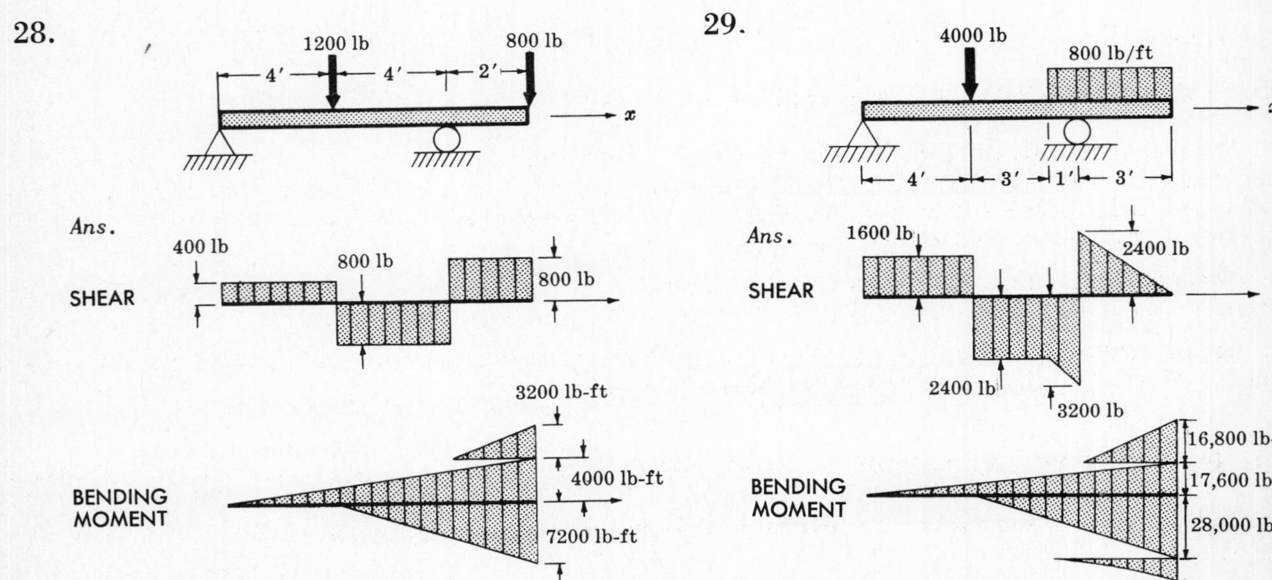

CHAPTER 7

Centroids and Moments of Inertia of Plane Areas

THE FIRST MOMENT OF AN ELEMENT OF AREA about any axis in the plane of the area is given by the product of the area of the element and the perpendicular distance between the element and the axis. For example, in the figure the first moment dQ_x of the element da about the x-axis is given by

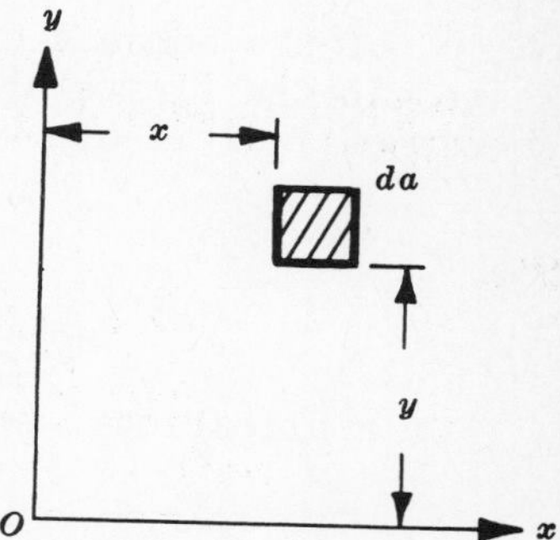

$$dQ_x = y\,da$$

About the y-axis the first moment is

$$dQ_y = x\,da$$

For applications, see Problem 1.

THE FIRST MOMENT OF A FINITE AREA about any axis in the plane of the area is given by the summation of the first moments about that same axis of all the elements of area contained in the finite area. This is frequently evaluated by means of an integral. If the first moment of the finite area is denoted by Q_x, then

$$Q_x = \int dQ_x$$

For applications, see Problems 3, 4, 5, 13, 15.

THE CENTROID OF AN AREA is defined by the equations

$$\bar{x} = \frac{\int x\,da}{A} = \frac{Q_y}{A}, \qquad \bar{y} = \frac{\int y\,da}{A} = \frac{Q_x}{A}$$

where A denotes the area. For applications, see Problems 1-5, 13, 15-17.

The centroid of an area is the point at which the area might be considered to be concentrated and still leave unchanged the first moment of the area about any axis. For example, a thin metal plate will balance in a horizontal plane if it is supported at a point directly under its center of gravity.

The centroids of a few areas are obvious. In a symmetrical figure such as a circle or square, the centroid coincides with the geometric center of the figure.

It is common practice to denote a centroid distance by a bar over the coordinate distance. Thus $\bar{x}$ indicates the x-coordinate of the centroid.

THE SECOND MOMENT, OR MOMENT OF INERTIA, OF AN ELEMENT OF AREA about any axis in the plane of the area is given by the product of the area of the element and the square of the perpendicular distance between the element and the axis. In the above figure, the moment of inertia dI_x of the element about the x-axis is

$$dI_x = y^2\,da$$

About the y-axis the moment of inertia is

$$dI_y = x^2\,da$$

THE SECOND MOMENT, OR MOMENT OF INERTIA, OF A FINITE AREA about any axis in the plane of the area is given by the summation of the moments of inertia about that same axis of all of the elements of area contained in the finite area. This too, is frequently found by means of an integral. If the moment of inertia of the finite area about the x-axis is denoted by I_x then we have

$$I_x = \int dI_x = \int y^2\,da$$

$$I_y = \int dI_y = \int x^2\,da$$

For applications, see Problems 6,8,9,11.

UNITS of moment of inertia are the fourth power of a length, perhaps in^4 or ft^4.

THE PARALLEL AXIS THEOREM states that the moment of inertia of an area about any axis is equal to the moment of inertia about a parallel axis through the centroid of the area plus the product of the area and the square of the perpendicular distance between the two axes. For the area shown below, the axes x_G and y_G pass through the centroid of the plane area. The x and y axes are parallel axes located at distances x_1 and y_1 from the centroidal axes. Let A denote the area of the figure, I_{xG} and I_{yG} the moments of inertia about the axes through the centroid, and I_x and I_y the moments of inertia about the x and y axes. Then we have

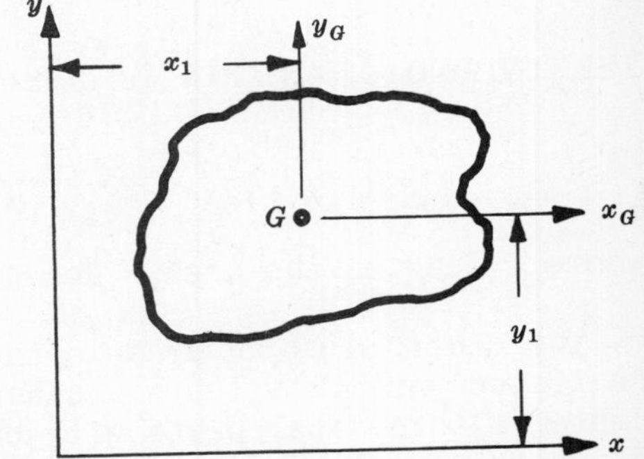

$$I_x = I_{xG} + A(y_1)^2$$

$$I_y = I_{yG} + A(x_1)^2$$

This relation is derived in Problem 7. For applications, see Problems 8,10,13-18.

COMPOSITE AREAS. The moment of inertia of a composite area is the summation of the moments of inertia of the component areas making up the whole. This frequently eliminates the necessity for integration if the area can be broken down into rectangles, triangles, circles, etc., for each of which the moment of inertia is known. See Problems 12,13,15-18.

RADIUS OF GYRATION. If the moment of inertia of an area A about the x-axis is denoted by I_x, then the radius of gyration r_x is defined by

$$r_x = \sqrt{I_x/A}$$

Similarly, the radius of gyration with respect to the y-axis is given by

$$r_y = \sqrt{I_y/A}$$

Since I is in units of length to the fourth power, and A is in units of length to the second power, then the radius of gyration has the units of length, say in. or ft. It is frequently useful for comparative purposes but has no physical significance. See Problems 15, 16.

SOLVED PROBLEMS

1. Locate the centroid of a triangle.

Let us introduce the coordinate system shown.

The y-coordinate of the centroid is defined by the equation

$$\bar{y} = \frac{\int y\,da}{A}$$

It is simplest to choose an element such that y is constant for all points in the element. The horizontal shaded area satisfies this condition and the area da of the element is $s\,dy$. Thus

$$\bar{y} = \frac{\int y\,s\,dy}{A}$$

The product $y\,s\,dy$ represents the first moment of the shaded element about the x-axis.

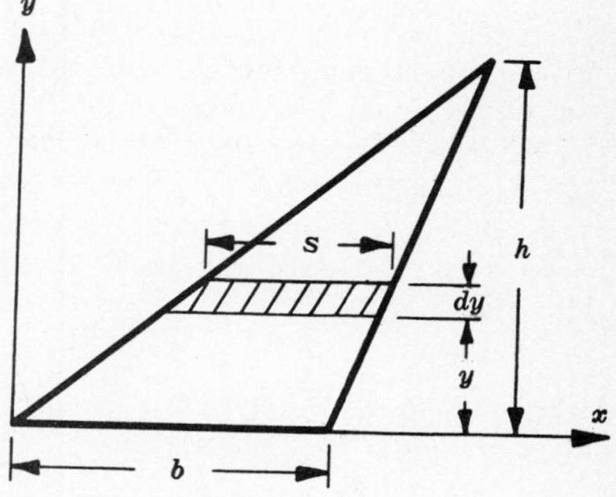

From similar triangles, $\frac{s}{b} = \frac{h-y}{h}$. Substituting this value of s in the above integral,

$$\bar{y} = \frac{\int_0^h y\frac{b}{h}(h-y)\,dy}{\frac{1}{2}bh} = \frac{2}{h^2}\int_0^h (hy-y^2)\,dy$$

$$= \frac{2}{h^2}\left\{h[y^2/2]_0^h - [y^3/3]_0^h\right\} = \frac{2}{h^2}\left(\frac{h^3}{2}-\frac{h^3}{3}\right) = \frac{1}{3}h$$

Note that the altitude h is measured perpendicular to the base of length b.

2. Locate the centroid of a semi-circle.

The polar coordinate system shown will be a logical choice for such a contour.

The shaded element of area is approximately a rectangle and its area is given by $\rho\,d\theta\,d\rho$. The y-coordinate of the centroid is given by the equation

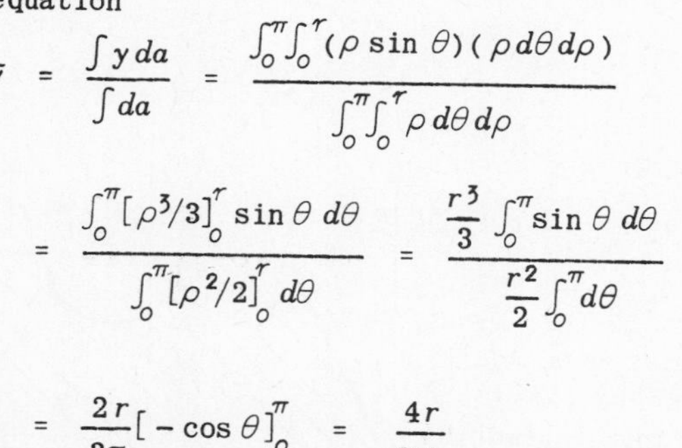

$$\bar{y} = \frac{\int y\,da}{\int da} = \frac{\int_0^\pi\int_0^r(\rho\sin\theta)(\rho\,d\theta\,d\rho)}{\int_0^\pi\int_0^r \rho\,d\theta\,d\rho}$$

$$= \frac{\int_0^\pi[\rho^3/3]_0^r\sin\theta\,d\theta}{\int_0^\pi[\rho^2/2]_0^r\,d\theta} = \frac{\frac{r^3}{3}\int_0^\pi\sin\theta\,d\theta}{\frac{r^2}{2}\int_0^\pi d\theta}$$

$$= \frac{2r}{3\pi}[-\cos\theta]_0^\pi = \frac{4r}{3\pi}$$

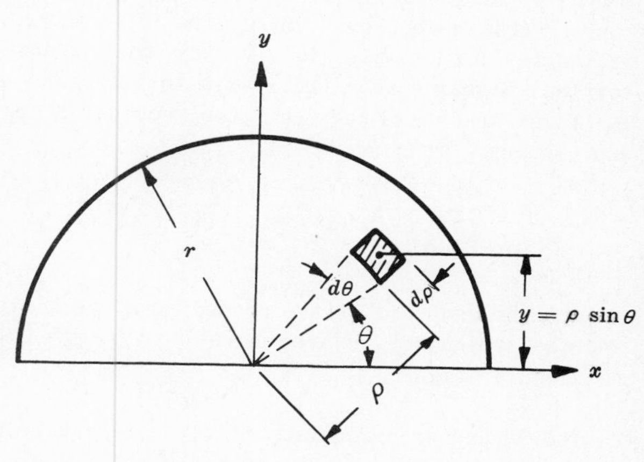

3. Locate the centroid of the shaded area remaining after the semi-circle of radius 2 in. has been removed from the semi-circular area of radius 5 in.

In this case there is no need to integrate. The shaded area may be regarded as consisting of the difference of areas between the 5 in. semi-circle and the 2 in. semi-circle. The y-coordinate of the centroid of the shaded area is given by

$$\bar{y} = \frac{\int y\,da}{A}$$

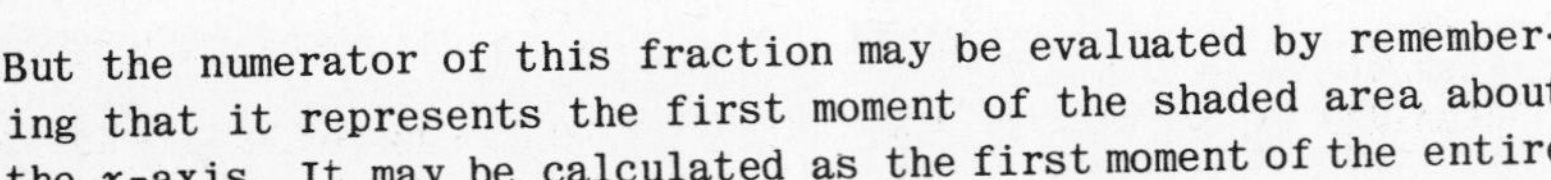

But the numerator of this fraction may be evaluated by remembering that it represents the first moment of the shaded area about the x-axis. It may be calculated as the first moment of the entire 5 in. semi-circular area minus the first moment of the 2 in. semi-circular area about the x-axis. The first moment of the 5 in. semi-circular area about the x-axis is given by the product of its area and the vertical distance from the x-axis to the centroid of this area. Similarly for the first moment of the 2 in. semi-circular area. The location of the centroid of each of these areas was found in Problem 2. Thus

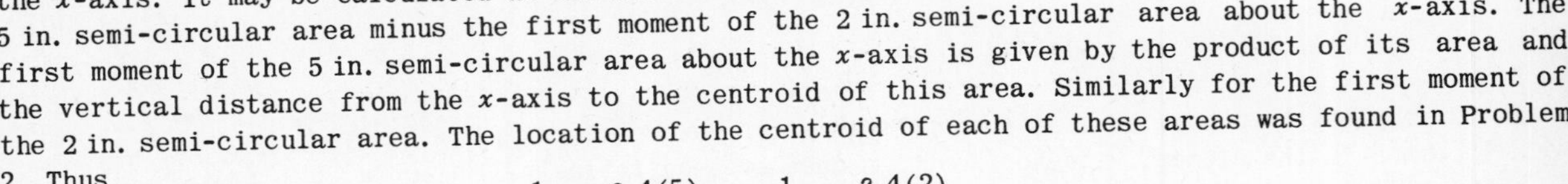

$$\bar{y} = \frac{\frac{1}{2}\pi(5)^2\,\frac{4(5)}{3\pi} - \frac{1}{2}\pi(2)^2\,\frac{4(2)}{3\pi}}{\frac{1}{2}\pi(5)^2 - \frac{1}{2}\pi(2)^2} = 2.36 \text{ in.}$$

From symmetry this point lies on the y-axis.

4. Locate the centroid of the shaded area remaining after the two rectangles shown have been removed from the original rectangle.

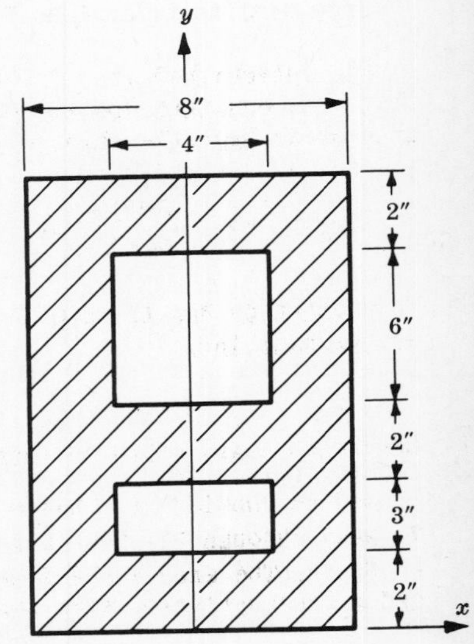

The y-axis is selected as the vertical axis of symmetry and the x-axis is taken to coincide with the base of the figure. The y-coordinate of the centroid is given by $\bar{y} = \frac{\int y\,da}{A}$ and because of symmetry it lies on the y-axis.

The shaded area may be regarded as consisting of the original rectangle 8 in. × 15 in. minus the two smaller rectangles. Integration is not necessary in this case because the location of the centroid of each of the three rectangles is known from symmetry. The numerator of the above fraction represents the first moment of the shaded area about the x-axis and it may be calculated as the first moment of the 8 in. × 15 in. rectangle minus the first moment of each of the two rectangles that have been removed. The first moment of the 8 in.×15 in. rectangle, for example, is given by the product of its area by the vertical distance from the x-axis to its centroid, which is 7.5 in. The first moments of the other two rectangles are given by analogous expressions. Thus

$$\bar{y} = \frac{(8)(15)(7.5) - (4)(3)(3.5) - (4)(6)(10)}{(8)(15) - (4)(3) - (4)(6)} = 7.35 \text{ in.}$$

5. Locate the centroid of the shaded area remaining after one corner and the semi-circular area have been removed from the originally rectangular figure.

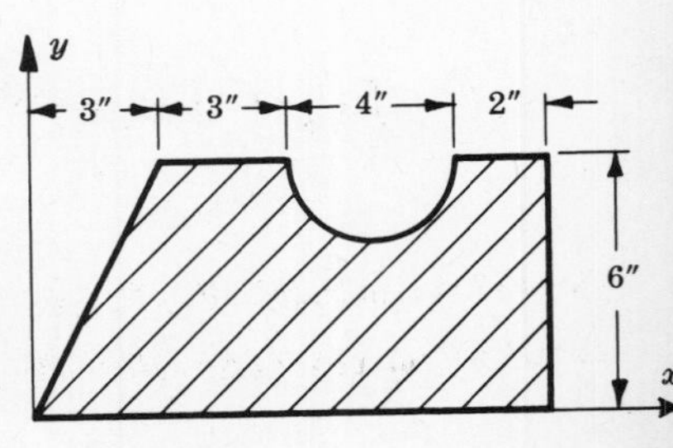

The shaded area consists of (1) a rectangle 6 in. × 12 in., minus (2) a triangle 6 in. × 3 in., minus (3) a semi-circular area. Again, since the centroids of (2) and (3) were determined in Problems 1 and 2 respectively, integration is not necessary and a finite summation may be used.

The y-coordinate of the centroid is given by $\bar{y} = \dfrac{\int y\,da}{A}$. The numerator, representing the first moment of the shaded area about the x-axis, may be evaluated as the first moment of the rectangle, minus that of the triangle, minus that of the semicircle. Thus

$$\bar{y} = \frac{(12)(6)(3) - \frac{1}{2}(3)(6)(4) - \frac{1}{2}\pi(2)^2[6 - \frac{4(2)}{3\pi}]}{(12)(6) - \frac{1}{2}(3)(6) - \frac{1}{2}\pi(2)^2} = 2.60 \text{ in.}$$

Similarly, the x-coordinate of the centroid may be located by $\bar{x} = \dfrac{\int x\,da}{A}$. The numerator here represents the first moment of the rectangle, minus that of the triangle, minus that of the semicircle about the y-axis. Thus

$$\bar{x} = \frac{(12)(6)(6) - \frac{1}{2}(3)(6)(1) - \frac{1}{2}\pi(2)^2(8)}{(12)(6) - \frac{1}{2}(3)(6) - \frac{1}{2}\pi(2)^2} = 6.58 \text{ in.}$$

6. Determine the moment of inertia of a rectangle about an axis through the centroid and parallel to the base.

Let us introduce the coordinate system shown. The moment of inertia I_{xG} about the x-axis passing through the centroid is given by

$$I_{xG} = \int y^2 da$$

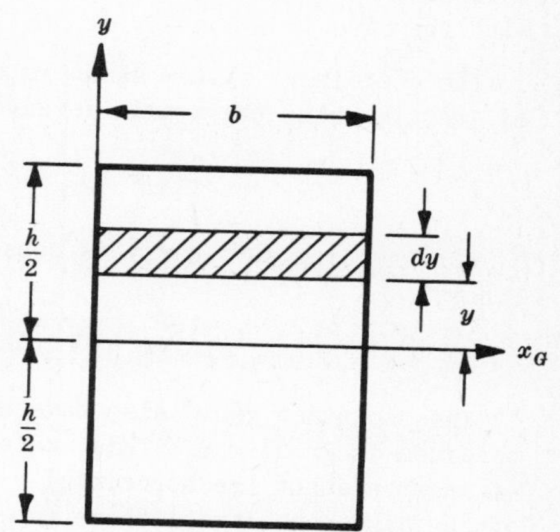

For convenience it is logical to select an element such that y is constant for all points in the element. The shaded area shown has this characteristic.

$$I_{xG} = \int_{-h/2}^{h/2} y^2 b\,dy = b\left[\frac{y^3}{3}\right]_{-h/2}^{h/2} = \frac{1}{12}bh^3$$

This quantity has the dimension of a length to the fourth power, perhaps in^4.

7. Derive the parallel-axis theorem for plane areas.

Let us consider the plane area A shown. The axes x_G and y_G pass through its centroid, whose location is presumed to be known. The axes x and y are located at known distances y_1 and x_1 respectively, from the axes through the centroid.

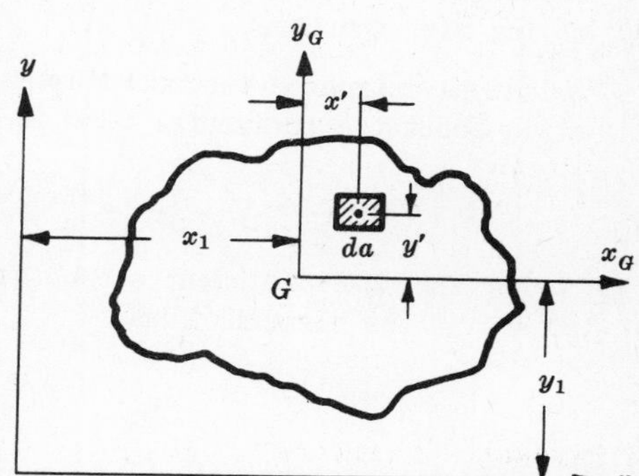

For the element of area da the moment of inertia about the x-axis is given by

$$dI_x = (y_1 + y')^2 da$$

For the entire area A the moment of inertia about the x-axis is

$$I_x = \int dI_x = \int (y_1 + y')^2 da$$

$$= \int (y_1)^2 da + 2\int y_1 y' da + \int (y')^2 da$$

The first integral on the right is equal to $y_1^2 \int da = y_1^2 A$ because y_1 is a constant. The second integral on the right is equal to $2y_1 \int y' da = 2y_1(0) = 0$ because the axis from which y' is measured passes through the centroid of the area. The third integral on the right is equal to I_{xG}, i.e. the moment of inertia of the area about the horizontal axis through the centroid. Thus

$$I_x = I_{xG} + A(y_1)^2$$

A similar consideration in the other direction would show that

$$I_y = I_{yG} + A(x_1)^2$$

This is the parallel-axis theorem for plane areas. It is to be noted that one of the axes involved in each equation must pass through the centroid of the area. In words, this may be stated as follows: The moment of inertia of an area with reference to an axis not through the centroid of the area is equal to the moment of inertia about a parallel axis through the centroid of the area plus the product of the same area and the square of the distance between the two axes.

The moment of inertia always has a positive value, with a minimum value for axes through the centroid of the area in question.

8. Determine the moment of inertia of a rectangle about an axis coinciding with the base.

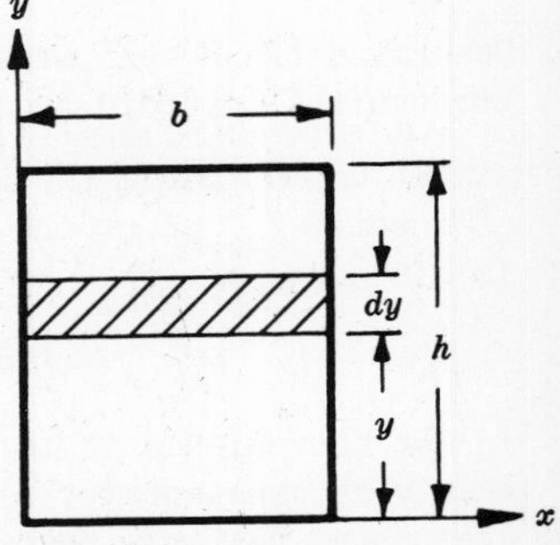

The coordinate system shown is convenient. By definition the moment of inertia about the x-axis is given by

$$I_x = \int y^2 da$$

For the element shown y is constant for all points in the element. Hence

$$I_x = \int_0^h y^2 b\, dy = b[y^3/3]_0^h = \frac{1}{3} b h^3$$

This solution could also have been obtained by applying the parallel axis theorem to the result obtained in Problem 6. This states that the moment of inertia about the base is equal to the moment of inertia about the horizontal axis through the centroid plus the product of the area and the square of the distance between these two axes. Thus

$$I_x = I_{xG} + A(y_1)^2 = \frac{1}{12} b h^3 + b h (\frac{h}{2})^2 = \frac{1}{3} b h^3$$

9. Determine the moment of inertia of a triangle about an axis coinciding with the base.

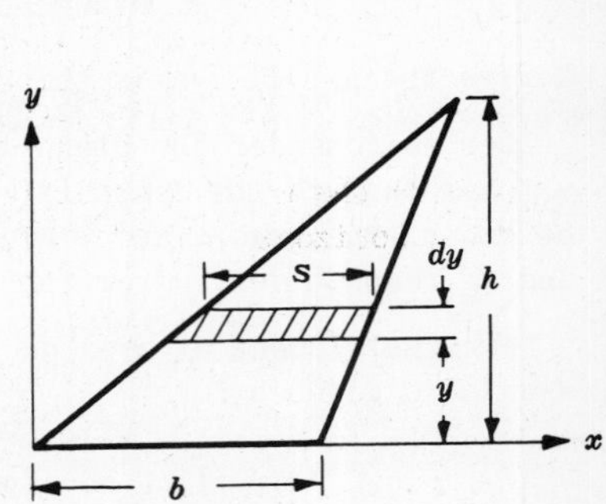

Let us introduce the coordinate system shown. The moment of inertia about the horizontal base is

$$I_x = \int y^2 da$$

For the shaded element shown the quantity y is constant for all points in the element. Thus

$$I_x = \int_0^h y^2 s\, dy$$

From the similar triangles involved, we have

$$\frac{s}{b} = \frac{h-y}{h}$$

Substituting this value of s in the integral, we find

$$I_x = \int_0^h y^2 \frac{b}{h}(h-y)dy = \frac{b}{h}[h\int_0^h y^2 dy - \int_0^h y^3 dy] = \frac{1}{12} b h^3$$

10. Determine the moment of inertia of a triangle about an axis through the centroid and parallel to the base.

Let the x_G-axis pass through the centroid and take the x-axis to coincide with the base as shown.

From Problem 1 the x_G-axis is located a distance of $h/3$ above the base. Also, the parallel axis theorem tells us that

$$I_x = I_{xG} + A(y_1)^2$$

But I_x was determined in Problem 9, and A and y_1 $(= h/3)$ are known. Hence we may solve for the desired unknown, I_{xG}. Substituting,

$$\frac{1}{12}bh^3 = I_{xG} + \frac{1}{2}bh(\frac{h}{3})^2 \qquad \text{or} \qquad I_{xG} = \frac{1}{36}bh^3$$

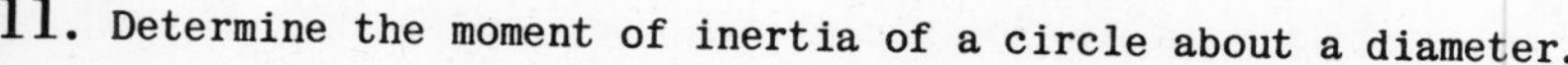

11. Determine the moment of inertia of a circle about a diameter.

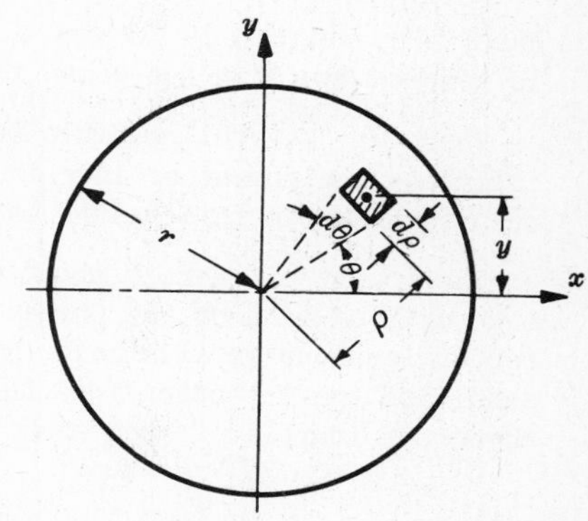

Let us select the shaded element of area shown and work with the polar coordinate system. The radius of the circle is r.

To find I_x we have the definition $I_x = \int y^2 da$

But $y = \rho \sin\theta$ and $da = \rho\, d\theta\, d\rho$. Hence

$$I_x = \int_0^{2\pi}\int_0^r \rho^2 \sin^2\theta \; \rho \; d\theta \, d\rho$$

$$= \int_0^{2\pi} \sin^2\theta \, d\theta \, [\tfrac{1}{4}\rho^4]_0^r = \frac{r^4}{4}\int_0^{2\pi}\sin^2\theta \, d\theta = \frac{\pi r^4}{4}$$

If D denotes the diameter of the circle, then $D = 2r$ and $I_x = \dfrac{\pi D^4}{64}$. This is half the value of the polar moment of inertia of a solid circular area.

Hence the moment of inertia of a semicircular area about an axis coinciding with its base is

$$I_x = \frac{1}{2}\,\frac{\pi D^4}{64} = \frac{\pi D^4}{128}$$

12. Determine the moment of inertia of the hollow rectangular area about a horizontal axis through the centroid.

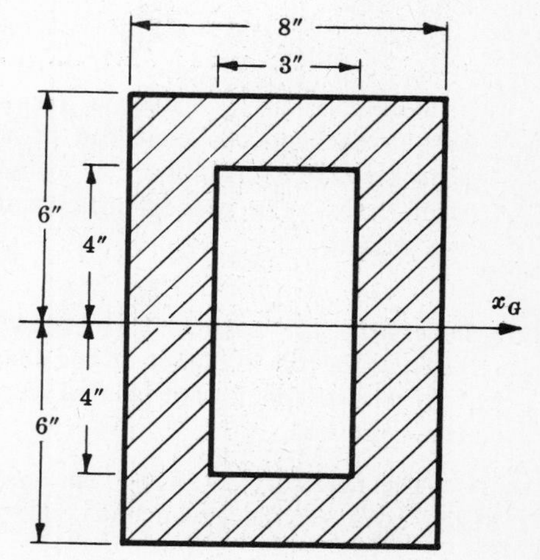

The x_G-axis passes through the centroid of the figure. Perhaps the simplest technique is to compute the moment of inertia of the large 8 in. × 12 in. rectangle about the x_G-axis and from that subtract the moment of inertia about the same axis of the 3 in. × 8 in. rectangle.

From Problem 6, the moment of inertia of a rectangle about an axis through the centroid and parallel to the base is given by

$$I_{xG} = bh^3/12$$

Thus for the shaded area we have

$$I_{xG} = \frac{1}{12}(8)(12)^3 - \frac{1}{12}(3)(8)^3 = 1024 \text{ in}^4$$

13. Determine the moment of inertia of the T-section shown about a horizontal axis passing through the centroid.

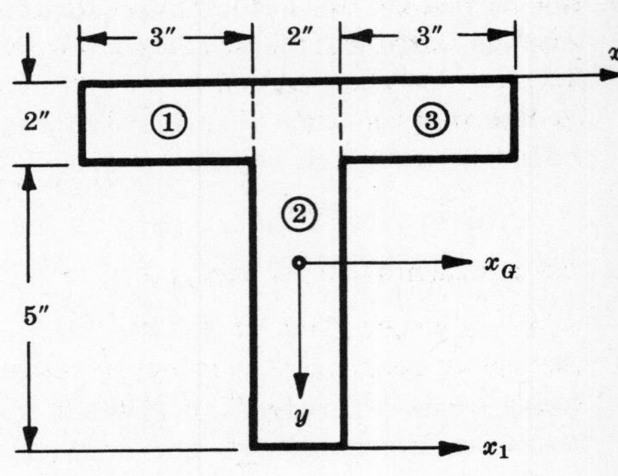

It is first necessary to locate the centroid of the area. To do this, we introduce the x-y coordinate system shown. By definition, the y-coordinate of the centroid is given by

$$\bar{y} = \frac{\int y\,da}{A}$$

The numerator of this expression represents the first moment of the entire area about the x-axis. This may be calculated by multiplying the area of each of the three component rectangles 1, 2, and 3 by the distance from the x-axis to the centroid of the particular rectangle. Thus

$$\bar{y} = \frac{(3)(2)(1) + (7)(2)(3.5) + (3)(2)(1)}{(3)(2) + (7)(2) + (3)(2)} = +2.35 \text{ in.}$$

Hence the centroid is located 2.35 in. below the x-axis. The horizontal axis passing through this point is denoted by x_G in the above figure.

There are several possible techniques for determining the required moment of inertia. One is to calculate the moment of inertia of the entire area about the x-axis, then use the parallel axis theorem to transfer this result to the x_G axis.

The moment of inertia about the x-axis is found as the sum of the moments of inertia about this same axis of each of the three component rectangles. The expression for the moment of inertia of a rectangle about an axis coinciding with its base was derived in Problem 8. Note that it is easiest to subdivide the T-section into the three rectangles shown, rather than in any other manner, because the moment of inertia of each about the x-axis is known from Problem 8. Thus

$$I_x = (1/3)(3)(2)^3 + (1/3)(2)(7)^3 + (1/3)(3)(2)^3 = 245 \text{ in}^4$$

The parallel axis theorem may now be used to find the moment of inertia of the entire figure about the x_G-axis. Thus

$$I_x = I_{xG} + A(y_1)^2, \quad 245 = I_{xG} + 26(2.35)^2 \quad \text{and} \quad I_{xG} = 101 \text{ in}^4$$

14. Determine the moment of inertia of the T-section of Problem 13 about a horizontal axis x_1 through its lower extremity.

This axis is located $(7-2.35) = 4.65$ in. below the horizontal axis through the centroid. The parallel axis theorem may be used to transfer the known moment of inertia from the x_G-axis to the x_1-axis. Thus

$$I_{x_1} = I_{xG} + A(y_1)^2 = 101 + 26(4.65)^2 = 664 \text{ in}^4$$

Note carefully that the parallel axis theorem can only be used if one of the two axes concerned passes through the centroid of the area. For example, it is *not* permissible to transfer from the x-axis to the x_1-axis merely by adding the product of the area and the square of the distance between these axes. The reason this is not valid is that neither of these axes passes through the centroid of the figure.

15. Determine the moment of inertia and also the radius of gyration of the channel section shown about a horizontal axis through the centroid.

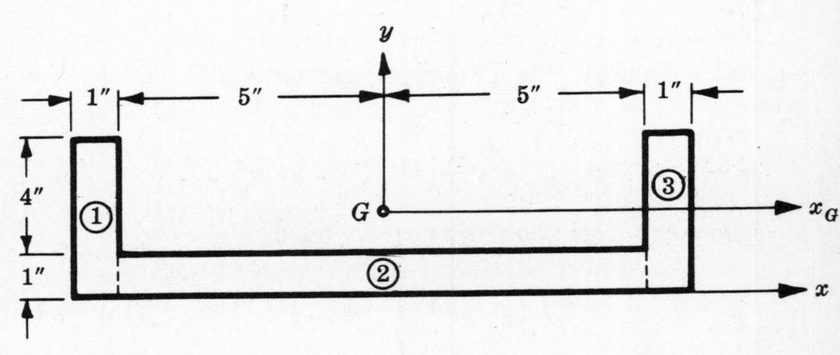

The centroid lies on the y-axis and its location is given by

$$\bar{y} = \frac{\int y\,da}{A}$$

The numerator of this expression represents the first moment of the area about the x-axis. The entire area is composed of the three component rectangles shown. The first moment of each of these rectangles about the x-axis is given by the product of its area and the perpendicular distance from its centroid to the x-axis. Thus

$$\bar{y} = \frac{(1)(5)(2.5) + (10)(1)(0.5) + (1)(5)(2.5)}{(1)(5) + (10)(1) + (1)(5)} = 1.5 \text{ in.}$$

The horizontal axis passing through the centroid is denoted by x_G in the above figure.

It is convenient to first determine the moment of inertia with respect to the x-axis. For each of the three component rectangles the moment of inertia about an axis through its base was found in Problem 8 to be $I_x = bh^3/3$. For the entire figure,

$$I_x = \frac{1}{3}(1)(5)^3 + \frac{1}{3}(10)(1)^3 + \frac{1}{3}(1)(5)^3 = 86.6 \text{ in}^4$$

From the parallel axis theorem,

$$I_x = I_{xG} + A(y_1)^2, \quad 86.6 = I_{xG} + 20(1.5)^2 \quad \text{and} \quad I_{xG} = 41.6 \text{ in}^4$$

The radius of gyration with respect to the x_G-axis is $r_{xG} = \sqrt{I_{xG}/A} = \sqrt{41.6/20} = 1.45$ in.

16. Determine the moment of inertia and also the radius of gyration of the I-section shown about a horizontal axis passing through the centroid.

To locate the centroid, which lies on the y-axis, we have

$$\bar{y} = \frac{\int y\,da}{A}$$

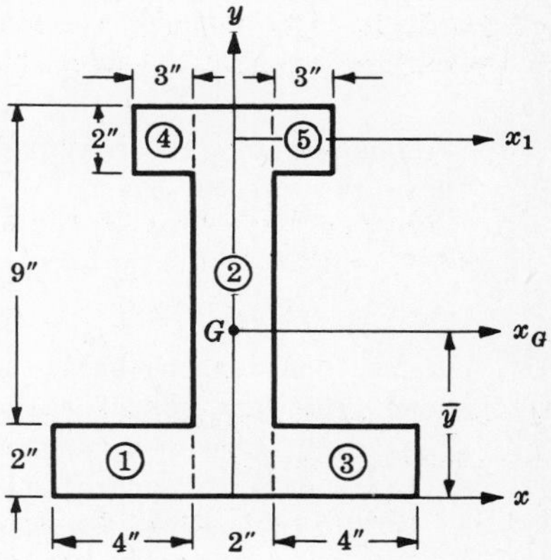

The entire section is divided into the five component rectangles shown and the numerator of the above fraction may then be evaluated by a numerical summation. Thus

$$\bar{y} = \frac{4(2)(1) + 11(2)(5.5) + 4(2)(1) + 3(2)(10) + 3(2)(10)}{(4)(2) + (11)(2) + (4)(2) + (3)(2) + (3)(2)}$$

$$= 5.14 \text{ in.}$$

The horizontal axis passing through the centroid is denoted by x_G in the figure.

We shall first determine the moment of inertia with respect to the x-axis. For the rectangles 1, 2, and 3 the moment of inertia about this axis is given by

$$I_x = \frac{1}{3}bh^3$$

For the rectangles 4 and 5 it is first necessary to determine the moment of inertia about a horizontal axis x_1 passing through the centroid of these rectangles, then apply the parallel axis theorem to transfer this result to the x-axis.

For the entire figure we thus have

$$I = (1/3)(4)(2)^3 + (1/3)(2)(11)^3 + (1/3)(4)(2)^3 + [(1/12)(3)(2)^3 + (3)(2)(10)^2]2$$

$$= 2113 \text{ in}^4$$

From the parallel axis theorem, $I_x = I_{xG} + A(y_1)^2$, $2113 = I_{xG} + 50(5.14)^2$ and $I_{xG} = 793$ in^4.

The radius of gyration with respect to the x_G axis is

$$r_{xG} = \sqrt{I_{xG}/A} = \sqrt{793/50} = 3.99 \text{ in.}$$

17. Determine the moment of inertia of the hollow rectangular area about a horizontal axis through its centroid.

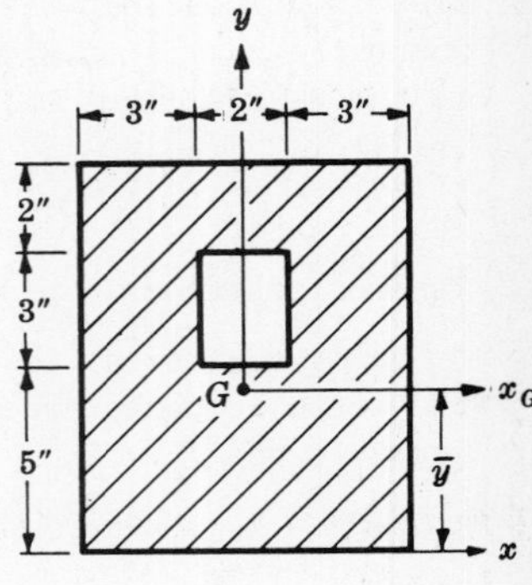

The centroid lies on the y-axis and its location is given by

$$\bar{y} = \frac{\int y\,da}{A}$$

The numerator may be evaluated as the first moment of the entire 8 in. × 10 in. rectangle about the x-axis minus the first moment of the 2 in. × 3 in. rectangle that has been removed. Thus

$$\bar{y} = \frac{(8)(10)(5) - (2)(3)(6.5)}{(8)(10) - (2)(3)} = 4.88 \text{ in.}$$

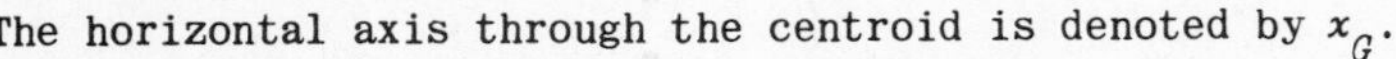

The horizontal axis through the centroid is denoted by x_G.

We shall first compute the moment of inertia of the entire 8 in. × 10 in. rectangle about the x_G-axis. This is done by finding its moment of inertia about a horizontal axis through the centroid of this rectangle (assuming that the 2 in.×3 in. hole is not present), then transferring this result to the x_G-axis. For the entire 8 in.×10 in. rectangle this application of the parallel axis theorem gives

$$I'_{xG} = (1/12)(8)(10)^3 + (8)(10)(5 - 4.88)^2 = 668 \text{ in}^4$$

Similarly for the 2 in.×3 in. rectangle that has been removed, the moment of inertia of it with respect to the x_G-axis is found by computing the moment of inertia with respect to a horizontal axis through the centroid of this 2 in.×3 in. area then transferring this result to the x_G-axis. This yields

$$I''_{xG} = (1/12)(2)(3)^3 + (2)(3)(6.5 - 4.88)^2 = 20.3 \text{ in}^4$$

Consequently the moment of inertia of the hollow rectangular area is given by the difference of these two values. Thus

$$I_x = 668 - 20.3 = 647.7 \text{ in.}^4$$

18. Let us consider the built-up I-beam with the cross-section shown. The four structural angles are identical and from a steel manufacturer's handbook the moment of inertia of each angle about a horizontal axis through its centroid is known to be 6.7 in^4 and the cross-sectional area is given as 4.61 in^2. The centroid of the angle is located 1.23 in. from the 4 in. face of the angle. Determine the moment of inertia of the entire cross-sectional area about a horizontal axis through the centroid of the area.

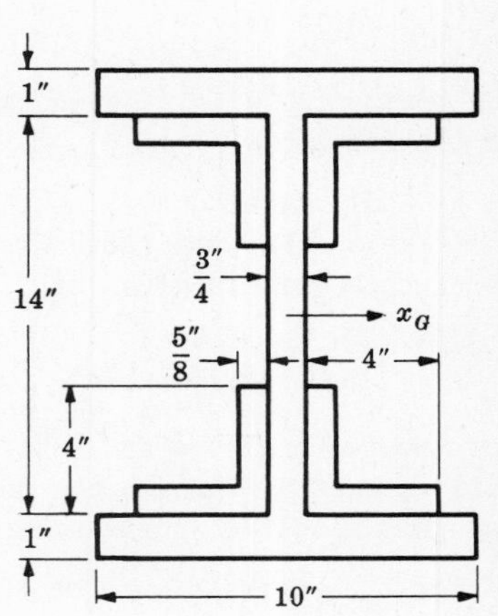

The horizontal axis x_G through the centroid is an axis of symmetry. The moment of inertia of the vertical web ($\frac{3}{4}$ in.×14 in.) about the x_G-axis is

$$I'_{xG} = (1/12)bh^3 = (1/12)(3/4)(14)^3 = 171 \text{ in}^4$$

The moment of inertia of each angle about the x_G-axis is equal to the moment of inertia about the horizontal axis through the centroid of the angle only plus the product of the area of the angle and the square of the distance between these two axes. Thus according to the parallel axis theorem, for all four angles we have

$$I''_{xG} = 4[6.7 + 4.61(7 - 1.23)^2] = 640 \text{ in}^4$$

The moment of inertia of the horizontal flanges, each 1 in.×10 in., about the x_G-axis may also be found by the parallel axis theorem. Thus

$$I'''_{xG} = 2[(1/12)(10)(1)^3 + (10)(1)(7.5)^2] = 1127 \text{ in}^4$$

Consequently the moment of inertia of the entire area about the x_G-axis is

$$I_x = 171 + 640 + 1127 = 1938 \text{ in}^4$$

This result neglects the effect of any welds or rivets.

SUPPLEMENTARY PROBLEMS

19. Locate the centroid of the shaded area shown in Fig. (*a*) below where the rectangular portion has been removed from the semicircle. *Ans.* $\bar{x} = 0$, $\bar{y} = 2.80$ in.

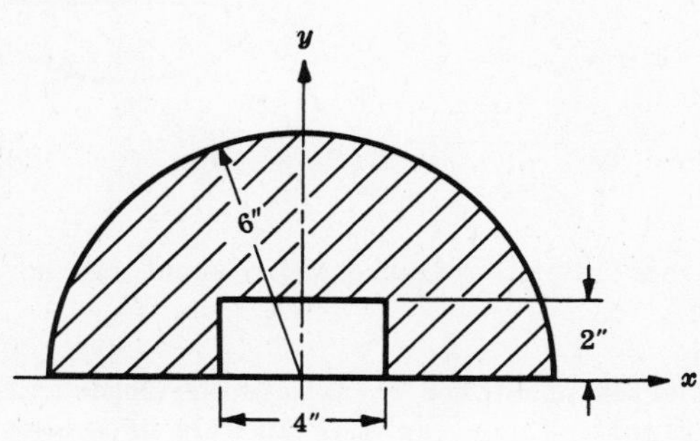

Fig. (*a*) Prob. 19

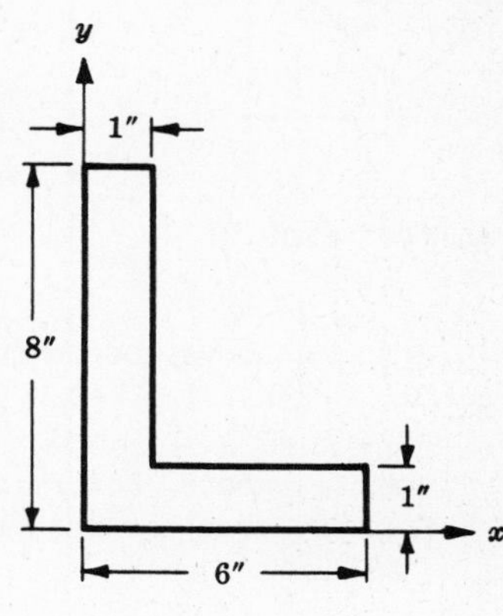

Fig. (*b*) Prob. 20

20. Find the centroid of the angle-section shown in Fig. (*b*) above. *Ans.* $\bar{x} = 1.65$ in., $\bar{y} = 2.65$ in.

21. Locate the centroid of the shaded area shown in Fig. (*c*) below. *Ans.* $\bar{x} = 0$, $\bar{y} = 4.19$ in.

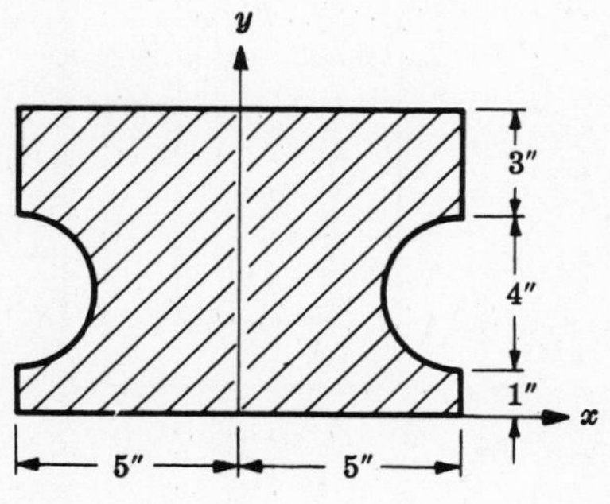

Fig. (*c*) Prob. 21

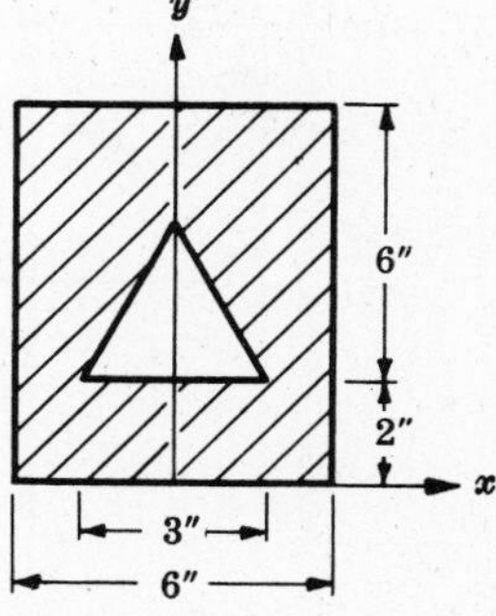

Fig. (*d*) Prob. 22

22. Locate the centroid of the shaded area remaining after the equilateral triangle has been removed from the rectangle as shown in Fig. (*d*) above. *Ans.* $\bar{x} = 0$, $\bar{y} = 4.10$ in.

23. Determine the moment of inertia of a rectangle having a base of 3 in. and a height of 8 in. about an axis through its centroid and parallel to the base. *Ans.* $I_{xG} = 128 \text{ in}^4$

24. Determine the moment of inertia of an equilateral triangle 6 in. on a side about an axis through its centroid and parallel to the base. *Ans.* $I_{xG} = 23.4 \text{ in}^4$

25. Determine the moment of inertia of a circle of diameter 5 in. about a diameter. *Ans.* $I_{xG} = 30.7\ \text{in}^4$

26. Determine the moment of inertia of one quadrant of a circle of radius 2 in. about a diameter coinciding with one side of the quadrant. *Ans.* $I = 3.14\ \text{in}^4$

27. Determine the moment of inertia of the diamond-shaped figure shown in Fig. (*a*) below with respect to the horizontal axis of symmetry. *Ans.* $I_{xG} = 85.4\ \text{in}^4$

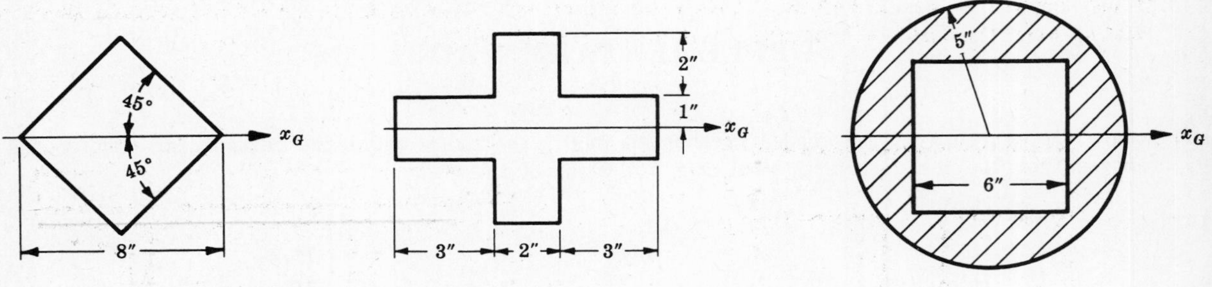

Fig. (*a*) Prob. 27 Fig. (*b*) Prob. 28 Fig. (*c*) Prob. 29

28. Refer to Fig. (*b*) above. Determine the moment of inertia of the figure shown about the horizontal axis of symmetry. *Ans.* $I_{xG} = 40\ \text{in}^4$

29. Refer to Fig. (*c*) above. Determine the moment of inertia about the x_G-axis of the shaded area remaining after the square area has been removed from the circle. The x_G-axis is an axis of symmetry. *Ans.* $I_{xG} = 383\ \text{in}^4$

30. Refer to Fig. (*d*) below. Determine the moment of inertia of the wide-flange section shown about a horizontal axis of symmetry. Also, determine the radius of gyration about this same axis. *Ans.* $I_{xG} = 1033\ \text{in}^4$, $r_{xG} = 3.74$ in.

31. Determine the moment of inertia of the wide-flange section shown in Problem 30 about the vertical axis of symmetry. *Ans.* $I_{yG} = 2775\ \text{in}^4$

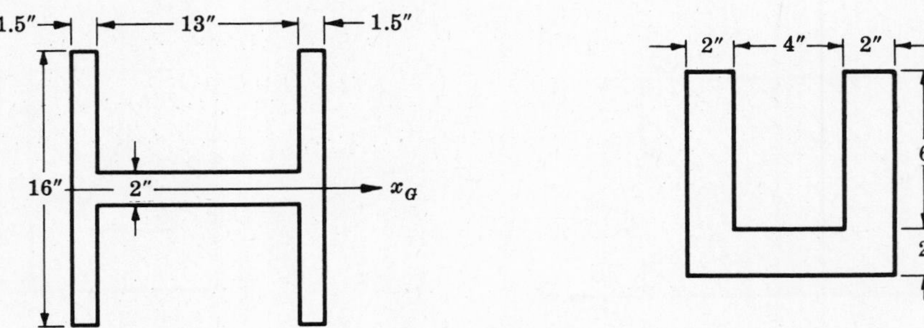

Fig. (*d*) Prob. 30 Fig. (*e*) Prob. 32

32. Determine the moment of inertia of the channel-type section about a horizontal axis through the centroid. Refer to Fig. (*e*) above. What is the radius of gyration about this same axis? *Ans.* $I_{xG} = 231\ \text{in}^4$, $r_{xG} = 2.40$ in.

33. Locate the centroid of the channel-type section shown below and determine the moment of inertia of the cross-sectional area about a horizontal axis through the centroid. *Ans.* $\bar{y} = 1.53$ in., $I_{xG} = 84\ \text{in}^4$

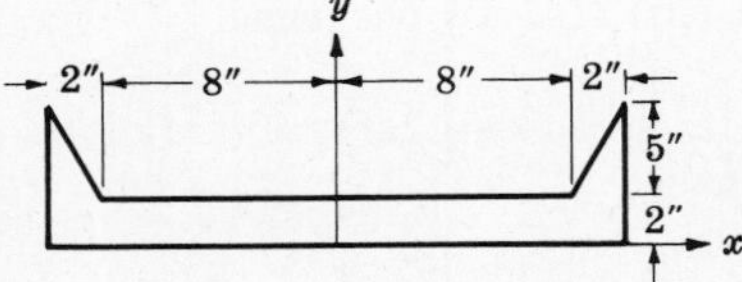

34. Consider the T-section shown in Fig. (*a*) below. Determine the width b so that the centroid lies along the lower edge of the flange, i.e. along the line a-a. For this value of b, calculate the moment of inertia about the horizontal axis through the centroid. *Ans.* b = 2.5 in., I_{xG} = 80.0 in^4

35. Built-up sections of the type shown in Fig. (*b*) below are frequently used for upper chords of bridge trusses. The section shown consists of two channels and a cover plate. The channels each have a cross-sectional area of 5.86 in^2 and a moment of inertia about a horizontal axis through the centroid of the channel alone of 78.5 in^4. Determine the moment of inertia of the entire section about a horizontal axis through its centroid. *Ans.* I_{xG} = 296 in^4

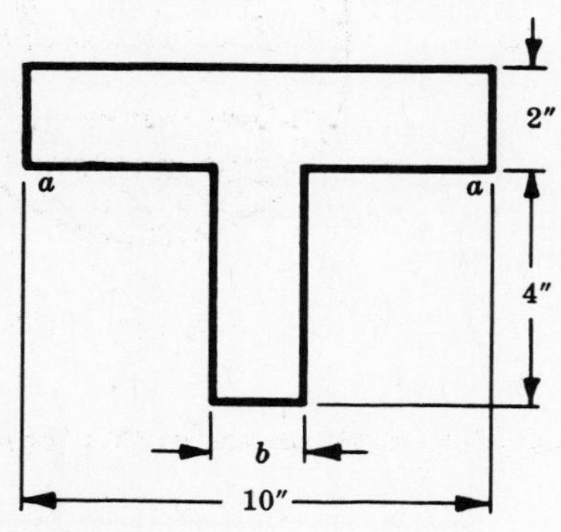

Fig. (*a*) Prob. 34

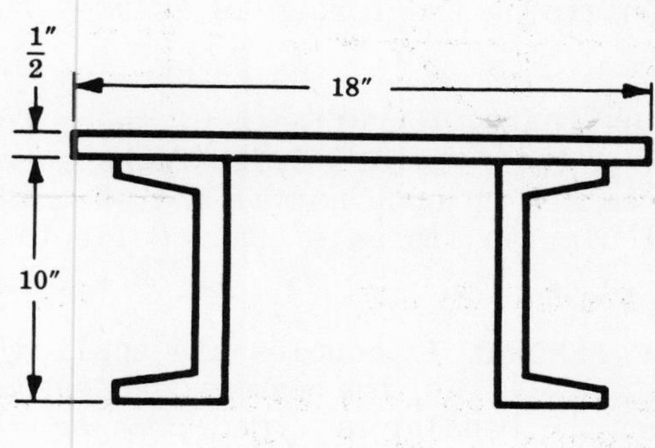

Fig. (*b*) Prob. 35

CHAPTER 8

Stresses in Beams

TYPES OF LOADS ACTING ON BEAMS. Either forces or couples that lie in a plane containing the longitudinal axis of the beam may act upon the member. The forces are understood to act perpendicular to the longitudinal axis, and the plane containing the forces is assumed to be a plane of symmetry of the beam.

EFFECTS OF LOADS. The effects of these forces and couples acting on a beam are (*a*) to impart deflections perpendicular to the longitudinal axis of the bar, and (*b*) to set up both normal and shearing stresses on any cross-section of the beam perpendicular to its axis. Beam deflections will be considered in Chapters 9 and 10.

TYPES OF BENDING. If couples are applied to the ends of the beam and no forces act on the bar, then the bending is termed *pure bending*. For example, in the beam shown the portion of the beam between the two downward forces is subject to pure bending. Bending produced by forces that do not form couples is called *ordinary bending*. A beam subject to pure bending has only normal stresses with no shearing stresses set up in it; a beam subject to ordinary bending has both normal and shearing stresses acting within it.

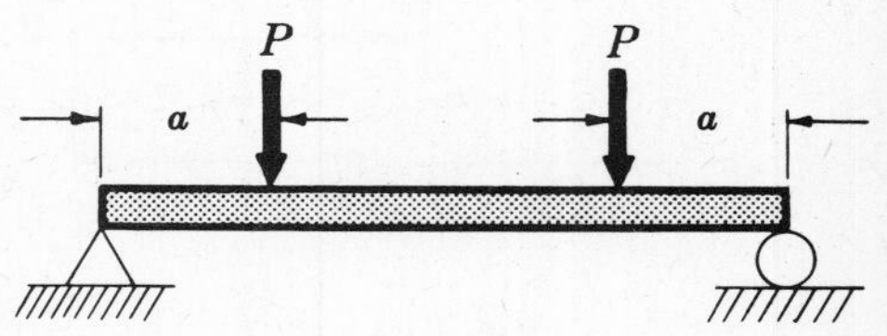

NATURE OF BEAM ACTION. It is convenient to imagine a beam to be composed of an infinite number of thin longitudinal rods or fibers. Each longitudinal fiber is assumed to act independently from every other fiber, i.e. there are no lateral pressures or shearing stresses between the fibers. The beam shown above, for example, will deflect downward and the fibers in the lower part of the beam undergo extension, while those in the upper part are shortened. These changes in the lengths of the fibers set up stresses in the fibers. Those that are extended have tensile stresses acting on the fibers in the direction of the longitudinal axis of the beam, while those that are shortened are subject to compressive stresses.

NEUTRAL SURFACE. There always exists one surface in the beam containing fibers that do not undergo any extension or compression, and thus are not subject to any tensile or compressive stress. This surface is called the neutral surface of the beam.

NEUTRAL AXIS. The intersection of the neutral surface with any cross-section of the beam perpendicular to its longitudinal axis is called the neutral axis. All fibers on one side of the neutral axis are in a state of tension, while those on the opposite side are in compression.

BENDING MOMENT. The algebraic sum of the moments of the external forces to one side of any cross-section of the beam about an axis through that section is called the bending moment at that section. This concept was discussed in Chapter 6.

NORMAL STRESSES IN BEAMS. For any beam having a longitudinal plane of symmetry and subject to a bending moment M at a certain cross-section, the normal stress acting on a longitudinal fiber at a distance y from the neutral axis of the beam is given by

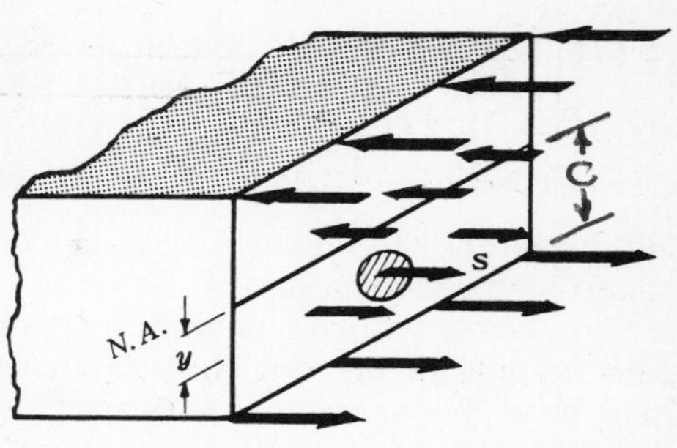

$$s = \frac{My}{I}$$

where I denotes the moment of inertia of the cross-sectional area about the neutral axis. This quantity was discussed in Chapter 7. The derivation of this equation is discussed in detail in Problem 1. For applications see Problems 2-14, 18-20. These stresses vary from zero at the neutral axis of the beam to a maximum at the outer fibers as shown. The stresses are tensile on one side of the neutral axis, compressive on the other. These stresses are also called bending, flexural, or fiber stresses.

LOCATION OF THE NEUTRAL AXIS. The neutral axis always passes through the centroid of the cross-section. Hence, the moment of inertia I appearing in the above equation for normal stress is the moment of inertia of the cross-sectional area about an axis through the centroid of the cross-section of the beam.

SECTION MODULUS. At the outer fibers of the beam the value of the coordinate y is frequently denoted by the symbol c. In that case the maximum normal stresses are given by

$$s = \frac{Mc}{I} \quad \text{or} \quad s = \frac{M}{I/c}$$

The ratio I/c is called the section modulus and is usually denoted by the symbol Z. The units are (inches, or feet)^3. The maximum bending stresses may then be represented in the form

$$s = \frac{M}{Z}$$

This form is convenient because values of Z are available in handbooks for a wide range of standard structural steel shapes. See Problems 5, 10, 15, 16.

ASSUMPTIONS. In the derivation of the above expression for normal stresses it is assumed that a plane section of the beam normal to its longitudinal axis prior to loading remains plane after the forces and couples have been applied. Further, it is assumed that the beam is initially straight and of uniform cross-section and that the moduli of elasticity in tension and compression are equal.

SHEARING FORCE. The algebraic sum of all the vertical forces to one side of any cross-section of the beam is called the shearing force at that section. This concept was discussed in Chapter 6.

SHEARING STRESSES IN BEAMS. For any beam subject to a shearing force V (expressed in pounds) at a certain cross-section both vertical and horizontal shearing stresses s_s are set up. The magnitudes of the vertical shearing stresses at any cross-section are such that these stresses have the shearing force V as a resultant. In the cross-section of the beam shown below the vertical plane of symmetry contains the applied forces and the neutral axis passes through the centroid of the section. The coordinate y is measured from the neutral axis. The moment of inertia of the *entire* cross-sectional area about the neutral axis is denoted by I. The shearing stress on all fibers a distance y_o from the neutral axis is given by the formula

$$s_s = \frac{V}{Ib}\int_{y_o}^{c} y\,da$$

where b denotes the width of the beam at the location where the shearing stress is being calculated. This expression is derived in Problem 21. For applications see Problems 22-28. The integral $\int_{y_0}^{c} y\,da$ represents the first moment of the shaded area of the cross-section about the neutral axis. This quantity was discussed in detail in Chapter 7. More generally, the integral always represents the first moment about the neutral axis of that part of the cross-sectional area of the beam between the horizontal plane on which the shearing stress s_s occurs and the outer face of the beam, i.e. the area between y_o and c.

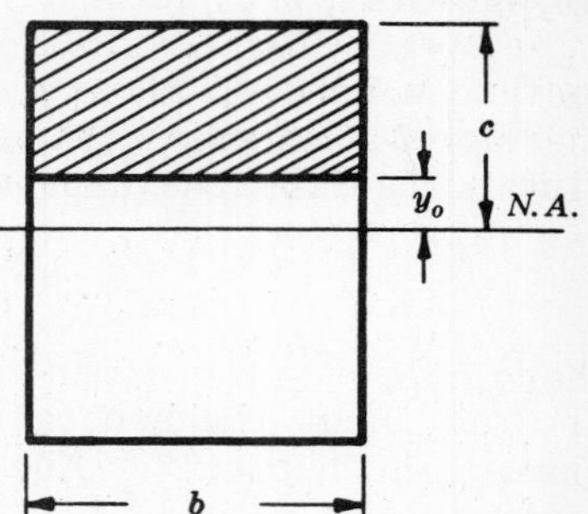

From the above expression it is evident that the maximum shearing stress always occurs at the neutral axis of the beam, whereas the shearing stress at the outer fibers is always zero. This is in contrast to the distribution of normal stress over the cross-section, since that varies from zero at the neutral axis to a maximum at the outer fibers.

In a beam of rectangular cross-section the above equation for shearing stress becomes

$$s_s = \frac{V}{2I}\left(\frac{h^2}{4} - y_o^2\right)$$

where s_s denotes the shearing stress on a fiber at a distance y_o from the neutral axis and h denotes the depth of the beam. The distribution of vertical shearing stress over the rectangular cross-section is thus parabolic, varying from zero at the outer fibers to a maximum at the neutral axis. This expression is derived in Problem 22. For applications see Problems 23, 24, 25, 26.

All of the above equations for shearing stress give both the vertical and also the horizontal shearing stresses at a point, as discussed in Problem 21, since the intensities of shearing stresses in these two directions are always equal.

SOLVED PROBLEMS

1. Derive an expression for the relationship between the bending moment acting at any section in a beam and the bending stress at any point in this same section.

Fig. (a) Fig. (b)

The beam shown in Fig. (a) above is loaded by the two couples M and consequently is in static equilibrium. Since the bending moment has the same value at all points along the bar, the beam is said to be in a condition of *pure bending*. To determine the distribution of bending stress in the beam, let us cut the beam by a plane passing through it in a direction perpendicular to the geometric axis of the bar. In this manner the forces under investigation become external to the new body formed, even though they were internal effects with regard to the original uncut body.

The free-body diagram of the portion of the beam to the left of this cutting plane now appears as in Fig. (b) above. Evidently a moment M must act over the cross-section cut by the plane so that the left portion of the beam will be in static equilibrium. The moment M acting on the cut section represents the effect of the right portion of the beam on the left portion. Since the right portion has been removed, it must be replaced by its effect on the left portion and this effect is represented by the the moment *M*. This moment is the resultant of the moments of forces acting perpendicular to the cut cross-section and in the plane of the page. It is now necessary to make certain assumptions in order to determine the nature of the variation of these forces over the cross-section.

It is convenient to consider the beam to be composed of an infinite number of thin longitudinal rods or fibers. It is assumed that every longitudinal fiber acts independently of every other fiber, that is, there are no lateral pressures or shearing stresses between adjacent fibers. Thus each fiber is subject only to axial tension or compression. Further, it is assumed that a plane section of the beam normal to its axis before loads are applied remains plane and normal to the axis after loading. Lastly, it is assumed that the material follows Hooke's law and that the moduli of elasticity in tension and compression are equal.

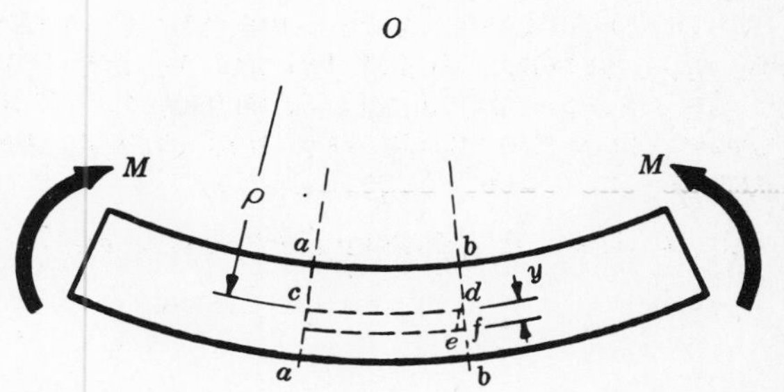

Let us next consider two adjacent cross-sections *aa* and *bb* marked on the side of the beam, as shown in the adjoining diagram. Prior to loading, these sections are parallel to each other. After the applied moments have acted on the beam, these sections are still planes but they have rotated with respect to each other to the positions shown, where *O* represents the center of curvature of the beam. Evidently the fibers on the upper surface of the beam are in a state of compression, while those on the lower surface have been extended slightly and are thus in tension. The line *cd* is the trace of the surface in which the fibers do not undergo any strain during bending and this surface is called the neutral surface, and its intersection with any cross-section is called the neutral axis. The elongation of the longitudinal fiber at a distance y (measured positive downward) may be found by drawing line *de* parallel to *aa*. If ρ denotes the radius of curvature of the bent beam, then from the similar triangles *cOd* and *edf* we find the strain of this fiber to be

1)
$$\epsilon = \frac{ef}{cd} = \frac{de}{cO} = \frac{y}{\rho}$$

Thus, the strains of the longitudinal fibers are proportional to the distance y from the neutral axis.

Since Hooke's law holds, and therefore $E = s/\epsilon$, or $s = E\epsilon$ it immediately follows that the stresses existing in the longitudinal fibers are proportional to the distance y from the neutral axis, or

2)
$$s = \frac{Ey}{\rho}$$

Let us consider a beam of rectangular cross-section, although the derivation actually holds for any cross-section which has a longitudinal plane of symmetry. In this case, these longitudinal, or bending stresses appear as in the adjoining diagram.

Let dA represent an element of area of the cross-section at a distance y from the neutral axis. The stress acting on dA is given by the above expression and consequently the force on this element is the product of the stress and the area dA, i.e.

3)
$$dF = \frac{Ey}{\rho}\,dA$$

However, the resultant longitudinal force acting over the cross-section is zero (for the case of pure bending) and this condition may be expressed by the summation of all forces dF over the cross-section. This is done by integration:

4)
$$\int \frac{Ey}{\rho}\,dA = \frac{E}{\rho}\int y\,dA = 0$$

Evidently $\int y\,dA = 0$. However, this integral represents the first moment of the area of the cross section with respect to the neutral axis, since y is measured from that axis. But, from Chapter 7 we may write $\int y\,dA = \bar{y}A$, where $\bar{y}$ is the distance from the neutral axis to the centroid of the cross-sectional area. From this, $\bar{y}A = 0$; and since A is not zero, then $\bar{y} = 0$. Thus the neutral axis always passes through the centroid of the cross-section.

The moment of the elemental force dF about the neutral axis is given by

5) $$dM = y\,dF = y(\frac{Ey}{\rho}\,dA)$$

The resultant of the moments of all such elemental forces summed over the entire cross-section must be equal to the bending moment M acting at that section and thus we may write

6) $$M = \int \frac{Ey^2}{\rho}\,dA$$

But $I = \int y^2\,dA$ and thus we have

7) $$M = \frac{EI}{\rho}$$

It is to be carefully noted that this moment of inertia of the cross-sectional area is computed with respect to the axis through the centroid of the cross-section. But previously we had

8) $$s = \frac{Ey}{\rho}$$

Eliminating ρ from these last two equations, we obtain

9) $$s = \frac{My}{I}$$

This formula gives the so-called bending or flexural stresses in the beam. In it, M is the bending moment at any section, I the moment of inertia of the cross-sectional area about an axis through the centroid of the cross-section, and y the distance from the neutral axis (which passes through the centroid) to the fiber on which the stress s acts.

The value of y at the outer fibers of the beam is frequently denoted by c and at these fibers the bending stresses are maximum and there we may write

10) $$s = \frac{Mc}{I}$$

2. A beam is loaded by a couple of 12,000 lb-in at each of its ends, as shown in the adjoining diagram. The beam is steel and of rectangular cross-section 1 in. wide by 2 in. deep. Determine the maximum bending stress in the beam and indicate the variation of bending stress over the depth of the beam.

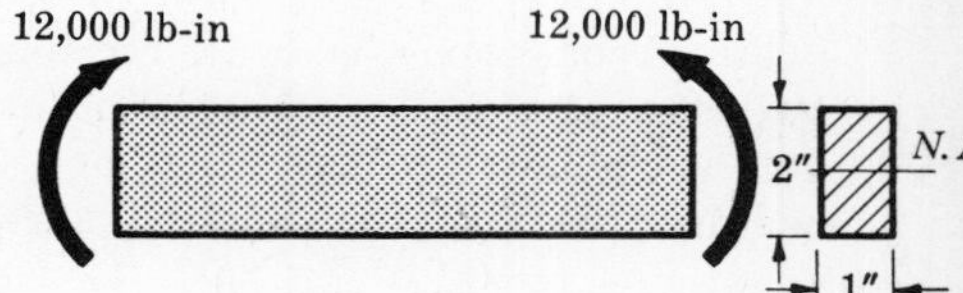

From Problem 1, bending takes place about the horizontal neutral axis denoted by $N.A.$ This axis passes through the centroid of the cross-section. The moment of inertia of the shaded rectangular cross-section about this axis was found in Problem 6, Chapter 7 to be

$$I = \frac{1}{12}bh^3 = \frac{1}{12}(1)(2)^3 = 0.667\text{ in}^4$$

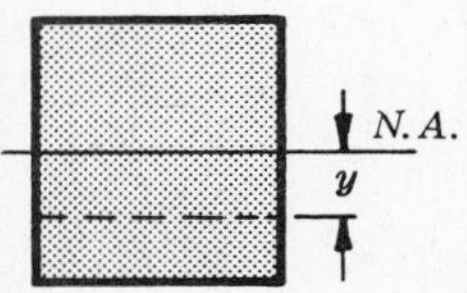

Also from problem 1, the bending stress at a distance y from the neutral axis is given by $s = My/I$, where y is illustrated in the adjacent diagram. Thus, all longitudinal fibers of the beam at the distance y from the neutral axis are subject to the same bending stress given by the above formula.

Since M and I are constant along the length of the bars, evidently the maximum bending stress occurs on those fibers where y takes on its maximum value. These are the fibers along the upper and lower surfaces of the beam, and from inspection it is obvious that for the direction of loading shown the upper fibers are in compression and the lower fibers in tension. For the lower fibers, $y = 1$ in. and the maximum bending stress is

$$s = \frac{12,000(1)}{0.667} = 18,000 \text{ lb/in}^2$$

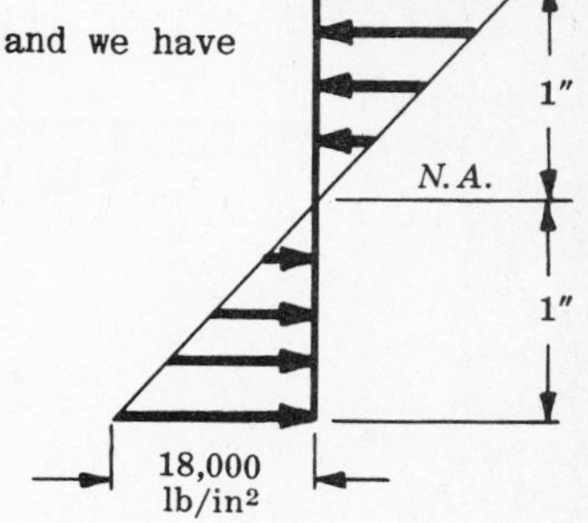

For the fibers along the upper surface y may be considered to be negative and we have

$$s = \frac{12,000(-1)}{0.667} = -18,000 \text{ lb/in}^2$$

Thus the peak stresses are 18,000 lb/in^2 in tension for all fibers along the lower surface of the beam and 18,000 lb/in^2 in compression for all fibers along the upper surface. According to the formula $s = My/I$, the bending stress varies linearly from zero at the neutral axis to a maximum at the outer fibers and hence the variation over the depth of the beam may be plotted as in the adjacent diagram.

3. A beam of circular cross-section is 7 in. in diameter. It is simply supported at each end and loaded by two concentrated loads of 20,000 lb each applied 12 in. from the end of the beam. Determine the maximum bending stress in the beam.

Here the moment is not constant along the length of the beam, as it was in Problem 2. The loading is illustrated in the adjoining diagram together with the bending moment diagram obtained by the methods of Chapter 6. It is to be noted that the portion of the beam between the two downward loads of 20,000 lb is in a condition termed *pure bending* and everywhere in that region the bending moment is equal to

$$(20,000)(12) = 240,000 \text{ lb-in.}$$

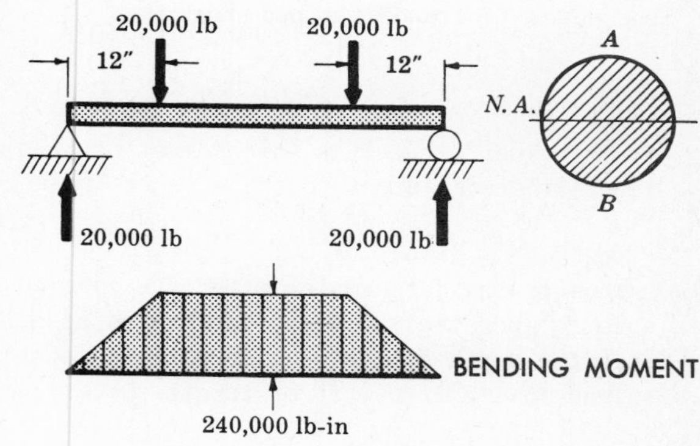

From Problem 11 of Chapter 7 the moment of inertia of the shaded circular cross-section about the neutral axis, which passes through the centroid of the circle, is

$$I = \pi D^4/64 = \pi(7)^4/64 = 118 \text{ in}^4$$

The bending stress at a distance y from the horizontal neutral axis shown is $s = My/I$. Evidently the maximum bending stresses occur along the fibers located at the ends of a vertical diameter and designated as A and B. This maximum stress is the same at all such points between the applied loads. At point B, $y = 3.5$ in. and the stress becomes

$$s = \frac{240,000(3.5)}{118} = 7120 \text{ lb/in}^2 \text{ tension.}$$

At point A the stress is 7120 lb/in^2 compression.

4. A steel cantilever beam 16 ft 8 in. in length is subjected to a concentrated load of 320 lb acting at the free end of the bar. The beam is of rectangular cross-section, 2 in. wide by 3 in. deep. Determine the magnitude and location of the maximum tensile and compressive bending stresses in the beam.

The bending moment diagram for this type of loading has been determined in Problem 1, Chapter 6. It is triangular with a maximum ordinate at the supporting wall, as shown below in Fig. (a). The maximum bending moment is merely the moment of the 320 lb force about an axis through point B and perpendicular

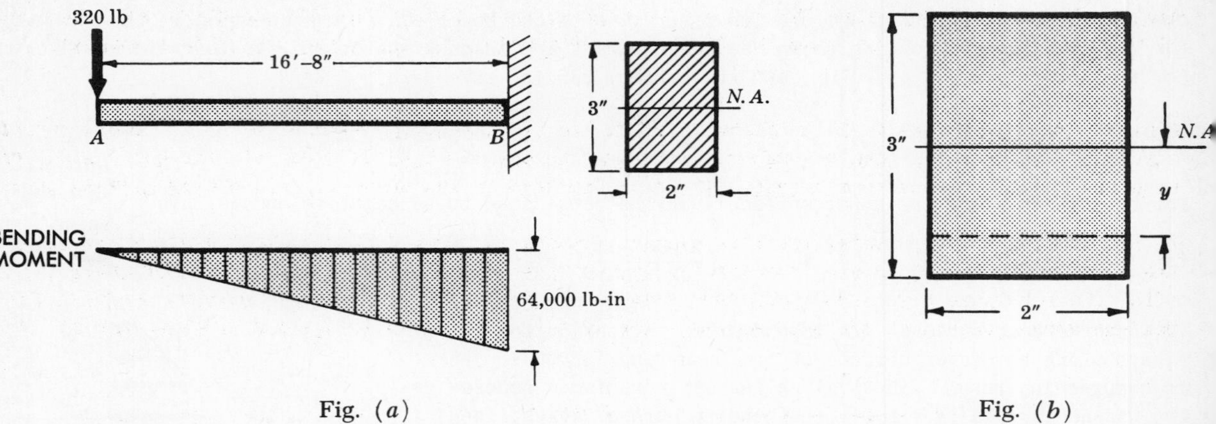

Fig. (a) Fig. (b)

to the plane of the page. It is 320(200) = 64,000 lb-in.

The bending stress at a distance y from the neutral axis, which passes through the centroid of the cross-section, is $s = My/I$ where y is illustrated in Fig. (b) above. In this expression I denotes the moment of inertia of the cross-sectional area about the neutral axis and is given by

$$I = \frac{1}{12}bh^3 = \frac{1}{12}(2)(3)^3 = 4.50 \text{ in}^4$$

Thus at the supporting wall where the bending moment is maximum the peak tensile stress occurs at the upper fibers of the beam and is

$$s = \frac{My}{I} = \frac{64,000(1.5)}{4.50} = 21,400 \text{ lb/in}^2$$

It is evident that this stress must be tension because all points of the beam deflect downward. At the lower fibers adjacent to the wall the peak compressive stress occurs and is equal to 21,400 lb/in^2.

5. Let us reconsider Problem 4 for the case where the rectangular beam is replaced by a commercially available rolled steel section, designated as a 6 WF 15-1/2. This standard manner of designation indicates that the depth of the section is 6 in., that it is a so-called *wide flange* section, and lastly that it weighs 15-1/2 lb per ft of length. Determine the maximum tensile and compressive bending stresses.

Such a beam has the symmetric cross-section shown in the adjacent diagram and bending takes place about the horizontal neutral axis passing through the centroid. Extensive handbooks listing properties of all available rolled steel shapes are available to designers and an abridged table is presented at the end of this chapter. From that table the moment of inertia about the neutral axis is found to be 28.1 in^4.

The bending stress at a distance y from the neutral axis is given by $s = My/I$. At the outer fibers, $y = c$ and

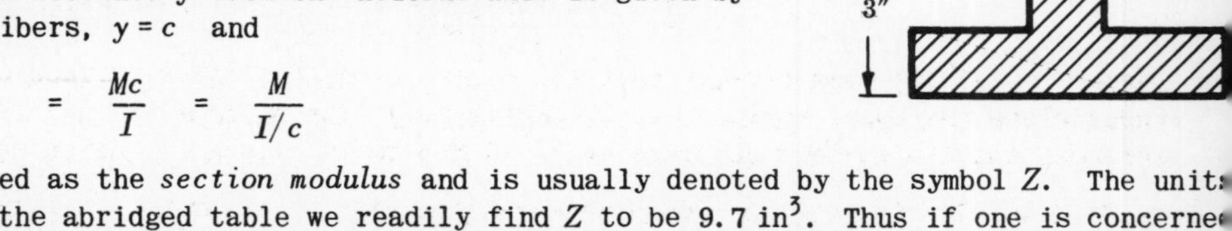

$$s = \frac{Mc}{I} = \frac{M}{I/c}$$

The ratio I/c is designated as the *section modulus* and is usually denoted by the symbol Z. The units are obviously in^3. From the abridged table we readily find Z to be 9.7 in^3. Thus if one is concerned only with bending stresses occuring at the outer fibers, which is frequently the case since we are often interested only in maximum stresses, then the section modulus is a convenient quantity to work with, particularly for standard structural shapes.

The stresses in the extreme fibers at the section of the beam immediately adjacent to the wall are thus given by

$$s = \frac{M}{I/c} = \frac{M}{Z} = \frac{64,000}{9.7} = 6600 \text{ lb/in}^2$$

Again, since the fibers along the top of the beam are stretching, the stress there will be tension. Along the lower face of the beam the fibers are shortening and there the stress is compressive.

5. A simply supported beam is 10 ft long and 3 × 5 in. in cross-section. At a point 4 ft from one support it carries a concentrated load of 6000 lb. Determine the maximum bending stress in the beam. Also, determine the bending stress in the outer fibers of the beam at the cross-section midway between supports.

The bending moment diagram for this type of loading has been discussed in Problem 4 of Chapter 6. The loaded beam, together with the bending moment diagram, may be represented as shown in Fig. (*a*) below, after the reactions have been found from statics. Inspection of the moment diagram reveals that the maximum moment occurs at the section where the 6000 lb load is applied. The bending moment at that

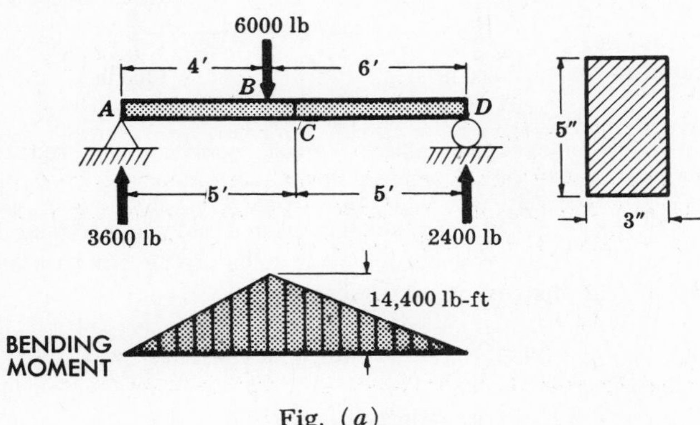

Fig. (*a*)

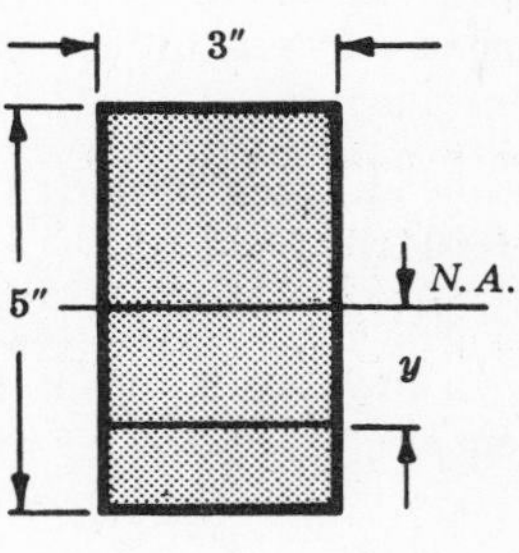

Fig. (*b*)

section is the moment of the 3600 lb reaction about an axis through B and perpendicular to the plane of the page. This is

$$3600(4) = 14{,}400 \text{ lb-ft} = 172{,}800 \text{ lb-in}$$

The same value could, of course, have been obtained by calculating the moment of the 2400 lb reaction about point B. The bending stress at any distance y from the neutral axis, which passes through the centroid of the cross-section, is $s = My/I$ where y is illustrated in Fig. (*b*) above. For a rectangular cross-section,

$$I = \frac{1}{12}bh^3 = \frac{1}{12}(3)(5)^3 = 31.2 \text{ in}^4$$

Thus, under the concentrated load the stress at any fiber a distance y from the neutral axis is

$$s = \frac{My}{I} = \frac{172{,}800\,y}{31.2}$$

At the lower fibers $y = 2.5$ in. and there the stress assumes the maximum value of

$$s = \frac{172{,}800(2.5)}{31.2} = 13{,}800 \text{ lb/in}^2$$

Inspection of the beam reveals that all fibers along the lower surface of the bar are in tension and those along the upper surface are in compression. The maximum tension is 13,800 lb/in² along the lower surface, and the maximum compression is 13,800 lb/in² along the upper surface.

To determine the stress midway between supports, it is first necessary to calculate the bending moment at section C. This is readily given by the moment of the 2400 lb reaction about an axis through C and perpendicular to the plane of the page. It is $2400(5) = 12{,}000$ lb-ft or 144,000 lb-in. The moment of inertia is, of course, 31.2 in⁴ and thus the bending stress at any fiber a distance y from the neutral axis is

$$s = \frac{My}{I} = \frac{144{,}000\,y}{31.2}$$

At the lower fibers y takes on its maximum value of 2.5 in. and there

$$s = \frac{144,000(2.5)}{31.2} = 11,500 \text{ lb/in}^2$$

For all fibers along the lower surface of the beam at the central section C the tension is 11,500 lb/in². A compression of equal magnitude exists in all fibers along the upper surface at this same section.

7. A beam 8 ft in length is simply supported at each end and bears a uniformly distributed load of 400 lb per ft of length. The cross-section of the bar is rectangular, 3 × 6 in. Determine the magnitude and location of the maximum bending stress in the beam. Also, determine the bending stress at a point 1 in. below the upper surface of the beam at the section midway between supports.

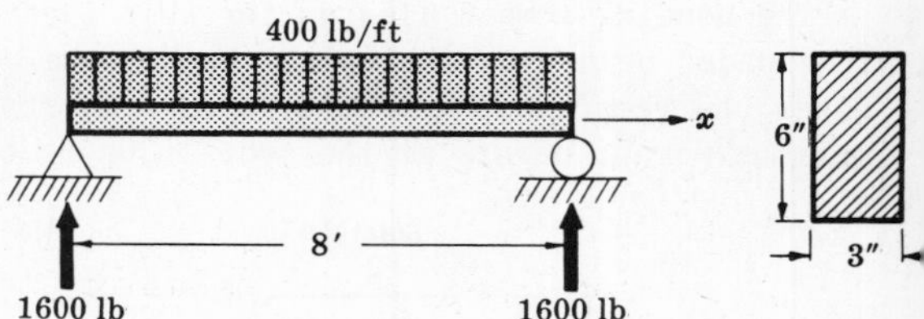

The beam, together with the distributed load, is sketched above. From symmetry the reactions are each 1600 lb.

The bending moment diagram for a uniformly distributed load acting on a simply supported beam was shown in Problem 6, Chapter 6 to be parabolic, varying from zero at the ends of the bar to a maximum at the center. The value of the bending moment at the center of this beam is

$$M_{x=4} = 1600(4) - 1600(2) = 3200 \text{ lb-ft} = 38,400 \text{ lb-in}$$

The bending moment diagram thus appears as in the adjoining diagram.

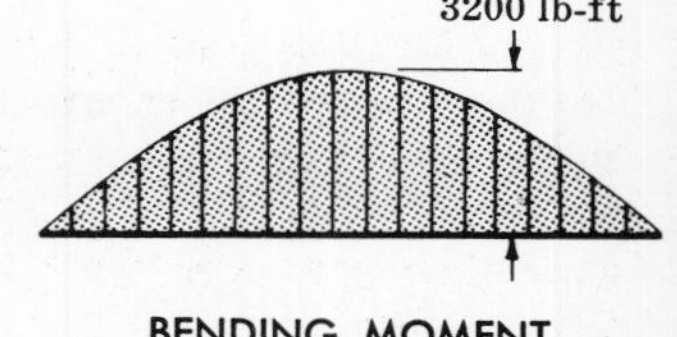

BENDING MOMENT

Since the maximum bending moment occurs at the center of the beam, then the maximum bending stress must also occur at the center of the beam, at $x = 4$ ft. At any fiber located a distance y from the neutral axis of that section, the bending stress is

$$s = My/I = 38,400\, y/I$$

The moment of inertia was found in Problem 6, Chapter 7, to be

$$I = \frac{1}{12}bh^3 = \frac{1}{12}(3)(6)^3 = 54 \text{ in}^4$$

Thus, at the center section $$s = \frac{38,400\,y}{54}$$

The maximum bending stress occurs at either the upper or lower extreme fibers. At the lower fibers, $y = 3$ in. and

$$s = 38,400(3)/54 = 2130 \text{ lb/in}^2$$

From inspection, the fibers along the lower surface of the beam are extended in length, consequently the stress in these fibers is tensile. In the upper fibers, $y = -3$ in., a compressive stress of equal magnitude exists.

At a point 1 in. below the upper surface of the beam at this central section the bending stress is

$$s = 38,400(-2)/54 = -1420 \text{ lb/in}^2$$

It is to be noted that y was taken to be negative in this calculation, since the point in question lies above the neutral axis. Consequently the bending stress is compressive as indicated by the final negative sign.

8. Let us reconsider the uniformly loaded beam of Problem 7. Determine the maximum bending stress in the beam if now the weight of the beam is considered in addition to the load of 400 lb per ft of length. The beam is steel and weighs 0.283 lb/in³.

Since the cross-section of the beam is 3×6 in., the volume of a 1 ft length of the beam is (3)(6)(12) = 216 in^3 and the weight of 1 ft length of the beam is 216(0.283) = 61.2 lb.

For design purposes the weight of the beam is called the dead load, abbreviated D.L. This weight of the beam per ft of length may be considered to act in addition to the load of 400 lb/ft. This applied load is called the live load, abbreviated L.L. The 61.2 lb acts uniformly over every ft of length of the beam and thus the resultant load is 400 + 61.2 = 461.2 lb/ft.

The total load over the entire beam is 461.2(8) = 3690 lb. Each end reaction is consequently 1845 lb and the bending moment at the center of the beam is

$$M_{x=4} \quad = \quad 1845(4) - 1845(2) \quad = \quad 3690 \text{ lb-ft} \quad = \quad 44{,}300 \text{ lb-in}$$

The bending moment diagram has the same appearance as in Problem 7 but with a maximum ordinate at the center of 3690 lb-ft.

The maximum bending stress occurs at the outer fibers of the beam midway between supports and is given by $s = My/I$ with $y = 3$ in. as before. Substituting,

$$s \quad = \quad 44{,}300(3)/54 \quad = \quad 2460 \text{ lb/in}^2$$

The previously obtained value of 2130 lb/in^2 which neglected the weight of the beam was 13.4% lower than this value. In actual practice it is almost always necessary to take the weight of the beam into account.

9. A cantilever beam 10 ft long is subjected to a uniformly distributed load of 1500 lb per ft of length. The allowable working stress in either tension or compression is 20,000 lb/in^2. If the cross-section is to be rectangular, determine the dimensions if the height is to be twice as great as the width.

The bending moment diagram for a uniform load acting over a cantilever beam was determined in Problem 2, Chapter 6. It was found to be parabolic, varying from zero at the free end of the beam to a maximum at the supporting wall. The loaded beam and the accompanying bending moment diagram are shown at the right. The maximum moment at the wall is given by

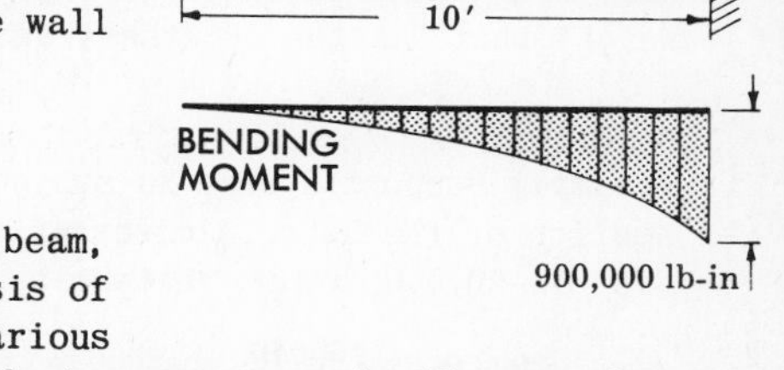

$$M_{x=10} \quad = \quad 15{,}000(5) \quad = \quad 75{,}000 \text{ lb-ft} \quad = \quad 900{,}000 \text{ lb-in}$$

It is to be noted that this problem involves the design of a beam, whereas all former problems in this chapter called for the analysis of stresses acting in beams of known dimensions and subject to various loadings. The only cross-section that need be considered for design purposes is the one where the bending moment is a maximum, i.e. at the supporting wall. Thus we wish to design a rectangular beam to resist a bending moment of 900,000 lb-in with a maximum bending stress of 20,000 lb/in^2.

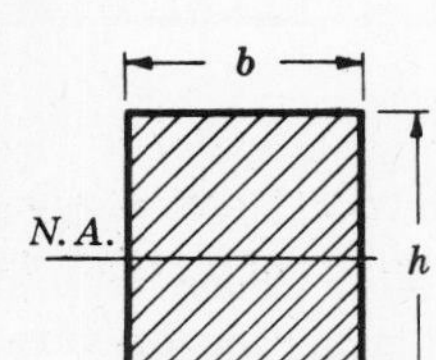

Since the cross-section is to be rectangular it will have the appearance shown in the diagram at the left where the width is denoted by b and the height by $h = 2b$, in accordance with the specifications. The moment of inertia about the neutral axis, which passes through the centroid of the section, is given by

$$I \quad = \quad \frac{1}{12}bh^3 \quad = \quad \frac{1}{12}b(2b)^3 \quad = \quad \frac{2}{3}b^4$$

At the cross-section of the beam adjacent to the supporting wall the bending stress in the beam is given by $s = My/I$. The maximum bending stress in tension occurs along the upper surface of the beam, since these fibers elongate slightly, and at this surface $y = b$ and $s = 20{,}000$ lb/in^2. Then

$$s \quad = \quad \frac{My}{I} \qquad \text{or} \qquad 20{,}000 \quad = \quad \frac{900{,}000b}{\frac{2}{3}b^4}$$

from which $b = 4.06$ in., and $h = 2b = 8.12$ in.

10. Select a suitable wide-flange section to carry the loading on the cantilever beam described in Problem 9. The working stress in either tension or compression is 20,000 lb/in^2.

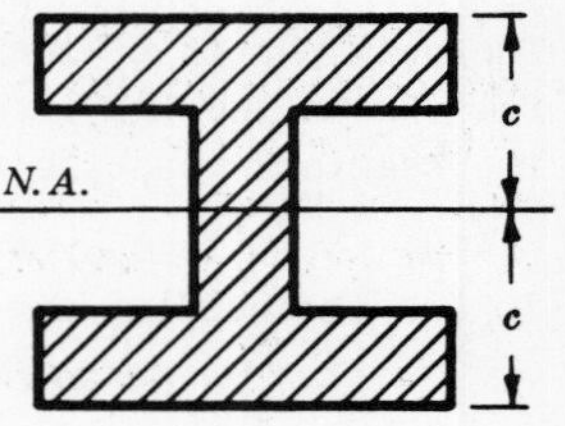

The bending moment diagram is of course the same as that shown in Problem 9. The maximum bending moment occurs at the supporting wall and as before is 900,000 lb-in.

For any wide-flange section, the bending stress on a fiber located a distance y from the neutral axis of the section is given by $s = My/I$. It is presumed that the beam is oriented as in the adjacent diagram. The maximum bending stresses obviously occur when y takes on its maximum value, which is at the outer fibers of the beam. Let us denote this maximum value of y by c, i.e. c is half the depth of the section. Then the maximum bending stress may be written in the form

$$s_{max} = \frac{Mc}{I} = \frac{M}{I/c} = \frac{M}{Z}$$

where Z is the section modulus of the beam.

From the last equation we have $Z = M/S$. Thus the required section modulus is given simply as the maximum bending moment divided by the allowable working stress. For the cantilever beam in question this becomes

$$Z = \frac{M}{S} = \frac{900,000}{20,000} = 45 \text{ in}^3$$

Consequently a suitable beam will be one having a section modulus of at least 45 in^3. It is of course unlikely that any standard rolled section will have exactly this section modulus and it is customary to select a section having either this Z, if possible, or a greater value. In this manner the working stress will not exceed the allowable value of 20,000 lb/in^2.

Reference to the abridged table of properties of wide-flange sections offered at the end of this chapter indicates that a 12 WF 36 section will be suitable for the design. It has a section modulus of 45.9 in^3, which is in excess of the required value of 45 in^3. This is the wide-flange section of least weight that has the necessary section modulus.

11. A simply supported beam is subject to the three concentrated forces shown in Fig. (a) below. The cross-section of the beam is circular and the maximum allowable working stress in either tension or compression is 20,000 lb/in^2. Determine the required diameter of the bar.

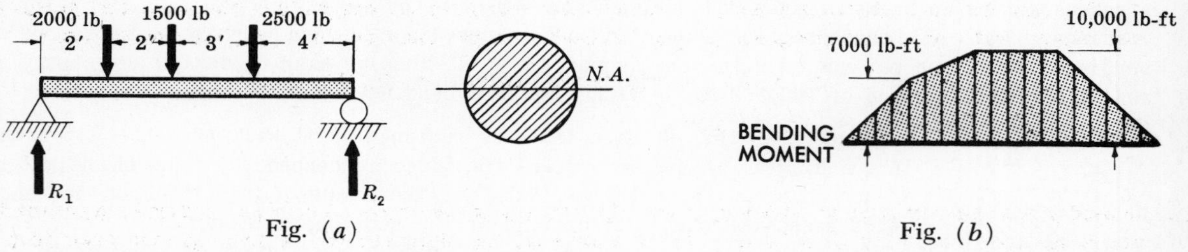

Fig. (a) Fig. (b)

This loading has been discussed in Problem 5, Chapter 6. There the reactions were found from statics to be R_1 = 3500 lb and R_2 = 2500 lb. The bending moment diagram consisted of a series of straight lines as shown in Fig. (b) above.

Since this is a design problem, it is only necessary to consider that cross-section of the beam where the bending moment is a maximum. From the above diagram it is evident that the maximum value the bending moment attains is 10,000 lb-ft and it happens that this value prevails at all sections between the 1500 lb and the 2500 lb loads. Hence, it is desired to design a beam of circular cross-section to withstand a bending moment of 10,000 lb-ft, or 120,000 lb-in, with a maximum allowable bending stress of 20,000 lb/in^2.

The cross-section of the beam is shown below. Evidently from the direction of the three applied loads the lower fibers of the beam are in tension and the upper fibers in compression. The maximum tension occurs along fiber A, since it is farthest from the neutral axis. Maximum compression exists in fiber B.

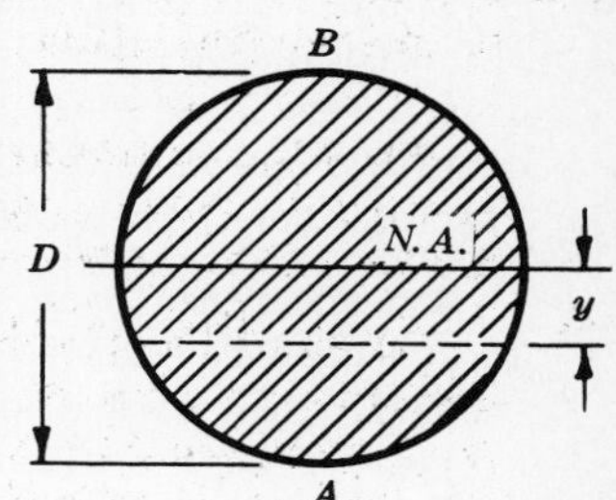

For any fiber a distance y from the neutral axis, which passes through the centroid of the circle, the bending stress is $s = My/I$. For fiber A, y has the value $D/2$. In Problem 11, Chapter 7, it was shown that the moment of inertia of a circle about a diameter is $I = \pi D^4/64$. Thus at point A, the point of maximum tension, the bending stress is 20,000 lb/in^2 and

$$s = \frac{My}{I} \quad \text{becomes} \quad 20{,}000 = \frac{120{,}000(D/2)}{\pi D^4/64} \quad \text{and} \quad D = 3.94 \text{ in.}$$

Consideration of the maximum compression would lead to the same required diameter.

12. If a steel wire 1/64 in. in diameter is coiled around a pulley 15 in. in diameter, determine the maximum bending stress set up in the wire. Take $E = 30 \cdot 10^6$ lb/in^2.

Since the radius of curvature of the wire is constant, 7.5 in., it is evident from Equation 7 of Problem 1 of this chapter, namely $M = EI/\rho$, that the bending moment M must be constant everywhere in the wire. Thus the wire acts as a beam subject to pure bending. An enlarged sketch of a portion of the wire is shown in the adjacent diagram. For any fiber in the wire at a distance y from the neutral axis, the normal strain ϵ was found in Equation (1) of Problem 1 of this chapter to be

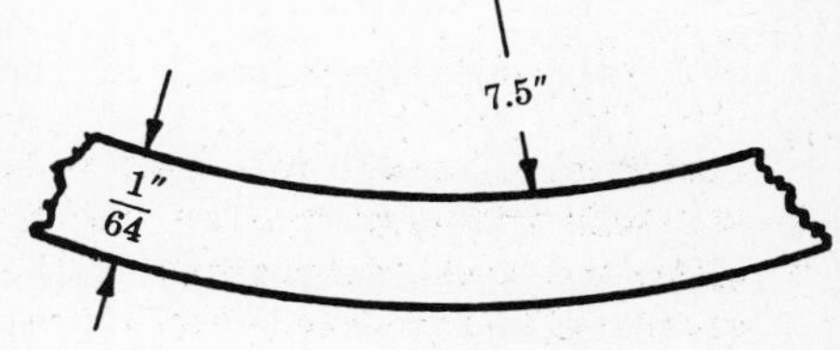

$$\epsilon = \frac{y}{\rho}$$

where ρ denotes the radius of curvature of the beam at that point.

The maximum strain occurs at the fibers where y assumes its maximum value, i.e. $\frac{1}{2}(1/64)$ in. from the neutral axis. The radius of curvature is approximately 7.5 in. More accurately, this radius should be measured to the neutral surface of the wire, but the value in that case would only differ from 7.5 in. by $\frac{1}{2}(1/64)$ in. and this quantity may reasonably be neglected.

Thus the maximum strain at the outer fibers of the wire is $\epsilon = \dfrac{\frac{1}{2}(1/64)}{7.5} = 0.00104.$

The longitudinal fibers are subject to tensile stresses on one side of the wire and compressive on the other, with no other stresses acting. Hooke's Law may then be used to find the stress s:

$$s = E\epsilon = (30 \cdot 10^6)(0.00104) = 31{,}200 \text{ lb/in}^2$$

This is the maximum stress in the wire.

13. Consider the simply supported beam, shown in Fig. (a) below, subject to the uniformly varying load having a maximum intensity of w lb per ft of length at the right end of the bar. If the beam is a 10 WF 49 section (i.e. a commercially available wide-flange section having a depth of 10 in. and weighing 49 lb/ft of length), determine the maximum load intensity w that may be applied if the working stress is 18,000 lb/in^2 in either tension or compression.

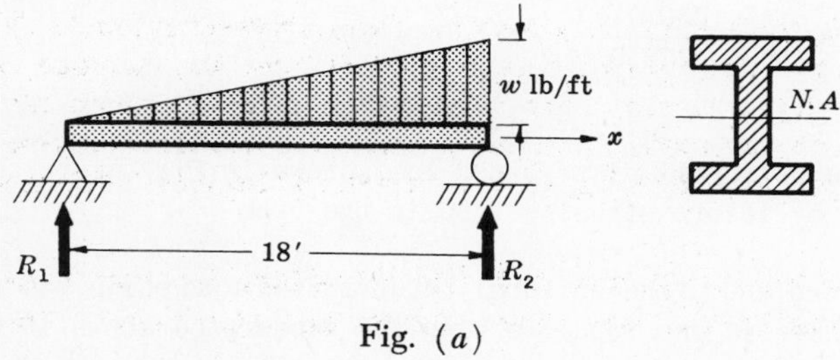

Fig. (a)

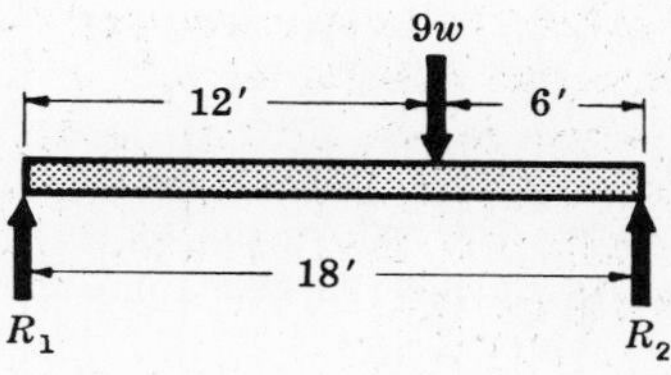

Fig. (b)

The reactions R_1 and R_2 may readily be determined in terms of the unknown w by replacing the distributed load by its resultant. Since the average value of the distributed load is $w/2$ lb/ft acting over a length of 18 ft, the resultant is a force of magnitude $18(w/2) = 9w$ lb acting through the centroid of the triangular loading diagram, i.e., 12 ft to the right of R_1. This resultant thus appears as in Fig. (*b*) above. From statics we immediately have $R_1 = 3w$ lb and $R_2 = 6w$ lb.

The shearing force and bending moment diagrams for this type of loading were discussed in Problem 10, Chapter 6. Let us introduce an x-axis coinciding with the beam and having its origin at the left support. Then at a distance x to the right of the left reaction, the intensity of load is found from similar triangle relationships to be $(x/18)w$ lb/ft. This portion of the loaded beam between R_1 and the section x appears as in Fig. (*c*) below. In accordance with the procedure explained in Problem 10, Chap-

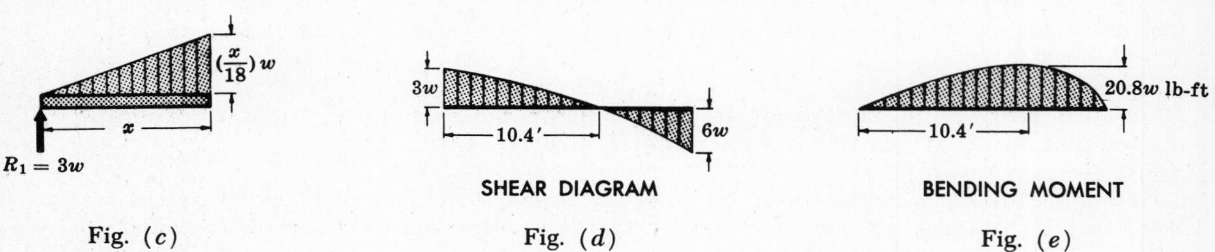

Fig. (*c*) Fig. (*d*) Fig. (*e*)

ter 6, the shearing force V at the section a distance x from the left support is given by

$$V = 3w - \frac{1}{2}(\frac{x}{18})w \cdot x = 3w - \frac{1}{36}wx^2$$

This equation holds for all values of x and from it the shear diagram is readily plotted, as shown in Fig. (*d*) above.

The point of zero shear is found by setting

$$3w - \frac{1}{36}wx^2 = 0 \qquad \text{from which} \qquad x = \sqrt{108} = 10.4 \text{ ft}$$

This is also the point where the bending moment assumes its maximum value.

The bending moment M at the section a distance x from the left support is given by

$$M = 3wx - \frac{1}{2}(\frac{x}{18})wx \cdot \frac{x}{3} = 3wx - \frac{1}{108}wx^3$$

Again, this equation holds for all values of x and from it the bending moment diagram may be plotted as in Fig. (*e*) above. At the point of zero shear, $x = 10.4$ ft, the bending moment is found by substitution in the above equation to be

$$M_{x=10.4} = 3w(\sqrt{108}) - \frac{1}{108}w(\sqrt{108})^3 = 2w\sqrt{108} \text{ lb-ft} = 250w \text{ lb-in}$$

This is the maximum bending moment in the beam.

The bending stress on any fiber a distance y from the neutral axis of the beam is given by $s = My/I$. The moment of inertia I of the beam is found from the table at the end of this chapter to be 272.9 in^4. The maximum tensile stress occurs at the lower fibers of the beam where $y = 5$ in. at the section where the bending moment is a maximum. This stress is 18,000 lb/in^2 and consequently

$$s = My/I \quad \text{becomes} \quad 18{,}000 = \frac{(250w)(5)}{272.9} \quad \text{or} \quad w = 3930 \text{ lb/ft}$$

14. Consider the over-hanging beam subject to the uniformly distributed load shown in Fig. (*a*) below. The beam is an 8 WF 19 section. What is the maximum flexure stress set up in the beam?

The bending moment diagram for this beam subject to the uniformly distributed load shown has already been determined in Problem 13, Chapter 6. There it was found to have the appearance shown in Fig. (*b*)

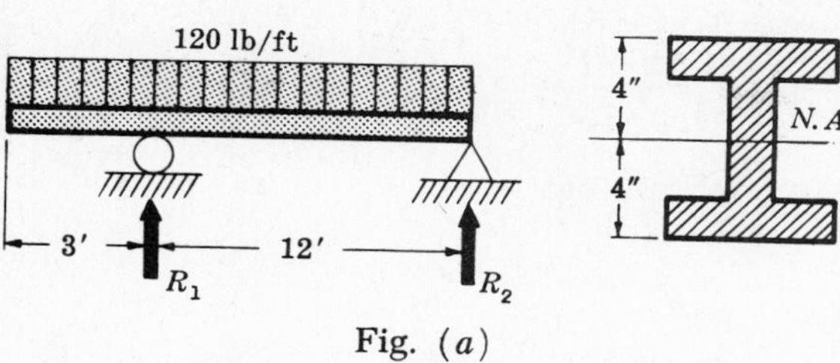

Fig. (a)

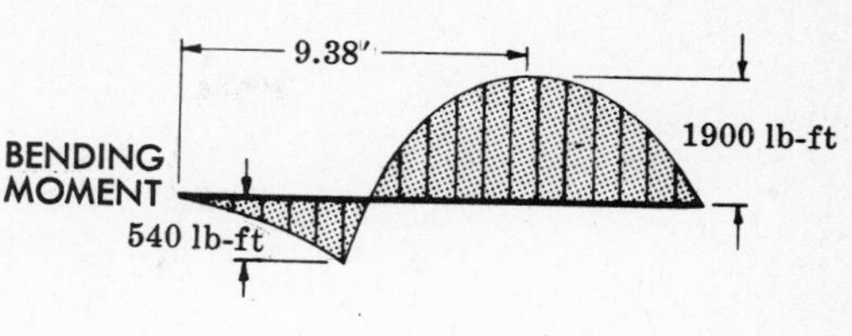

Fig. (b)

above. Thus the maximum bending moment in the beam is 1900 lb-ft or 22,800 lb-in.

The bending, or flexural stress on any longitudinal fiber a distance y from the neutral axis, which passes through the centroid of the cross-section, is given by $s = My/I$ where M denotes the bending moment at the cross-section under consideration. For this beam the moment of inertia I of the cross-sectional area about the axis through the centroid is found from the table at the end of this chapter to be 64.7 in^4.

Thus at the region of maximum bending moment, $x = 9.38$ ft, the maximum stress occurs at the outer fibers of the beam where $y = 4$ in. Then

$$s = My/I = 22,800(4)/64.7 = 1410 \text{ lb/in}^2$$

From the fact that the bending moment is positive at this section, we know that the bar must be concave upward there, according to the assumed sign convention for bending moments (see Chapter 6); hence at this section the lower fibers are subject to a tension of 1410 lb/in^2 and the upper fibers to a compression of 1410 lb/in^2.

15. Determine the section modulus of a beam of rectangular cross-section.

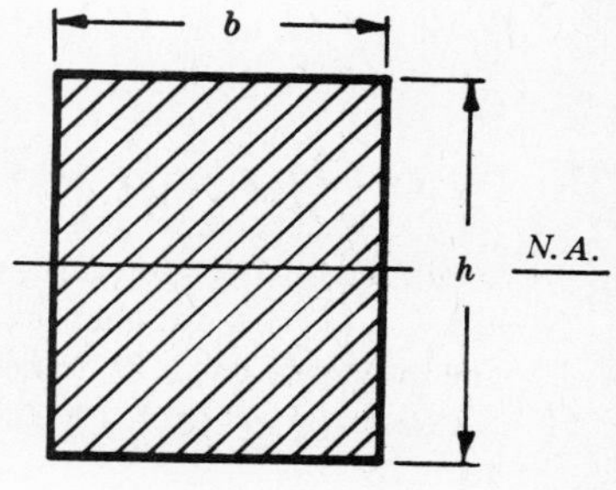

Let h denote the depth of the beam and b its width. Bending is assumed to take place about the neutral axis through the centroid of the cross-section. The moment of inertia about the neutral axis is $I = bh^3/12$.

At the outer fibers the distance to the neutral axis is $h/2$, and this is commonly denoted by c. The maximum bending stresses at these outer fibers are given by

$$s_{max} = \frac{Mc}{I} = \frac{M}{I/c}$$

The ratio I/c is called the section modulus and is usually denoted by Z. Then $s_{max} = M/Z$. For the beam of rectangular cross-section,

$$Z = \frac{I}{c} = \frac{bh^3/12}{h/2} = \frac{bh^2}{6}$$

The section modulus Z has units of (inches, or feet)^3.

6. A beam of rectangular cross-section is to be cut from a circular log of diameter D. What should be the ratio of the depth of the beam to its width for maximum strength in pure bending?

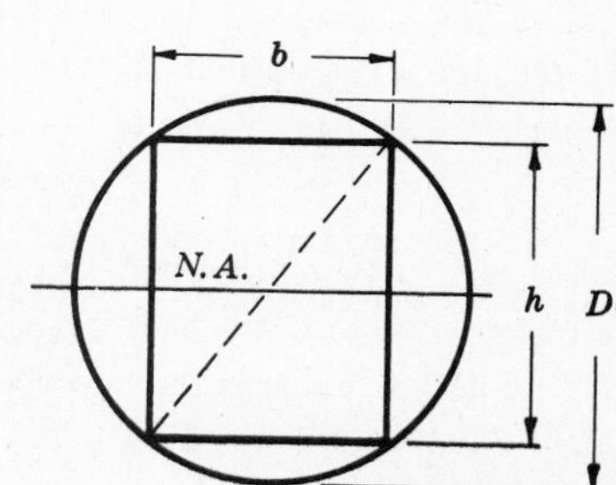

A sketch of the beam cross-section appears at the right, where h denotes the height of the beam and b its width.

The bending is considered to take place about the horizontal neutral axis shown. The maximum bending stresses occur at the outer fibers of the rectangular section at a distance $h/2$ above or below the neutral axis. Any fiber a distance y from the neutral axis is subject to a bending stress given by $s = My/I$, where I denotes the moment of inertia of the rectangular cross-section about the neutral axis, i.e. $bh^3/12$. At the outer fibers, $y = h/2$ and the maximum stresses there become

$$(1) \qquad s_{max} = \frac{M(h/2)}{bh^3/12} = \frac{M}{bh^2/6}$$

For any given value of the maximum bending stress, the maximum moment M that may be applied to th bar is found from the last equation to be

$$(2) \qquad M = \frac{1}{6}(s_{max})bh^2$$

For the condition of maximum strength, i.e. maximum moment M, the product bh^2 must be made a maximu since s_{max} is constant for a given material.

To maximize the quantity bh^2 we realize that it must be expressed in terms of one independent variable, say b, and we may do this from the right triangle relationship: $b^2 + h^2 = D^2$. Then

$$bh^2 = b(D^2 - b^2) = bD^2 - b^3$$

To maximize bh^2 we take the first derivative of the expression with respect to b and set it equal t zero, as follows:

$$\frac{d(bh^2)}{db} = \frac{d}{db}(bD^2 - b^3) = D^2 - 3b^2 = (b^2 + h^2) - 3b^2 = h^2 - 2b^2 = 0$$

Solving, $\frac{h}{b} = \sqrt{2}$. This is the desired ratio in order that the beam will carry a maximum moment M.

It is to be noted that the expression appearing in the denominator of the right side of equati (1), i.e. $bh^2/6$ is the section modulus of a rectangular bar. Thus the section modulus is actually t quantity to be maximized for greatest strength of the beam.

17. Two $\frac{1}{2} \times 8$ in. cover plates are welded to two channels 10 in. high and weighing 20 lb/ft of length to form the cross-section of a beam as shown. The loads are in a vertical plane and bending takes place about a horizontal axis. If the maximum allowable bending stress is 18,000 lb/in², determine the maximum bending moment that may be developed in the beam. The moment of inertia of each channel about a horizontal axis through its centroid is 78.5 in⁴.

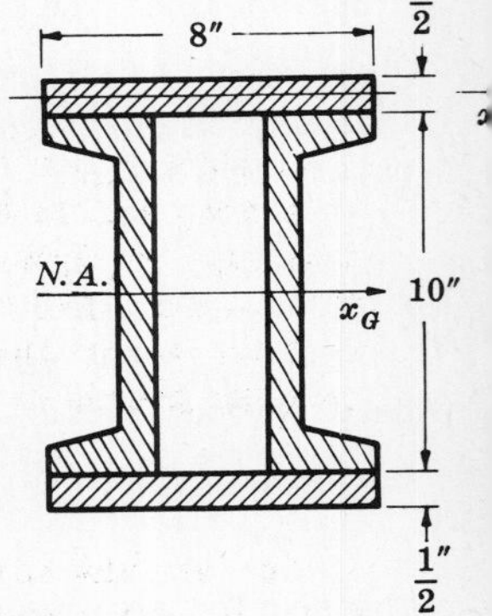

It is first necessary to calculate the moment of inertia of the entire beam cross-section about the horizontal neutral axis through the centroid denoted by x_G. The moment of inertia of each cover plate about the neutral axis x_G is equal to the moment of inertia about the horizontal axis denoted by x_1 passing through the centroid of the cover plate plus the product of the area of the plate and the square of the distance between the x_1 and x_G axes. Thus for each cover plate the moment of inertia about the neutral axis is

$$I_1 = \frac{1}{12}(8)(\frac{1}{2})^3 + 8(\frac{1}{2})(5\frac{1}{4})^2 = 110 \text{ in}^4$$

The moment of inertia of each channel about the x_G axis is given as 78.5 in⁴. Consequently, for t entire cross-section, the moment of inertia about the neutral axis is found by summation to be

$$I = 2(78.5) + 2(110) = 377 \text{ in}^4$$

The maximum bending stress in the beam occurs at the outer fibers of the cover plates and is giv by $s = Mc/I$, where I denotes the moment of inertia of the entire cross-section about the neutral ax and c represents the distance to the outer fibers of the cover plate, i.e. 5.5 in. If the maximum stre at these outer fibers is 18,000 lb/in², then

$$s = Mc/I \quad \text{becomes} \quad 18{,}000 = \frac{M(5.5)}{377} \quad \text{or} \quad M = 1{,}232{,}000 \text{ lb-in}$$

This is the maximum bending moment to which the beam should be subjected so as not to exceed the stre of 18,000 lb/in².

18. A beam is loaded by one couple at each of its ends as shown, the magnitude of each couple being 50,000 lb-in. The beam is steel and of T-type cross-section with the dimensions indicated. Determine (*a*) the maximum tensile stress in the beam and its location and (*b*) the maximum compressive stress and its location.

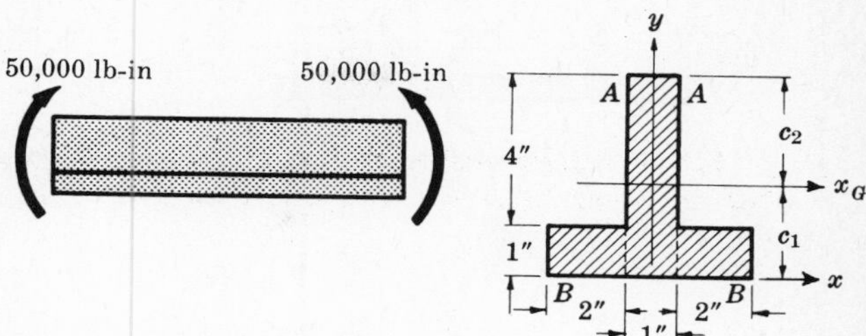

It is first necessary to locate the centroid of the cross-sectional area since the neutral axis is known to pass through the centroid. To do this we introduce the x-y coordinate system shown and proceed as in Problem 13, Chapter 7. The y-coordinate of the centroid is defined by

$$\bar{y} = \frac{\int y\,da}{A}$$

where the numerator of the right side represents the first moment of the entire area about the x-axis. The T-section may be considered to consist of the three rectangles indicated by the dotted lines and this expression becomes

$$\bar{y} = \frac{5(1)(2.5) + [2(1)(1/2)]2}{5(1) + [2(1)]2} = 1.61 \text{ in.}$$

Thus, the centroid is located 1.61 in. above the x-axis. The horizontal axis passing through this point is denoted by x_G as shown.

The moment of inertia about the x-axis is given by the sum of the moments of inertia about this same axis of each of the three component rectangles comprising the cross-section. Thus

$$I_x = \frac{1}{3}(2)(1)^3 + \frac{1}{3}(1)(5)^3 + \frac{1}{3}(2)(1)^3 = 43 \text{ in}^4$$

The moment of inertia about the x_G-axis may now be found by use of the parallel axis theorem. Thus

$$I_x = I_{xG} + A(\bar{y})^2, \quad 43 = I_{xG} + 9(1.61)^2 \quad \text{and} \quad I_{xG} = 19.7 \text{ in}^4$$

Evidently for the loading shown, the fibers below the x_G-axis are in tension, while the fibers above this axis are in compression. Let c_1 and c_2 denote the distances of the extreme fibers from the neutral axis (x_G) as shown. Obviously $c_1 = 1.61$ in. and $c_2 = 3.39$ in. The maximum tensile stress occurs in those fibers along B-B and is given by $s = Mc_1/I$, where I denotes the moment of inertia of the entire cross-section about the neutral axis passing through the centroid of the cross-section. Thus the maximum tensile stress is given by

$$s = Mc_1/I = 50{,}000(1.61)/19.7 = 4100 \text{ lb/in}^2$$

The maximum compressive stress occurs in those fibers along A-A and is given by $s = Mc_2/I$. To provide a consistent system of algebraic signs, it is desirable to assign a negative value to c_2 since it lies on the side of the x_G-axis opposite to that of c_1. Hence

$$s = Mc_2/I = 50{,}000(-3.39)/19.7 = -8600 \text{ lb/in}^2$$

The negative sign indicates that the stress is compressive.

19. A simply supported beam is loaded by the couple of 1000 lb-ft as shown in Fig. (*a*) below. The beam has a channel-type cross-section as illustrated. Determine the maximum tensile and compressive stresses in the beam.

The bending moment diagram for this particular loading has been determined in Problem 9, Chapter 6. There, it was found to have the appearance as in Fig. (*b*) below.

It is next necessary to locate the centroid of the cross-section since the neutral axis passes

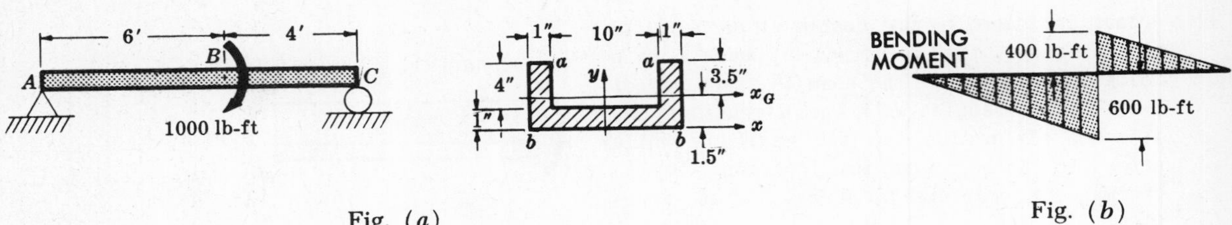

Fig. (a)

Fig. (b)

through that point. This has already been determined for this channel-type section in Problem 15, Chapter 7. There, the centroid was found to lie 1.5 in. above the x-axis and consequently the x_G-axis through the centroid lies 1.5 in. above the x-axis. In that same problem the moment of inertia of the entire cross-section about the x_G-axis was found to be 41.6 in^4.

In this problem it is necessary to distinguish carefully between positive and negative bending moments. One method of attack is to consider a cross-section of the beam slightly to the left of point B where the 1000 lb-ft couple is applied. According to the bending moment diagram the moment there is -600 lb-ft and, according to the sign convention adopted in Chapter 6, since the moment is negative the beam is concave downward at that section, as shown at the right. Thus the upper fibers are in tension and the lower fibers in compression. Along the upper fibers a-a the bending stress is given by $s = My/I$. Then

$$s_a \;=\; 600(12)(3.5)/41.6 \;=\; 605\ \text{lb/in}^2$$

Along the lower fibers b-b the value of y in the above formula for bending stress must be taken to be negative since these fibers lie on the other side of the neutral axis and there we have

$$s_b \;=\; 600(12)(-1.5)/41.6 \;=\; -260\ \text{lb/in}^2$$

It is next necessary to investigate the bending stresses at a section slightly to the right of point B. There the bending moment is 400 lb-ft and according to the usual sign convention the beam is concave upward at that section, as shown at the right. Here the upper fibers are in compression and the lower fibers in tension. Along the upper fibers a-a the bending stress is

$$s'_a \;=\; 400(12)(-3.5)/41.6 \;=\; -400\ \text{lb/in}^2$$

Along the lower fibers b-b we have

$$s'_b \;=\; 400(12)(1.5)/41.6 \;=\; 170\ \text{lb/in}^2$$

The maximum tensile and compressive stresses must now be selected from the above four values. Evidently the maximum tension is 605 lb/in^2 occurring in the upper fibers just to the left of point B and the maximum compression is 400 lb/in^2 occurring in the upper fibers also but just to the right of point B.

20. Consider the beam with overhanging ends loaded by the three concentrated forces shown in Fig. (a). The beam is simply supported and of T-type cross-section placed as shown. The material is gray cast iron having an allowable working stress in tension of 5000 lb/in^2 and in compression of 20,000 lb/in^2. Determine the maximum allowable value of P.

From symmetry each of the reactions denoted by R is equal to $P/2$. The bending moment diagram for this type of loading was discussed in Problem 12, Chapter 6. As shown there, the diagram consists of

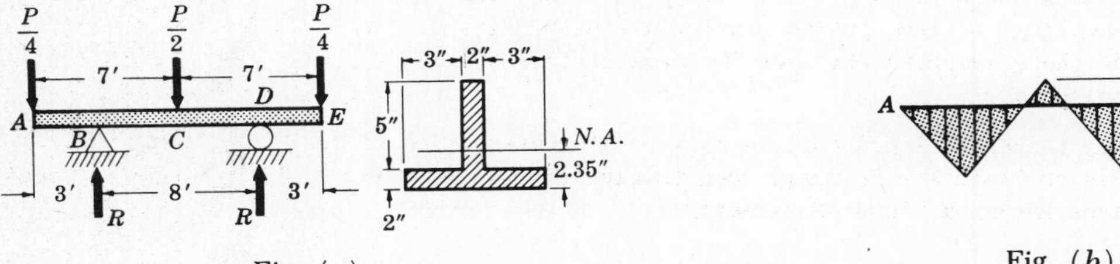

Fig. (a)

Fig. (b)

a series of straight lines connecting the ordinates representing bending moments at the points A, B, C, D and E. At B the bending moment is given by the moment of the force $P/4$ acting at A about an axis through B. Thus

$$M_B = -(P/4)(3) = -3P/4 \text{ lb-ft}$$

At C the bending moment is given by the sum of the moments of the forces $P/4$ and $R = P/2$ about an axis through C. Thus

$$M_C = -(P/4)(7) + (P/2)(4) = P/4 \text{ lb-ft}$$

The bending moment at D is equal to that at B by symmetry and the moment at each of the ends A and E is zero. Hence, the bending moment diagram plots as in Fig. (b) above.

The properties of the T-section used here have already been determined in Problem 13, Chapter 7. There the distance from the outer fibers of the flange to the centroid was found to be 2.35 in. and the moment of inertia about the neutral axis passing through the centroid was 101 in^4.

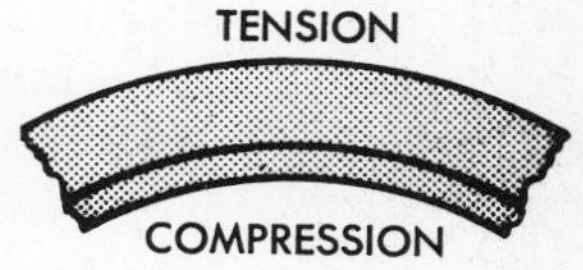

It is perhaps simplest to calculate four values of P based upon the various maximum tensile and compressive stresses that may exist at each of the points B and C and then select the minimum of these values. Let us first examine point B. Since the bending moment there is negative, the beam is concave downward at that point, as shown in the adjoining diagram. Evidently the upper fibers are in tension and the lower fibers are subject to compression. We shall first calculate a value of P assuming that the allowable tensile stress of 5000 lb/in^2 is realized in the upper fibers. Applying the flexure formula $s = My/I$ to these upper fibers, we find

$$5000 = (3P/4)(12)(4.65)/101 \qquad \text{or} \qquad P = 12,000 \text{ lb}$$

Next we shall calculate a value of P, assuming that the allowable compressive stress of 20,000 lb/in^2 is set up in the lower fibers. Again applying the flexure formula, we find

$$20,000 = (3P/4)(12)(2.35)/101 \qquad \text{or} \qquad P = 95,500 \text{ lb}$$

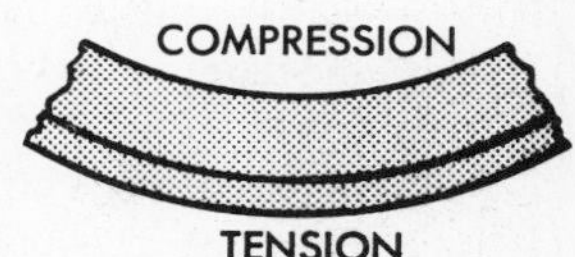

We shall now examine point C. Since the bending moment there is positive, the beam is concave upward at that point and appears as in the adjoining diagram. Here, the upper fibers are in compression and the lower fibers are subject to tension. First we will calculate a value of P assuming that the allowable tension of 5000 lb/in^2 is set up in the lower fibers. From the flexure formula we find

$$5000 = (P/4)(12)(2.35)/101 \qquad \text{or} \qquad P = 71,800 \text{ lb}$$

Lastly, we shall assume that the allowable compression of 20,000 lb/in^2 is set up in the upper fibers. Applying the flexure formula, we have

$$20,000 = (P/4)(12)(4.65)/101 \qquad \text{or} \qquad P = 145,000 \text{ lb}$$

The minimum of these four values is $P = 12,100$ lb. Thus the tensile stress at the points B and D is the controlling factor in determining the maximum allowable load.

21. In the case of a beam loaded by transverse forces acting perpendicular to the axis of the beam, not only are bending stresses parallel to the axis of the bar produced but shearing stresses also act over cross-sections of the beam perpendicular to the axis of the bar. Derive an expression for the intensity of these shearing stresses in terms of the shearing force at the section and the properties of the cross-section.

The theory to be developed applies only to a cross-section of rectangular shape. However, the results of this analysis are commonly used to give approximate values of the shearing stress in other cross-sections having a plane of symmetry.

Let us consider an element of length dx cut from a beam as shown in the adjoining diagram. We shall denote the bending moment at the left side of the element by M and that at the right side by $(M+dM)$, since in general the bending moment changes slightly as we move from one section to an adjacent section of the beam. If y is measured upward from the neutral axis, then the bending stress at the left section a-a is given by

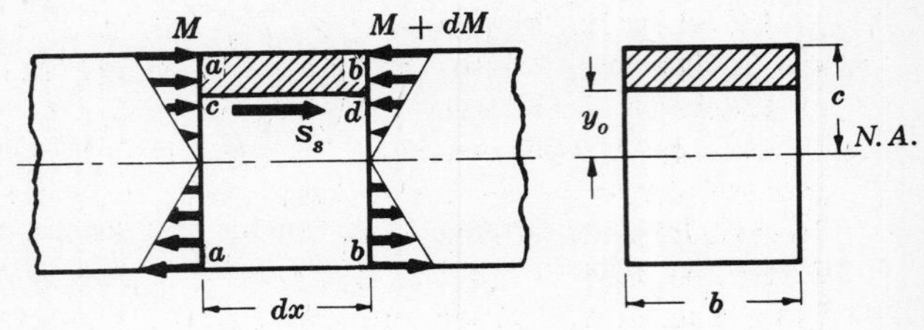

$$s = \frac{My}{I}$$

where I denotes the moment of inertia of the entire cross-section about the neutral axis. This stress distribution is illustrated above. Similarly, the bending stress at the right section b-b is

$$s' = \frac{(M + dM)y}{I}$$

Let us now consider the equilibrium of the shaded element $acdba$. The force acting on an area dA of the face ac is merely the product of the intensity of the force and the area; thus

$$s\,da = \frac{My}{I}\,dA$$

The sum of all such forces over the left face ac is found by integration to be

$$\int_{y_0}^{c} \frac{My}{I}\,dA$$

Likewise, the sum of all normal forces over the right face db is given by

$$\int_{y_0}^{c} \frac{(M + dM)y}{I}\,dA$$

Evidently, since these two integrals are unequal, some additional horizontal force must act on the shaded element to maintain equilibrium. Since the top face ab is assumed to be free of any externally applied horizontal forces, then the only remaining possibility is that there exists a horizontal shearing force along the lower face cd. This represents the action of the lower portion of the beam on the shaded element. Let us denote the shearing stress along this face by s_s as shown. Also, let b denote the width of the beam at the position where s_s acts. Then the horizontal shearing force along the face cd is

$$s_s\, b\, dx$$

For equilibrium of the element $acdba$ we have

$$\Sigma F_h = \int_{y_0}^{c} \frac{My}{I}\,da - \int_{y_0}^{c} \frac{(M+dM)y}{I}\,dA + s_s\, b\, dx = 0$$

Solving,

$$s_s = \frac{1}{Ib} \cdot \frac{dM}{dx} \int_{y_0}^{c} y\,dA$$

But from Problem 7, Chapter 6 we have $V = dM/dx$, where V represents the shearing force (in pounds) at the section a-a. Substituting,

$$s_s = \frac{V}{Ib} \int_{y_0}^{c} y\,dA$$

The integral in this last equation represents the first moment of the shaded cross-sectional area about the neutral axis of the beam. This area is always the portion of the cross-section that is above the level at which the desired shear stress acts. This first moment of area is sometimes denoted by Q, in which case the above formula becomes

$$s_s = \frac{VQ}{Ib}$$

The units of $\int y\,dA$ or of Q are in^3.

The shearing stress s_s just determined acts horizontally as shown above. However, let us consider the equilibrium of a thin element *mnop* of thickness t cut from any body and subject to a shearing stress s_1 on its lower face, as shown in the adjoining diagram. The total horizontal force on the lower face is $s_1 t\,dx$. For equilibrium of forces in the horizontal direction, an equal force but acting in the opposite direction must act on the upper face, hence the shear stress intensity there too is s_1. These two forces give rise to a couple of magnitude $s_1 t\,dx\,dy$. The only way in which equilibrium of the element can be maintained is for another couple to act over the vertical faces. Let the shear stress intensity on these faces be denoted by s_2. The total force on either vertical face is $s_2 t\,dy$. For equilibrium of moments about the center of the element we have

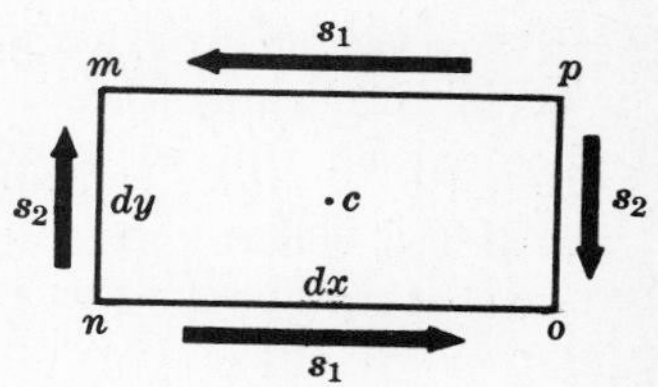

$$\Sigma M_C = s_1 t\,dx\,dy - s_2 t\,dy\,dx = 0 \qquad \text{or} \qquad s_1 = s_2$$

Thus we have the interesting conclusion that the shearing stresses on any two perpendicular planes through a point on a body are equal. Consequently, not only are there shearing stresses s_s acting horizontally at any point in the beam, but shearing stresses of an equal intensity also act vertically at that same point.

In summary, when a beam is loaded by transverse forces, both horizontal and vertical shearing stresses arise in the beam. The vertical shearing stresses are of such magnitudes that their resultant at any cross-section is exactly equal to the shearing force V at that same section.

22. Using the expression for shearing stress derived in Problem 21, determine the distribution of shearing stress in a beam of rectangular cross-section. What is the maximum shearing stress in a rectangular bar?

In Problem 21 the shearing stress s_s at a distance y_o from the neutral axis of the beam was found to be

$$s_s = \frac{V}{Ib}\int_{y_o}^{c} y\,dA$$

where V denotes the shearing force at the cross-section and b represents the width of the beam at the position where s_s is acting.

It is necessary to evaluate the above integral for a rectangular cross-section. Let h denote the height of the cross-section and b its width, as shown in the adjacent diagram.

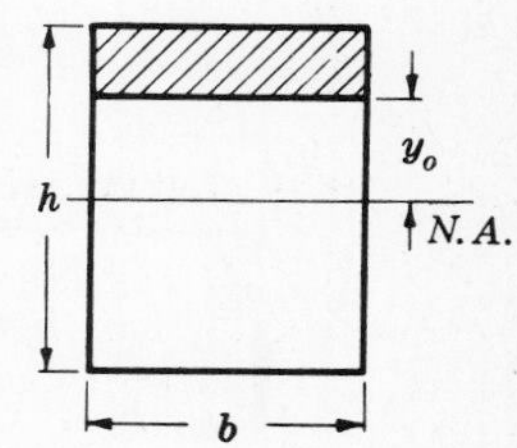

The integral represents the first moment of the shaded area about the neutral axis. It is to be noted that this area extends from the level at which the desired shear stress s_s acts to the extreme outer fibers of the beam. In this manner we find the shear stress s_s on all fibers a distance y_o from the neutral axis. Actually it is not necessary to integrate in such a simple case. Since the integral is known to represent the first moment of the shaded area about the neutral axis, we may calculate this first moment according to the definition given in Chapter 7. That is, the first moment of the shaded area is simply the product of the area and the perpendicular distance between the centroid of the area and the neutral axis. The area is given by $b(h/2 - y_o)$, and the distance from the centroid of the shaded region to the neutral axis is $\frac{1}{2}(h/2 + y_o)$. Consequently the value of the integral representing the first moment of area is

$$\int_{y_o}^{c} y\,da = \frac{1}{2}b\left(\frac{h}{2} + y_o\right)\left(\frac{h}{2} - y_o\right) = \frac{1}{2}\left(\frac{h^2}{4} - y_o^2\right)b$$

and the shearing stress s_s at a distance y_o from the neutral axis becomes

$$s_s = \frac{V}{Ib}\left[\frac{1}{2}b\left(\frac{h^2}{4} - y_o^2\right)\right] = \frac{V}{2I}\left(\frac{h^2}{4} - y_o^2\right)$$

From this it may be seen that the shearing stress over the cross-section varies in a parabolic manner from a maximum at the neutral axis ($y_o = 0$) to zero at the outer fibers of the beam ($y_o = h/2$). This variation may be plotted as in the adjoining diagram.

At the neutral axis, $y_o = 0$, the maximum shearing stress is found by substitution in the above formula to be

$$(s_s)_{max} = Vh^2/8I$$

But for a rectangular cross-section $I = bh^3/12$. Substituting,

$$(s_s)_{max} = \frac{Vh^2}{8(bh^3/12)} = \frac{3}{2}(\frac{V}{bh})$$

Thus the maximum shearing stress in the case of a rectangular cross-section is 50 percent greater than the average shearing stress obtained by dividing the shearing force V by the cross-sectional area bh.

23. A beam is of rectangular cross-section, 6 in. wide and 8 in. deep. It is subject to a system of transverse forces that gives rise to a maximum vertical shearing force of 4000 lb. Determine the maximum shearing stress in the beam.

The maximum shearing stress in a beam of rectangular cross-section was shown in Problem 22 to be 50% greater than the average shearing stress. The average shearing stress over the section where the shearing force is 4000 lb is obtained by dividing the shearing force by the cross-sectional area of 48 in^2. This gives

$$(s_s)_{av} = 4000/48 = 83.3 \text{ lb/in}^2$$

The maximum shearing stress occurs at the neutral axis of the beam and is

$$(s_s)_{max} = (3/2)(83.3) = 125 \text{ lb/in}^2$$

24. A beam of rectangular cross-section is simply supported at the ends and subject to the single concentrated force shown in Fig. (a) below. Determine the maximum shearing stress in the beam. Also, determine the shearing stress at a point one inch below the top of the beam at a section one foot to the right of the left reaction.

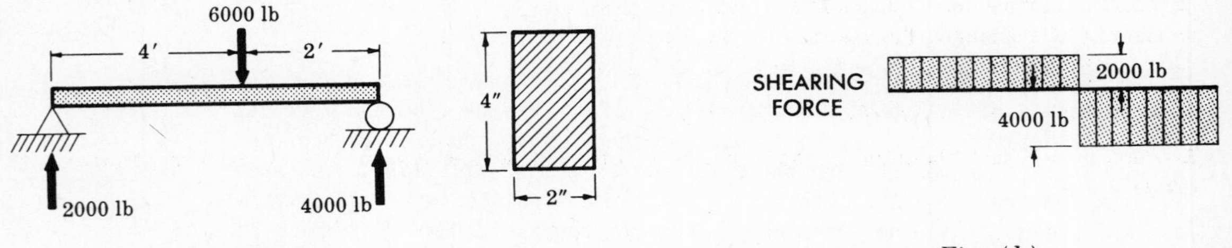

Fig. (a) Fig. (b)

The reactions are readily found from statics to be 2000 lb and 4000 lb as shown. The shearing force diagram for this type of loading was discussed in Problem 4, Chapter 6. For this beam the shear diagram has the appearance as in Fig. (b) above.

Inspection of the above shear diagram reveals that the maximum value of the shearing force is 4000 lb occurring at all cross-sections to the right of the 6000 lb load. The average value of the shearing stress acting over any cross-section in this region is simply the shearing force divided by the cross-sectional area, i.e.

$$(s_s)_{av} = 4000/2(4) = 500 \text{ lb/in}^2$$

By Problem 22 the maximum shearing stress is 50% greater than the average value. Hence

$$(s_s)_{max} = (3/2)(500) = 750 \text{ lb/in}^2$$

This maximum shearing stress occurs at all points along the neutral axis of the beam to the right of the 6000 lb load.

From the shear diagram, the shearing force acting over a section 1 ft to the right of the left reaction is 2000 lb. The shearing stress s_s at any point in this section a distance y_o from the neutral axis was shown in Problem 22 to be

$$s_s = \frac{V}{2I}\left(\frac{h^2}{4} - y_o^2\right)$$

At a point 1 in. below the top fibers of the beam, $y_o = 1$ in. Also, we have $h = 4$ in. and $I = bh^3/12 = 2(4)^3/12 = 10.7 \text{ in}^4$. Substituting,

$$s_s = \frac{2000}{2(10.7)}\left(\frac{16}{4} - 1\right) = 280 \text{ lb/in}^2.$$

25. A simply supported wood beam of rectangular cross-section is loaded as shown in Fig. (*a*). Determine the magnitude and location of the maximum shearing stress in the beam. Also, determine the maximum bending stress in the beam.

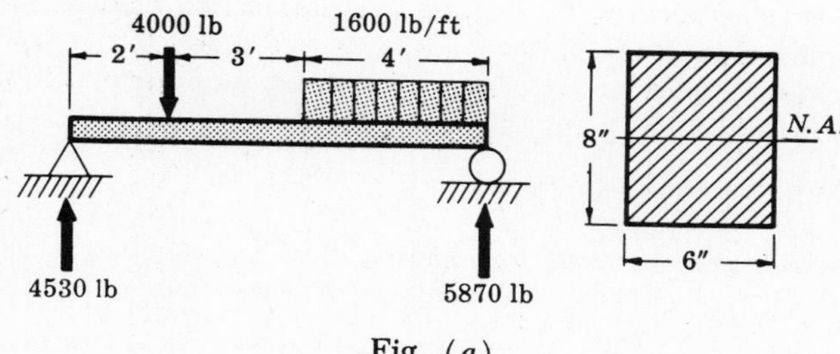

Fig. (*a*)

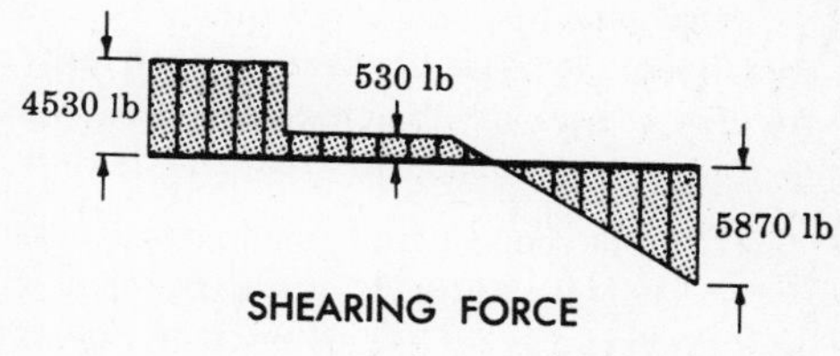

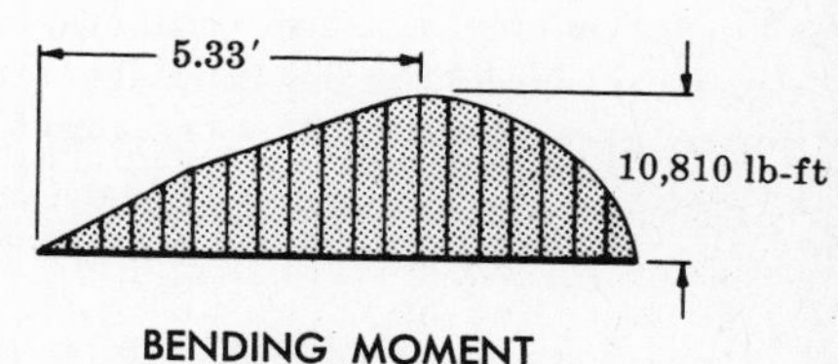

Fig. (*b*)

The shearing force and bending moment diagrams for this beam were determined in Problem 8, Chapter 6. There they were found to have the appearance as in Fig. (*b*).

Inspection of the shear diagram reveals that the maximum shearing force V is 5870 lb, occurring adjacent to the right support. The average shearing stress over that cross-section just to the left of the support is

$$(s_s)_{av} = 5870/6(8) = 122 \text{ lb/in}^2$$

Since the maximum shearing stress in a beam of rectangular cross-section is 50% greater than the average value,

$$(s_s)_{max} = (3/2)(122) = 183 \text{ lb/in}^2$$

Thus the maximum shearing stress in the beam is 183 lb/in^2, occurring at the neutral axis of the beam just to the left of the right support.

The bending moment diagram reveals that the maximum bending moment in the beam is 10,810 lb-ft or 130,000 lb-in. The maximum bending stress occurs at the outer fibers of the beam at this section of maximum bending moment and is given by

$$s = Mc/I$$

Here $c = 4$ in., i.e. the distance from the neutral axis to the outer fibers. Also, $I = 6(8)^3/12 = 256 \text{ in}^4$. Substituting,

$$s_{max} = 130{,}000(4)/256 = 2030 \text{ lb/in}^2$$

Thus the maximum bending stress is 2030 lb/in^2 occurring at the outer fibers of the beam at the section 5.33 ft to the right of the left reaction. This stress is tensile at the lower surface of the beam and compressive at the upper surface.

26. A simply supported redwood beam of rectangular cross-section is loaded by a single concentrated force P, as shown in Fig. (a). The maximum allowable stress in bending is 2000 lb/in^2 and the allowable horizontal shearing stress is 125 lb/in^2. Determine the maximum value the load P may have.

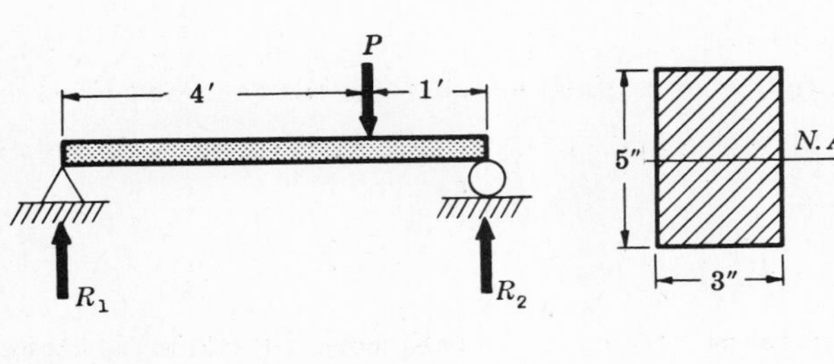

Fig. (a)

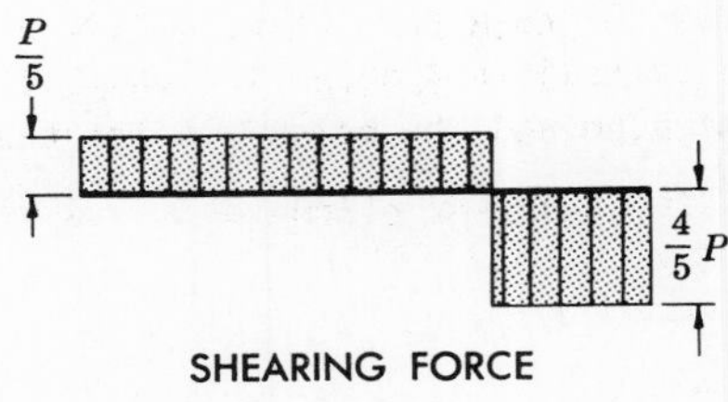

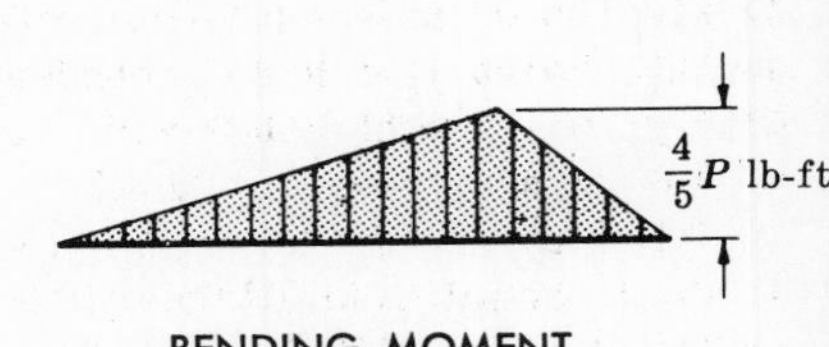

Fig. (b)

From statics the reactions are readily found to be $R_1 = P/5$ and $R_2 = 4P/5$. The shearing force and bending moment diagrams for this type of loading were discussed in Problem 4, Chapter 6. For this particular loading they appear as in Fig. (b).

It is perhaps simplest first to investigate the maximum value that P may have assuming that the shear stress is the controlling factor, then next to investigate the maximum value of P assuming that the bending stress is the controlling factor. The desired maximum value of P is then the minimum of these two values.

Let us first assume that the allowable shearing stress of 125 lb/in^2 is set up in the beam. From the shear diagram the maximum shearing force V is $4P/5$. The maximum shear stress occurs at the neutral axis at all sections to the right of the load P and was found in Problem 22 to be

$$(s_s)_{max} = \frac{3V}{2bh}$$

where b and h denote the width and height of the beam respectively. Substituting,

$$125 = \frac{3(4P/5)}{2(3)(5)} \quad \text{or} \quad P = 1560 \text{ lb}$$

We shall next assume that the allowable bending stress of 2000 lb/in^2 is set up in the beam. The maximum bending stress occurs at the extreme upper and lower fibers of the beam just under the load, since the bending moment is a maximum there. This maximum bending moment is seen from the moment diagram to be $4P/5$ lb-ft or $48P/5$ lb-in. The maximum bending stress is given by $s = Mc/I$ where $c = 2.5$ in. denotes the distance from the neutral axis to the outer fibers of the beam. At the lower surface of the beam the fibers are in tension and if the allowable stress of 2000 lb/in^2 is set up there, we have

$$2000 = \frac{(48P/5)(2.5)}{3(5)^3/12} \quad \text{or} \quad P = 2600 \text{ lb}$$

The maximum value the load P may have is thus the minimum of these two values, or 1560 lb. Thus the shearing stress governs the maximum allowable load.

27. Consider the cantilever beam subject to the concentrated load shown in the diagram below. The cross-section of the beam is of T-shape. Determine the maximum shearing stress in the beam and also determine the shearing stress 1 in. from the top surface of the beam at a section adjacent to the supporting wall.

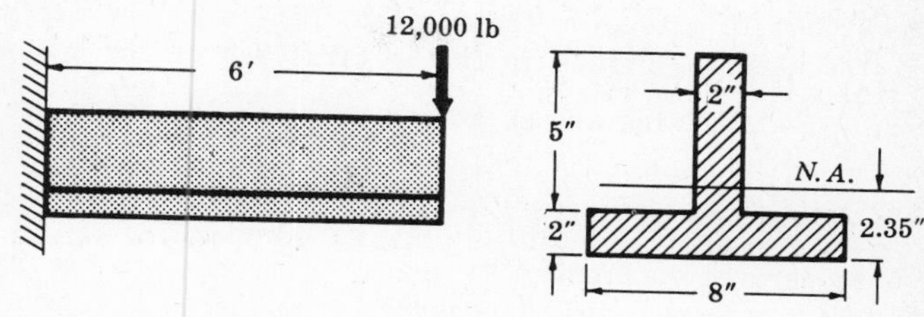

The shearing force diagram for this type of loading was discussed in Problem 1, Chapter 6. The shear force has a constant value of 12,000 lb at all points along the length of the beam. Because of this simple, constant value the shear diagram need not be drawn.

Further, the location of the centroid and the moment of inertia about the centroidal axis for this particular cross-section were determined in Problem 13, Chapter 7. The centroid was found to be 2.35 in. above the lower surface of the beam and the moment of inertia about a horizontal axis through the centroid was found to be 101 in^4.

The shearing stress at a distance y_o from the neutral axis through the centroid was found in Problem 21 to be

$$s_s = \frac{V}{Ib}\int_{y_o}^{c} y\,dA$$

Inspection of this equation reveals that the shearing stress is a maximum at the neutral axis, since at that point $y_o = 0$ and consequently the integral assumes the largest possible value. It is not necessary to integrate, however, since the integral is known in this case to represent the first moment of the area between the neutral axis and the outer fibers of the beam about the neutral axis. This area is represented by the shaded region shown to the right. The value of the integral could also, of course, be found by taking the first moment of the unshaded area below the neutral axis about the line, but that calculation would be somewhat more difficult.

Thus the first moment of the shaded area about the neutral axis is

$$2(4.65)(2.33) = 21.6 \text{ in}^3$$

and the shearing stress at the neutral axis, where $b = 2$ in., is found by substitution in the above general formula to be

$$s_s = \frac{12,000}{101(2)}(21.6) = 1290 \text{ lb/in}^2$$

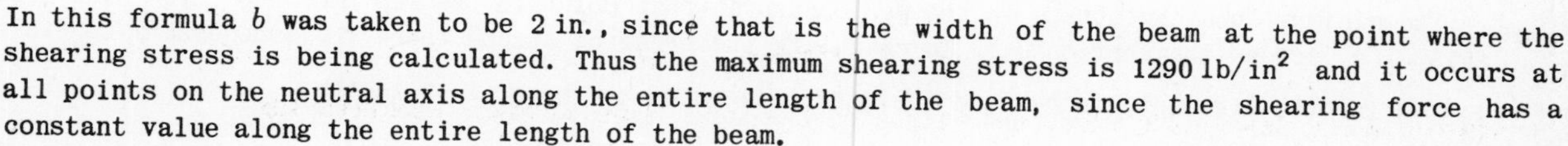

In this formula b was taken to be 2 in., since that is the width of the beam at the point where the shearing stress is being calculated. Thus the maximum shearing stress is 1290 lb/in^2 and it occurs at all points on the neutral axis along the entire length of the beam, since the shearing force has a constant value along the entire length of the beam.

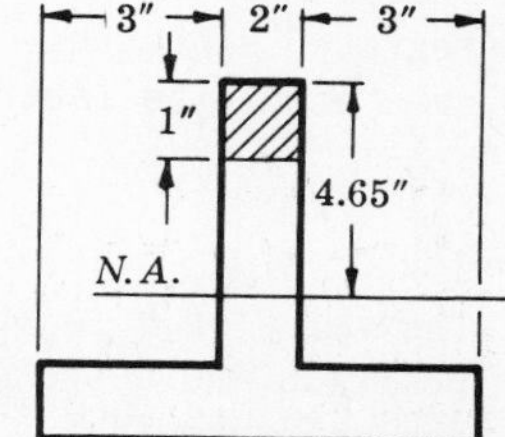

The shearing stress 1 in. from the top surface of the beam is again given by the formula

$$s_s = \frac{V}{Ib}\int_{y_o}^{c} y\,dA$$

Now, the integral represents the first moment of the new shaded area shown at the right about the neutral axis. Again it is not necessary to integrate to evaluate the integral, since the coordinate of the centroid of this shaded area is known. It is 4.15 in. above the neutral axis. Thus the first moment of this shaded area about the neutral axis is $2(1)(4.15) = 8.30$ in^3, and the shearing stress 1 in. below the top fibers is

$$s_s = \frac{12,000}{101(2)}(8.30) = 495 \text{ lb/in}^2$$

Again, b was taken to be 2 in. since that is the width of the beam at the point where the shearing stress is being evaluated. Since the shearing force is equal to 12,000 lb everywhere along the length of the beam, the shearing stress 1 in. below the top fibers is 495 lb/in^2 everywhere along the beam.

28. Consider a beam having an I-type cross-section as shown in the diagram below. A shearing force V of

32,000 lb acts over the section. Determine the maximum and minimum values of the shearing stress in the vertical web of the section.

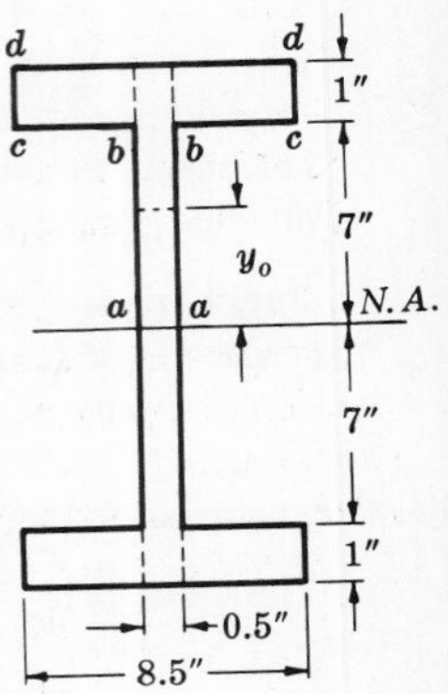

The shearing stress at any point in the cross-section is given by

$$s_s = \frac{V}{Ib}\int_{y_o}^{c} y\,dA$$

as derived in Problem 21. Here, y_o represents the location of the section on which s_s acts, and is measured from the neutral axis as shown. In this expression, I represents the moment of inertia of the entire cross-section about the neutral axis which passes through the centroid of the section. I is readily calculated by dividing the section into rectangles, as indicated by the dotted lines, and we have

$$I = \frac{1}{12}(0.5)(16)^3 + 2[\frac{1}{12}(8.0)(1)^3 + (8.0)(1)(7.5)^2] = 1072 \text{ in}^4$$

Inspection of the general formula for shearing stress reveals that this stress has a maximum value when $y_o = 0$, i.e. at the neutral axis, since at that point the integral takes on its largest possible value. It is not necessary to integrate to obtain the value of $\int_{y_o}^{c} y\,dA$, since this integral is shown to represent the first moment of the area between $y_o = 0$ (i.e. the neutral axis) and the outer fibers of the beam. This area is shaded in the adjoining diagram. For this area we have, taking its first moment about the neutral axis,

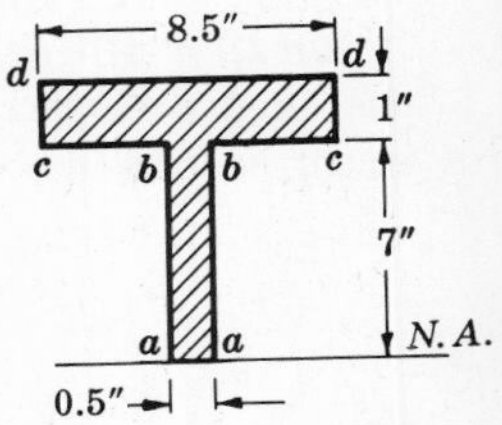

$$\int_0^8 y\,dA = 7(0.5)(3.5) + 8.5(1)(7.5) = 75.9 \text{ in}^3$$

Consequently the maximum shearing stress in the web occurs at the section a-a along the neutral axis and is found by substituting in the general formula for shearing stress to be

$$(s_s)_{max} = \frac{32,000}{(1072)(0.5)}(75.9) = 4540 \text{ lb/in}^2$$

The minimum shearing stress in the web occurs at that point in the web farthest from the neutral axis, i.e. across the section b-b. To calculate the shearing stress there, it is necessary to evaluate $\int_{y_o}^{c} y\,dA$ for the area between b-b and the outer fibers of the beam. This is the shaded area shown in the adjoining sketch. Again, it is not necessary to integrate, since this integral merely represents the first moment of this shaded area about the neutral axis. It is

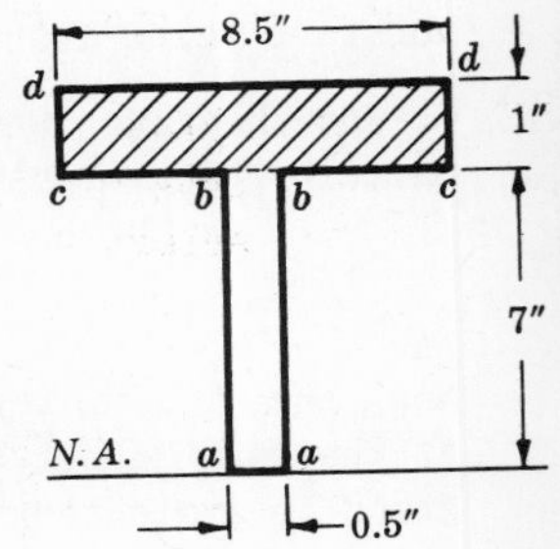

$$\int_7^8 y\,dA = (8\tfrac{1}{2})(1)(7\tfrac{1}{2}) = 63.7 \text{ in}^3$$

The value of b is still 0.5 in., since that is the width of the beam at the position where the shearing stress is being calculated. Substituting in the general formula

$$(s_s)_{min} = \frac{32,000}{(1072)(0.5)}(63.7) = 3800 \text{ lb/in}^2$$

It is to be noted that there is not too great a difference between the maximum and minimum values of shearing stress in the web of the beam. In fact, it is customary to calculate only an approximate value of the shearing stress in the web of such an I-beam. This value is obtained by dividing the total shearing force V by the cross-sectional area of the web alone. This approximate value becomes

$$(s_s)_{av} = \frac{32,000}{16(0.5)} = 4000 \text{ lb/in}^2$$

A more advanced analysis of shearing stresses in an I-beam reveals that the vertical web resists nearly all of the shearing force V and that the horizontal flanges resist only a small portion of this force. The shear stress in the web of an I-beam is specified by various codes at rather low values. Thus some codes specify 10,000 lb/in^2, others 13,000 lb/in^2.

SUPPLEMENTARY PROBLEMS

29. A cypress beam has a cross-section 4 in. × 8 in. The beam bends about an axis parallel to the 4 in. face. If the maximum bending stress developed is 7200 lb/in^2, determine the maximum bending moment. *Ans.* 307,200 lb-in

30. A cantilever beam 9 ft long carries a concentrated force of 8000 lb at its free end. The material is structural steel and the maximum bending stress is not to exceed 18,000 lb/in^2. Determine the required diameter if the bar is to be circular. *Ans.* 7.86 in.

31. An oak beam 12 ft long is simply supported at the ends and loaded at the center by a concentrated force of 1500 lb. The proportional limit of the wood is 7900 lb/in^2 and a safety factor of 4 is sufficient. Determine the cross-section of the beam if (*a*) it is to be square, and (*b*) if the depth is to be $1\frac{1}{2}$ times the width. *Ans.* *a*) 5.46 in., *b*) 4.17 in. × 6.27 in.

32. A simply supported pine beam is 10 ft long and carries a uniformly distributed load of 40 lb per foot of length. The maximum bending stress is not to exceed 1500 lb/in^2. If the depth of the beam is to be $1\frac{1}{4}$ times the width, determine the required cross-section. *Ans.* 2.5 in. × 3.1 in.

33. A 10 WF 29 section (see table at end of this chapter for properties) is used as a cantilever beam. The beam is 5 ft long and the allowable bending stress is 18,000 lb/in^2. Determine the maximum allowable intensity of uniform load that may be carried along the entire length of the beam. *Ans.* 3800 lb/ft

34. A steel beam 50 in. in length is simply supported at each end and carries a concentrated load of 25,000 lb acting 20 in. from one of the supports. Determine the maximum bending stresses set up in the beam if the cross-section is rectangular, 4 in. wide by 6 in. deep. *Ans.* 12,500 lb/in^2

35. Determine the maximum bending stresses for a bar loaded as in the preceding problem if the beam is a 10 WF 23 section. *Ans.* 12,500 lb/in^2

36. A strip of steel 0.04 in. thick is bent into an arc of a circle of 30 in. radius. Determine the maximum bending stress. Take $E = 30 \cdot 10^6$ lb/in^2. *Ans.* 20,000 lb/in^2

37. The maximum bending moment existing in a steel beam is 750,000 lb-in. Select the most economical wide-flange section to resist this moment if the working stress in both tension and compression is 20,000 lb/in^2. *Ans.* 12 WF 32

38. The beam shown in Fig. (*a*) below is simply supported at the ends and carries the two symmetrically placed loads of 15,000 lb each. If the working stress in either tension or compression is 18,000 lb/in^2, select the most economical wide-flange section to support these loads. *Ans.* 10 WF 21

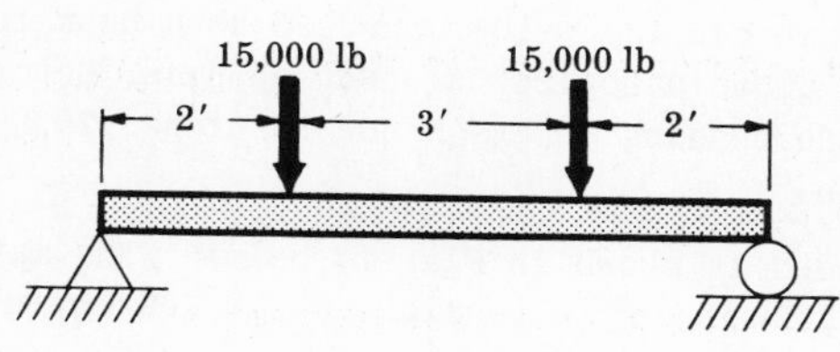

Fig. (*a*) Prob. 38

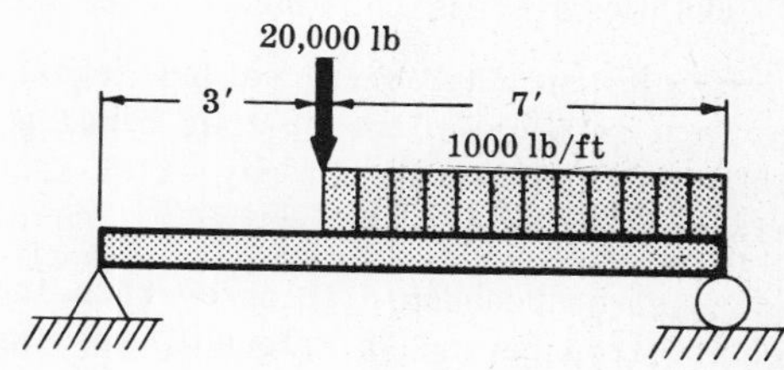

Fig. (*b*) Prob. 39

39. Consider the simply supported beam carrying the concentrated and uniform loads shown in Fig. (*b*) above. Select a suitable wide-flange section to resist these loads based upon a working stress in either tension or compression of 20,000 lb/in^2. *Ans.* 12 WF 25

40. The two distributed loads are carried by the simply supported beam as shown in Fig. (*a*) below. The cross-section of the beam is an 8 WF 28 section. Determine the magnitude and location of the maximum bending stress in the beam. *Ans.* 9000 lb/in^2, 5.5 ft from the right support

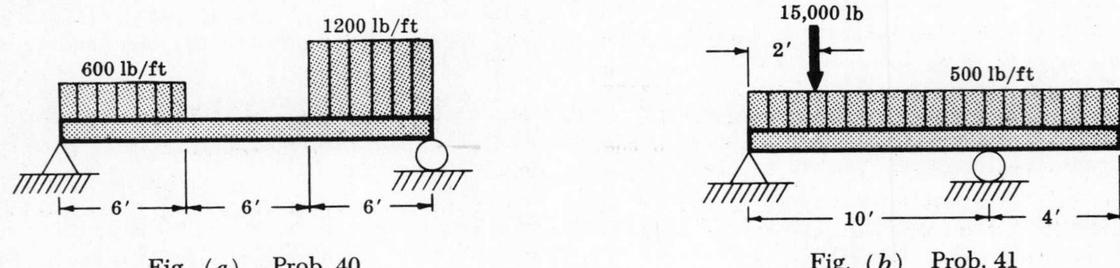

Fig. (*a*) Prob. 40

Fig. (*b*) Prob. 41

41. The overhanging beam shown in Fig. (*b*) above is of circular cross-section 6 in. in diameter. Determine (*a*) the maximum bending stress in the bar and its location, (*b*) the bending stress at the outer fibers of the bar at the section midway between supports.
Ans. *a*) 15,400 lb/in^2, under the concentrated load; *b*) 10,850 lb/in^2

42. Select the most economical wide-flange section to carry the loading described in the preceding problem. Use a working stress in both tension and compression of 18,000 lb/in^2. *Ans.* 10 WF 21

43. Refer to Fig. (*c*) below. A T-beam having the cross-section shown projects five feet from a wall as a cantilever beam and carries a uniformly distributed load of 600 lb/ft including its own weight. Determine the maximum tensile and compressive bending stresses in the beam.
Ans. +6400 lb/in^2, −15,200 lb/in^2

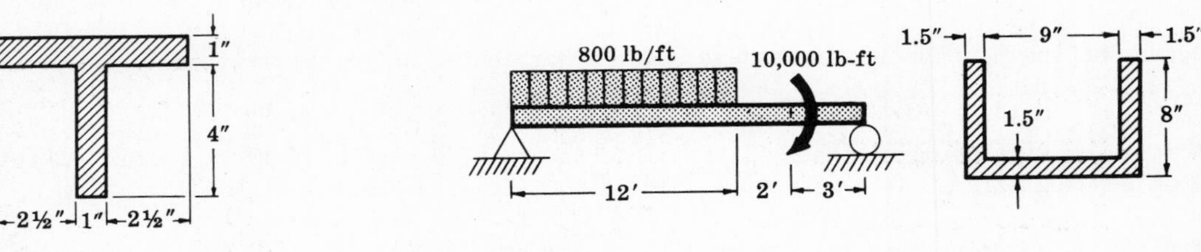

Fig. (*c*) Prob. 43

Fig. (*d*) Prob. 44

44. The simply supported steel beam is loaded by both the uniformly distributed load and the couple shown in Fig. (*d*) above. The beam is of channel-type cross-section as illustrated. Determine the maximum tensile and compressive stresses set up in the beam.
Ans. 3020 lb/in^2 tension, 5520 lb/in^2 compression

45. Two 5 × 5 × 3/8 in. angles are welded together as shown in Fig. (*e*) below, and used as a beam to support loads in a vertical plane so that bending takes place about a horizontal neutral axis. Determine the maximum bending moment that may exist in the beam if the bending stress is not to exceed 20,000 lb/in^2 in either tension or compression. *Ans.* 97,500 lb-in

46. The channel-shape beam with an overhanging end is loaded as shown in Fig. (*f*) below. The material is gray cast iron having an allowable working stress of 5000 lb/in^2 in tension and 20,000 lb/in^2 in compression. Determine the maximum allowable value of *P*. *Ans.* 2400 lb

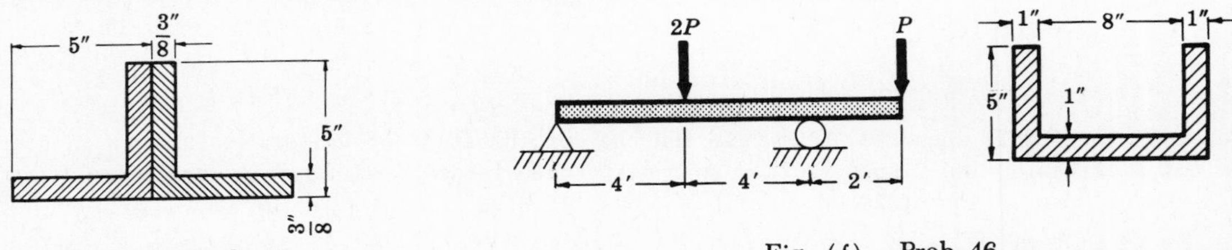

Fig. (*e*) Prob. 45

Fig. (*f*) Prob. 46

47. A wood beam 4 × 6 in. in cross-section is subject to a maximum transverse shearing force of 2000 lb. Determine the shearing stress at points one inch apart over the depth of the beam.
Ans. 0, 69.4 lb/in^2, 111 lb/in^2, 125 lb/in^2, 111 lb/in^2, 69.4 lb/in^2, 0

48. The simply supported beam of length 10 ft and cross-section 4 in. by 8 in. carries a uniform load of 200 lb/ft as shown in the adjoining diagram. Neglecting the weight of the beam, find: (*a*) the maximum normal stress in the beam, (*b*) the maximum shearing stress in the beam, (*c*) the shearing stress at a point 2 ft to the right of R_1 and 1 in. below the top surface of the beam.
Ans. *a*) 705 lb/in^2, *b*) 47 lb/in^2, *c*) 12.3 lb/in^2

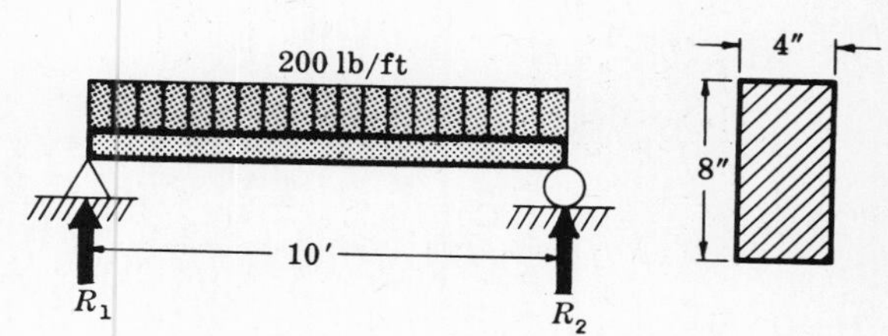

49. Determine (*a*) the maximum bending stress and (*b*) the maximum shearing stress in the beam shown in Fig. (*a*) below. The beam is simply supported and of rectangular cross-section.
Ans. *a*) 22,000 lb/in^2, *b*) 1660 lb/in^2

50. A rectangular beam of red cedar has a cross-section 6 × 8 in. The beam is simply supported at the ends and has a span of 8 ft. If the allowable bending stress is 2400 lb/in^2 and the allowable shearing stress is 90 lb/in^2, determine the intensity of uniform load that may be applied over the entire beam.
Ans. 720 lb/ft

51. A beam has the channel-type cross-section shown in Fig. (*b*) below. If the maximum shearing force along the length of the beam is 6000 lb, determine the maximum shearing stress in the beam.
Ans. 1110 lb/in^2

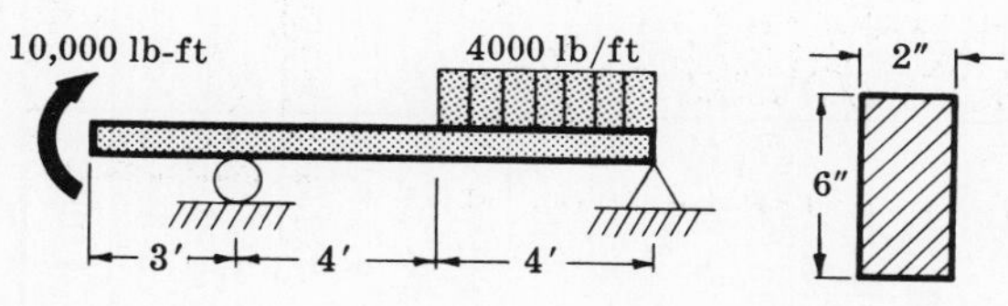

Fig. (*a*) Prob. 49

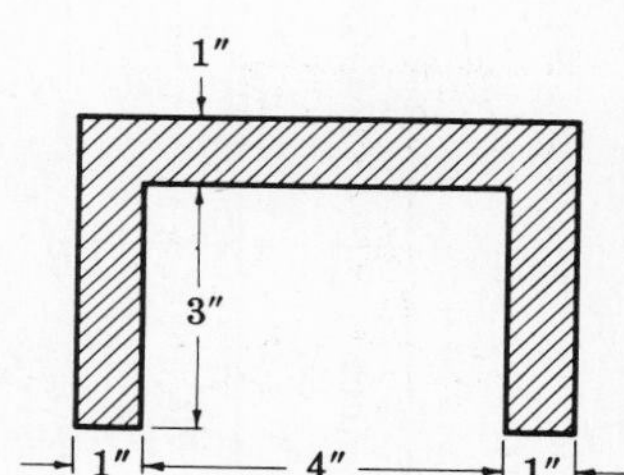

Fig. (*b*) Prob. 51

PROPERTIES OF WIDE FLANGE SECTIONS

(Abridged List)

Designation	Weight per foot (lb/ft)	Area (in^2)	I (about x-x axis) (in^4)	Z (in^3)	I (about y-y axis) (in^4)
18 WF 70	70.0	20.56	1153.9	128.2	78.5
18 WF 55	55.0	16.19	889.9	98.2	42.0
12 WF 72	72.0	21.16	597.4	97.5	195.3
12 WF 58	58.0	17.06	476.1	78.1	107.4
12 WF 50	50.0	14.71	394.5	64.7	56.4
12 WF 45	45.0	13.24	350.8	58.2	50.0
12 WF 40	40.0	11.77	310.1	51.9	44.1
12 WF 36	36.0	10.59	280.8	45.9	23.7
12 WF 32	32.0	9.41	246.8	40.7	20.6
12 WF 25	25.0	7.39	183.4	30.9	14.5
10 WF 89	89.0	26.19	542.4	99.7	180.6
10 WF 54	54.0	15.88	305.7	60.4	103.9
10 WF 49	49.0	14.40	272.9	54.6	93.0
10 WF 45	45.0	13.24	248.6	49.1	53.2
10 WF 37	37.0	10.88	196.9	39.9	42.2
10 WF 29	29.0	8.53	157.3	30.8	15.2
10 WF 23	23.0	6.77	120.6	24.1	11.3
10 WF 21	21.0	6.19	106.3	21.5	9.7
8 WF 40	40.0	11.76	146.3	35.5	49.0
8 WF 35	35.0	10.30	126.5	31.1	42.5
8 WF 31	31.0	9.12	109.7	27.4	37.0
8 WF 28	28.0	8.23	97.8	24.3	21.6
8 WF 27	27.0	7.93	94.1	23.4	20.8
8 WF 24	24.0	7.06	82.5	20.8	18.2
8 WF 19	19.0	5.59	64.7	16.0	7.9
6 WF $15\frac{1}{2}$	15.5	4.62	28.1	9.7	9.7

CHAPTER 9

Deflection of Beams - Double-Integration Method

INTRODUCTION. In Chapter 8 it was stated that lateral loads applied to a beam not only give rise to internal bending and shearing stresses in the bar, but also cause the bar to deflect in a direction perpendicular to its longitudinal axis. The stresses were examined in Chapter 8 and it is the purpose of this chapter and also Chapter 10 to examine two methods for calculating the deflections.

DEFINITION OF DEFLECTION OF A BEAM. The deformation of a beam is most easily expressed in terms of the deflection of the beam from its original unloaded position. The deflection is measured to the neutral surface of the deformed beam from the original neutral surface. The configuration assumed by the deformed neutral surface is known as the elastic curve of the beam. Figure 1, below, represents the beam in its original undeformed state and Figure 2 represents the beam in the deformed configuration it has assumed under the action of the loads.

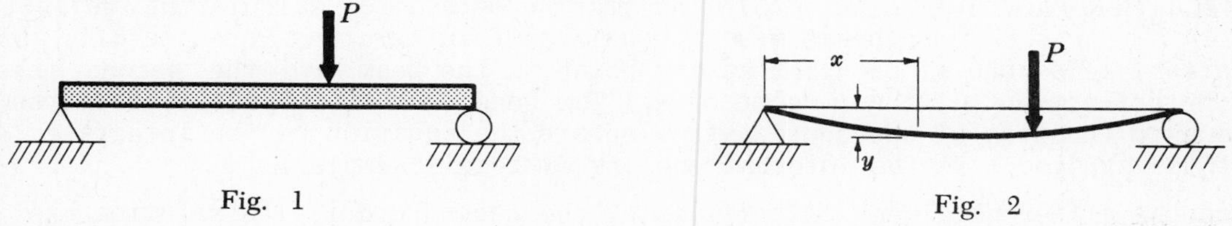

Fig. 1 Fig. 2

The displacement y is defined as the deflection of the beam. Usually it will be necessary to determine the deflection y for every value of x along the beam. This relation may be written in the form of an equation which is frequently called the equation of the deflection curve (or elastic curve) of the beam.

IMPORTANCE OF BEAM DEFLECTIONS. Specifications for the design of beams frequently impose limitations upon the deflections as well as the stresses. Consequently, in addition to the calculation of stresses as outlined in Chapter 8, it is essential that the designer be able to determine deflections. For example, in many building codes the maximum allowable deflection of a beam is not to exceed 1/300 of the length of the beam. Thus, a well designed beam must not only be able to carry the loads to which it will be subjected but it must not undergo undesirably large deflections. Also, the evaluation of reactions for statically indeterminate beams involves the use of various deformation relationships. These will be examined in detail in Chapter 11.

METHODS OF DETERMINING BEAM DEFLECTIONS. Numerous methods are available for the determination of beam deflections. The most commonly used are:

(*a*) The Double-Integration method

(*b*) The Moment-Area method

(*c*) Elastic Energy methods.

The first method is described in the present chapter. The Moment-Area technique will be examined in Chapter 10 and discussions of energy methods may be found in advanced books on strength of materials.

DOUBLE INTEGRATION METHOD. The differential equation of the deflection curve of the bent beam is

$$EI\frac{d^2y}{dx^2} = M \qquad (1)$$

where x and y are the coordinates shown in the above sketch of the deformed beam. That is, y is the deflection of the beam. This expression is derived in Problem 1. In this equation E denotes the modulus of elasticity of the beam and I represents the moment of inertia of the beam cross-section about the neutral axis, which passes through the centroid of the cross-section. Also, M represents the bending moment at the distance x from one end of the beam. This quantity was defined in Chapter 6 to be the algebraic sum of the moments of the external forces to one side of the section at a distance x from the end about an axis through this section. Usually, M will be a function of x and it will be necessary to integrate Equation *(1)* twice to obtain an algebraic equation expressing the deflection of y as a function of x.

Equation *(1)* is the basic differential equation that governs the deflection of all beams irrespective of the type of applied loading. For applications, see Problems 2, 6, 8, 10, 13, 15, 17, 20, 22, and 24.

THE INTEGRATION PROCEDURE. The double integration method for calculating deflections of beams merely consists of integrating Equation *(1)*. The first integration yields the slope dy/dx at any point in the beam and the second integration gives the deflection y for any value of x. The bending moment M must, of course, be expressed as a function of the coordinate x before the equation can be integrated. For the cases to be studied here the integrations are extremely simple.

Since the differential equation *(1)* is of the second order, its solution must contain two constants of integration. These two constants must be evaluated from known conditions concerning the slope or deflection at certain points in the beam. For example, in the case of a cantilever beam the constants would be determined from the conditions of zero change of slope as well as zero deflection at the built-in end of the beam.

Frequently two or more equations are necessary to describe the bending moment in the various regions along the length of a beam. This was emphasized in Chapter 6. In such a case, Equation *(1)* must be written for each region of the beam and integration of these equations yields two constants of integration for each region. These constants must then be determined so as to impose conditions of continuous deformations and slopes at the points common to adjacent regions. For example, see Problems 15, 17, 20, 22, and 24.

SIGN CONVENTIONS. The sign conventions for bending moment adopted in Chapter 6 will be retained here. The quantities E and I appearing in Equation *(1)* are, of course, positive. Thus, from this equation, if M is positive for a certain value of x, then d^2y/dx^2 is also positive. With the above sign convention for bending moments, it is necessary to consider the coordinate x along the length of the beam to be positive to the right and the deflection y to be positive upward. This will be explained in detail in Problem 1. With these algebraic signs the integration of Equation *(1)* may be carried out to yield the deflection y as a function of x, with the understanding that upward beam deflections are positive and downward deflections negative.

ASSUMPTIONS AND LIMITATIONS. In the derivation of Equation *(1)* it is assumed that deflections caused by shearing action are negligible compared to those caused by bending action. Also, it is assumed that the deflections are small compared to the cross-sectional dimensions of the beam. Further, the beam is presumed to be straight prior to the application of the loads. These conditions are in addition to the assumptions concerning beam theory that were listed in Chapter 8.

SOLVED PROBLEMS

1. Obtain the differential equation of the deflection curve of a beam loaded by lateral forces.

In Problem 1 of Chapter 8 the relationship

$$M = \frac{EI}{\rho} \tag{1}$$

was derived. In this expression M denotes the bending moment acting at a particular cross-section of the beam, ρ the radius of curvature to the neutral surface of the beam at this same section, E the modulus of elasticity, and I the moment of inertia of the cross-sectional area about the neutral axis passing through the centroid of the cross-section. In this book we will be concerned only with those beams for which E and I are constant along the entire length of the beam, but in general both M and ρ will be functions of x.

Equation *(1)* may be written in the form

$$\frac{1}{\rho} = \frac{M}{EI} \tag{2}$$

where the left side of equation *(2)* represents the curvature of the neutral surface of the beam. Since M will vary along the length of the beam, the deflection curve will be of variable curvature.

Let the heavy line in the adjacent sketch represent the deformed neutral surface of the bent beam. Originally the beam coincided with the x-axis prior to loading and the coordinate system that is usually found to be most convenient is shown in the sketch. The deflection y is taken to be positive in the upward direction; hence for the particular beam shown, all deflections are negative.

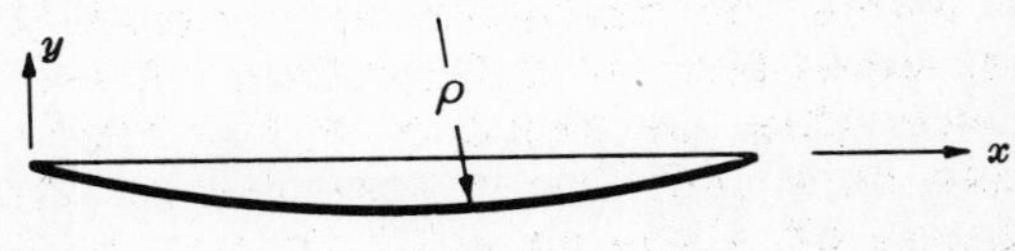

An expression for the curvature at any point along the curve representing the deformed beam is readily available from differential calculus. The exact formula for curvature is

$$\frac{1}{\rho} = \frac{d^2y/dx^2}{[1+(dy/dx)^2]^{3/2}} \tag{3}$$

In this expression, dy/dx represents the slope of the curve at any point; and for small beam deflections this quantity and in particular its square are small in comparison to unity and may reasonably be neglected. This assumption of small deflections simplifies the expression for curvature to the form

$$\frac{1}{\rho} \approx \frac{d^2y}{dx^2} \tag{4}$$

Hence for small deflections, equation *(2)* becomes $d^2y/dx^2 = M/EI$ or

$$EI\frac{d^2y}{dx^2} = M \tag{5}$$

This is the differential equation of the deflection curve of a beam loaded by lateral forces. In any problem it is necessary to integrate this equation to obtain an algebraic relationship between the deflection y and the coordinate x along the length of the beam. This will be carried out in the following problems.

2. Determine the deflection at every point of the cantilever beam subject to the single concentrated force P, as shown in the adjoining figure.

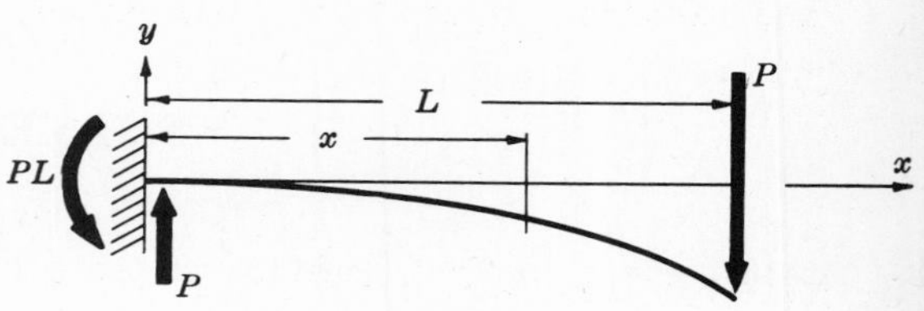

The x-y coordinate system shown is introduced, where the x-axis coincides with the original unbent position of the beam. The deformed beam has the appearance indicated by the heavy line. It is first necessary to find the reactions exerted by the supporting wall upon the bar; and as discussed in Problem 1, Chapter 6 these are easily found from statics to be a vertical force reaction P and a moment PL as shown.

The bending moment at any cross-section a distance x from the wall is given by the sum of the moments of these two reactions about an axis through this section. Evidently the upward force P produces a positive bending moment Px, and the couple PL if acting alone would produce curvature of the bar as shown. According to the sign convention of Chap. 6, this constitutes negative bending. Hence the bending moment M at the section x is

$$M = -PL + Px$$

The differential equation of the bent beam is

$$EI\frac{d^2y}{dx^2} = M$$

where E denotes the modulus of elasticity of the material and I represents the moment of inertia of the cross-section about the neutral axis. Substituting,

$$1) \qquad EI\frac{d^2y}{dx^2} = -PL + Px$$

This equation is readily integrated once to yield

$$2) \qquad EI\frac{dy}{dx} = -PLx + \frac{Px^2}{2} + C_1$$

which represents the equation of the slope, where C_1 denotes a constant of integration. This constant may be evaluated by use of the condition that the slope dy/dx of the beam at the wall is zero since the beam is rigidly clamped there. Thus $(dy/dx)_{x=0} = 0$. Equation *(2)* is true for all values of x and y, and if the condition at $x = 0$ is substituted we obtain $0 = 0 + 0 + C_1$ or $C_1 = 0$.

Next, integration of equation *(2)* yields

$$3) \qquad EIy = -PL\frac{x^2}{2} + \frac{Px^3}{6} + C_2$$

where C_2 is a second constant of integration. Again, the condition at the supporting wall will determine this constant. There, at $x = 0$, the deflection y is zero since the bar is rigidly clamped. Substituting $(y)_{x=0} = 0$ in equation *(3)*, we find $0 = 0 + 0 + C_2$ or $C_2 = 0$.

Thus equations *(2)* and *(3)* with $C_1 = C_2 = 0$ give the slope dy/dx and deflection y at any point x in the beam. The deflection is a maximum at the right end of the beam ($x = L$) under the load P, and from equation *(3)* it is found to be

$$EI(y)_{max} = -\frac{PL^3}{3}$$

where the negative value denotes that this point on the deflection curve lies below the x-axis. If only the magnitude of the maximum deflection at $x = L$ is desired, it is usually denoted by Δ and we have

4) $$\Delta_{max} = \frac{PL^3}{3EI}$$

3. The cantilever beam shown in Problem 2 is 10 ft in length and loaded by a force P of 10,000 lb. The beam is an 18 WF 47 steel section, having a moment of inertia about the neutral axis of 736.4 in^4. Determine the maximum deflection of the beam. Take $E = 30 \cdot 10^6$ lb/in^2.

The maximum deflection occurs at the free end of the beam under the concentrated load and was found in Problem 2 to be

$$\Delta_{max} = \frac{PL^3}{3EI} = \frac{10,000(120)^3}{3(30 \cdot 10^6)(736.4)} = 0.261 \text{ in.}$$

This deflection is downward as indicated in the drawing in Problem 2. In the derivation of this deflection formula it was assumed that the material of the beam follows Hooke's law. Actually, from the above calculation alone there is no assurance that the material is not stressed beyond the proportional limit. If it were, then the basic beam bending equation $EI(d^2y/dx^2) = M$ would no longer be valid and the above numerical value would be meaningless. Consequently, in every problem involving beam deflections it is to be emphasized that it is necessary to determine that the maximum bending stress in the beam is below the proportional limit of the material. This is easily done by use of the flexure formula derived in Problem 1, Chapter 8. According to this formula

$$s = Mc/I$$

where s denotes the bending stress, M the bending moment, c the distance from the neutral axis to the outer fibers of the beam, and I the moment of inertia of the beam cross-section about the neutral axis. The maximum bending moment in this problem occurs at the supporting wall and is given by $M_{max} = 10,000(120) = 1,200,000$ lb-in. Substituting in the formula for bending stress, we have

$$s_{max} = 1,200,000(9)/736.4 = 14,700 \text{ lb/in}^2$$

Since this value is below the proportional limit of steel, which is approximately 30,000 lb/in^2 for low-carbon structural steel, the use of the beam deflection equation was justifiable.

4. Determine the slope of the right end of the cantilever beam loaded as shown in Problem 2. For the beam described in Problem 3, determine the value of this slope.

In Problem 2 the equation of the slope was found to be

$$EI\frac{dy}{dx} = -PLx + \frac{Px^2}{2}$$

At the free end, $x = L$, and we have $$EI\left(\frac{dy}{dx}\right)_{x=L} = -PL^2 + \frac{PL^2}{2}$$

The slope at the free end is thus $$\left(\frac{dy}{dx}\right)_{x=L} = -\frac{PL^2}{2EI}.$$

For the beam described in Problem 3, this becomes $$\left(\frac{dy}{dx}\right)_{x=L} = -\frac{10,000(120)^2}{2(30 \cdot 10^6)(736.4)}$$

$$= -0.0326 \text{ radian.}$$

5. Determine the deflection at the midpoint, x = 5 ft, of the beam described in Problem 3.

In Problem 2 the equation of the deflection curve was found to be $EIy = -PL\frac{x^2}{2} + \frac{Px^3}{6}$.

At the midpoint, x = 60 in. and the other parameters have the same values as in Prob. 3. Substituting

$$(y)_{x=60\text{ in.}} = \frac{1}{(30\cdot 10^6)(736.4)}\left[-10{,}000(120)\frac{(60)^2}{2} + \frac{10{,}000(60)^3}{6}\right] = -0.0816\text{ in.}$$

The negative sign indicates that this point on the deflected beam lies below the x-axis.

6. Determine the deflection at every point of the cantilever beam subject to the uniformly distributed load of w lb per unit length as shown.

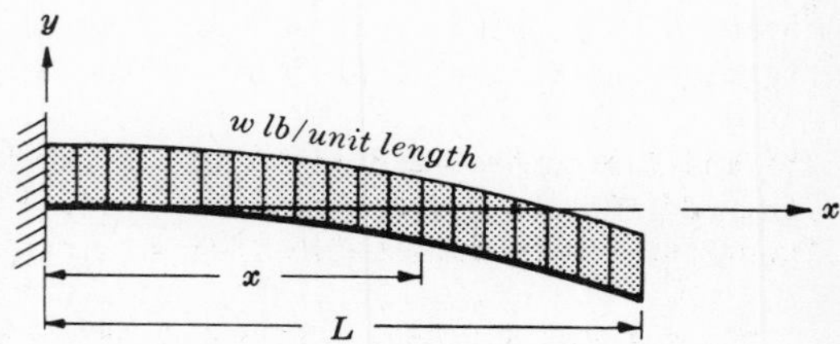

The x-y coordinate system shown is introduced, where the x-axis coincides with the original unbent position of the beam. The deformed beam has the appearance indicated by the heavy line. The equation for the bending moment could be determined in a manner analogous to that used in Problem 2, but instead let us seek a slight simplification of that technique. Let us determine the bending moment at the section a distance x from the wall by considering the forces to the right of this section rather than those to the left.

The force of w lb/unit length acts over the length $(L-x)$ to the right of this section and hence the resultant force is $w(L-x)$ lb. This force acts at the midpoint of this length of beam to the right of x and thus its moment arm from x is $\frac{1}{2}(L-x)$. The bending moment at the section x is thus given by

MOMENT EQ.

$$M = -\frac{w}{2}(L-x)^2$$

the negative sign being necessary since downward loads produce negative bending.

The differential equation describing the bent beam is thus

1) DIF EQ.
$$EI\frac{d^2y}{dx^2} = -\frac{w}{2}(L-x)^2$$

The first integration yields

2)
$$EI\frac{dy}{dx} = \frac{w}{2}\left[\frac{(L-x)^3}{3}\right] + C_1$$

where C_1 denotes a constant of integration.

This constant may be evaluated by realizing that the left end of the beam is rigidly clamped. At that point, $x = 0$, we have no change of slope and hence $(dy/dx)_{x=0} = 0$. Substituting these values in equation *(2)*, we find $0 = wL^3/6 + C_1$ or $C_1 = -wL^3/6$. We thus have

2)
$$EI\frac{dy}{dx} = \frac{w}{6}(L-x)^3 - \frac{wL^3}{6}$$

The next integration yields

3)
$$EIy = -\frac{w}{6}\left[\frac{(L-x)^4}{4}\right] - \frac{wL^3}{6}x + C_2$$

where C_2 represents a second constant of integration.

At the clamped end, $x = 0$, of the beam the deflection is zero and since equation *(3)* holds for all values of x and y, it is permissible to substitute this pair of values in it. Doing this, we obtain

$$0 = -wL^4/24 + C_2 \quad\text{or}\quad C_2 = wL^4/24.$$

The final form of the deflection curve of the beam is thus

$$EIy = -\frac{w}{24}(L-x)^4 - \frac{wL^3}{6}x + \frac{wL^4}{24} \tag{3'}$$

The deflection is a maximum at the right end of the bar ($x=L$) and at that point we have from equation *(3')*

$$EI(y)_{max} = -\frac{wL^4}{6} + \frac{wL^4}{24} = -\frac{wL^4}{8}$$

where the negative value denotes that this point on the deflection curve lies below the x-axis. If only the magnitude of the maximum deflection is desired, it is usually denoted by Δ and this becomes

$$\boxed{\Delta_{max} = \frac{wL^4}{8EI}} \tag{4}$$

7. The cantilever beam described in Problem 6 is of rectangular cross-section, 4 × 6 in. The bar is 6 ft long and carries a uniform load of 1000 lb/ft. The material is steel, for which $E = 30 \cdot 10^6$ lb/in^2. Determine the maximum deflection of the beam.

The maximum deflection of the beam occurs at the free end and was found in Problem 6 to be

$$\Delta_{max} = \frac{wL^4}{8EI}$$

Here, I for a rectangular cross-section is $bh^3/12 = 4(6)^3/12 = 72$ in^4. It is important to retain consistent units when substituting in such an equation. One manner of doing this is to take w in units of lb/in, L in inches, E in lb/in^2, and I in in^4. Doing this we obtain

$$\Delta_{max} = \frac{(1000/12)(72)^4}{8(30 \cdot 10^6)(72)} = 0.130 \text{ in.}$$

Again, the maximum stress in the bar should be investigated. The maximum bending moment occurs at the supporting wall and is

$$M_{max} = 1000(6)(3) = 18{,}000 \text{ lb-ft}$$

The maximum bending stress occurs at the outer fibers of the beam at this section adjacent to the wall and is given by $s = Mc/I$ where $c = 3$ in. Substituting,

$$s_{max} = \frac{Mc}{I} = \frac{(18{,}000 \times 12)(3)}{72} = 9000 \text{ lb/in}^2$$

Since this maximum bending stress is well below the proportional limit of the material, which is at least 30,000 lb/in^2, the use of the above deflection formula is valid.

8. Obtain an expression for the deflection curve of the simply supported beam subject to the uniformly distributed load of w lb per unit length as shown.

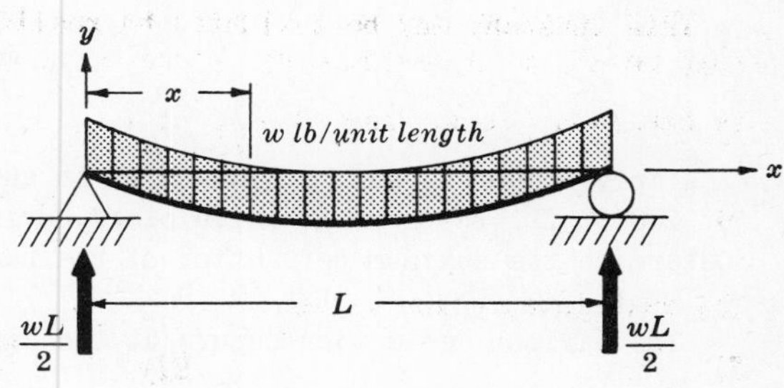

The x-y coordinate system shown is introduced, where the x-axis coincides with the original unbent position of the beam. The deformed beam has the appearance indicated by the heavy line. The total load acting on the beam is wL lb and because of symmetry, each of the end reactions is $wL/2$ lb. Because of the symmetry of loading, it is evident that the deflected beam is symmetric about the midpoint of the bar at $x = L/2$.

The equation for the bending moment at any section of a beam loaded and supported as this one is was discussed in Problem 6, Chapter 6. According to the method indicated there, the portion of the uniform load to the left of the section a distance x from the left support is replaced by its resultant acting at the midpoint of the section of length x. The resultant is wx lb acting downward and hence giving rise to a negative bending moment. The reaction $wL/2$ gives rise to a positive bending moment. Consequently, for any value of x, the bending moment is

$$M = \frac{wL}{2}x - wx\left(\frac{x}{2}\right)$$

The differential equation of the bent beam is $EI(d^2y/dx^2) = M$. Substituting,

1) $$EI\frac{d^2y}{dx^2} = \frac{wL}{2}x - \frac{wx^2}{2}$$

Integrating,

2) $$EI\frac{dy}{dx} = \frac{wL}{2}\left(\frac{x^2}{2}\right) - \frac{w}{2}\left(\frac{x^3}{3}\right) + C_1$$

It is to be noted that dy/dx represents the slope of the beam. Since the deflected beam is symmetric about the center of the span, i.e. about $x = L/2$, it is evident that the slope must be zero there. That is, the tangent to the deflected beam is horizontal at the midpoint of the beam. This condition enables us to determine C_1. Substituting this condition in *(2)* we obtain $(dy/dx)_{x=L/2} = 0$,

$$0 = \frac{wL}{4}\left(\frac{L^2}{4}\right) - \frac{w}{6}\left(\frac{L^3}{8}\right) + C_1 \quad \text{or} \quad C_1 = -\frac{wL^3}{24}$$

The slope dy/dx at any point is thus given by

2') $$EI\frac{dy}{dx} = \frac{wL}{4}x^2 - \frac{w}{6}x^3 - \frac{wL^3}{24}$$

Integrating again, we find

3) $$EIy = \frac{wL}{4}\left(\frac{x^3}{3}\right) - \frac{w}{6}\left(\frac{x^4}{4}\right) - \frac{wL^3}{24}x + C_2$$

This second constant of integration C_2 is readily determined by the fact that the deflection y is zero at the left support. Substituting $(y)_{x=0} = 0$ in *(3)*, we find $0 = 0 - 0 - 0 + C_2$ or $C_2 = 0$.

The final form of the deflection curve of the beam is thus

3') $$EIy = \frac{wL}{12}x^3 - \frac{w}{24}x^4 - \frac{wL^3}{24}x$$

The maximum deflection of the beam occurs at the center because of symmetry. Substituting $x = L/2$ in equation *(3')*, we obtain

$$EI(y)_{max} = -\frac{5wL^4}{384}$$

Or, without regard to algebraic sign the maximum deflection of a uniformly loaded simply supported beam is

4) $$\Delta_{max} = \frac{5}{384}\cdot\frac{wL^4}{EI}$$

9. A simply supported beam of length 10 ft and rectangular cross-section 4 × 8 in. carries a uniform load of 200 lb/ft. The beam is white pine, having a proportional limit of 6000 lb/in² and $E = 1.3\cdot10^6$ lb/in². Determine the maximum deflection of the beam.

The maximum deflection occurs at the center of the beam and is (see Prob. 8) $\Delta_{max} = \frac{5}{384}\cdot\frac{wL^4}{EI}$.

For the rectangular cross-section we have $I = \frac{1}{12}bh^3 = \frac{1}{12}(4)(8)^3 = 171$ in⁴.

Let us take the uniform load to be 200/12 lb/in and the length 120 in. Substituting these values,

$$\Delta_{max} = \frac{5}{384}\cdot\frac{(200/12)(120)^4}{(1.3\cdot10^6)(171)} = 0.202 \text{ in.}$$

The maximum bending stress in this beam was investigated in Problem 48, Chapter 8 and found to be 705 lb/in^2. Since this value is below the proportional limit of the material, the use of the above deflection formula is valid.

0. Obtain an equation for the deflection curve of the simply supported beam subject to the concentrated load P applied at the center of the beam as shown.

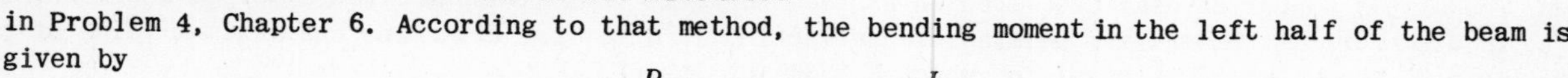

The x-y coordinate system shown is introduced. The deformed beam has the appearance indicated by the heavy line. Because of symmetry each end reaction is obviously $P/2$.

The equation for the bending moment at any section of a beam loaded as this one is was discussed in Problem 4, Chapter 6. According to that method, the bending moment in the left half of the beam is given by

$$M = \frac{P}{2}x \quad \text{for} \quad 0 < x < \frac{L}{2}$$

The differential equation of the bent beam is $EI \dfrac{d^2y}{dx^2} = M$. Substituting,

$$EI\frac{d^2y}{dx^2} = \frac{P}{2}x \quad \text{for} \quad 0 < x < \frac{L}{2} \tag{1}$$

The first integration of this equation yields

$$EI\frac{dy}{dx} = \frac{P}{2}\left(\frac{x^2}{2}\right) + C_1 \tag{2}$$

The slope of the beam is represented by dy/dx. Since the beam is loaded at its midpoint, the deflections are symmetric about the center of the beam, i.e. about the section $x = L/2$. This condition of symmetry tells us that the slope must be zero at $x = L/2$, i.e. the tangent to the deflected beam is horizontal there. Substituting this condition $(dy/dx)_{x=L/2} = 0$ in equation *(2)*, we obtain

$$0 = \frac{P}{4}\left(\frac{L^2}{4}\right) + C_1 \quad \text{or} \quad C_1 = -\frac{PL^2}{16}$$

Thus the slope dy/dx at any point in the beam is given by

$$EI\frac{dy}{dx} = \frac{P}{4}x^2 - \frac{PL^2}{16} \tag{2'}$$

Integrating again, we find

$$EIy = \frac{P}{4}\left(\frac{x^3}{3}\right) - \frac{PL^2}{16}x + C_2 \tag{3}$$

The second constant of integration C_2 is determined by the fact that the deflection y of the beam is zero at the left support. Thus $(y)_{x=0} = 0$. Substituting in *(3)*, we obtain $0 = 0 - 0 + C_2$ or $C_2 = 0$.

Thus the deflection curve of the left half of the beam is given by

$$EIy = \frac{P}{12}x^3 - \frac{PL^2}{16}x \tag{3'}$$

At this point it is to be carefully noted that it is *not* permissible to make use of the condition that the deflection y is zero at the right support, i.e. $(y)_{x=L} = 0$. This is because the bending moment equation, $M = (P/2)x$, is valid only for values of x less than $L/2$, i.e. to the left of the applied load P. To the right of force P the bending moment equation contains one additional term, and it would be necessary to work with the bending moment equation in the right half of the beam if the condition $(y)_{x=L} = 0$ were to be used. Actually there is no need to examine deflections to the right of the load since it is known that the deflection curve of the beam is symmetric about $x = L/2$. Briefly, in determining constants of integration it is permissible to use only those conditions on deflection or slope that pertain to the interval of the beam for which the bending moment equation was written.

Evidently the maximum deflection of the beam occurs at the center by virtue of symmetry. At this point the deflection is

$$EI(y)_{max} = -\frac{PL^3}{48}$$

Or, without regard to algebraic sign the maximum deflection of a simply supported beam subject to a centrally applied load P is

$$4) \qquad \Delta_{max} = \frac{PL^3}{48EI}$$

11. The simply supported beam described in Problem 10 is 14 ft long and of circular cross-section, 4 in. in diameter. If the maximum permissible deflection is 0.20 in., determine the maximum value of the load P. The material is steel for which $E = 30 \cdot 10^6$ lb/in^2.

The maximum deflection was found in equation *(4)* of Problem 10 to be $\Delta_{max} = \dfrac{PL^3}{48EI}$.

For a beam of circular cross-section (see Prob. 11, Chap. 7), $I = \pi D^4/64 = \pi 4^4/64 = 12.6$ in^4.

Also, L = 14 ft = 168 in. Substituting, $0.20 = \dfrac{P(168)^3}{48(30 \cdot 10^6)(12.6)}$ or $P = 765$ lb.

With this load applied at the center of the beam the reaction at each end is 383 lb and the bending moment at the center of the beam is 383(7) = 2681 lb-ft. This is the maximum bending moment in the beam and the maximum bending stress occurs at the outer fibers of the beam at this central section. The maximum bending stress is given by $s = \dfrac{Mc}{I}$. Then $s_{max} = \dfrac{2681(12)(2)}{12.6} = 5100$ lb/in^2. This is below the proportional limit of the material; hence the use of the deflection equation was permissible.

12. Consider again the simply supported beam described in Problem 11. Determine the slope of the beam at the left support.

From equation *(2')*, Problem 10, the slope dy/dx at any section at a distance x from the left support is given by

$$1) \qquad EI\frac{dy}{dx} = \frac{P}{4}x^2 - \frac{PL^2}{16}$$

At the left support $x = 0$ and as before, we have $E = 30 \cdot 10^6$ lb/in^2, $I = 12.6$ in^4, $P = 765$ lb, and $L = 168$ in. Substituting,

$$2) \qquad (30 \cdot 10^6)(12.6)\left(\frac{dy}{dx}\right)_{x=0} = -\frac{765(168)^2}{16} \quad \text{or} \quad \left(\frac{dy}{dx}\right)_{x=0} = 0.00358 \text{ radian}$$

The slope dy/dx actually represents the tangent of the angle of inclination of the deflection curve. For very small deformations such as we are considering in this chapter, the value of the angle expressed in units of radians is very nearly equal to the tangent of the angle; hence the slope $(dy/dx)_{x=0}$ is expressed as 0.00358 radian. Inspection of the units in equation *(2)* reveals that dy/dx must be dimensionless, and the radian is indeed a dimensionless unit of angular measure.

13. Determine the equation of the deflection curve for a simply supported beam loaded by a couple M_1 at the right end of the bar as shown.

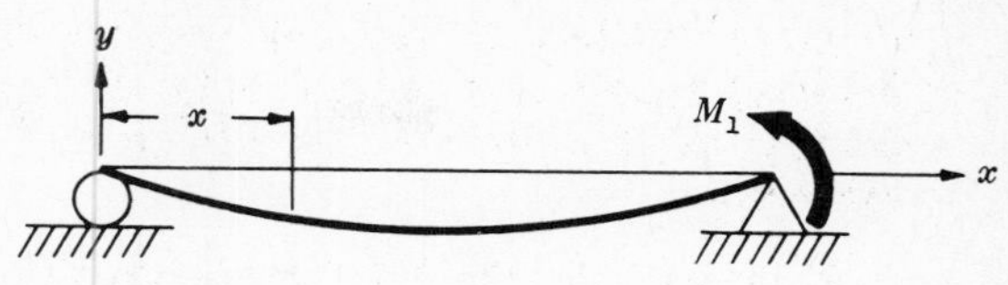

It is first necessary to determine the reactions acting on the beam. Since the applied couple M_1 can be held in equilibrium only by the action of another couple, it is apparent that the end reactions must be forces of equal magnitude R, but opposite in direction as indicated below. To find their magnitude we may write the statics equation

$$\Sigma M_O = -M_1 + RL = 0 \quad \text{or} \quad R = M_1/L$$

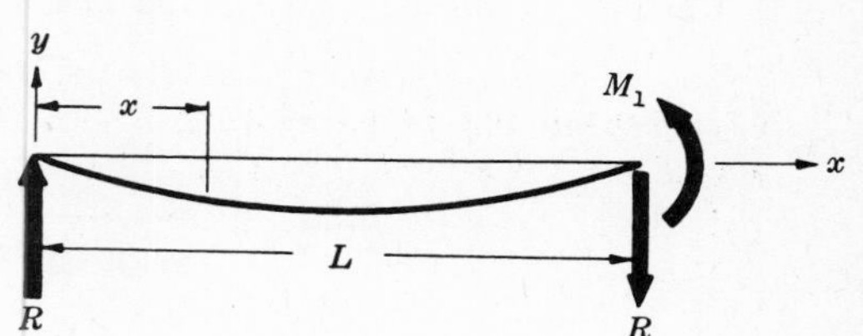

The heavy line indicates the configuration of the deflected beam. The bending moment at any section a distance x from the left reaction is

$$M = Rx = \frac{M_1}{L}x$$

This equation is valid for all values of x.

The differential equation of the deformed beam is

$$1) \qquad EI\frac{d^2y}{dx^2} = \frac{M_1}{L}x$$

Integrating once, we obtain

$$2) \qquad EI\frac{dy}{dx} = \frac{M_1}{L}\left(\frac{x^2}{2}\right) + C_1$$

There is no information available concerning the slope of the beam, hence it is not possible to determine C_1 at this stage. It is to be noted that there is no symmetry to the loading, hence there is no reason to expect the slope to be zero at the midpoint of the beam. We integrate again and obtain

$$3) \qquad EIy = \frac{M_1}{2L}\left(\frac{x^3}{3}\right) + C_1x + C_2$$

At this stage we are able to determine the constants of integration C_1 and C_2. It is evident that the deflection y is zero at the left support, i.e. $(y)_{x=0} = 0$. Substituting these values of x and y in equation *(3)*, we obtain $0 = 0 + 0 + C_2$ or $C_2 = 0$.

Also, the deflection y is zero at the right support, i.e. $(y)_{x=L} = 0$. Substituting these values of x and y in *(3)*, we find $0 = \frac{M_1}{6L}L^3 + C_1L$ or $C_1 = -\frac{M_1L}{6}$.

The deflection curve of the beam is consequently

$$3') \qquad EIy = \frac{M_1x^3}{6L} - \frac{M_1L}{6}x$$

The maximum deflection of the beam occurs at that point where the slope is zero, i.e. at that point where the tangent to the deflection curve is horizontal. The coordinate x of this point is readily found by setting the left side of *(2)* equal to zero. Doing this we get $0 = \frac{M_1x^2}{2L} - \frac{M_1L}{6}$ or $x = \frac{L}{\sqrt{3}}$.

The maximum deflection of the beam thus occurs at a distance $L/\sqrt{3}$ from the left reaction. The value of this deflection is found by substituting $x = L/\sqrt{3}$ in equation *(3′)*. This yields

$$4) \qquad EI(y)_{max} = \frac{M_1}{6L}\cdot\frac{L^3}{3\sqrt{3}} - \frac{M_1L}{6}\cdot\frac{L}{\sqrt{3}} = -\frac{M_1L^2\sqrt{3}}{27}$$

14. A simply supported beam is loaded by a couple M_1 as shown in Problem 13. The beam is 6 ft long and of square cross-section, 2 in. on a side. If the maximum permissible deflection in the beam is 0.20 in. and the allowable bending stress is 20,000 lb/in^2, find the maximum allowable load M_1. Take $E = 30 \cdot 10^6$ lb/in^2.

It is perhaps simplest to determine two values of M_1: one based upon the assumption that the deflection of 0.20 in. is realized, the other based on the assumption that the maximum bending stress in the bar is 20,000 lb/in^2. The true value of M_1 is then the minimum of these two values.

Let us first consider that the maximum deflection in the beam is 0.20 in. According to equation *(4)*, Problem 13, we have

$$0.20 = \frac{M_1 (72)^2 \sqrt{3}}{27(30 \cdot 10^6)\ \frac{1}{12}(2)(2)^3} \qquad \text{or} \qquad M_1 = 24,000 \text{ lb-in}$$

We shall now assume that the allowable bending stress of 20,000 lb/in^2 is set up in the outer fibers of the beam at the section of maximum bending moment. The bending moment diagram is shown at the right. From this it is seen that the maximum bending moment in the beam is M_1. Using the usual flexure formula, $s = Mc/I$, we have at the outer fibers of the bar at the right end, i.e. at the section of maximum bending moment,

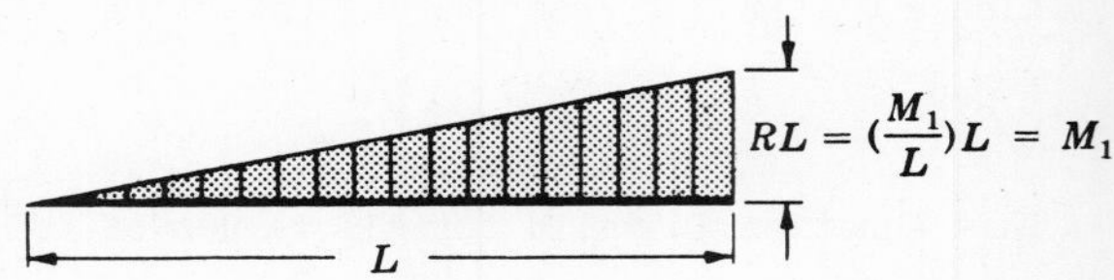

$$20,000 = \frac{M_1(1)}{\frac{1}{2}(2)(2)^3} \qquad \text{or} \qquad M_1 = 26,700 \text{ lb-in}$$

Thus the maximum allowable load is $M_1 = 24,000$ lb-in.

15. Determine the deflection curve of the simply supported beam subject to the concentrated force P applied as shown.

The x-y coordinate system is introduced as shown. The heavy line indicates the configuration of the deformed beam. From statics the reactions are easily found to have the values $R_1 = Pb/L$ and $R_2 = Pa/L$.

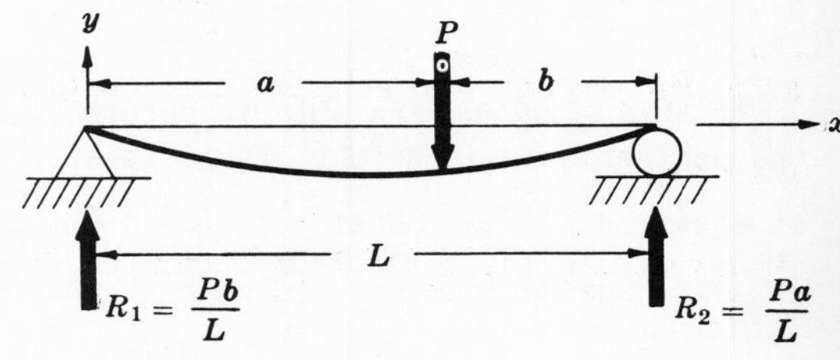

This problem presents one feature that distinguishes it from the other problems solved thus far in this chapter. That feature is that it is essential to consider two different equations describing the bending moment in the beam. One equation is valid to the left of the load P, the other holds to the right of this force. The integration of each equation gives rise to two constants of integration and thus there are four constants of integration to be determined rather than two as all other problems encountered thus far have offered.

In the region to the left of the force P we have the bending moment $M = (Pb/L)x$ for $0 < x < a$. The differential equation of the bent beam thus becomes

$$1) \qquad EI\frac{d^2y}{dx^2} = \frac{Pb}{L}x \qquad \text{for} \quad 0 < x < a$$

The first integration yields

$$2) \qquad EI\frac{dy}{dx} = \frac{Pb}{L}(\frac{x^2}{2}) + C_1$$

No definite information is available about the slope dy/dx at any point in this region. Since the load is not applied at the center of the beam there is no reason to believe that the slope is zero at $x = L/2$. However, we may say that the slope of the beam under the point of application of the force P is denoted by

$$3)\qquad EI\left(\frac{dy}{dx}\right)_{x=a} = \frac{Pba^2}{2L} + C_1$$

The next integration of equation (2) yields

$$4)\qquad EIy = \frac{Pb}{2L}\left(\frac{x^3}{3}\right) + C_1x + C_2$$

At the left support, $y = 0$ when $x = 0$. Substituting these values in equation (4) we immediately find $C_2 = 0$. It is to be noted that it is not permissible to use the condition $y = 0$ at $x = L$ in (4) since equation (1) is not valid in that region. We may denote the deflection under the point of application of the force P by

$$5)\qquad EI(y)_{x=a} = \frac{Pba^3}{6L} + C_1a$$

In the region to the right of the force P the bending moment equation is $M = (Pb/L)x - P(x - a)$ for $a < x < L$. Thus we have

$$6)\qquad EI\frac{d^2y}{dx^2} = \frac{Pb}{L}x - P(x-a) \qquad \text{for } a < x < L$$

The first integration of this equation yields

$$7)\qquad EI\frac{dy}{dx} = \frac{Pb}{L}\left(\frac{x^2}{2}\right) - \frac{P(x-a)^2}{2} + C_3$$

Although nothing definite may be said about the slope in this portion of the beam, we may denote the slope under the point of application of the force P by

$$8)\qquad EI\left(\frac{dy}{dx}\right)_{x=a} = \frac{Pba^2}{2L} + C_3$$

Under the concentrated load P the slope as given by equation (3) must be equal to that given by equation (8). Consequently the right sides of these two equations must be equal and we have

$$\frac{Pba^2}{2L} + C_1 = \frac{Pba^2}{2L} + C_3 \quad \text{or} \quad C_1 = C_3$$

Equation (7) may now be integrated to give

$$9)\qquad EIy = \frac{Pb}{2L}\left(\frac{x^3}{3}\right) - \frac{P(x-a)^3}{6} + C_3x + C_4$$

We may denote the deflection under the concentrated load by

$$10)\qquad EI(y)_{x=a} = \frac{Pba^3}{6L} + C_3a + C_4$$

The deflection at $x = a$ as given by (5) must equal that given by (10). Thus the right sides of these two equations are equal and we have $\frac{Pba^3}{6L} + C_1a = \frac{Pba^3}{6L} + C_3a + C_4$. Since $C_1 = C_3$, we have $C_4 = 0$.

The condition that $y = 0$ when $x = L$ may now be substituted in equation (9). Doing this we obtain

$$0 = \frac{PbL^2}{6} - \frac{Pb^3}{6} + C_3L \qquad \text{or} \qquad C_3 = \frac{Pb}{6L}(b^2 - L^2)$$

In this manner all four constants of integration are determined. These values may now be substituted in equations (4) and (9) to give

$$4')\qquad EIy = \frac{Pb}{6L}[x^3 - (L^2 - b^2)x] \qquad \text{for } 0 < x < a$$

and

$$9')\qquad EIy = \frac{Pb}{6L}\left[x^3 - \frac{L}{b}(x-a)^3 - (L^2 - b^2)x\right] \qquad \text{for } a < x < L$$

These two equations are necessary to describe the deflection curve of the bent beam. Each equation is valid only in the region indicated and it is not possible to replace these two equations by one equation containing the variable x to various powers that will be valid over the entire length of the beam.

It is to be noted that the deflections indicated by equations *(4')* and *(9')* are valid no matter where the load P is applied, i.e. it does not matter if P lies to the right or the left of the center line of the beam.

16. Consider the simply supported beam described in Problem 15. If the cross-section is rectangular, 2×4 in., and $P = 4000$ lb with $a = 4$ ft, $b = 2$ ft, determine the maximum deflection of the beam. The beam is steel, for which $E = 30 \cdot 10^6$ lb/in^2.

Since $a > b$ it is evident that the maximum deflection must occur to the left of the load P. It occurs at that point where the slope of the beam is zero.

Differentiating *(4')*, the slope in this region is given by $EI\dfrac{dy}{dx} = \dfrac{Pb}{6L}[3x^2 - (L^2 - b^2)]$.

Setting the slope equal to zero, we find $x = \sqrt{(L^2 - b^2)/3}$. This is the location of the point where the deflection is maximum. The deflection at this point is found by substituting this value of x in equation *(4')*. The maximum deflection is thus

$$EI(y)_{max} = -\frac{Pb\sqrt{3}}{27L}(L^2 - b^2)^{3/2}$$

For the rectangular section we have $I = 2(4^3)/12 = 10.7$ in^4. Also, $P = 4000$ lb, $b = 24$ in., $L = 72$ in., and $E = 30 \cdot 10^6$ lb/in^2. Substituting,

$$y_{max} = -\frac{4000(24)\sqrt{3}}{30 \cdot 10^6(10.7)(27)(72)}(72^2 - 24^2)^{3/2} = -0.0834 \text{ in.}$$

The negative sign indicates that this point on the bent beam lies below the x-axis.

By application of the formula $s = Mc/I$ the maximum bending stress, which occurs under the load P, is found to be 12,000 lb/in^2. This is below the proportional limit of steel, so use of the above deflection equations is valid.

17. Determine the equation of the deflection curve for the cantilever beam loaded by a uniformly distributed load of w lb per unit length over the portion of the beam shown.

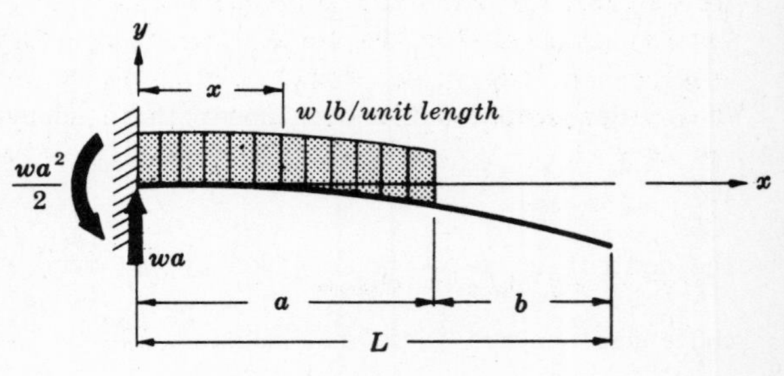

It is first necessary to determine the reactions exerted by the supporting wall upon the beam. These are easily found from statics to be a vertical force of magnitude wa lb together with a couple of magnitude $wa^2/2$. Again, two equations are required to describe the bending moment along the length of the bar.

For any point under the uniform load the bending moment at a distance x from the wall is given by

$$M = wax - \frac{wa^2}{2} - \frac{wx^2}{2}$$

In obtaining this equation the portion of the uniform load to the left of the section x was replaced by its resultant of wx lb acting downward at a distance $x/2$ from the wall. Also, according to the sign convention adopted in Chapter 6 the couple $wa^2/2$ produces negative bending. The differential equation of the loaded portion of the beam becomes

1) $$EI\frac{d^2y}{dx^2} = wax - \frac{wa^2}{2} - \frac{wx^2}{2} \qquad \text{for} \quad 0 < x < a$$

Integrating the first time we obtain

2) $$EI\frac{dy}{dx} = wa(\frac{x^2}{2}) - \frac{wa^2}{2}x - \frac{w}{2}(\frac{x^3}{3}) + C_1$$

Since the bar is clamped at the left end, $x = 0$, we know that the slope dy/dx must be zero there. Substituting these values in equation *(2)*, we find $C_1 = 0$. Integrating again, we find

3) $$EIy = \frac{wa}{2}(\frac{x^3}{3}) - \frac{wa^2}{2}(\frac{x^2}{2}) - \frac{w}{6}(\frac{x^4}{4}) + C_2$$

The deflection y of the beam is zero at the wall where $x = 0$. Substituting in *(3)*, we obtain $C_2 = 0$. Thus the equation of the bent beam in the loaded region is

4) $$EIy = \frac{wa}{6}x^3 - \frac{wa^2}{4}x^2 - \frac{w}{24}x^4$$

From equation *(4)* the deflection y at $x = a$ is given by

5) $$EI(y)_{x=a} = -wa^4/8$$

Also, from equation *(2)* the slope dy/dx at $x = a$ is given by

6) $$EI(dy/dx)_{x=a} = -wa^3/6$$

At any section in the unloaded region of the beam, i.e. $a < x < L$, the bending moment is zero. This is most easily seen by considering the moments of the forces to the right of such a section about an axis through this section and perpendicular to the plane of the page. Since there are no loads to the right of the section, the bending moment is zero everywhere in this region. Thus in this region we have

$$EI\frac{d^2y}{dx^2} = 0 \qquad \text{for} \quad a < x < L$$

Integrating once we obtain

7) $$EI\frac{dy}{dx} = C_3$$

The constant C_3 may be evaluated by realizing that the slope dy/dx at $x = a$ is the same for both the loaded and also the unloaded regions of the beam. That is, at the point $x = a$ the slope as given by the equation for the loaded region is equal to the slope as given by the equation for the unloaded region. For the loaded region the slope at $x = a$ was found in equation *(6)*. For the unloaded region the slope according to equation *(7)* is a constant C_3. Equating the right sides of these two equations we have $C_3 = -wa^3/6$. The slope in the unloaded region is thus

7′) $$EI\frac{dy}{dx} = -\frac{wa^3}{6}$$

Integrating this we obtain

8) $$EIy = -\frac{wa^3}{6}x + C_4$$

The constant C_4 may be evaluated by realizing that at the point $x = a$ the deflection y as given by equation *(5)* must be equal to the deflection as given by *(8)* for the unloaded region. Equating the right sides of these two equations at the common point $x = a$, we have $C_4 = wa^4/24$.

Thus two equations are necessary to describe the deflection curves of the loaded and unloaded regions of the beam. They are

4′) $$EIy = \frac{wa}{6}x^3 - \frac{wa^2}{4}x^2 - \frac{w}{24}x^4 \qquad \text{for} \quad 0 < x < a$$

8′) $$EIy = -\frac{wa^3}{6}x + \frac{wa^4}{24} \qquad \text{for} \quad a < x < L$$

Inspection of equation *(7′)* reveals that the slope of the beam is constant in the unloaded region. Thus the deflected beam is straight in that region.

18. Determine the equation of the deflection curve for the cantilever beam loaded by a uniformly distributed load of w lb per unit length as well as a concentrated force P applied as shown.

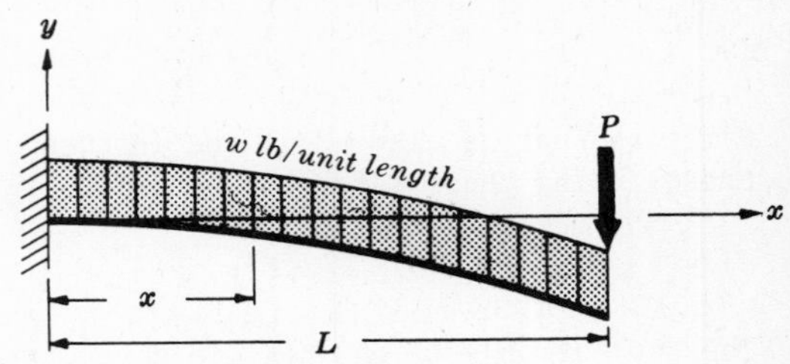

The deformed beam has the configuration indicated by the heavy line. The x-y coordinate system is introduced as shown. One logical approach to this problem is to determine the reactions at the wall, then write the differential equation of the bent beam, integrate this equation twice and determine the constant of integration from the conditions of zero slope and zero deflection at the wall.

Actually this procedure has already been carried out in Problem 2 if only the concentrated load alone acts on the beam, and in Problem 6 if only the uniformly distributed load alone is acting. Due to the concentrated force only the deflection y was found in equation *(3)* of Problem 2 to be

1) $$EIy = -PL\frac{x^2}{2} + \frac{Px^3}{6}$$

Due to the uniformly distributed load only the deflection y was found in *(3′)* of Problem 6 to be

2) $$EIy = -\frac{w}{24}(L-x)^4 - \frac{wL^3}{6}x + \frac{wL^4}{24}$$

It is possible to obtain the resultant effect of these two loads when they act simultaneously merely by adding together the effects of each as they act separately. This is called the *method of superposition*. It is useful in determining deflections of beams subject to a combination of loads, such as we have here. Essentially it consists of utilizing the results of simpler beam deflection problems to build up the solutions of more complicated problems. Thus it is not an independent method of determining beam deflections.

According to this method the deflection at any point of a beam subject to a combination of loads can be obtained as the sum of the deflections produced at this point by each of the loads acting separately. The final deflection equation resulting from the combination of loads is then obtained by adding the deflection equations for each load.

For this beam the final deflection equation is given merely by adding equations *(1)* and *(2)*. The resultant deflection y due to both loads is given by

3) $$EIy = -PL\frac{x^2}{2} + \frac{Px^3}{6} - \frac{w}{24}(L-x)^4 - \frac{wL^3}{6}x + \frac{wL^4}{24}$$

The slope dy/dx at any point in the beam is merely found by differentiating both sides of equation *(3)* with respect to x.

The method of superposition is valid in all cases where there is a linear relationship between each separate load and the separate deflection which it produces.

19. The cantilever beam in Problem 18 is an 8 WF 35 steel section 10 ft in length. The concentrated load at the free end of the beam is 5000 lb and the beam also supports a uniform load of 480 lb/ft, including the weight of the beam. Determine the deflection at a point 8 ft from the supporting wall. Also, determine the maximum stress in the beam. Take $E = 30 \cdot 10^6$ lb/in^2.

From the table in Chapter 8 we find the moment of inertia about the neutral axis of this section to be 126.5 in^4. In terms of the coordinate system used in Problem 18, we wish to evaluate the deflection y at x = 8 ft = 96 in. Also, we have L = 120 in., P = 5000 lb, w = 480 lb/ft = 40 lb/in. Substituting these values in equation *(3)* of Problem 18, we find

$$30\cdot10^6(126.5)[y]_{x=8\text{ ft}} = -5000(120)\frac{(96)^2}{2} + \frac{5000(96)^3}{6} - \frac{40}{24}(120-96)^4 - \frac{40(120)^3(96)}{6} + \frac{40(120)^4}{24}$$

Solving, $[y]_{x=8\text{ ft}} = -0.73$ in. The negative sign indicates that the deformed beam lies below the x-axis, which coincides with the original unbent configuration of the bar.

The maximum bending moment is readily determined by inspection of Problems 1 and 2, Chapter 6. According to Problem 1 the maximum bending moment in this beam due to the concentrated force of 5000 lb only occurs at the supporting wall and is 5000(10) = 50,000 lb-ft. According to Problem 2 the maximum bending moment due to the distributed load only also occurs at the wall and its value is 480(10)(5) = 24,000 lb-ft. The method of superposition may also be applied in the determination of stresses, and according to this principle the resultant bending moment at the wall is the sum of the bending moments due to the various forces acting separately. Thus the maximum bending moment is 50,000 + 24,000 = 74,000 lb-ft. The maximum bending stress occurs at the outer fibers of this section adjacent to the wall where the bending moment is maximum and is given by $s = Mc/I$. Substituting,

$$s_{max} = \frac{74{,}000(12)(4)}{126.5} = 28{,}100\text{ lb/in}^2$$

Since this maximum stress is below the proportional limit of steel, the use of the deflection equations was valid.

10. Determine the deflection curve of the simply supported beam loaded by the couple M_1 as shown in Fig. (a).

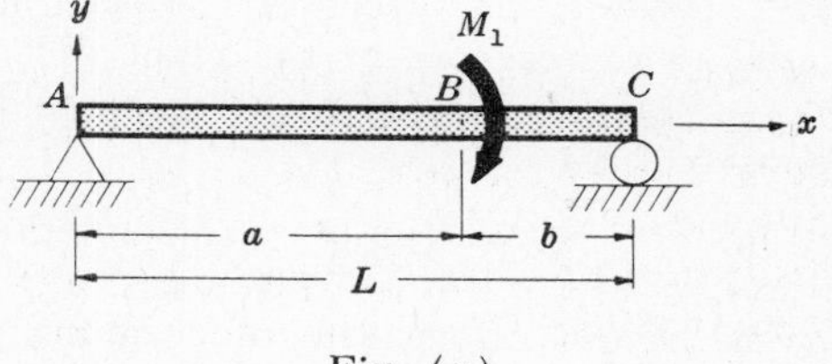

Fig. (a)

The reactions and bending moment equations for this type of loading have been discussed in Prob. 9, Chap. 6. As demonstrated there, the reactions must constitute a couple as shown in Fig. (b). From statics we have

$$\Sigma M_A = M_1 - RL = 0 \qquad \text{or} \qquad R = M_1/L$$

The configuration of the bent beam is indicated by the heavy line. The x-axis coincides with the original unbent position of the bar. The bending moment in the region to the left of the load M_1 is evidently

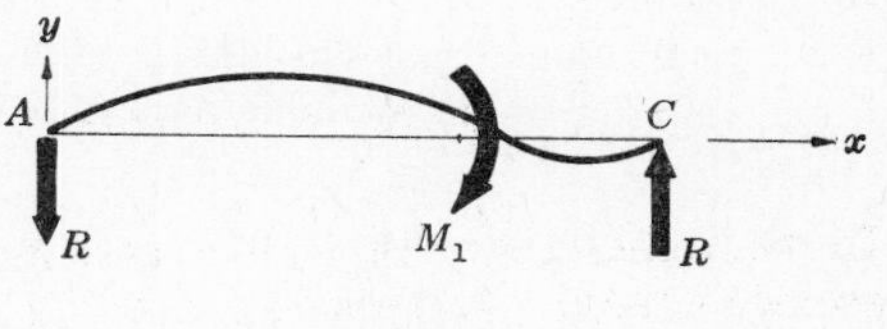

Fig. (b)

1) $$M = -Rx \qquad \text{for} \quad 0 < x < a$$

while in the region to the right of M_1 the bending moment is given by

2) $$M = -Rx + M_1 \qquad \text{for} \quad a < x < L$$

The couple M_1 produces a positive bending moment, since if it alone acted upon the region BC it would produce bending as shown in the adjoining sketch. According to the sign convention of Chapter 6 this constitutes positive bending, hence in equation (2) M_1 bears a positive sign.

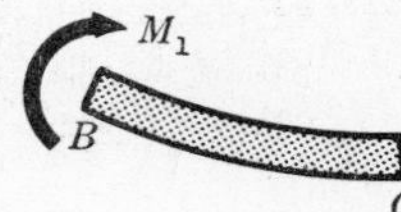

The differential equation of the region of the bent beam to the left of M_1 is

3) $$EI\frac{d^2y}{dx^2} = -Rx \qquad \text{for} \quad 0 < x < a$$

Integrating once we obtain

4) $$EI\frac{dy}{dx} = -R\frac{x^2}{2} + C_1$$

Since we have no definite information concerning the slope in this region, we are unable to evaluate C_1 immediately. However, we may say that the slope of the beam under the point of application of the couple M_1 is

5) $$EI\left(\frac{dy}{dx}\right)_{x=a} = -R\frac{a^2}{2} + C_1$$

Integration of equation *(4)* yields

$$6)\qquad EIy = -\frac{R}{2}\left(\frac{x^3}{3}\right) + C_1x + C_2$$

It is evident that the deflection y is zero at the left support where $x = 0$. Substituting this value, $(y)_{x=0} = 0$, in equation *(6)* we obtain $0 = 0 + 0 + C_2$ or $C_2 = 0$.

The differential equation of the region of the bent beam to the right of M_1 is

$$7)\qquad EI\frac{d^2y}{dx^2} = -Rx + M_1 \qquad \text{for} \quad a < x < L$$

Integrating the first time we find

$$8)\qquad EI\frac{dy}{dx} = -R\frac{x^2}{2} + M_1x + C_3$$

Again no definite information is available concerning the slope in this region, but we may say that the slope under the point of application of M_1 is

$$9)\qquad EI\left(\frac{dy}{dx}\right)_{x=a} = -R\frac{a^2}{2} + M_1a + C_3$$

However, the slope of the beam under the point of application of M_1 has only one value, its value as represented by the right sides of equations *(5)* and *(9)*. Equating the right sides of these two equations so as to force these two expressions for slope to be equal at this common point, we have

$$10)\qquad -R\frac{a^2}{2} + C_1 = -R\frac{a^2}{2} + M_1a + C_3 \qquad \text{or} \qquad C_1 = M_1a + C_3$$

The second integration of equation *(8)* yields

$$11)\qquad EIy = -\frac{R}{2}\left(\frac{x^3}{3}\right) + M_1\frac{x^2}{2} + C_3x + C_4$$

It is evident that the deflection y is zero at the right support where $x = L$. Substituting this value, $(y)_{x=L} = 0$, in equation *(11)* we obtain

$$12)\qquad 0 = -\frac{RL^3}{6} + M_1\frac{L^2}{2} + C_3L + C_4$$

One additional relationship is needed to determine all of the constants of integration. This relation is that the deflection of the beam under the point of application of M_1 is the same no matter if it is calculated from the equation for the left region or the right region of the beam. It is to be emphasized that there is no reason for assuming the deflection to be zero under the point of application of the couple. Substituting $x = a$ in *(6)* and *(11)* and equating the right sides, we obtain

$$13)\qquad -\frac{Ra^3}{6} + C_1a = -\frac{Ra^3}{6} + M_1\frac{a^2}{2} + C_3a + C_4 \qquad \text{or} \qquad C_1a = M_1\frac{a^2}{2} + C_3a + C_4$$

Solving equations *(10)*, *(12)* and *(13)* simultaneously, we find

$$C_1 = -\frac{M_1L}{3} + M_1a - \frac{M_1a^2}{2L}, \qquad C_3 = -\frac{M_1L}{3} - \frac{M_1a^2}{2L}, \qquad C_4 = \frac{M_1a^2}{2}$$

Substituting these values in equations *(6)* and *(11)*, we obtain the two equations required to describe the deflection curve of the bent beam:

$$14)\qquad EIy = -\frac{M_1x^3}{6L} - \frac{M_1Lx}{3} + M_1ax - \frac{M_1a^2x}{2L} \qquad \text{for} \quad 0 < x < a$$

and

$$15)\qquad EIy = -\frac{M_1x^3}{6L} + \frac{M_1x^2}{2} - \frac{M_1Lx}{3} - \frac{M_1a^2x}{2L} + \frac{M_1a^2}{2} \qquad \text{for} \quad a < x < L$$

To summarize, two equations were required to define the bending moment along the entire length of the beam. Two second order differential equations then had to be integrated and two constants of integration arose from the solution of each of these two equations. Thus we had four constants of integration and it was necessary to use four boundary conditions to determine them. These conditions were:

a) $y = 0$ when $x = 0$.
b) $y = 0$ when $x = L$.
c) When $x = a$, the deflections given by equations *(6)* and *(11)* are equal.
d) When $x = a$, the slopes given by equations *(4)* and *(8)* are equal.

21. In the simply supported beam of Problem 20 take M_1 = 1000 lb-ft, a = 6 ft, and b = 4 ft. The bar is of rectangular cross-section, 1 × 2 in. The material is steel for which $E = 30 \cdot 10^6$ lb/in². Determine:
a) The deflection under the point of application of M_1.
b) The deflection at x = 3 ft.
c) The deflection at x = 8 ft.

a) To calculate the deflection under the point of application of the 1000 lb-ft couple, we may use either equation *(14)* or *(15)* of Problem 20 with x = 6 ft. Since equation *(14)* is simpler, we shall work with it. Substituting the values x = 6 ft = 72 in., a = 6 ft = 72 in., L = 10 ft = 120 in., M_1 = 1000 lb-ft = 12,000 lb-in and $I = 1(2^3)/12 = 2/3$ in⁴ in equation *(14)*, we obtain

$$30 \cdot 10^6 (\tfrac{2}{3}) [y]_{x=6\text{ ft}} = -\frac{12{,}000(72)^3}{6(120)} - \frac{12{,}000(120)(72)}{3} + 12{,}000(72)^2 - \frac{12{,}000(72)^3}{2(120)}$$

Solving, $[y]_{x=6\text{ ft}} = 0.133$ in.

b) Equation *(14)* of Problem 20 may be used to calculate the deflection at x = 3 ft. In this equation we substitute x = 3 ft = 36 in., a = 72 in., L = 120 in., M_1 = 12,000 lb-in and I = 2/3 in⁴. This equation then becomes

$$30 \cdot 10^6 (\tfrac{2}{3}) [y]_{x=3\text{ ft}} = -\frac{12{,}000(36)^3}{6(120)} - \frac{12{,}000(120)(36)}{3} + 12{,}000(72)(36) - \frac{12{,}000(72)^2(36)}{2(120)}$$

Solving, $[y]_{x=3\text{ ft}} = 0.185$ in.

c) Equation *(15)* of Problem 20 may be used to calculate the deflection at x = 8 ft. In this equation we substitute x = 8 ft = 96 in., a = 72 in., L = 120 in., M_1 = 12,000 lb-in and I = 2/3 in⁴. The equation then becomes

$$30 \cdot 10^6 (\tfrac{2}{3}) [y]_{x=8\text{ ft}} = -\frac{12{,}000(96)^3}{6(120)} + \frac{12{,}000(96)^2}{2} - \frac{12{,}000(120)(96)}{3} - \frac{12{,}000(72)^2(96)}{2(120)} + \frac{12{,}000(72)^2}{2}$$

Solving, $[y]_{x=8\text{ ft}} = 0.024$ in.

The bending moment diagram for this beam has already been determined in Problem 9 of Chapter 6. There, the maximum bending moment was found to be 600 lb-ft acting on a section immediately to the left of the applied couple. The maximum bending stress in the beam occurs at the outer fibers of this section and is given by

$$s = \frac{Mc}{I}$$

$$s_{max} = \frac{600(12)(1)}{2/3} = 10{,}800 \text{ lb/in}^2$$

Since this stress is below the proportional limit of steel, the use of the deflection equations was permissible.

22. Determine the equation of the deflection curve for the overhanging beam loaded by the two equal forces P as shown.

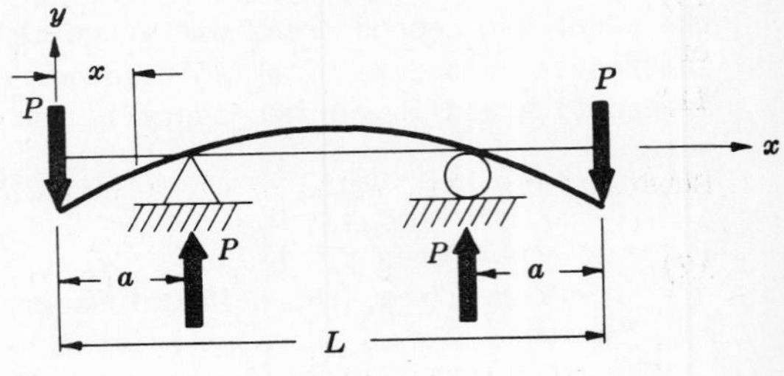

The x-y coordinate system is introduced as shown with the x-axis coinciding with the original unbent position of the bar. The fact that the left end of the bar deflects from the coordinate system presents no additional difficulties. From the conditions of symmetry it is evident that each support exerts a force P upon the bar.

The bending moment in the left overhanging region is given by

$$M = -Px \qquad \text{for} \quad 0 < x < a$$

and the differential equation of the bent beam in that region is

1) $$EI\frac{d^2y}{dx^2} = -Px \qquad \text{for} \quad 0 < x < a$$

The first integration of this equation yields

2) $$EI\frac{dy}{dx} = -P\frac{x^2}{2} + C_1$$

Nothing definite is known about the slope dy/dx in this region. In particular it is to be emphasized that there is no justification for assuming the slope to be zero at the point of support, $x = a$. We may denote the slope there by the notation

3) $$EI\left(\frac{dy}{dx}\right)_{x=a} = -P\frac{a^2}{2} + C_1$$

The next integration yields

4) $$EIy = -\frac{P}{2}\left(\frac{x^3}{3}\right) + C_1x + C_2$$

Since the beam is hinged at the support, it is known the deflection y is 0 there. Thus $(y)_{x=a} = 0$. Substituting $y = 0$ when $x = a$ in equation *(4)*, we find

5) $$0 = -\frac{Pa^3}{6} + C_1a + C_2$$

The bending moment in the central region of the beam between supports is $M = -Pa$ and the differential equation of the bent beam in the central region is

6) $$EI\frac{d^2y}{dx^2} = -Pa \qquad \text{for} \quad a < x < (L-a)$$

Integrating we obtain

7) $$EI\frac{dy}{dx} = -Pax + C_3$$

Because of the symmetry of loading it is evident that the slope dy/dx must be zero at the midpoint of the bar. Thus $(dy/dx)_{x=L/2} = 0$. Substituting these values of x and dy/dx in *(7)*, we find

8) $$0 = -Pa\frac{L}{2} + C_3 \qquad \text{or} \qquad C_3 = \frac{PaL}{2}$$

Also, from equation *(7)* we may say that the slope of the beam over the left support, $x = a$, is given by substituting $x = a$ in this equation. This yields

9) $$EI\left(\frac{dy}{dx}\right)_{x=a} = -Pa^2 + \frac{PaL}{2}$$

But the slope dy/dx as given by this expression must be equal to that given by equation *(3)* since the bent bar at that point must have the same slope, no matter which equation is being considered. Equating the right sides of equations *(3)* and *(9)*, we obtain

10) $$-\frac{Pa^2}{2} + C_1 = -Pa^2 + \frac{PaL}{2}$$

or

11) $$C_1 = -\frac{Pa^2}{2} + \frac{PaL}{2}$$

Substituting this value of C_1 in equation *(5)*, we find

12) $$0 = -\frac{Pa^3}{6} - \frac{Pa^3}{2} + \frac{Pa^2L}{2} + C_2 \quad \text{or} \quad C_2 = \frac{2Pa^3}{3} - \frac{Pa^2L}{2}$$

The next integration of equation *(7)* yields

13) $$EIy = -Pa\frac{x^2}{2} + \frac{PaL}{2}x + C_4$$

Again, it may be said that the deflection y is zero at the left support where $x = a$. Although this same condition was used previously in obtaining equation *(5)*, there is no reason why it should not be used again. In fact, it is essential to use it in order to solve for the constant C_4 in equation *(13)*. Thus, substituting the values $(y)_{x=a} = 0$ in equation *(13)*, we obtain

14) $$0 = -\frac{Pa^3}{2} + \frac{Pa^2L}{2} + C_4 \quad \text{or} \quad C_4 = \frac{Pa^3}{2} - \frac{Pa^2L}{2}$$

Thus two equations were required to define the bending moment in the left and central regions of the beam. Each equation was used in conjunction with the second-order differential equation describing the bent beam, and thus two constants of integration arose from the solution of each of these two equations. It was necessary to utilize four conditions concerning slope and deflection in order to determine these four constants. These conditions were:

a) When $x = a$, $y = 0$ for the overhanging portion of the beam.
b) When $x = a$, $y = 0$ for the central portion of the beam.
c) When $x = L/2$, $dy/dx = 0$ for the central portion of the beam.
d) When $x = a$, the slope dy/dx is the same for the deflection curve on either side of the support.

Finally, the equations of the bent beam may be written in the forms

15) $$EIy = -\frac{P}{6}x^3 - \frac{Pa^2x}{2} + \frac{PaLx}{2} + \frac{2Pa^3}{3} - \frac{Pa^2L}{2} \quad \text{for} \quad 0 < x < a$$

and

16) $$EIy = -\frac{Pax^2}{2} + \frac{PaLx}{2} + \frac{Pa^3}{2} - \frac{Pa^2L}{2} \quad \text{for} \quad a < x < (L-a)$$

Because of the symmetry there is no need to write the equation for the deformed beam in the right overhanging region.

23. In the overhanging beam described in Problem 22, each force P is 4000 lb. The distance a is 3 ft and the length L is 16 ft. The bar is steel and of circular cross-section, 4 in. in diameter. Determine the deflection under each load P and also the deflection at the center of the beam. Take $E = 30 \cdot 10^6$ lb/in^2.

The moment of inertia is given by $I = \frac{\pi}{64}(4)^4 = 12.6$ in^4 according to Problem 11, Chapter 7. Also, we have $a = 3$ ft $= 36$ in., $L = 16$ ft $= 192$ in. The deflection anywhere in the left overhanging region is given by equation *(15)* of Problem 22. Under the concentrated force P we have $x = 0$, and substituting these values in equation *(15)* we obtain

$$30 \cdot 10^6 (12.6)[y]_{x=0} = \frac{2(4000)(36)^3}{3} - \frac{4000(36)^2(192)}{2} \quad \text{or} \quad [y]_{x=0} = -0.98 \text{ in.}$$

The deflection anywhere in the central portion between supports is given by equation *(16)* of Problem 22. At the center of the beam we have $x = 8$ ft $= 96$ in. and as before $a = 36$ in., $L = 192$ in., and $P = 4000$ lb. Substituting in equation *(16)*, we find

$$30\cdot 10^6(12.6)[y]_{x=8\text{ ft}} = -\frac{4000(36)(96)^2}{2} + \frac{4000(36)(192)(96)}{2} + \frac{4000(36)^3}{2} - \frac{4000(36)^2(192)}{2}$$

Solving, $[y]_{x=8\text{ ft}} = 0.69$ in.

The maximum bending stress occurs at the outer fibers of the bar everywhere between the supports, since the bending moment has the constant value of 4000(3) = 12,000 lb-ft in this region. This maximum stress is given by

$$s = \frac{Mc}{I} = \frac{12,000(12)(2)}{12.6} = 22,800\text{ lb/in}^2$$

This is less than the proportional limit of the material.

24. Determine the deflection curve of the overhanging beam subject to a uniform load of w lb per unit length and supported as shown.

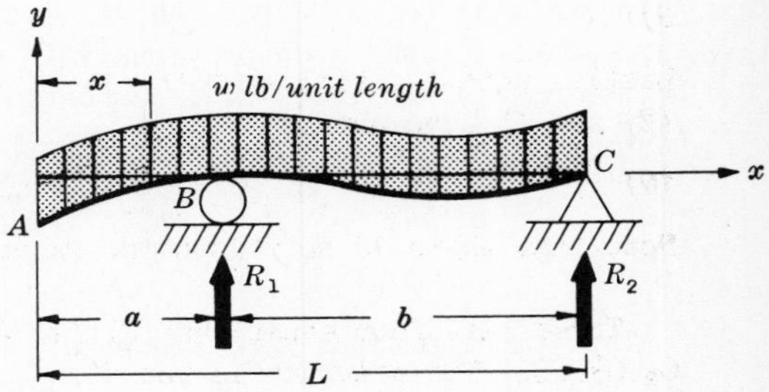

The determination of reactions and the bending moment diagram for such a beam was discussed in Problem 13, Chapter 6. As in that problem we shall replace the entire distributed load by its resultant of wL lb acting at the midpoint of the length L. Taking moments about the right reaction, we have

$$\Sigma M_C = R_1 b - \frac{wL^2}{2} = 0 \quad\text{or}\quad R_1 = \frac{wL^2}{2b}$$

Summing forces vertically, we find $\Sigma F_v = \dfrac{wL^2}{2b} + R_2 - wL = 0$ or $R_2 = wL - \dfrac{wL^2}{2b}$.

The bending moment equation in the left overhanging region is $M = -\dfrac{wx^2}{2}$ for $0 < x < a$. Consequently the differential equation of the bent beam in that region is

1) $$EI\frac{d^2y}{dx^2} = -\frac{wx^2}{2} \quad\text{for}\quad 0 < x < a$$

Two successive integrations yield

2) $$EI\frac{dy}{dx} = -\frac{w}{2}\left(\frac{x^3}{3}\right) + C_1$$

3) $$EIy = -\frac{w}{6}\left(\frac{x^4}{4}\right) + C_1x + C_2$$

The bending moment equation in the region between supports is $M = -\dfrac{wx^2}{2} + R_1(x-a)$. The differential equation of the bent beam in that region is thus

4) $$EI\frac{d^2y}{dx^2} = -\frac{wx^2}{2} + \frac{wL^2}{2b}(x-a) \quad\text{for}\quad a < x < L$$

Two integrations of this equation yield

5) $$EI\frac{dy}{dx} = -\frac{w}{2}\left(\frac{x^3}{3}\right) + \frac{wL^2}{2b}\left[\frac{(x-a)^2}{2}\right] + C_3$$

6) $$EIy = -\frac{w}{6}\left(\frac{x^4}{4}\right) + \frac{wL^2}{4b}\left[\frac{(x-a)^3}{3}\right] + C_3x + C_4$$

Since we started with two second-order differential equations, *(1)* and *(4)*, and two constants of integration arose from each, we have four constants C_1, C_2, C_3 and C_4 to evaluate from known conditions concerning slopes and deflections. These conditions are:

a) When $x = a$, $y = 0$ in the overhanging region.
b) When $x = a$, $y = 0$ in the region between supports.
c) When $x = L$, $y = 0$ in the region between supports.
d) When $x = a$, the slope given by equation (2) must be equal to that given by equation (5). Consequently the right sides of these equations must be equal when $x = a$.

Substituting condition (a) in equation (3), we obtain

7) $$0 = -wa^4/24 + C_1 a + C_2$$

Substituting condition (b) in equation (6), we find

8) $$0 = -wa^4/24 + C_3 a + C_4$$

Substituting condition (c) in equation (6), we get

9) $$0 = -wL^4/24 + wL^2 b^2/12 + C_3 L + C_4$$

Lastly, equating slopes at the left reaction by substituting $x = a$ in the right sides of equations (2) and (5), we obtain

10) $$-wa^3/6 + C_1 = -wa^3/6 + C_3$$

Note that there is no reason for assuming the slope to be zero at the left support, $x = a$.

These last four equations (7),(8),(9),(10) may now be solved for the four unknown constants C_1, C_2,C_3,C_4. The solution is found to be

11) $$C_1 = C_3 = \frac{w(L^4 - a^4)}{24b} - \frac{wL^2 b}{12}$$

12) $$C_2 = C_4 = \frac{wa^4}{24} - \frac{w(L^4 - a^4)a}{24b} + \frac{wL^2 ab}{12}$$

The two equations describing the deflection curve of the bent bar are found by substituting these values of the constants in equations (3) and (6). These equations may be written in the final forms

3') $$EIy = -\frac{wx^4}{24} + \frac{w(L^4 - a^4)x}{24b} - \frac{wL^2 bx}{12} + \frac{wa^4}{24} - \frac{w(L^4 - a^4)a}{24b} + \frac{wL^2 ab}{12} \quad \text{for } 0 < x < a$$

6') $$EIy = -\frac{wx^4}{24} + \frac{wL^2 (x-a)^3}{12b} + \frac{w(L^4 - a^4)x}{24b} - \frac{wL^2 bx}{12} + \frac{wa^4}{24} - \frac{w(L^4 - a^4)a}{24b} + \frac{wL^2 ab}{12} \quad \text{for } a < x < L$$

25. For the overhanging beam discussed in Problem 24 consider the uniform load to be 120 lb/ft, $a = 3$ ft and $b = 12$ ft. The bar is of rectangular cross-section 3×4 in. Determine the maximum deflection of the beam. Take $E = 30 \cdot 10^6$ lb/in^2.

An approximate representation of the deflected beam is shown in Problem 24. The point where the maximum deflection occurs is not immediately evident. It may be at the extreme left end of the beam, where $x = 0$, or it may be at some intermediate point between the supports. If the maximum does occur between supports, it is unlikely that it occurs midway between them since there is no symmetry to the system. In the event that it occurs between supports, the location of the point may be determined by finding the point where the slope of the beam is zero. The slope anywhere in the region between supports is given by equation (5) of Problem 24, and if we set the slope dy/dx in that equation equal to zero and use the value of C_3 given by equation (11) we find

$$0 = -\frac{wx^3}{6} + \frac{wL^2 (x-a)^2}{4b} + \frac{w(L^4 - a^4)}{24b} - \frac{wL^2 b}{12}$$

Substituting $w = 120$ lb/ft = 10 lb/in, $a = 3$ ft = 36 in., $b = 12$ ft = 144 in. and $L = 180$ in., we find

$$0 = -\frac{10x^3}{6} + \frac{10(180)^2 (x-36)^2}{4(144)} + \frac{10[(180)^4 - (36)^4]}{24(144)} - \frac{10(180)^2 (144)}{12}$$

Solving by trial-and-error, x = 110.4 in. = 9.20 ft. This locates the point where the slope is zero.

The deflection at x = 110.4 in. may be found by substituting this value in equation *(6')* of Problem 24. This leads to the relation

$$(30\cdot 10^6)\frac{1}{12}(3)(4^3)[y]_{x=110.4\text{ in.}} = -\frac{10(110.4)^4}{24} + \frac{10(180)^2(110.4-36)^3}{12(144)} + \frac{10[(180)^4-(36)^4]110.4}{24(144)}$$

$$-\frac{10(180)^2(144)(110.4)}{12} + \frac{10(36)^4}{24} - \frac{10[(180)^4-(36)^4](36)}{24(144)} + \frac{10(180)^2(36)(144)}{12}$$

Solving, $[y]_{x=110.4\text{ in.}} = -0.10$ in.

The calculus technique of equating the first derivative dy/dx to zero for the purpose of determining the location of a point where the value of a function is maximum fails to detect any maximum deflection that may exist at such a point as x = 0. Hence it is necessary to investigate the deflection at that point. Substituting x = 0 in equation *(3')* of Problem 24, we find

$$(30\cdot 10^6)\frac{1}{12}(3)(4^3)[y]_{x=0} = \frac{10(36)^4}{24} - \frac{10[(180)^4-(36)^4](36)}{24(144)} + \frac{10(180)^2(36)(144)}{12}$$

Solving, $[y]_{x=0} = +0.065$ in.

Thus the assumed form of the deflection curve shown in Problem 24 is incorrect in the overhanging region for this particular beam. Actually, the beam bends upward in this region. For other values of a and b it would be possible for the beam to bend into the configuration shown there.

The maximum deflection of the beam is consequently 0.10 in. downward at a point 9.20 ft from the left end.

The bending moment diagram for this beam was investigated in Problem 13 of Chapter 6. There the maximum bending moment was found to be 1900 lb-ft. The maximum bending stress is given by

$$s_{max} = \frac{Mc}{I} = \frac{1900(12)(2)}{\frac{1}{12}(3)(4^3)} = 2840\ \text{lb/in}^2$$

The use of the above deflection equations is thus justified.

It is to be noted that the section where the bending stress is a maximum is not the section where the deflection is a maximum.

SUPPLEMENTARY PROBLEMS

26. Consider the cantilever beam loaded as shown in Problem 2. The load P is 5000 lb, L = 12 ft, the moment of inertia of the cross-section is 250 in^4, and $E = 30\cdot 10^6$ lb/in^2. Find the maximum deflection of the beam. *Ans.* −0.670 in.

27. Consider the uniformly loaded cantilever beam discussed in Problem 6. The total load is 6000 lb, the length of the beam is 10 ft and the moment of inertia of the cross-section is 200 in^4. Determine the deflection and slope at the free end of the beam. Take $E = 30\cdot 10^6$ lb/in^2.
Ans. −0.216 in., −0.0024 rad

28. A 10 WF 45 section is used as a simply supported beam. It is 14 ft long and carries a uniformly distributed load of 15,000 lb. Determine the maximum deflection of the beam. Take $E = 30\cdot 10^6$ lb/in^2. Also, find the maximum stress in the beam. *Ans.* −0.124 in., 6400 lb/in^2

29. Consider the simply supported beam loaded by the central load P as discussed in Prob. 10. The length of the beam is 20 ft, the force P is 25,000 lb, I = 1064 in^4, and $E = 30\cdot 10^6$ lb/in^2. Determine the maximum deflection of the beam. *Ans.* −0.225 in.

30. Consider the simply supported beam loaded as shown in Problem 15. The length of the beam is 20 ft, a = 15 ft, the load P = 1000 lb, and I = 150 in^4. Determine the deflection at the center of the beam. Take $E = 30 \cdot 10^6$ lb/in^2. *Ans.* −0.044 in.

31. Refer to Fig. (a) below. Determine the deflection at every point of the cantilever beam subject to the single moment M_1 shown. *Ans.* $EIy = -M_1x^2/2$

32. The cantilever beam described in Problem 31 is of circular cross-section, 5 in. in diameter. The length of the beam is 10 ft and the applied moment is 5000 lb-ft. Determine the maximum deflection of the beam. Take $E = 30 \cdot 10^6$ lb/in^2. *Ans.* −0.469 in.

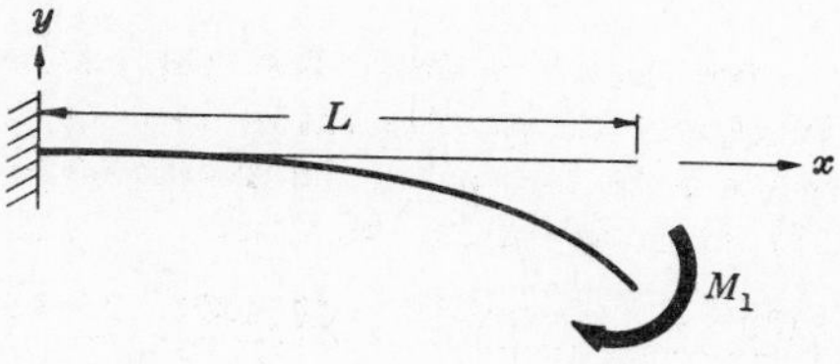

Fig. (a) Prob. 31

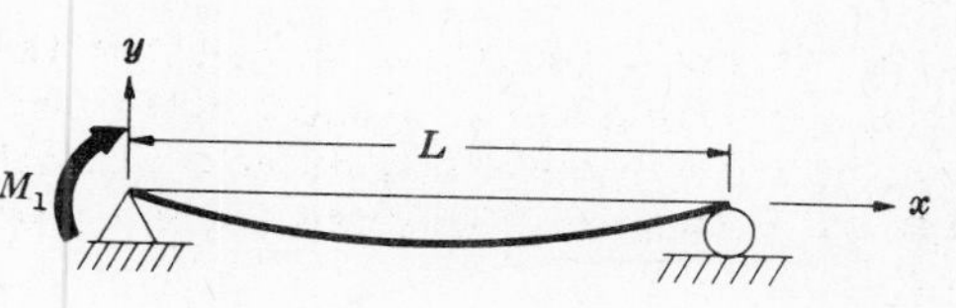

Fig. (b) Prob. 33

33. Refer to Fig. (b) above. Determine the equation of the deflection curve for a simply supported beam loaded by a couple M_1 at the left end of the bar as shown. *Ans.* $EIy = -\dfrac{M_1}{6L}x^3 + \dfrac{M_1}{2}x^2 - \dfrac{M_1L}{3}x$

34. The cross-section of the beam described in Problem 33 is an 8 WF 27 section. The length of the beam is 10 ft and the applied moment is 30,000 lb-ft. Determine the location of the point of maximum deflection of the beam and the value of the maximum deflection.
Ans. −0.118 in. occurring 4.20 ft from left end

35. Determine the equation of the deflection curve for the simply supported beam loaded by a uniformly distributed load of w lb/unit length as well as a concentrated force P applied at the midpoint as shown in Fig. (c) below. *Ans.* $EIy = \dfrac{wLx^3}{12} - \dfrac{wx^4}{24} - \dfrac{wL^3x}{24} + \dfrac{Px^3}{12} - \dfrac{PL^2x}{16}$ for $0 < x < \dfrac{L}{2}$

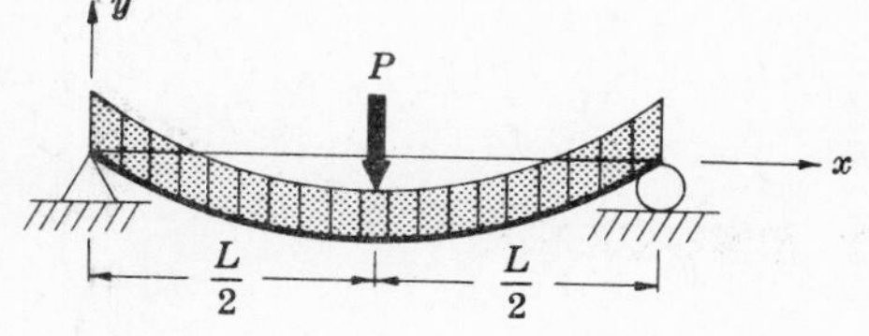

Fig. (c) Prob. 35

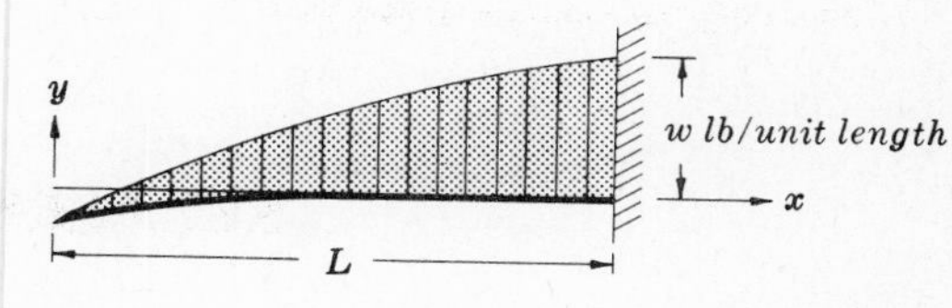

Fig. (d) Prob. 36

36. Refer to Fig. (d) above. Find the equation of the deflection curve for the cantilever beam subject to the uniformly varying load shown. *Ans.* $EIy = -\dfrac{wx^5}{120L} + \dfrac{wL^3x}{24} - \dfrac{wL^4}{30}$

37. The cross-section of the cantilever beam loaded as shown in Problem 36 is rectangular, 2 × 3 in. The bar is aluminum for which $E = 10 \cdot 10^6$ lb/in^2 and the length of the beam is 3 ft. Determine the permissible maximum intensity of loading if the maximum deflection is not to exceed 0.15 in. and the maximum stress is not to exceed 8000 lb/in^2. *Ans.* w = 1333 lb/ft

38. Refer to the adjoining Fig. (e). Determine the equation of the deflection curve for the simply supported beam supporting the load of uniformly varying intensity.
Ans. $EIy = \dfrac{wL}{2}\left(-\dfrac{x^5}{60L^2} + \dfrac{x^3}{18} - \dfrac{7L^2x}{180}\right)$

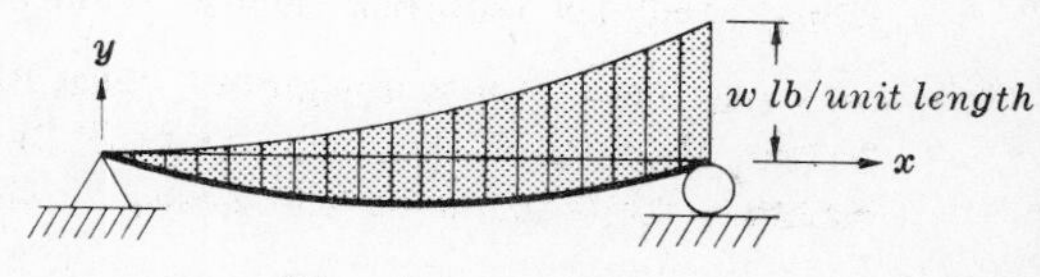

Fig. (e) Prob. 38

39. The cross-section of the beam described in Problem 38 has a moment of inertia of 150 in^4. The beam is 8 ft long and the intensity of loading at the right end is 2500 lb/ft. Determine the deflection at a point 2 ft from the right support. Assume $E = 30 \cdot 10^6$ lb/in^2. *Ans.* −0.019 in.

40. Determine the equation of the deflection curve for the cantilever beam loaded by the concentrated force P as shown in Fig. (f) below.

Ans. $EIy = -\dfrac{P}{6}(a-x)^3 - \dfrac{Pa^2}{2}x + \dfrac{Pa^3}{6}$ for $0 < x < a$; $EIy = -\dfrac{Pa^2}{2}x + \dfrac{Pa^3}{6}$ for $a < x < L$

41. For the cantilever beam described in Problem 40, take $P = 1000$ lb, $a = 6$ ft, and $b = 4$ ft. The beam is of equilateral triangular cross-section 6 in. on a side. The cross-section of the beam has a vertical axis of symmetry. Determine the maximum deflection of the beam. Take $E = 30 \cdot 10^6$ lb/in^2. *Ans.* −0.355 in.

Fig. (f) Prob. 40

Fig. (g) Prob. 42

42. Refer to Fig. (g) above. Find the equation of the deflection curve for the cantilever beam loaded by the couple M_1. *Ans.* $EIy = -\dfrac{M_1 x^2}{2}$ for $0 < x < a$; $EIy = -M_1 ax + \dfrac{M_1 a^2}{2}$ for $a < x < L$

43. Refer to Fig. (h) below. A simply supported beam is subject to the two symmetrically placed loads shown. Determine the deflection curve of the bent beam.

Ans. $EIy = \dfrac{Px^3}{6} + \left(\dfrac{Pa^2}{2} - \dfrac{PaL}{2}\right)x$ for $0 < x < a$; $EIy = \dfrac{Pax^2}{2} - \dfrac{PaLx}{2} + \dfrac{Pa^3}{6}$ for $a < x < (a+b)$

44. The symmetrically loaded beam described in Problem 43 is an 8 WF 19 section. The length of the beam is 20 ft, $a = 4$ ft, and $P = 3000$ lb. Determine the deflection under the point of application of each force P. Take $E = 30 \cdot 10^6$ lb/in^2. *Ans.* −0.35 in.

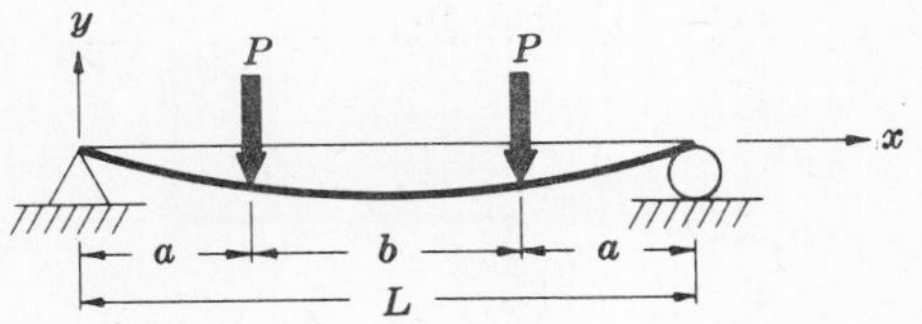

Fig. (h) Prob. 43

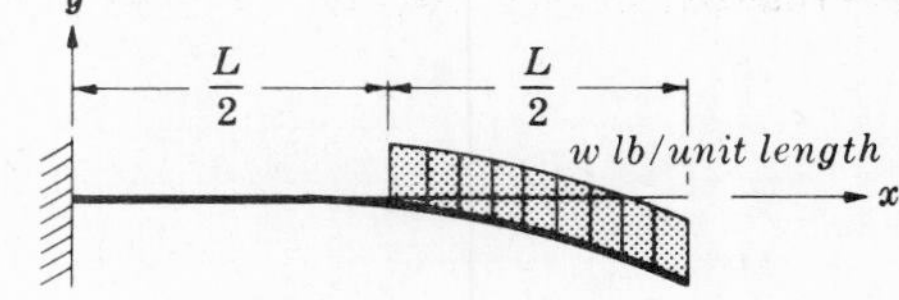

Fig. (i) Prob. 45

45. Refer to Fig. (i) above. Find the equation of the deflection curve for the cantilever beam loaded over half its length by a uniformly distributed load of w lb per unit length as shown. Using this equation, determine the maximum deflection.

Ans. $EIy = \dfrac{wLx^3}{12} - \dfrac{3wL^2x^2}{16}$ for $0 < x < \dfrac{L}{2}$

$EIy = -\dfrac{w(L-x)^4}{24} - \dfrac{7wL^3x}{48} + \dfrac{15wL^4}{384}$ for $\dfrac{L}{2} < x < L$ $\qquad \Delta_{max} = \dfrac{41}{384}\left(\dfrac{wL^4}{EI}\right)$

46. The cantilever beam loaded as shown in Problem 45 is a 10 WF 21 section. The length of the beam is 12 ft and the uniform loading is 800 lb/ft. Determine the maximum deflection of the beam. Assume $E = 30 \cdot 10^6$ lb/in^2. *Ans.* −0.96 in.

47. Refer to Fig. (j) below. A simply supported overhanging beam supports the uniform load shown. Find the equation of the deflection curve of the bent beam. Take the origin of the coordinate system at the level of the supports.

Ans. $EIy = -\frac{wx^4}{24} + \frac{wL^3x}{48} - \frac{wLx}{4}(\frac{L}{2} - a)^2 + \frac{wa^4}{24} - \frac{waL^3}{48} + \frac{wLa}{4}(\frac{L}{2} - a)^2$ for $0 < x < a$

$EIy = -\frac{wx^4}{24} + \frac{wL(x-a)^3}{12} + \frac{wL^3x}{48} - \frac{wLx}{4}(\frac{L}{2} - a)^2 + \frac{wa^4}{24} - \frac{waL^3}{48} + \frac{wLa}{4}(\frac{L}{2} - a)^2$ for $a < x < (a+b)$

48. The symmetrically supported beam described in Problem 47 is 30 ft in length and the distance between supports is 20 ft. The moment of inertia of the cross-section is 400 in^4 and the uniform load is 800 lb/ft. Find the deflection at the center of the beam. Assume $E = 30 \cdot 10^6$ lb/in^2. *Ans.* -0.166 in.

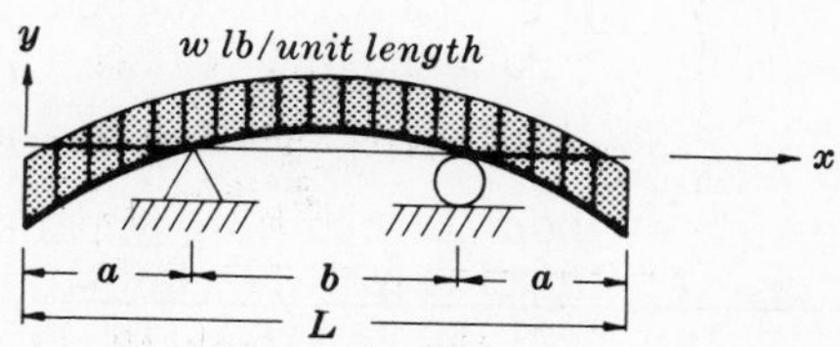

Fig. (j) Prob. 47

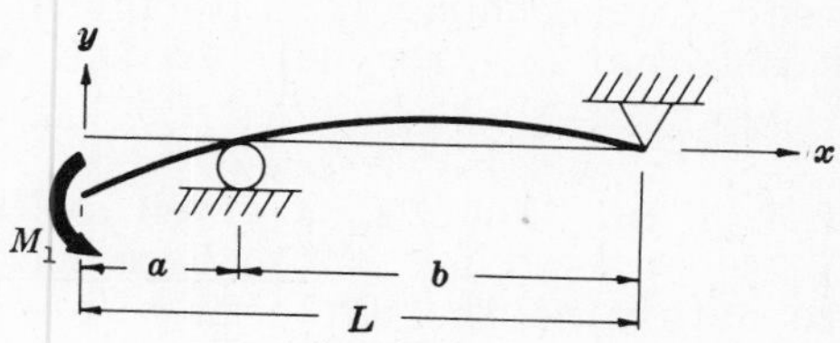

Fig. (k) Prob. 49

49. Refer to Fig. (k) above. A beam with an overhanging end is loaded by a couple M_1 as shown. Determine the equation of the deflection curve for the bent beam. Take the origin of coordinates at the level of the supports.

Ans. $EIy = -\frac{M_1x^2}{2} + M_1ax + \frac{M_1(L-a)x}{3} - \frac{M_1a^2}{2} - \frac{M_1a(L-a)}{3}$ for $0 < x < a$

$EIy = -\frac{M_1(L-x)^3}{6(L-a)} - \frac{M_1x(L-a)}{6} + \frac{M_1L(L-a)}{6}$ for $a < x < L$

50. The beam described in Problem 49 is a 10 WF 45 section, 12 ft in length. The supports are spaced such that $a = 3$ ft and the applied moment M_1 is 100,000 lb-ft. Determine the deflection under the point of application of the moment. Assume $E = 30 \cdot 10^6$ lb/in^2. *Ans.* -0.312 in.

CHAPTER 10

Deflection of Beams - Moment-Area Method

INTRODUCTION. In Chapter 9 it was mentioned that several methods are available for the determination of beam deflections. That chapter was devoted to an exposition of the Double-Integration Method. In the present chapter the second method, namely the Moment-Area Method, will be investigated in detail. It may be considered to constitute an alternative procedure to the double-integration process.

STATEMENT OF THE PROBLEM. A given system of loads acts upon a beam. The dimensions of the beam and the modulus of elasticity are known. It is desired to determine the deflection of any point in the bent beam from its original position.

FIRST MOMENT-AREA THEOREM. In the accompanying figure, *AB* represents a portion of the deflection curve of a bent beam. The shaded diagram immediately below *AB* represents the corresponding portion of the bending moment diagram. The construction of this diagram for many different types of loadings was examined in detail in Chapter 6. Tangents to the deflection curve are drawn at each of the points *A* and *B* as indicated.

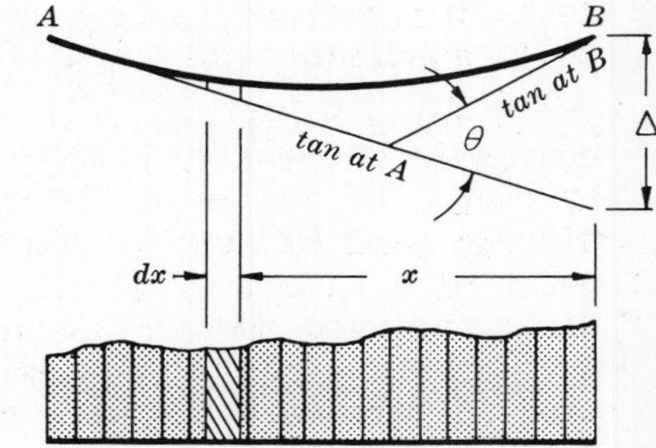

The First Moment-Area Theorem states that: The angle between the tangents at *A* and *B* is equal to the area of the bending moment diagram between these two points, divided by the product *EI*.

If θ denotes the angle between the tangents as shown in the diagram, then this theorem may be stated in equation form as follows:

$$\theta = \int_A^B \frac{M\,dx}{EI}$$

This theorem is derived in Problem 1. For applications, see Problems 5 and 13.

In this equation, *E* represents the modulus of elasticity of the beam, and *I* denotes the moment of inertia of the beam cross-section about the neutral axis, which passes through the centroid of the cross-section. Also, *M* represents the bending moment at the distance *x* from the point *B* as shown.

SECOND MOMENT-AREA THEOREM. Let us consider the vertical distance between the point *B* on the deflection curve shown above and the tangent to this curve drawn at point *A*. This vertical distance is denoted by Δ in the diagram.

The Second Moment-Area Theorem states that: The vertical distance of point *B* on a deflection curve from the tangent drawn to the curve at A is equal to the moment with respect to the vertical through B of the area of the bending moment diagram between A and B, divided by the product EI.

This theorem may be stated in equation form as follows:

$$\Delta = \int_A^B \frac{Mx\,dx}{EI}$$

This theorem is derived in Problem 2. For applications, see Problems 4, 6-12, 14-17.

SIGN CONVENTIONS. In the use of the first theorem, areas corresponding to a positive bending moment diagram are considered positive, those arising from a negative moment diagram are taken as negative. With reference to the above deflection curve AB and accompanying tangents, a positive net area implies that the tangent at B makes a positive or counterclockwise angle with the tangent drawn at A. In the use of the second theorem, the moments of areas of positive bending moment diagrams are considered to be positive and such positive products of areas and moment arms give rise to positive deflections. Positive deflections are taken to be those where point B lies *above* the tangent drawn at point A. It is to be noted that this sign convention bears no relation to that used in the double-integration method.

THE MOMENT-AREA PROCEDURE. The determination of the deflection of a specified point on a loaded beam is made in accordance with the following procedure.

1. The reactions of the beam are determined. In the particular case of a cantilever beam this step may frequently be omitted.
2. An approximate deflection curve is drawn. This curve must be consistent with the known conditions at the supports, such as zero slope or zero deflection.
3. The bending moment diagram is drawn for the beam. The procedure for this was discussed in Chapter 6. Frequently it is convenient to construct the moment diagram by *parts*, as discussed in Problems 14 and 15, of Chapter 6. Actually, the M/EI diagram must be used in connection with either of the above theorems, but for beams of constant cross-section, the M/EI diagram has the same shape as the ordinary bending moment diagram, except that each ordinate is divided by the product EI. Accordingly it is permissible in the case of beams of constant cross-section to work directly with the bending moment diagram and then divide the computed areas or moment-areas by EI. Or, equivalently, the angles or deflections may be multiplied by EI when areas or moment-areas of the ordinary moment diagram are used. This is actually the procedure commonly followed and is used in all illustrative problems in this chapter.
4. Convenient points A and B are selected and a tangent is drawn to the assumed deflection curve at one of these points, say A.
5. The deflection of point B from the tangent at A is then calculated by the Second Moment-Area Theorem.

In certain simple cases, particularly those involving cantilever beams, this deflec- ion of B from the tangent at A may actually be the desired deflection. In many cases, owever, it will be necessary to apply the Second Moment-Area Theorem to another point n the beam and then examine the geometric relationship between these two calculated eflections in order to obtain the desired deflection. No general statements regarding his phase of the procedure may be made. Specific examples of this technique may be ound in Problems 15, 16, 17.

COMPARISON OF MOMENT-AREA AND DOUBLE-INTEGRATION METHODS. If the deflection of only a single point of a beam is desired, the Moment-Area Method is usually more convenient than the Double-Integration Method. On the other hand, if the equation of the deflection curve of the entire beam is desired there is usually no procedure superior to Double-Integration. In the particular case of cantilever beams, the Moment-Area method is usually to be preferred. However, in any particular case the selection of one method or the other lies with the worker. Occasionally, there is little preference of one method over the other and more often the preference is entirely a personal one.

ASSUMPTIONS AND LIMITATIONS. As explained, in Problems 1 and 2, the Moment-Area method may be derived from the equation relating bending moment at a point in a beam and the curvature of the neutral surface at that same point. This same relation was used in deriving the Double-Integration procedure. Hence both methods are based upon the same fundamental relationship and thus both are subject to the same limitations. These are mentioned in the corresponding section in Chapter 9.

EI Δ = area of moment diagram × centroid of same.

SOLVED PROBLEMS

1. Derive the First Moment-Area Theorem.

In the figure, let AB represent a portion of the deflection curve of a bent beam. Let us consider an element of the beam of length ds. The radius of curvature of this element is denoted by ρ, and the bending moment in the beam at this point is denoted by M. From Problem 1, Chapter 8, we have the relationship given in equation *(7)*

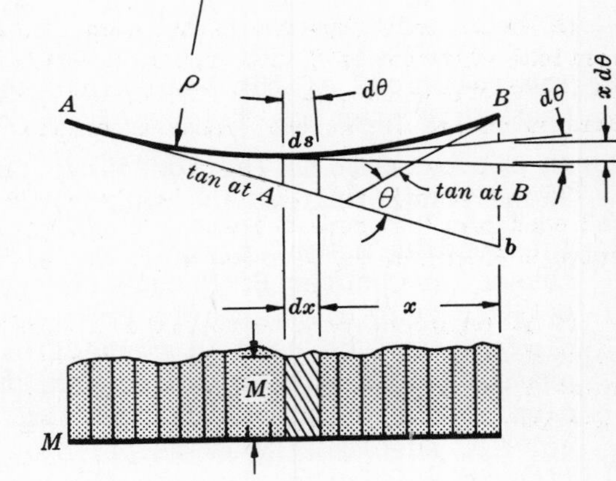

$$1) \qquad M = \frac{EI}{\rho}$$

where E represents the modulus of elasticity of the material and I denotes the moment of inertia of the cross-sectional area of the beam about its neutral axis.

The figure immediately below AB represents the bending moment diagram corresponding to the length AB of the beam. Construction of this diagram was discussed in Chapter 6.

The element of length ds is considered to subtend an angle $d\theta$, measured with respect to the center of curvature of the element ds, as shown. It is evident that $ds = \rho\, d\theta$; hence $\rho = ds/d\theta$. Substituting in equation *(1)*,

$$d\theta = \frac{M}{EI}\, ds$$

Since we are concerned only with very small lateral deflections of beams, it will be satisfactory to replace ds by its horizontal projection dx. Thus

$$d\theta = \frac{M\, dx}{EI}$$

This angle $d\theta$ may also be thought of as the angle between tangents to the deflection curve at the ends of the element of length ds, as shown above. This is true because the sides of these two angles are perpendicular. The angle θ between tangents to the deflection curve at the points A and B may now be found by summing all such angles $d\theta$, that is,

$$\theta = \int d\theta = \int_A^B \frac{M\,dx}{EI}$$

This is called the First Moment-Area Theorem. In words, this theorem is: The angle between the tangents at two points A and B of the deflection curve of a beam is equal to the area under the bending moment diagram between these two points, divided by EI.

For a sign convention, we shall take positive areas to be those arising from positive moment diagrams. A positive net area will be taken to denote that the right hand tangent at B makes a counterclockwise angle with the left hand tangent at A.

2. Derive the Second Moment-Area Theorem.

Let us again refer to the figure in Problem 1. It is desired to calculate the vertical distance of point B on the deflection curve from the tangent drawn to this curve at a point A. This distance is represented by the line segment Bb shown there. The contribution to this length Bb made by the bending of the element of length ds is the vertical element $x\,d\theta$ shown there. However, in Problem 1 it was shown that $d\theta = M\,dx/EI$. Hence

$$x\,d\theta = \frac{Mx\,dx}{EI}$$

With reference to the figure shown in Problem 1, the right side of this equation represents the moment of the shaded area $M\,dx$ about a vertical line through B, divided by EI. Integration yields

$$Bb = \int_A^B \frac{Mx\,dx}{EI}$$

In words, this equation states that if A and B are points on the deflection curve of a beam, the vertical distance of B from the tangent drawn to the curve at A is equal to the moment with respect to the vertical through B of the area of the bending moment diagram between A and B, divided by EI. This is called the Second Moment-Area Theorem.

For a sign convention, we shall take moments of areas of positive moment diagrams to be positive and such positive moment-areas will give rise to positive deflections. We shall further define positive deflections to be those where the point B lies above the tangent drawn at A.

It is important to note that this theorem indicates relative deflections, i.e. the deflection of point B with respect to the tangent drawn to the curve at A. The true or absolute deflection of point B may be zero, as in the case of a point directly over one support of a beam, yet in such a case there may be a non-zero relative deflection with respect to the tangent at A.

3. Determine the areas and locate the centroids of the figures commonly occurring in bending moment diagrams drawn by parts.

We need be concerned primarily with only three geometric figures: the rectangle, the triangle, the parabola. For the rectangle, the area is of course equal to the product of the lengths of two adjacent sides, and the centroid lies at the geometric center of the rectangle. The area of a triangle is half the product of its base and its altitude, and the centroid of a triangle was located in Problem 1, Chapter 7.

Let us now consider the case of the parabola shown in the adjoining figure. Note carefully that the parabola is placed so that its vertex lies at the origin of the coordinate system.

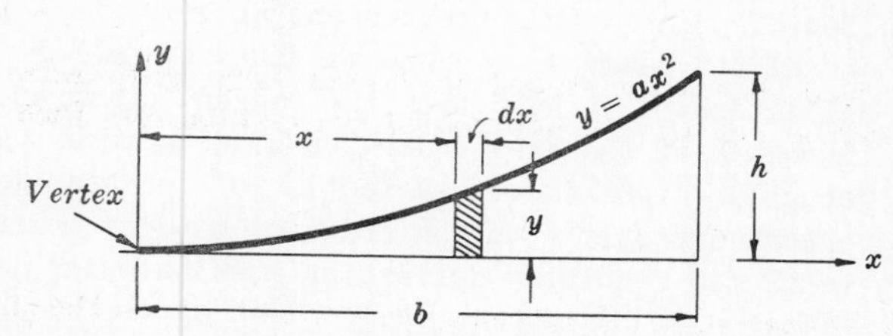

To determine the area, we consider first the area of the small shaded element of width dx and altitude y. Evidently its area is $y\,dx$.

To find the area under the entire parabola shown, we must sum the areas of all such elements by means of an integral, viz.:

$$A = \int y\,dx = \int_0^b ax^2\,dx = \frac{1}{3}a\,[x^3]_0^b = \frac{1}{3}a\,b^3$$

But when $x = b$, $y = h$; hence $a = \dfrac{h}{b^2}$ and thus $A = \dfrac{1}{3}bh$. = Area PARABOLA

To locate the x-coordinate of the centroid of this parabolic area, we employ the definition of the centroid given in Chapter 7, viz.:

$$\bar{x} = \frac{\int x\,da}{A}$$

Then $$\bar{x} = \frac{\int x(y\,dx)}{bh/3} = \frac{\int_0^b x(ax^2)dx}{bh/3} = \frac{a[x^4/4]_0^b}{bh/3} = \frac{(h/b^2)(b^4/4)}{bh/3} = \frac{3}{4}b.$$ = centroid PARABOLA

The rectangle, triangle, and parabola will be the only geometric figures occurring in the moment area treatments of beams subject to applied couples, concentrated forces, and uniformly distributed loads.

Thus the moments of areas under the bending moment diagrams, used in the Second Moment-Area Theorem, may be found in many cases by use of the above expressions for areas and centroids. The first moment of the area is equal to the product of its centroidal distance from the moment axis and its area.

4. The cantilever beam shown is subject to the concentrated force P applied at the free end of the beam. Determine the deflection under the point of application of the load.

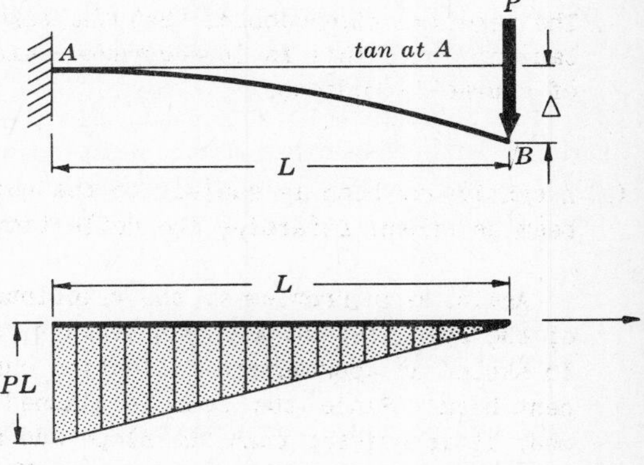

In the case of a cantilever beam, the reactions at the wall need not be determined although their determination is, of course, extremely simple. It is known that the slope and deflection at the clamped end A are each zero by definition of a cantilever beam. Hence the heavy curved line represents a realistic deflection curve.

The bending moment diagram is most easily drawn by working from right to left across the beam and is shown at the right. The construction of the moment diagram for this type of loading was discussed in Problem 1, Chapter 6.

Next, a tangent to the deflection curve is drawn at point A. In the case of a cantilever beam, this tangent coincides with the original unbent position of the bar and is represented by the straight line shown. Hence, in this particular case, the deflection of point B from the tangent at A is the actual desired deflection. The deflection of the free end of the beam, B, may now be found by use of the Second Moment-Area Theorem. By this theorem, the deflection of point B from the tangent drawn at A is given by the moment about the vertical line through B of the area under the bending moment diagram between A and B divided by the product EI. Actually, since the cross-section of the beam is constant along the length of the beam, it is easier to work directly with the ordinary bending moment diagram rather than with the M/EI diagram. In that event, the resulting deflection must be multiplied by the product EI.

Thus, from the Second Moment-Area Theorem, EI times the deflection of B from the tangent at A, denoted by Δ, is given by the moment of the shaded moment diagram about a vertical line through B. This moment of area may be calculated by multiplying the area by the distance of the centroid of the area from the vertical line through B. The area of the triangular moment diagram is $\frac{1}{2}(L)(-PL)$, where the negative sign is used because the bending moment is negative. The centroid of the moment diagram lies at a distance $2L/3$ from the right end. Hence the Moment-Area Theorem becomes

$$EI\Delta = \frac{1}{2}(L)(-PL)(\frac{2}{3}L) = -\frac{PL^3}{3} \quad \text{or} \quad \Delta = -\frac{PL^3}{3EI}$$

The negative sign implies that the final position of B lies below the tangent drawn at A.

5. Determine the slope at the right end of the cantilever beam discussed in Problem 4.

The heavy curved line representing the deflected beam has been sketched in Problem 4. Also, the moment diagram was presented there. These are reproduced here.

In the first of these diagrams tangents are drawn to the deflection curve of the bent beam at the clamped end A and the free end B. These are designated as the tangents at A and B respectively. According to the First Moment-Area Theorem the angle θ between these two tangents is equal to the area under the bending moment diagram between A and B, divided by EI. Thus from this theorem EI times the angle θ is given by the area under the shaded moment diagram, and we have

$$EI\theta = \tfrac{1}{2}(L)(-PL)$$

where the negative sign accompanying PL is used because the bending moment is negative. Solving,

$$\theta = -\frac{PL^2}{2EI}$$

The negative sign denotes that the right hand tangent at B makes a clockwise angle with the left hand tangent at A. This is in accordance with the sign convention discussed in Problem 1. The angle θ is of course in radians.

6. A cantilever beam is subject to the uniformly distributed load acting over the entire length of the beam as shown. Determine the deflection of the free end of the beam.

Again, as in Problem 4, the reactions at the end of the beam need not be found. It is first necessary to sketch an approximate deflection curve for the bent beam. Since the beam is clamped at the left end, it is evident that the slope and also the deflection at that end are each zero. Thus the heavy curved line represents a deflection curve that is in agreement with the known conditions of zero slope and zero deflection at the left end of the beam.

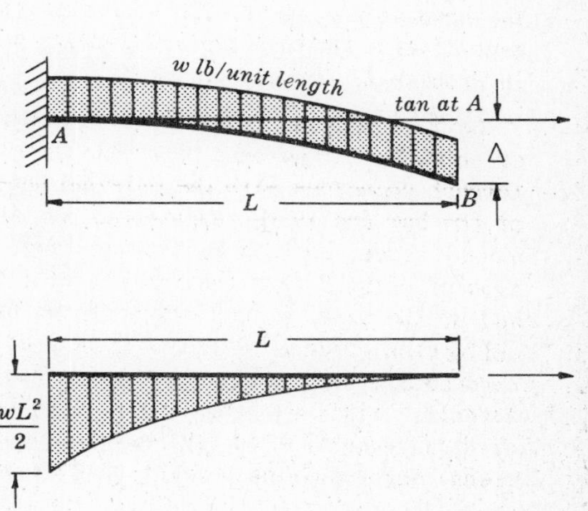

The bending moment diagram is best drawn by working from right to left across the beam. The construction of this bending moment diagram was discussed in Problem 2, Chapter 6. As mentioned in that problem, the maximum bending moment occurs at the supporting wall and has the value $wL^2/2$, where w represents the intensity of the uniform load per unit length of the beam. Also, in that problem, it was shown that the bending moment diagram is a parabola.

Next, a tangent to the deflection curve is drawn at point A. The free end of the beam is designated as B. Again, as in Problem 4, this tangent coincides with the original position of the beam and is represented by the straight line shown. Thus the deflection of point B from the tangent drawn at A represents the desired deflection. The deflection of point B may now be found by use of the Second Moment-Area Theorem. From this theorem, the deflection of point B from the tangent drawn at A is given by the moment about the vertical line through B of the area under the bending moment diagram between A and B, divided by EI. It is most convenient to work with the bending moment diagram shown above and then multiply the resulting deflection by the product EI.

Thus by the Second Moment-Area Theorem, EI times the deflection of B from the tangent at A, denoted by Δ, is given by the moment of the shaded moment diagram about a vertical line through B. This moment of area may be calculated by multiplying the area by the distance between the centroid of the area and the vertical line through B. The area of the parabolic moment diagram was shown in Problem 3 to be 1/3 the area of the rectangle enclosing the parabola. Thus the area under the moment diagram is given by $\frac{1}{3}(L)(-\frac{wL^2}{2})$, where the negative sign is used because the bending moment is negative. Also in Problem 3, it was shown that the centroid of the parabolic figure lies at a distance $3L/4$ from the right end. Hence the Second Moment-Area Theorem becomes

$$EI\Delta = \frac{1}{3}(L)(-\frac{wL^2}{2})(\frac{3L}{4}) = -\frac{wL^4}{8} \quad \text{or} \quad \Delta = -\frac{wL^4}{8EI}$$

The negative sign indicates that the final position of B lies below the tangent drawn at A.

7. A cantilever beam is subject to the uniformly distributed load extending from the midpoint of the beam to the extreme end as shown. Determine the deflection of the free end of the beam.

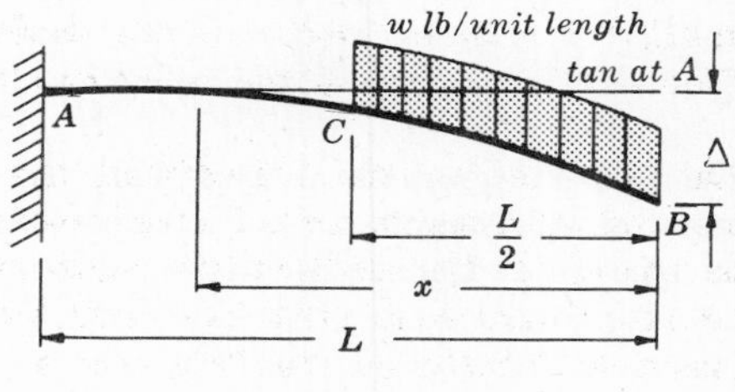

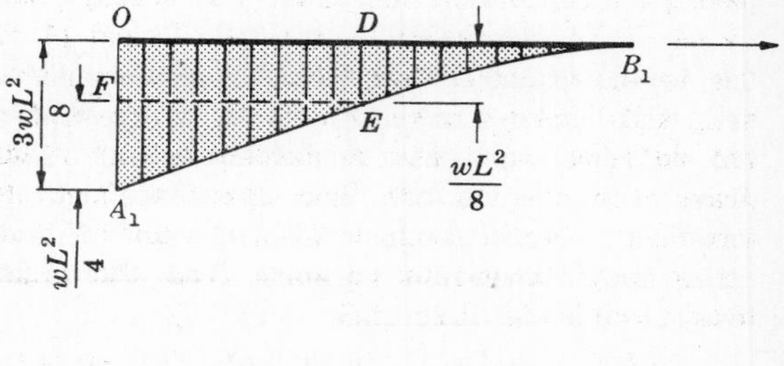

The reactions at the left end of the beam need not be calculated. The heavy curved line represents a reasonable approximation to the deflection curve of the beam. This curve is in agreement with the known conditions of zero slope and zero deflection at the left end of the beam. The bending moment diagram is most simply drawn by working from right to left across the beam. Under the uniform load, the moment must be parabolic, as discussed in Problem 2, Chapter 6. In constructing the portion of the moment diagram between A and C, the uniform load extending from C to B may be replaced by its resultant which is $wL/2$ lb acting downward. At any point between A and C, located a distance x from the end B, the bending moment is given by the moment of the resultant of the distributed load about an axis through this same point and perpendicular to the plane of the page. Thus the bending moment anywhere between A and C is given by $-\frac{wL}{2}(x - \frac{L}{4})$. This is a linear function, and hence the bending moment diagram plots as a straight line between A and C. Thus the bending moment diagram consists of a parabolic region DB_1E together with a trapezoidal region $ODEA_1$, as shown above.

Next, a tangent to the deflection curve is drawn at point A. The free end of the beam is designated as B. This tangent coincides with the original position of the beam and is represented by the straight line shown. Thus the deflection of point B from the tangent drawn at A represents the deflection of the free end of the beam. This deflection may be found by use of the Second Moment-Area Theorem. According to this theorem, the deflection of B from the tangent at A is given by the moment about the vertical line through B of the area under the bending moment diagram between A and B divided by EI. Again, it is convenient to work with the ordinary bending moment diagram shown above and then multiply the resulting deflection by EI.

The moment of the area OB_1A_1 about a vertical line through B may be evaluated most easily by considering the area to be divided into three portions: the parabolic region DB_1E, the rectangular region $ODEF$, and the triangular region FEA_1. The moment of the parabolic region is given by the product of its area (which is 1/3 of the area of the rectangle enclosing it) and the distance from B_1 to the centroid of the parabolic area (which is 3/4 of the length $L/2$). The moment of this area is thus

$$\frac{1}{3}\left(\frac{L}{2}\right)\left(\frac{-wL^2}{8}\right)\left(\frac{3}{4}\cdot\frac{L}{2}\right)$$

The moment of the rectangular region is given by the product of its area and the distance from B_1 to its centroid which is $3L/4$. The moment of this area is thus

$$\frac{L}{2}\left(\frac{-wL^2}{8}\right)\left(\frac{L}{2}+\frac{1}{2}\cdot\frac{L}{2}\right)$$

The moment of the triangular region is given by the product of its area and the distance from B_1 to its centroid which is $\left(\frac{L}{2}+\frac{2}{3}\cdot\frac{L}{2}\right)$. The moment of this area is thus

$$\frac{1}{2}\left(\frac{L}{2}\right)\left(\frac{-wL^2}{4}\right)\left(\frac{L}{2}+\frac{2}{3}\cdot\frac{L}{2}\right)$$

The moment of the entire area OB_1A_1 about the vertical line through B is equal to the sum of the moments of the three areas named above. Hence the Second Moment-Area Theorem becomes

$$EI\Delta = \frac{1}{3}\left(\frac{L}{2}\right)\left(\frac{-wL^2}{8}\right)\left(\frac{3}{4}\cdot\frac{L}{2}\right) + \frac{L}{2}\left(\frac{-wL^2}{8}\right)\left(\frac{L}{2}+\frac{1}{2}\cdot\frac{L}{2}\right) + \frac{1}{2}\left(\frac{L}{2}\right)\left(\frac{-wL^2}{4}\right)\left(\frac{L}{2}+\frac{2}{3}\cdot\frac{L}{2}\right) \quad \text{or} \quad \Delta = \frac{-41wL^4}{384EI}$$

The negative sign indicates that the final position of B lies below the tangent drawn at A.

8. A cantilever beam is loaded by the moment M_1 as well as the uniformly distributed load extending over half of its length as shown. Determine the deflection of the free end of the beam.

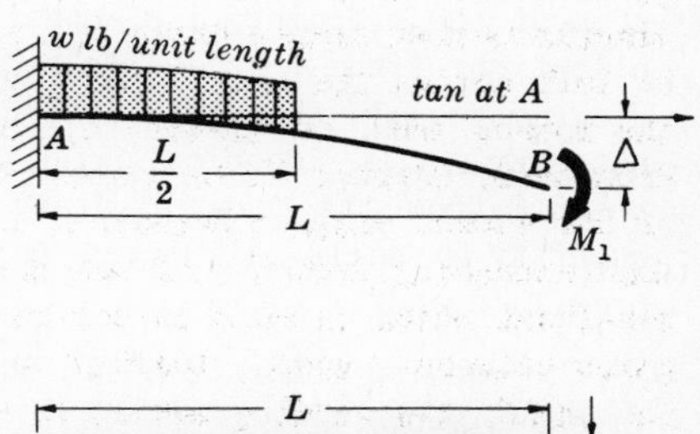

The reactions at the left end of the beam need not be calculated. The heavy curved line represents the approximate shape of the deflection curve of the beam. This curve is in agreement with the conditions of zero slope and zero deflection at the left end of the beam. In this case, since there are two loads acting on the beam, it is perhaps simplest to construct the bending moment diagram by *parts* as discussed in Problems 14 and 15 of Chapter 6. One bending moment diagram will be constructed to represent the bending moment due to M_1 alone without regard to the uniform load, and a second moment diagram will be drawn for the uniform load alone without regard to the moment M_1. In the construction of each of these diagrams, it is most convenient to work from the right end of the beam toward the left end.

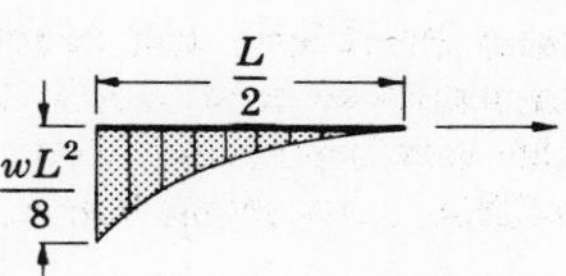

The first bending moment diagram due to M_1 alone is evidently a rectangle, since the bending moment due to M_1 is the same at all points along the beam. The moment M_1 causes the beam to bend into a configuration that is concave downward, which, according to the sign convention of Chapter 6, constitutes a negative bending moment.

The second bending moment diagram due to the uniform load is parabolic as discussed in Problem 6, except that the parabola here corresponds only to the portion of the beam that is subject to the uniform load, i.e. the left half of the beam.

A tangent to the deflection curve is now drawn at point A. The free end of the beam is designated as point B. This tangent coincides with the original position of the beam and is represented by the straight line shown. Thus the deflection of point B from the tangent drawn at A represents the desired deflection of the free end of the beam. According to the Second Moment-Area Theorem, this deflection of B from the tangent at A is given by the moment about the vertical line through B of the area under the entire bending moment diagram between A and B, divided by EI.

The moment of the area of the entire bending moment diagram between A and B about the vertical line through B is most easily found by adding the moments of the rectangular and parabolic areas about this vertical line. For each of these areas, the moment about the line through B is given by the product of the area and the distance from the centroid of the area to B. These are the same areas and centroidal distances used in previous problems. Thus the Second Moment-Area Theorem gives

$$EI\Delta = (-M_1)(L)(\frac{L}{2}) + \frac{1}{3}(\frac{L}{2})(\frac{-wL^2}{8})(\frac{L}{2} + \frac{3}{4}\cdot\frac{L}{2}) \quad \text{or} \quad \Delta = -\frac{M_1L^2}{2EI} - \frac{7wL^4}{384EI}$$

The negative signs indicate that the final position of B lies below the tangent drawn at A.

9. The simply supported beam is loaded by a concentrated force applied at its midpoint as shown. Find the maximum deflection of the beam.

The reactions at the ends of the beam are each $P/2$ by symmetry. The heavy line represents the deflection curve of the beam; evidently this curve must be symmetric about the midpoint of the beam since the load is centrally applied. Because of this symmetry, the tangent drawn at the midpoint of the deflected beam will be horizontal. The midpoint of the beam is denoted as point A, and the tangent drawn at A is shown by the horizontal line. The right end of the beam is designated as point B. Again, because of symmetry, the maximum deflection of the beam must occur at its midpoint.

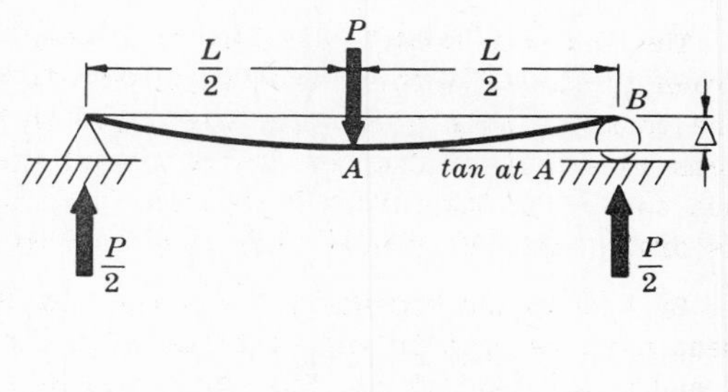

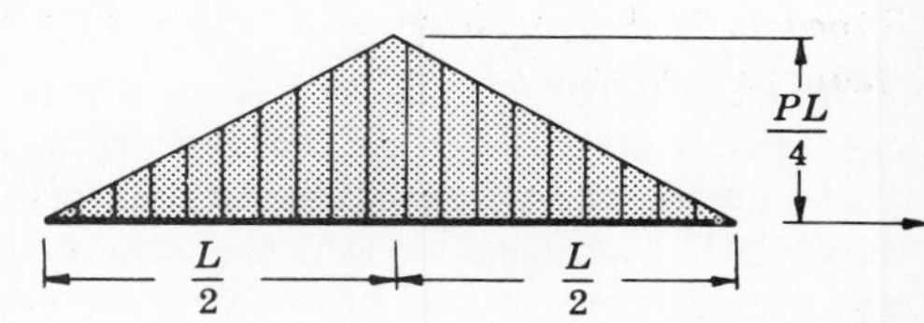

It is desired to find the deflection of the beam at the midpoint, i.e. under the point of application of the force P. Inspection of the above diagram reveals that this central deflection, denoted by Δ, is identical with the deflection of point B from the tangent drawn at A. This last quantity is easily calculated by the Second Moment-Area Theorem.

The bending moment diagram for this particular loading was discussed in Problem 4, Chapter 6, and is shown above.

In calculating the deflection of B from the tangent at A, it is necessary to evaluate the moment of the area under the moment diagram between these two points about a vertical line through B and divide this quantity by the product EI. The area to be considered is thus the right half of the above shaded moment diagram, i.e. the triangle of altitude $PL/4$ and base $L/2$. The distance from the centroid of this triangle to the vertical line through B is $(2/3)(L/2)$. Hence the Second Moment-Area Theorem applied between the points A and B gives the desired deflection:

$$EI\Delta = \frac{1}{2}(\frac{L}{2})(\frac{PL}{4})(\frac{2}{3}\cdot\frac{L}{2}) \quad \text{or} \quad \Delta = \frac{PL^3}{48EI}$$

The fact that this quantity is positive indicates that point B lies above the tangent at A.

It is to be noted that the Second Moment-Area Theorem always indicates relative deflections, i.e. the deflection of one point on a beam with respect to a tangent drawn at a second point on the beam. In this particular problem the true or absolute displacement of B is zero, but we may think of a deflection of B relative to the tangent drawn at A. Fortunately, this relative displacement is exactly equal to the desired maximum displacement because of symmetry.

0. The simply supported beam is loaded by the two symmetrically placed forces shown. Find the maximum deflection of the beam.

The reactions at the ends of the beam are each equal to P by symmetry. The heavy line represents the deflection curve of the beam and this curve must be symmetric about the midpoint of the beam because the loads are symmetrically applied. As in Problem 9, the tangent to the deflection curve at the midpoint of the beam, denoted by A, is horizontal. The right end of the beam is designated as point B. Because of symmetry, the maximum deflection of the beam must occur at its midpoint.

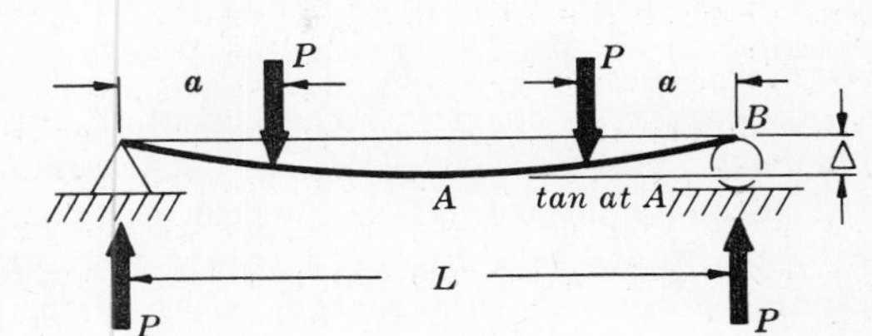

However, the diagram indicates that the deflection at the midpoint is equal to the deflection of point B from the tangent at A, both deflections being represented by Δ. It is simple to calculate this second deflection by use of moment-areas.

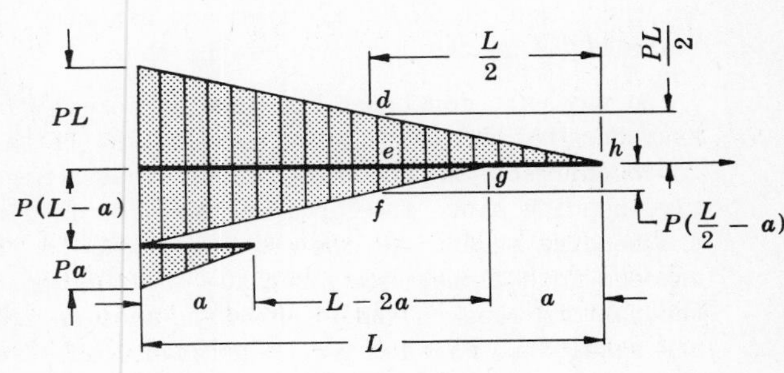

The bending moment diagram is perhaps best drawn in *parts*, working from right to left along the beam. In this procedure, discussed in Problems 14 and 15 of Chapter 6, the moment diagram due to each of the forces P applied individually is drawn with the result that the final moment diagram appears as shown above.

By the Second Moment-Area Theorem, the deflection of B from the tangent at A is equal to the moment with respect to the vertical line through B of the area of the bending moment diagram between A and B, divided by EI. This area consists of the two triangles *deh* and *efg* shown above. The locations of the centroids of these triangles were discussed in Problem 3. From the Second Moment-Area Theorem, we have

$$EI\Delta = \frac{1}{2}\left(\frac{L}{2}\right)\left(\frac{PL}{2}\right)\left(\frac{2}{3}\cdot\frac{L}{2}\right) + \frac{1}{2}\left(\frac{L-2a}{2}\right)\left[-\left(\frac{PL}{2} - Pa\right)\right]\left[a + \frac{2}{3}\left(\frac{L-2a}{2}\right)\right] = \frac{PL^2a}{8} - \frac{Pa^3}{6}$$

or

$$\Delta = \frac{PL^3}{24EI}\left(\frac{3a}{L} - \frac{4a^3}{L^3}\right)$$

The negative sign in the first of the quantities appearing in square brackets is used because that quantity represents the altitude of triangle *efg* and this triangle corresponds to a negative bending moment. In this particular problem the bending moment diagram could have been drawn in the conventional manner just as advantageously, rather than by parts as illustrated above.

11. The simply supported beam is loaded by the uniformly distributed load shown. Determine the maximum deflection of the beam.

The reactions at the ends of the beam are each $wL/2$ by symmetry. The heavy line represents the deflection curve of the beam. This curve must be symmetric about the midpoint of the beam since the loading and supports are symmetric. The midpoint is denoted as point A and because of symmetry the tangent to the deflection curve at A must be horizontal. The right end of the beam is designated as point B. Because of symmetry, the maximum deflection of the beam must occur at its midpoint.

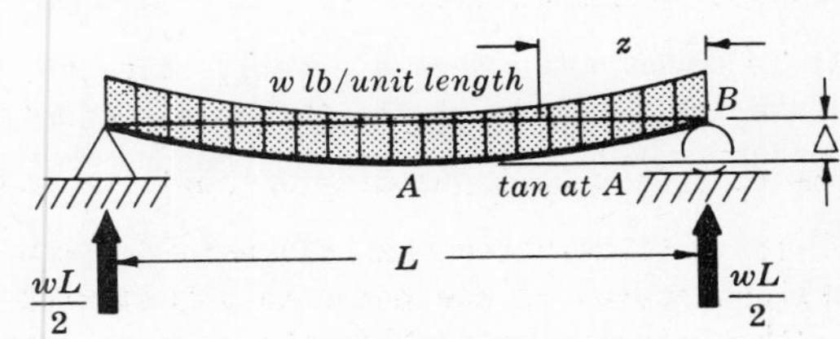

It is desired to find the deflection of the beam at its midpoint. From the above diagram, it is evident that this central deflection, denoted by Δ, is identical to the deflection of point B from the tangent drawn at A. This last quantity may, of course, be found by the moment-area technique.

The bending moment diagram is perhaps best drawn by parts so that the moment-area theorem may be applied in as simple a manner as possible. It is easiest to work from right to left along the beam.

Let us first construct the moment diagram corresponding to the reaction $wL/2$ at the right end of the beam. If this force only is considered to act on the beam, it will cause positive bending since the force is upward. If we introduce a coordinate z with origin at the right end of the beam and with the variable z directed toward the left end of the beam as shown in the above diagram, then the bending moment at this section z due to the reaction is given by $(wL/2)z$. This plots as the triangle shown in the adjoining Fig. (a), being zero at the right end of the beam and assuming the value $wL^2/2$ at the left end of the beam.

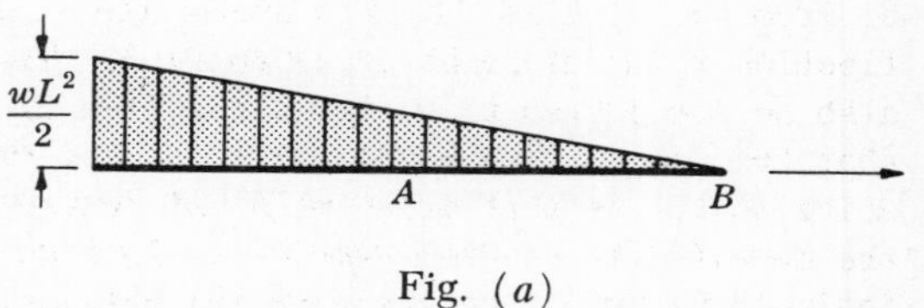

Fig. (a)

We may next construct the bending moment diagram due to the distributed load only. If this force only acts on the beam, it will cause negative bending since the force is downward. To find the bending moment at the section z due to the uniform load alone, we may replace the portion of the uniform load to the right of the section z by its resultant, which is a concentrated load of wz lb applied at a distance $z/2$ to the right of the section z. Hence the bending moment at this section due to the uniform load is $-wz^2/2$. This plots as the parabola shown in Fig. (b), being zero at the right end of the beam and assuming the value $wL^2/2$ at the left end of the beam.

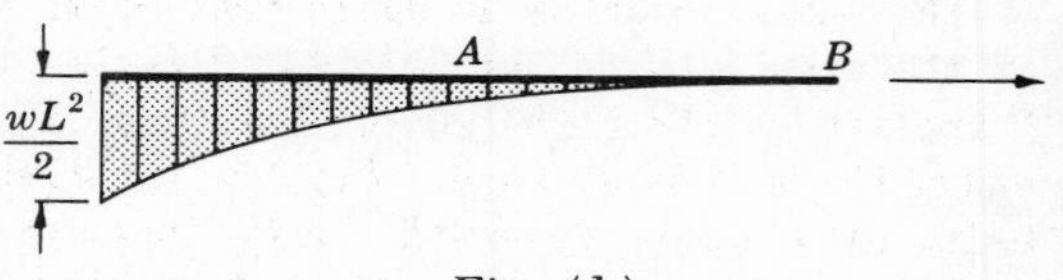

Fig. (b)

The entire bending moment diagram is composed of the two parts, as shown in Fig. (c).

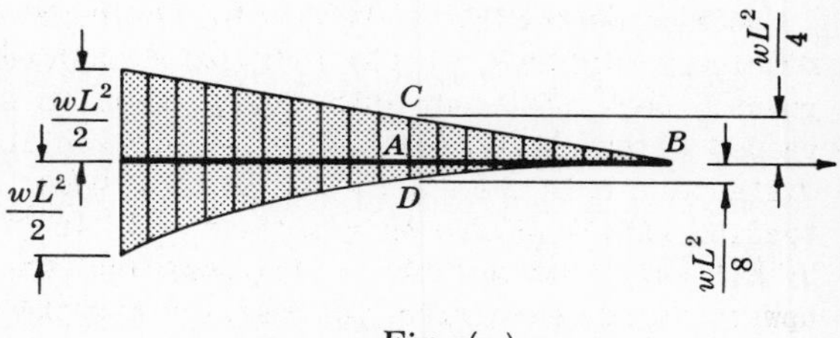

Fig. (c)

According to the Second Moment-Area Theorem, the deflection of B from the tangent at A is equal to the moment with respect to the vertical line through B of the area of the bending moment diagram between A and B, divided by EI. From the above bending moment diagram we can calculate the moment of the triangular area ABC about the vertical line through B to be given by the product of the area of the triangle and the distance from B to the centroid of the triangle, which is $(2/3)(L/2)$. This moment of area is consequently

$$\frac{1}{2}\left(\frac{L}{2}\right)\left(\frac{wL^2}{4}\right)\left(\frac{2}{3}\cdot\frac{L}{2}\right)$$

Also, the moment of the parabolic area ABD about the vertical line through B is given by the product of the area ABD (which is 1/3 the area of the surrounding rectangle) and the distance from B to the centroid of the area, which is $(3/4)(L/2)$. This moment of area is consequently

$$\frac{1}{3}\left(\frac{-wL^2}{8}\right)\left(\frac{L}{2}\right)\left(\frac{3}{4}\cdot\frac{L}{2}\right)$$

It is to be noted that a negative sign is used in conjunction with the term $wL^2/8$, since it represents the altitude of the parabola at the point A and this parabolic area corresponds to a negative bending moment. Thus, from the Second Moment-Area Theorem, we have

$$EI\Delta = \frac{1}{2}\left(\frac{L}{2}\right)\left(\frac{wL^2}{4}\right)\left(\frac{2}{3}\cdot\frac{L}{2}\right) + \frac{1}{3}\left(\frac{-wL^2}{8}\right)\left(\frac{L}{2}\right)\left(\frac{3}{4}\cdot\frac{L}{2}\right)$$

or

$$\Delta = \frac{5wL^4}{384EI}$$

2. The overhanging beam is loaded by two concentrated forces as shown. Find the deflection at the midpoint of the beam.

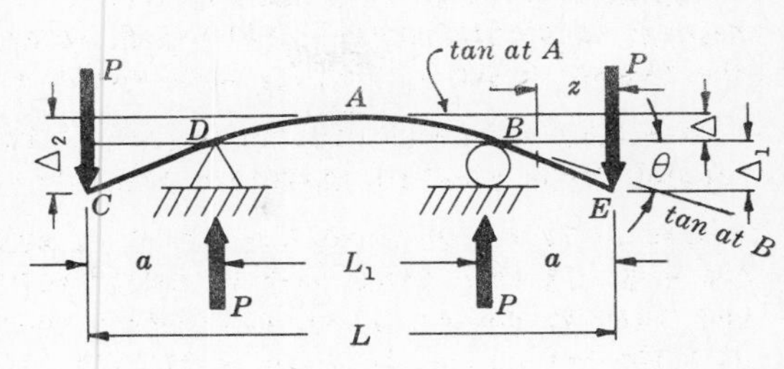

Each of the reactions is equal to P lb by symmetry. The deflection curve of the beam, shown by the heavy line, is symmetric about the midpoint of the beam and consequently the tangent to the deflection curve is horizontal at the midpoint. The midpoint of the beam is designated as point A and the point directly over the right support is B. From the diagram it is evident that the deflection of the midpoint is given by Δ, which may also be considered to be the deflection of B from the tangent drawn at A. Again, it is to be noted that the moment-area technique indicates relative deflections, in this case the deflection of B relative to the tangent drawn at A. Of course, the absolute deflection of B is zero. However, the use of the moment-area theorem enables us to find the desired deflection of the midpoint by the somewhat indirect method of determining the deflection of B relative to the tangent at A.

The moment diagram is again best drawn by parts. Let us introduce a coordinate z with the origin at the right end of the beam and directed toward the left, as shown above. Due to the downward force P applied at the right end of the beam, the bending moment at any section a distance z from the right end of the beam is given by $-Pz$, the negative sign being used since downward loads correspond to negative bending moments. The moment diagram for this load only is shown in the adjacent Fig. (a).

Working from right to left along the beam, the bending moment due to the reaction P applied at point B does not come into being until we have passed to the left of B. Then, at any section a distance z from the right end of the beam, the bending moment due to this force P only is given by $P(z-a)$, this quantity being positive because upward forces give rise to positive bending moments. The moment diagram for this load only plots as shown in Fig. (b).

Lastly, the bending moment diagram due to the reaction P applied at D comes into being after we have passed to the left of point D. At any section a distance z from the right end of the beam, the bending moment due to this force P only is given by $P[z-(L_1+a)]$. Thus the moment diagram for this loading only plots as shown in Fig. (c).

The entire bending moment diagram is consequently composed of the above three triangles, as shown in Fig. (d).

The deflection of point B from the tangent drawn to the deflection curve at point A is given by the moment of the area under the moment diagram between A and B about a vertical line through B, divided by EI. This area under the moment diagram consists of a triangle AFB and a trapezoid $ABHG$ as shown. The trapezoid may be treated most conveniently by dividing it into a rectangle of altitude Pa and a triangle of altitude $PL_1/2$. The area of each of these figures together with the

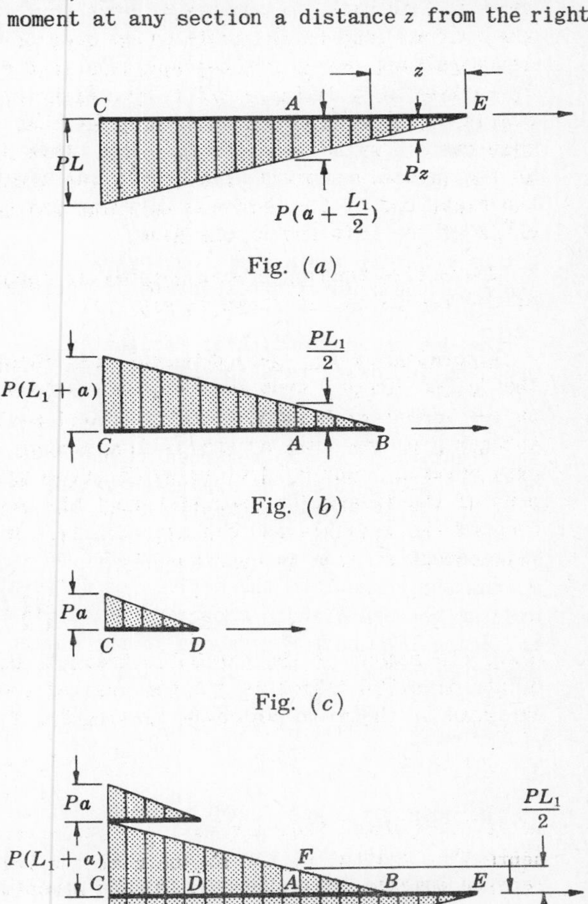

Fig. (a)

Fig. (b)

Fig. (c)

Fig. (d)

location of each centroid of area was discussed in Problem 3. According to the Moment-Area Theorem the desired deflection Δ is thus given by

$$EI\Delta = \frac{1}{2}(\frac{L_1}{2})(\frac{PL_1}{2})(\frac{2}{3}\cdot\frac{L_1}{2}) + (\frac{L_1}{2})(-PA)(\frac{L_1}{4}) + \frac{1}{2}(\frac{L_1}{2})(\frac{-PL_1}{2})(\frac{2}{3}\cdot\frac{L_1}{2}) \quad \text{or} \quad \Delta = -\frac{PaL_1^2}{8EI}$$

13. Determine the slope of the beam described in Problem 12 at a point directly over the support at B.

For the determination of slopes, the First Moment-Area Theorem is useful. Again, let us consider the two points A and B as described in Problem 12. As stated above, the tangent to the deflection curve of the beam is horizontal at the point A because of symmetry of both loading and support. If the change of slope of the deflection curve can be evaluated between A and B, then evidently this change will also be equal to the slope itself at B since the slope at A is zero. The First Moment-Area Theorem tells us that the angle between the tangents at A and B is equal to the area of the bending moment diagram between these two points divided by EI. Evidently the angle that the tangent at A makes with the horizontal is zero, and we shall denote the angle that the tangent at B makes with the horizontal by θ. Then, from the Moment-Area Theorem, we have merely to evaluate the area under the moment diagram between A and B. This area, as before, consists of the triangle AFB together with the trapezoid $ABHG$. From moment-areas we thus have

$$EI\theta = \frac{1}{2}(\frac{L_1}{2})(\frac{PL_1}{2}) + (\frac{L_1}{2})(-Pa) + \frac{1}{2}(\frac{L_1}{2})(\frac{-PL_1}{2})$$

or

$$\theta = -\frac{PaL_1}{2EI}$$

The negative sign indicates that the tangent at B makes a clockwise angle with the tangent drawn at A. This agrees with the sign convention adopted in Problem 1.

14. Let us consider again the overhanging beam described in Problem 12. Determine the deflection of the end E of the beam with respect to its original position.

The desired deflection is designated as Δ_1 in the figure of Problem 12. Evidently from that figure, the following relationship exists:

$$\Delta_1 = \Delta_2 - \Delta$$

The quantity Δ_2 may be regarded as the deflection of point E from the tangent to the deflection curve at A. As such, it may be evaluated by the Second Moment-Area Theorem. The quantity Δ has already been evaluated in Problem 12.

To compute Δ_2 we may employ the Second Moment-Area Theorem which states that the deflection of E from the tangent to the deflection curve at A is equal to the moment of the area under the moment diagram between A and E about a vertical line through E, divided by EI. This area consists of the triangles AFB and AGE as shown in the moment diagram drawn by parts in Problem 12. According to the Moment-Area Theorem, the deflection Δ_2 is given by

$$EI\Delta_2 = \frac{1}{2}(\frac{L_1}{2})(\frac{PL_1}{2})[a + \frac{2}{3}\cdot\frac{L_1}{2}] + \frac{1}{2}(a + \frac{L_1}{2})[-P(a + \frac{L_1}{2})][\frac{2}{3}(a + \frac{L_1}{2})] = -\frac{PaL_1^2}{8} - \frac{Pa^2L_1}{2} - \frac{Pa^3}{3}$$

The quantity Δ was found in Problem 12 to be $\Delta = -\dfrac{PaL_1^2}{8EI}$.

Finally, the desired end deflection Δ_1 may be found from the geometric relationship

$$\Delta_1 = \Delta_2 - \Delta$$

to be

$$\Delta_1 = -\frac{PaL_1^2}{8EI} - \frac{Pa^2L_1}{2EI} - \frac{Pa^3}{3EI} + \frac{PaL_1^2}{8EI} = -\frac{Pa^2L_1}{2EI} - \frac{Pa^3}{3EI}$$

15. The simply supported beam is loaded by the concentrated load shown. Determine the deflection at the midpoint of the beam.

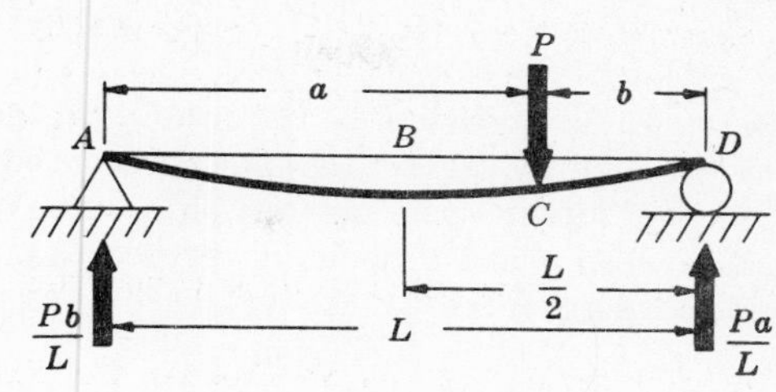

From statics the end reactions have the values indicated. The approximate form of the deflection curve is indicated by the heavy line. Evidently there is no symmetry of the deflection curve about the midpoint of the beam, and this feature renders this problem somewhat more difficult than the preceding ones. The left end of the beam is designated as point A, the midpoint as B, the point of application of the force P as C, and the right end as D.

Again, it is convenient to draw the bending moment diagram by parts. Working from left to right along the length of the beam, the moment due to the left reaction alone at any section a distance x from the left end is given by Pbx/L. This is represented by the triangle shown in Fig. (a).

As we proceed from left to right along the beam, the effect of the downward load P on the moment diagram does not become evident until we have passed to the right of its point of application. At any section a distance x from the point A and lying to the right of load P, the bending moment due to the force P alone is given by $-P(x-a)$. This may be represented by the triangle shown in Fig. (b).

The entire bending moment diagram drawn by parts appears in Fig. (c).

The desired deflection at the midpoint of the beam may be found by the following rather indirect use of the Moment-Area Theorem. A tangent to the deflection curve is drawn at the left end of the beam. This is designated as the tangent at point A, as shown in Fig. (d). It is now possible to calculate the deflection of point D from the tangent at A by use of the Second Moment-Area Theorem. This deflection is designated as Dd in Fig. (d). Again it is to be remembered that the moment-area technique indicates relative deflections, in this case the deflection of point D relative to the tangent drawn at A. The absolute deflection of point D is of course zero, since the beam is supported at that point.

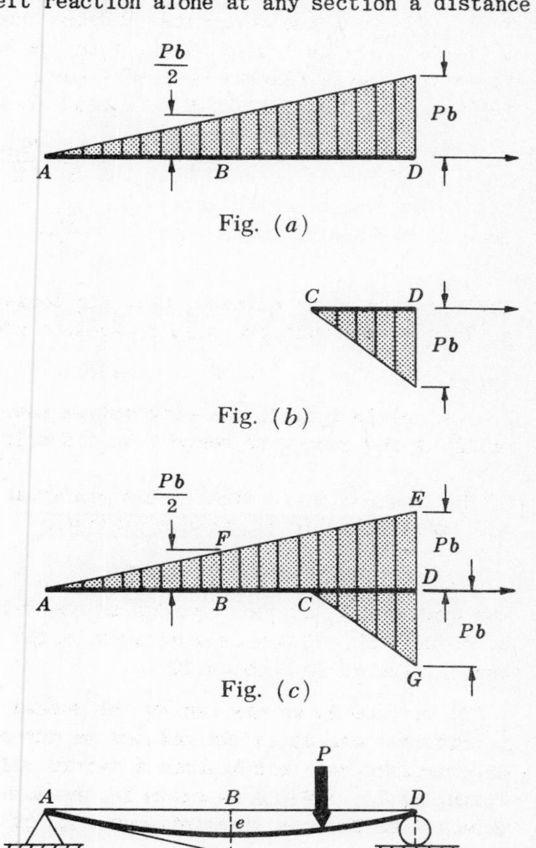

Fig. (a)

Fig. (b)

Fig. (c)

Fig. (d)

According to the Second Moment-Area Theorem, the deflection of D from the tangent drawn at A is given by the moment of the area under the moment diagram between A and D about a vertical line through D, divided by EI. Thus, taking the moment of the above triangles ADE and CDG about the vertical line through D, we obtain

$$EI(Dd) = \frac{1}{2}(L)(Pb)\left(\frac{L}{3}\right) + \frac{1}{2}(b)(-Pb)\left(\frac{b}{3}\right) = \frac{PbL^2}{6} - \frac{Pb^3}{6}$$

As stated previously, B represents the midpoint of the beam. Evidently from similar triangles, the line segment Bf shown in Fig. (d) must be exactly half the length of Dd and thus we may write

$$EI(Bf) = \frac{PbL^2}{12} - \frac{Pb^3}{12}$$

Next, it is possible to calculate the deflection of the midpoint of the beam from the tangent drawn at A. This deflection is represented by the line segment ef in the above figure. According to the Second Moment-Area Theorem this is given by the moment of the area under the bending moment diagram between A and B about a vertical line through B, divided by EI. This portion of the moment diagram is represented by the triangle ABF. Applying the Second Moment-Area Theorem we have

$$EI(ef) = \frac{1}{2}(\frac{L}{2})(\frac{Pb}{2})(\frac{1}{3}\cdot\frac{L}{2}) = \frac{PbL^2}{48}$$

From the above representation of the deflection curve of the beam it is apparent that the desired midpoint deflection is represented by the line segment Be. This may be found from the relationship

$$Be = Bf - ef$$

Substituting the above values in the right side of the equation, we find for the desired deflection of the midpoint

$$EI(Be) = \frac{PbL^2}{12} - \frac{Pb^3}{12} - \frac{PbL^2}{48} \quad \text{or} \quad Be = \frac{PbL^2}{48EI}(3 - \frac{4b^2}{L^2})$$

Note that this is not the maximum deflection of the beam except for the special case when $a = b = L/2$. Also, it is assumed that the load P lies to the right of the midpoint of the beam, otherwise the triangular moment diagram CDG would extend to the left of the midpoint of the beam and it would then be necessary to take a portion of it into account in calculating the deflection ef.

16. The overhanging beam is loaded by the concentrated force of 10,000 lb applied as shown in Fig. (*a*). The moment of inertia of the cross-section of the beam about its neutral axis is 1000 in^4 and $E = 30\cdot10^6$ lb/in^2. Determine the deflection under the point of application of the 10,000 lb force.

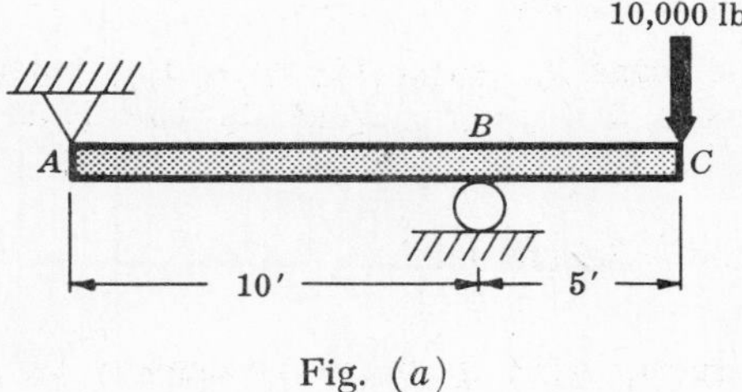

Fig. (*a*)

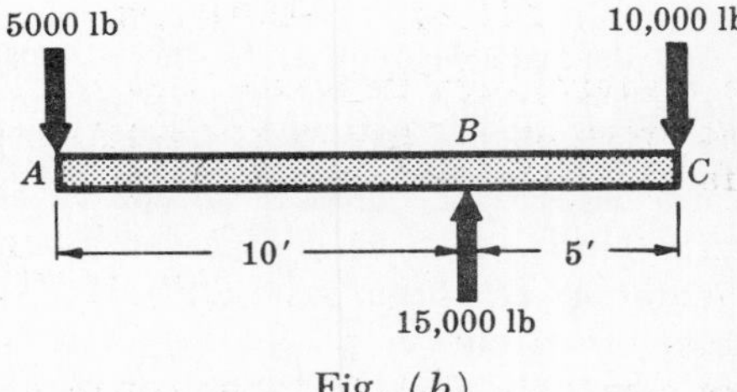

Fig. (*b*)

From statics the reactions are easily found to be a downward force of 5000 lb acting at A and an upward force of 15,000 lb applied at B. The free-body diagram of the beam is shown in Fig. (*b*).

The bending moment diagram for this problem may be drawn in any one of several equally convenient ways. Let us draw it in the conventional manner, in which case it appears as in Fig. (*c*).

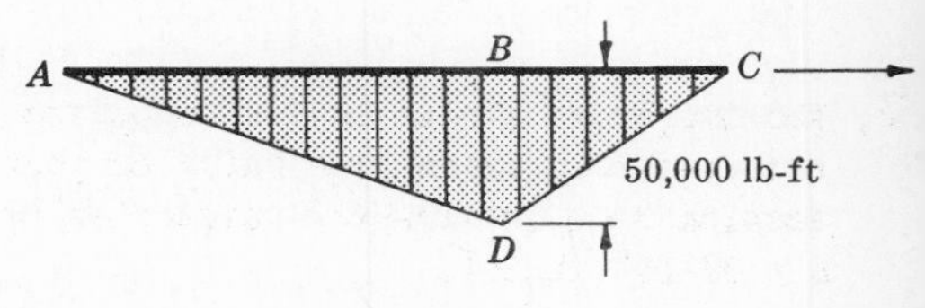

Fig. (*c*)

The approximate nature of the deflection curve is represented by the heavy line in Fig. (*d*). A tangent to the deflection curve is drawn at the point A, as indicated.

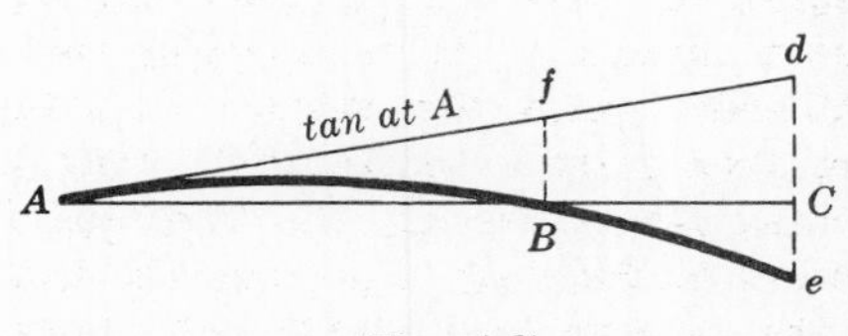

Fig. (*d*)

According to the Second Moment-Area Theorem the deflection of point C, whose final position is indicated by e, from the tangent at A is given by the moment of the area under the bending moment diagram between A and C about a vertical line through C divided by EI. Thus the moment of the area of the entire triangle ADC about a vertical line through C must be calculated. This gives for the deflection of C with respect to the tangent at A,

$$EI(de) = \frac{1}{2}(10)(-50{,}000)(5 + \frac{1}{3}\cdot 10) + \frac{1}{2}(5)(-50{,}000)(\frac{2}{3}\cdot 5) = -2{,}496{,}000$$

It is to be carefully noted that the units of the right side of this equation are lb-ft^3.

Next, it is necessary to calculate the deflection of point B from the tangent drawn at A. This is represented by the line segment fB in the above diagram. Again, it is to be remembered that the Moment-Area Theorem indicates relative deflections, in this case the deflection of B relative to the tangent at A. Actually, of course, the true or absolute deflection of point B is zero. This relative deflection may be found by taking the moment of the area of the triangle ABD about a vertical line through B and then dividing by EI. This gives

$$EI(fB) = \frac{1}{2}(10)(-50{,}000)(\frac{1}{3}\cdot 10) = -833{,}000$$

Again, the units of the right side of this equation are lb-ft^3.

From a consideration of the similar triangles AfB and AdC in Fig. (d), we may write $fB/10 = dC/15$. Hence from the above value of $EI(fB)$ we have

$$EI(dC) = \frac{15}{10}(-833{,}000) = -1{,}250{,}000 \text{ lb-ft}^3$$

Evidently the desired deflection of the point C, represented by the line segment Ce, is given by

$$Ce = de - dC$$

Substituting the above values in the right side of this equation, we find for the desired deflection of the point C:

$$EI(Ce) = -2{,}496{,}000 - (-1{,}250{,}000) = -1{,}246{,}000$$

The negative sign indicates that the final position of point C, designated as e, lies below the tangent drawn at A. Let us now substitute $I = 1000$ in^4 and $E = 30\cdot 10^6$ lb/in^2 in the last equation. This gives the desired deflection

$$Ce = \frac{-1{,}246{,}000(1728)}{30\cdot 10^6(1000)} = -0.0720 \text{ in.}$$

Note carefully that it is necessary to introduce the factor 1728 as shown to convert to consistent units, since the quantity $-1{,}246{,}000$ is in lb-ft^3 units whereas E and I are expressed in lb/in^2 and in^4 respectively.

7. The overhanging beam is loaded by the uniformly distributed load shown. The bar is of rectangular cross-section 3×4 in. Find the deflection of point A. Take $E = 30\cdot 10^6$ lb/in^2.

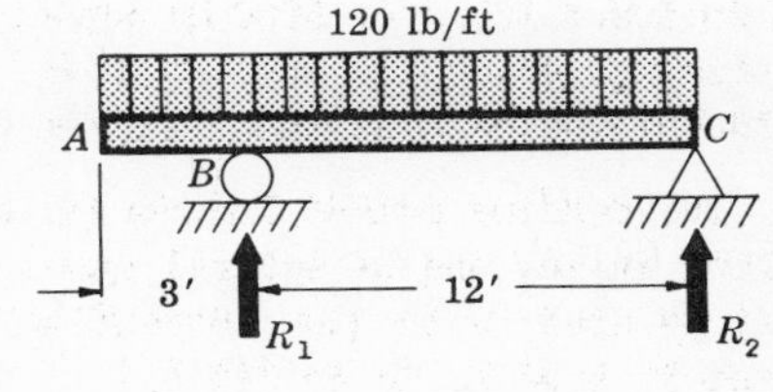

From statics the reactions are readily found to be $R_1 = 1125$ lb and $R_2 = 675$ lb. The bending moment diagram for this beam is best adapted to the Moment-Area technique if it is drawn by parts working from both ends toward the support at point B. We first work from point C to the left, considering first the bending moment due to the reaction R_2 only, which appears in Fig. (a); and then the bending moment due to the uniform load of 120 lb/ft only, which appears as the parabola in Fig. (b).

Fig. (a)

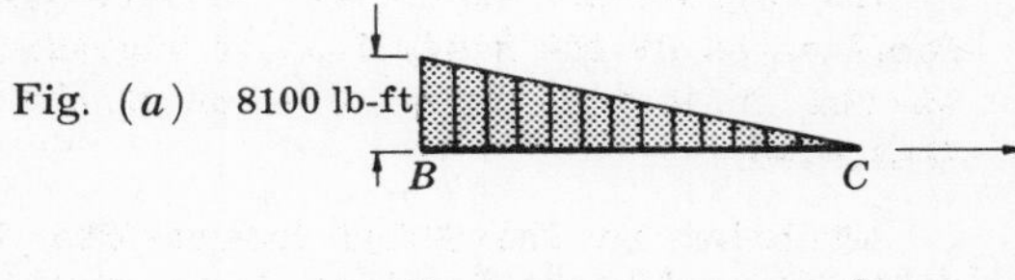

Fig. (b)

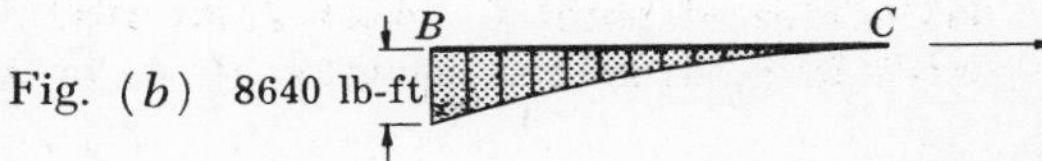

We next work from point A to the right and obtain the parabolic bending moment diagram due to the uniform load in the region AB, as shown in Fig. (c).

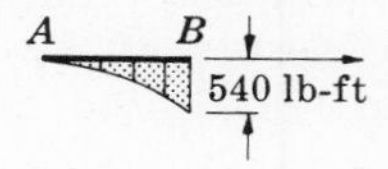

Fig. (c)

The complete bending moment diagram drawn by parts thus appears as shown in Fig. (d).

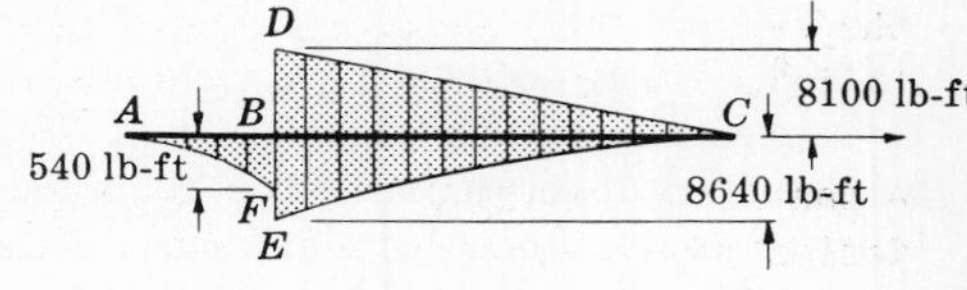

Fig. (d)

The approximate nature of the deflection curve is represented by the heavy line in Fig. (e). A tangent to the deflection curve is drawn at the point B as indicated.

The deflection represented by Af is desired. Let us begin by finding the deflection of A (whose final position is represented by f) from the tangent drawn at B. This is readily given by the Second Moment-Area Theorem to be the moment of the area under the bending moment diagram between A and B about a vertical line through A, divided by EI. This area under the moment diagram between A and B is represented by the parabola ABF. Applying the Second Theorem, we find

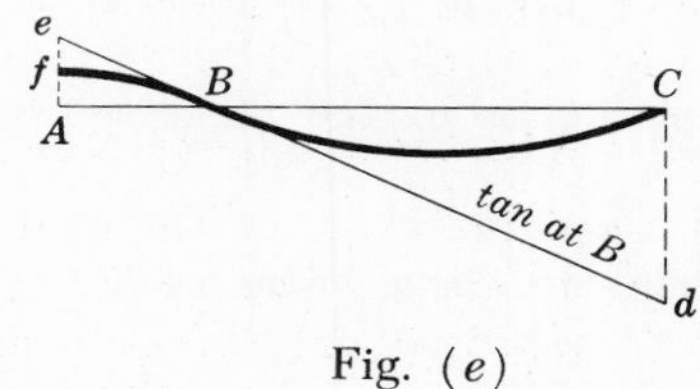

Fig. (e)

$$EI(ef) = \frac{1}{3}(3)(-540)(\frac{3}{4} \cdot 3) = -1216 \text{ lb-ft}^3$$

The negative sign indicates that the final position of point A (represented by f) lies below the tangent drawn at B.

We shall next calculate the deflection of C from the tangent drawn at B. This distance is represented by Cd and is readily given by the Second Moment-Area Theorem to be the moment about the vertical line through C of the area under the bending moment diagram between B and C, divided by EI. Thus we have

$$EI(Cd) = \frac{1}{2}(12)(8100)(\frac{2}{3} \cdot 12) + \frac{1}{3}(12)(-8640)(\frac{3}{4} \cdot 12) = 77,900 \text{ lb-ft}^3$$

Again, it is to be remembered that this is a relative deflection, since the absolute deflection of point C is zero.

From a consideration of the similar triangles BCd and ABe, we have $eA/3 = Cd/12$. Hence from the above value of $EI(Cd)$ we have

$$EI(eA) = \frac{3}{12}(77,900) = 19,500 \text{ lb-ft}^3$$

From Fig. (e) it is evident that the desired deflection (Af) is given by

$$Af = eA - ef$$

Substituting the above values in the right side of the equation, we find

$$EI(Af) = 19,500 - 1216 = 18,284 \text{ lb-ft}^3$$

Now substitute $E = 30 \cdot 10^6$ lb/in^2 and $I = 3(4^3)/12 \approx 16$ in^4 in the last equation and obtain

$$Af = \frac{18,284(1728)}{(30 \cdot 10^6)(16)} = 0.065 \text{ in.}$$

Again, it is necessary to introduce the factor 1728 since the right side of the equation was originally in units of lb-ft^3 whereas the values of E and I were stated in units of lb/in^2 and in^4 respectively. This value of deflection agrees with that found in Problem 24, Chapter 9 by the Double-Integration Method.

SUPPLEMENTARY PROBLEMS

18. A cantilever beam of length L is subject to a concentrated force P applied at a distance a from the fixed end of the beam. Determine the deflection at the free end of the beam. *Ans.* $Pa^2(3L-a)/6EI$

19. The cantilever beam described in Problem 18 is 10 ft long and a load of 1000 lb is applied 6 ft from the support. The cross-section of the beam is 2 in. by 6 in. The material is steel for which $E = 30 \cdot 10^6$ lb/in^2. Determine the maximum deflection of the beam. *Ans.* 0.230 in.

20. A cantilever beam of length L is loaded by a couple of magnitude M_1 at its free end. Determine the deflection of the free end of the beam. *Ans.* $M_1L^2/2EI$

21. The cantilever beam of Problem 20 is an 8 WF 24 section. The beam is 18 ft long. (*a*) Determine the maximum value of the moment M_1 so that the deflection at the free end of the bar does not exceed 0.25 in. (*b*) For this load find the maximum bending stress in the bar. Assume $E = 30 \cdot 10^6$ lb/in^2
Ans. *a*) 26,400 lb-in, *b*) 1270 lb/in^2

22. Consider the cantilever beam loaded as in Problem 6. Determine the slope of the beam at the end B.
Ans. $\theta_B = -wL^3/6EI$

23. A cantilever beam of length L is loaded by a couple of magnitude M_1 applied at the midpoint of the bar. Determine the deflection of the free end of the bar. *Ans.* $3M_1L^2/8EI$

24. If the cantilever beam described in Problem 23 is a 12 WF 58 section, determine the maximum moment M_1 that may be applied so as not to exceed a working flexural stress of 18,000 lb/in^2. The length of the beam is 10 ft. Also, determine the deflection of the free end of the beam due to this applied moment. Assume $E = 30 \cdot 10^6$ lb/in^2. *Ans.* 1,410,000 lb-in, 0.532 in.

25. A cantilever beam of length L is loaded by a uniform load of w lb/ft extending from the clamped end to the midpoint of the length of the bar. Determine the maximum deflection of the beam.
Ans. $7wL^4/384EI$

26. The cantilever beam described in Problem 25 is 12 ft long and is loaded by a uniform load of 1000 lb/ft extending from the clamped end to the midpoint of the length of the bar. The bar is steel for which $E = 30 \cdot 10^6$ lb/in^2 and the moment of inertia of the cross-sectional area is 100 in^4. Determine the maximum deflection of the beam. *Ans.* 0.218 in.

27. A cantilever beam is 10 ft in length and the moment of inertia of the cross-sectional area is 1000 in^4. At the free end of the beam a 10,000 lb load is applied, and at a point 6 ft from the fixed end another 10,000 lb load is applied. Determine the maximum deflection of the beam. Take $E = 30 \cdot 10^6$ lb/in^2. *Ans.* 0.274 in.

28. A steel beam of rectangular cross-section 3 in. by 5 in. is 10 ft long. It is simply supported at each end and is loaded by a uniformly distributed load of 100 lb/ft over its entire length as well as a concentrated force of 2000 lb applied at the midpoint of its length. Determine the maximum deflection of the beam. Take $E = 30 \cdot 10^6$ lb/in^2. *Ans.* 0.105 in.

29. A simply supported beam is loaded by the uniformly distributed loads shown in Fig. (*a*). Determine the maximum deflection of the beam.

$$Ans. \quad -\frac{wa^4}{24EI} + \frac{wa^2L^2}{16EI}$$

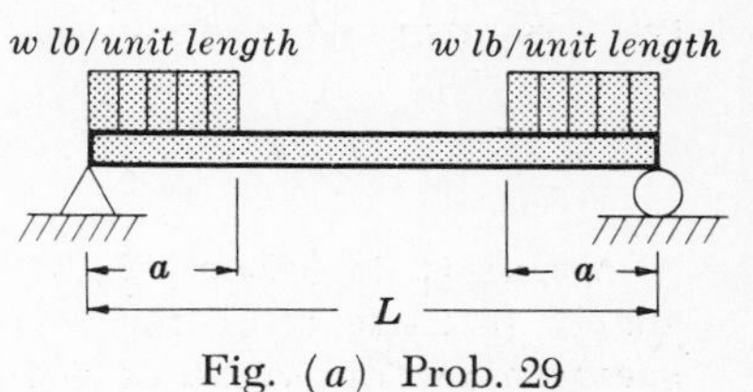

Fig. (*a*) Prob. 29

30. Consider the simply supported hickory beam subject to the three concentrated forces shown in Fig. (*b*). The beam is 6 in. by 12 in. in cross-section and the modulus of elasticity is $2.2 \cdot 10^6$ lb/in^2. Determine the maximum deflection of the beam. Also, determine the maximum bending stress.
Ans. 3.10 in., 5000 lb/in^2

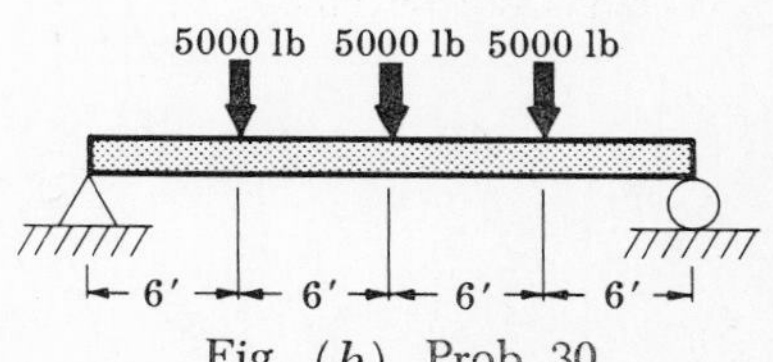

Fig. (*b*) Prob. 30

31. A simply supported beam with overhanging ends is loaded by the uniformly distributed loads shown in Fig. (*c*). Determine the deflection of the midpoint of the beam with respect to an origin at the level of the supports.

Ans. $\dfrac{wa^2(L-2a)^2}{16EI}$ (above level of supports)

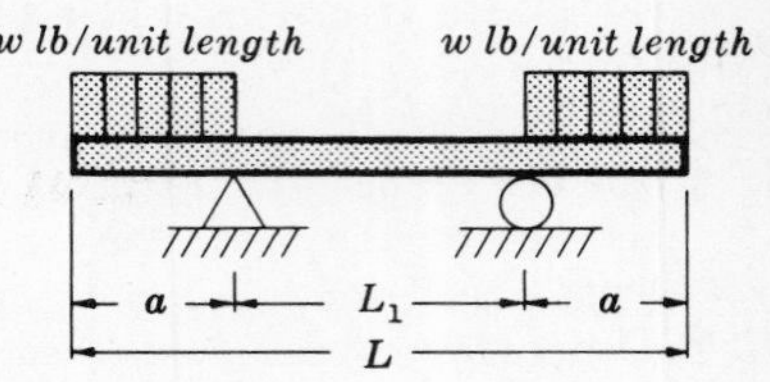

Fig. (*c*) Prob. 31

32. For the beam described in Problem 31, determine the deflection of one end of the beam with respect to an origin at the level of the supports.

Ans. $\dfrac{wa^3L}{4EI} - \dfrac{3wa^4}{8EI}$ (below level of supports)

33. Consider the simply supported beam described in Problem 49, Chapter 9. By use of the moment-area method, determine the deflection of the beam under the point of application of the applied moment M_1. Compare this result with that obtained by use of the double-integration method.

Ans. $\dfrac{M_1a^2}{2EI} + \dfrac{M_1a(L-a)}{3EI}$

34. The beam described in Problem 33 is of circular cross-section. The length of the beam is 12 ft and a = 3 ft. For an applied moment of 50,000 lb-in determine the required diameter of the beam, such that the deflection under the point of application of the applied moment is 0.30 in. Take $E = 30 \cdot 10^6$ lb/in^2. Also, determine the maximum bending stress in the bar. *Ans.* 3.86 in., 8950 lb/in^2

35. The overhanging beam is loaded by the uniformly distributed load as well as the concentrated force shown in Fig. (*d*) below. Determine the deflection of point *A* on the beam.

Ans. $-\dfrac{wa^3b}{3EI} + \dfrac{Pab^2}{4EI} - \dfrac{wa^4}{8EI}$ (below level of supports)

36. In the overhanging beam described in Problem 35, $a = b = 6$ ft, $P = 4000$ lb and $w = 800$ lb/ft. The beam is a 10 WF 21 section. Determine the deflection of point *A*. Take $E = 30 \cdot 10^6$ lb/in^2. *Ans.* 0.140 in.

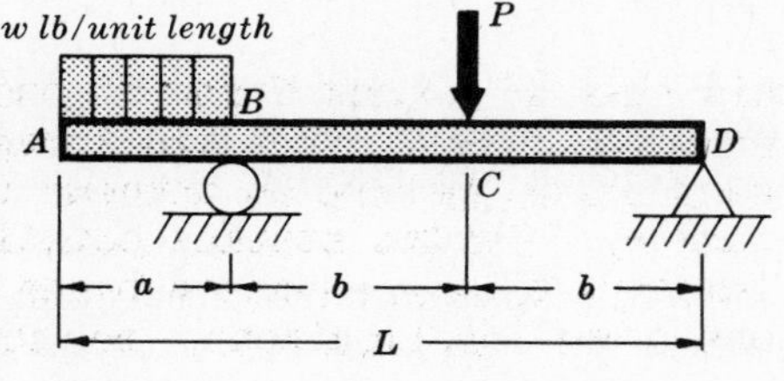

Fig. (*d*) Prob. 35

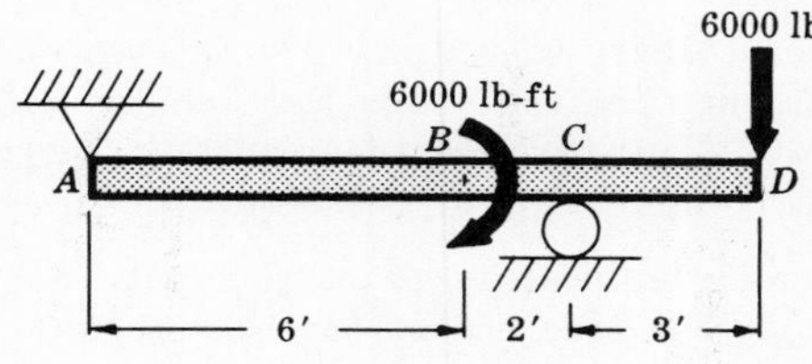

Fig. (*e*) Prob. 37

37. Consider the simply supported overhanging beam shown in Fig. (*e*) above. The loading consists of a moment of 6000 lb-ft and a force of 6000 lb applied as shown. The moment of inertia of the cross-sectional area of the beam is 42 in^4 and $E = 30 \cdot 10^6$ lb/in^2. Determine the deflection of point *B* where the moment is applied. *Ans.* 0.104 in.

CHAPTER 11

Statically Indeterminate Beams

STATICALLY DETERMINATE BEAMS. In Chapters 9 and 10 the deflections and stresses were determined for beams having various conditions of loading and support. In all of the cases treated it was always possible to completely determine the reactions exerted upon the beam merely by appling the equations of static equilibrium. In these cases the beams are said to be statically determinate.

STATICALLY INDETERMINATE BEAMS. In this chapter we shall consider those beams where the number of unknown reactions exceeds the number of equilibrium equations available for the system. In such a case it is necessary to supplement the equilibrium equations with additional equations stemming from the deformations of the beam. In these cases the beams are said to be statically indeterminate.

TYPES OF STATICALLY INDETERMINATE BEAMS. Several common types of statically indeterminate beams are illustrated below. Although a wide variety of such structures exists in practice the following three diagrams illustrate the nature of an indeterminate system. For the beams shown below the reactions of each constitute a parallel force system and hence there are two equations of static equilibrium available. Thus the determination of the reactions in each of these cases necessitates the use of additional equations arising from the deformation of the beam.

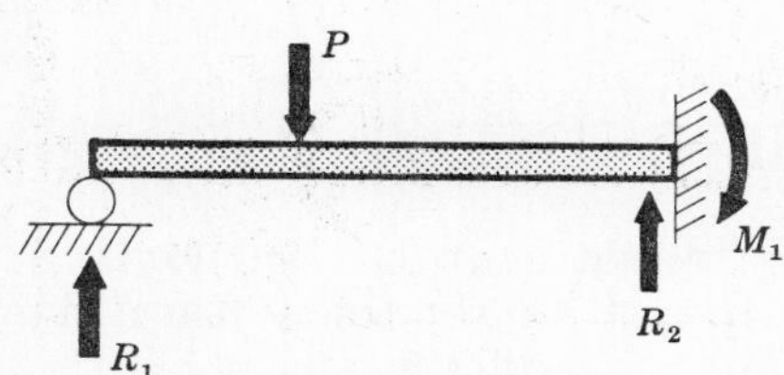

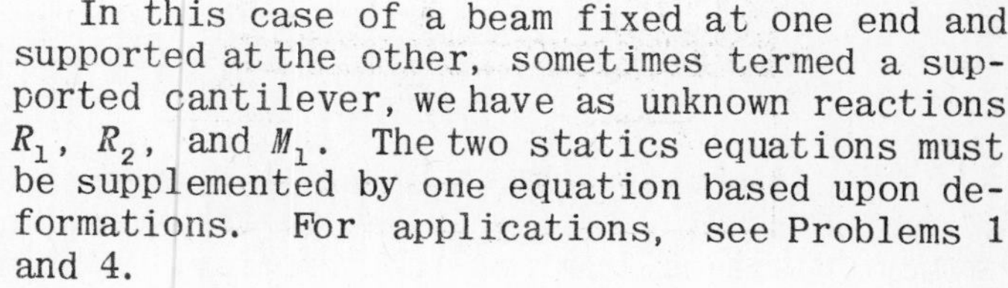

In this case of a beam fixed at one end and supported at the other, sometimes termed a supported cantilever, we have as unknown reactions R_1, R_2, and M_1. The two statics equations must be supplemented by one equation based upon deformations. For applications, see Problems 1 and 4.

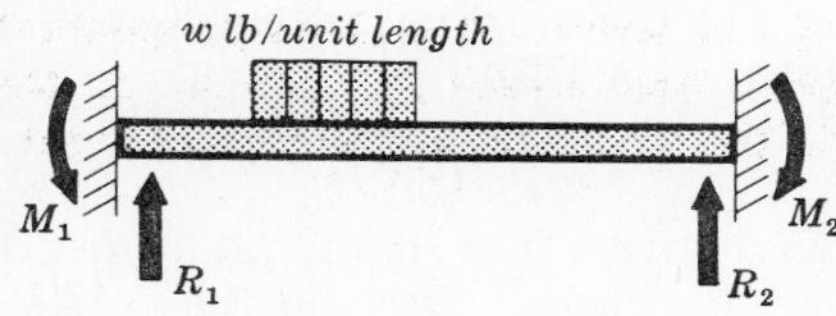

In this case of a beam fixed or clamped at both ends the unknown reactions are R_1, R_2, M_1, and M_2. The two statics equations must be supplemented by two equations arising from the deformations. For applications, see Problems 9, 12 and 15.

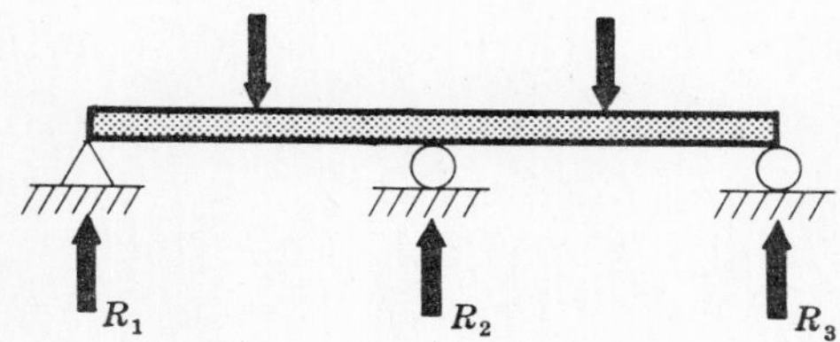

In this case the beam is supported on three supports at the same level. The unknown reactions are R_1, R_2, and R_3. The two statics equations must be supplemented by one equation based upon deformations. A beam of this type that rests on more than two supports is called a *continuous beam*. For applications, see Problems 19, 22, 23 and 24.

NATURE OF EQUATIONS ARISING FROM THE BEAM DEFORMATIONS. In the first of the above illustrations the equations based upon deformations are derived by making use of the fact that the deflection of the left end of the beam, the supported end, is zero. Using the Second Moment-Area Theorem this deflection is expressed as a function of R_1, then set equal to zero, and the resulting equation solved for R_1. The two equations of statics then yield values of R_2 and M_1.

In the second of the illustrations above the deformation equations are based upon the facts that (*a*) the change of slope between the tangents drawn at the two ends of the beam is zero, and (*b*) the deflection of the left end of the beam is zero. The First Moment-Area Theorem is used to express the condition (*a*) and leads to an equation containing the unknowns R_1 and M_1. The Second Moment-Area Theorem is then used to express the second condition (*b*) and this leads to another equation containing R_1 and M_1. These two equations are then solved simultaneously. The two statics equations then yield values for R_2 and M_2.

The continuous beam illustrated in the third of the above diagrams is usually investigated in a somewhat different manner by use of the Three-Moment Theorem. The derivation of this theorem is however based upon simple deformation principles. This Theorem is derived in Problem 19.

THREE-MOMENT THEOREM. A continuous beam is one that rests on more than two supports. The continuous two-span beam shown in the adjacent diagram is subject to a partial uniform load as well as several concentrated forces. It is convenient to consider the bending moments at the various supports as the unknowns (rather than the reactions themselves) and write deformation equations in terms of these bending moments. This leads to the Three-Moment Theorem:

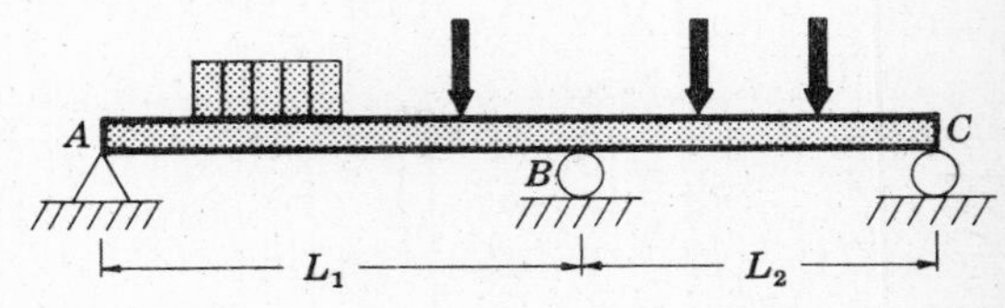

$$M_A L_1 + 2M_B(L_1 + L_2) + M_C L_2 = -\frac{6A_1 \bar{a}_1}{L_1} - \frac{6A_2 \bar{b}_2}{L_2}$$

In this equation M_A, M_B, and M_C designate bending moments at the supports A, B, and C respectively. L_1 and L_2 denote the span lengths, A_1 and A_2 represent the areas of the moment diagrams drawn on the temporary assumption that each of the spans of the beam is *simply supported*, and $\bar{a}_1$ and $\bar{b}_2$ designate the distances of the centroids of each of these moment diagrams from A and C respectively. The theorem in this form is applicable to all continuous beams having all supports at the same level. The derivation of the theorem is presented in Problem 19. For applications see Problems 20-24.

ASSUMPTIONS AND LIMITATIONS. The usual assumptions governing stresses and deflections of beams as presented in Chapters 8, 9, and 10 apply to the beams considered in this chapter. In addition, it is to be noted that the nature of the supports in contact with the beams is such that no horizontal reactions are exerted upon the beams.

SOLVED PROBLEMS

1. Consider the beam that is supported at the left end, clamped at the right end and subject to the concentrated load as shown in the adjoining diagram. Determine the reactions R_1, R_2, and M_1.

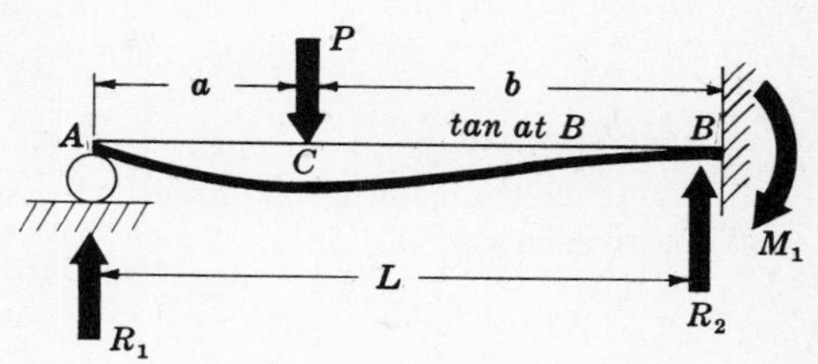

From statics we have

(1) $\Sigma M_B = R_1 L + M_1 - Pb = 0$

(2) $\Sigma F_v = R_1 + R_2 - P = 0$

This is a parallel force system and hence there are only two equations of equilibrium available. Thus any equilibrium equations in addition to the above two would not be independent equations. Yet these two equations contain the three unknowns R_1, R_2, and M_1. Thus we have a statically indeterminate system and it is necessary to supplement the statics equations with relationships arising from the deformations of the beam. In this case it is necessary to employ only one equation arising from deformations since we will then have three eqůations containing three unknowns.

To obtain this equation let us examine the deflected beam, indicated by the heavy curved line in the above figure. If a tangent to the deformed beam is drawn at B, i.e. the clamped end, then this tangent will coincide with the original unbent position of the beam. The deflection of the supported end A from this tangent at B is zero. Thus we may apply the Second Moment-Area Theorem developed in Problem 2, Chapter 10. The moment diagram when drawn by parts is shown in the adjoining diagram. From the Second Moment-Area Theorem, realizing that the deflection of A from the tangent at B is zero, and that this deflection is given by the moment of the area under the above moment diagram between A and B about a vertical line through A, we have

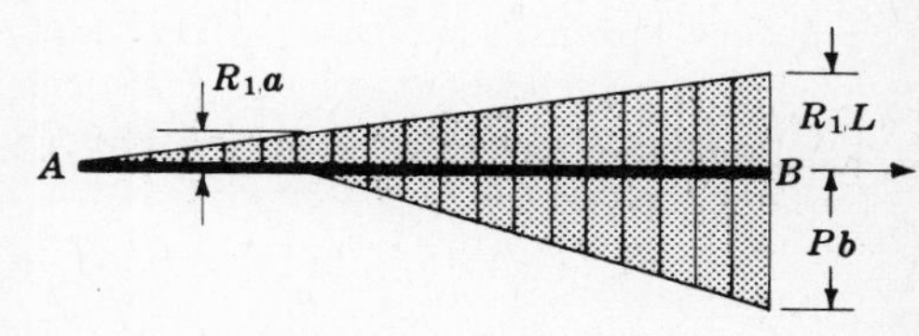

$$\frac{1}{2}(R_1 L)(L)(\frac{2}{3}L) + \frac{1}{2}(-Pb)(b)(a+\frac{2}{3}b) = 0 \quad \text{or} \quad R_1 = \frac{3Pb^2}{2L^3}(a+\frac{2}{3}b) = \frac{Pb^2}{2L^3}(2L+a) \qquad (3)$$

Substituting for R_1 in Equation (2),
$$R_2 = \frac{Pa}{2L^3}(3L^2 - a^2) \qquad (4)$$

Substituting these values in Equation (1),
$$M_1 = \frac{Pa}{2L^2}(L^2 - a^2) \qquad (5)$$

The unknowns reaction are thus completely determined.

2. The beam discussed in Problem 1 is an 8 WF 24 section. The load P is 5000 lb, $L = 20$ ft, and $a = 10$ ft. Determine the reactions and the maximum bending stresses in the beam.

Substituting in Equation (3) of Problem 1, $R_1 = \dfrac{5000(10)^2}{2(20)^3}(40+10) = 1560$ lb.

From Equation (4) of Problem 1 we have $R_2 = \dfrac{5000(10)}{2(20)^3}[3(400) - 100] = 3440$ lb.

Lastly, from Equation (5) we find $M_1 = \dfrac{5000(10)}{2(20)^2}(400-100) = 18,750$ lb-ft.

It is to be noted that these expressions are valid only if the proportional limit of the material is not exceeded at any point in the beam. This is because the Moment-Area Theorem was derived upon the assumption that this was true and this theorem was used in Problem 1 to determine the reactions.

Hence it is necessary to investigate the maximum bending stress in the beam. The only points that need be investigated are at the clamped end and also directly under the concentrated load. From the table at the end of Chapter 8 we have $I = 82.5\ \text{in}^4$ for this section.

At the clamped end the moment M_1 gives rise to a maximum bending stress in the outer fibers of

$$s = \frac{Mc}{I} = \frac{18,750(12)(4)}{82.5} = 10,900\ \text{lb/in}^2$$

Under the concentrated load the bending moment is $1560(10) = 15,600$ lb-ft. At the outer fibers the maximum bending stress is

$$s = \frac{15,600(12)(4)}{82.5} = 9100\ \text{lb/in}^2$$

The first value is the peak bending stress. Since it is less than the proportional limit of steel the use of the above expressions for reactions is justified.

3. For the beam described in Problem 2, determine the deflection under the point of application of the 5000 lb force.

This deflection is readily determined by use of the Second Moment-Area Theorem together with the bending moment diagram shown in Problem 1 and the reactions as found in Problem 2. According to the Second Moment-Area Theorem the deflection of point C (under the load) from the tangent drawn at B is equal to the moment of the area under the M/EI diagram between C and B about a vertical line through C. The area as shown in Problem 1 consists of a rectangle together with two triangles. The desired deflection Δ_c becomes

$$EI(\Delta_c) = 10(15,600)(5) + \tfrac{1}{2}(10)(15,600)(\tfrac{2}{3}\cdot 10) + \tfrac{1}{2}(10)(-50,000)(\tfrac{2}{3}\cdot 10) = -366,500$$

$$\Delta_c = -\frac{(366,500)(1728)}{(30\cdot 10^6)(82.5)} = -0.255\ \text{in.}$$

It is to be noted that the factor of 1728 is introduced to convert to consistant units, since the quantity $-366,500$ is in (lb-ft^3) units whereas E and I are expressed in lb/in^2 and in^4 respectively.

4. The beam shown in the adjoining figure is clamped at the left end, supported at the right end, and subject to a uniformly distributed load. Determine the reactions R_1, R_2, and M_1.

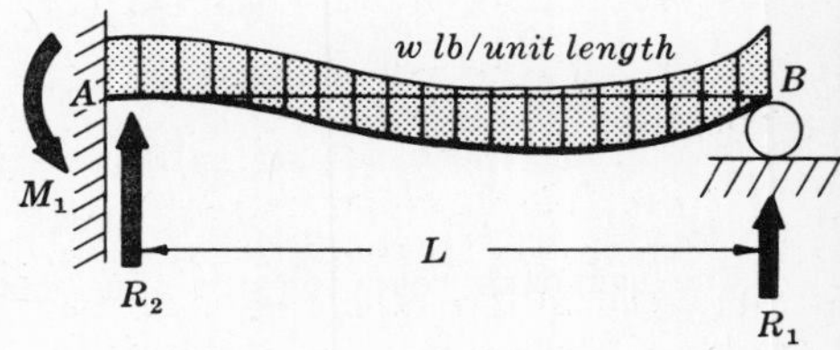

From statics we have

$$(1) \qquad \Sigma F_v = R_1 + R_2 - wL = 0$$

$$(2) \qquad \Sigma M_A = M_1 + R_1 L - wL^2/2 = 0$$

Again, as in Problem 1, there are only two statics equations available for this parallel force system. Since these two equations contain three unknowns the system is indeterminate and we must examine the deformations of the beam to obtain a third equation.

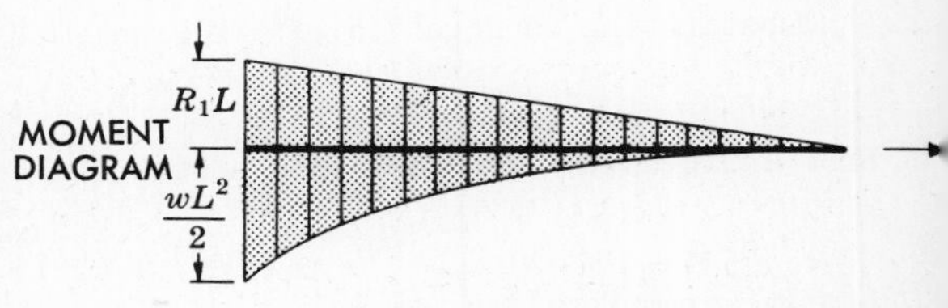

The deflected beam is indicated by the heavy line and the tangent to the beam at the clamped end coincides with the unbent configuration of the beam. The deflection of B from the tangent at A is zero. The moment diagram when drawn by parts has the appearance shown in the adjoining sketch.

From the Second Moment-Area Theorem, since the deflection of B from the tangent at A is zero, we have

$$(1/2)(R_1L)(L)(2L/3) + (-wL^2/2)(L)(1/3)(3L/4) = 0 \quad \text{or} \quad R_1 = (3/8)wL \tag{3}$$

Substituting this value in Equation (*1*) we find

$$R_2 = (5/8)wL \tag{4}$$

Finally, from Equation (*2*) we obtain

$$M_1 = (1/8)wL^2 \tag{5}$$

It is frequently convenient to represent the bending moment diagram in the following composite fashion rather than drawing it by parts as shown above. This may be done by first considering the parabolic moment diagram due to a uniformly distributed load acting on a simply supported beam. This moment diagram was obtained in Problem 6, Chapter 6. The maximum ordinate to this diagram was shown to be $wL^2/8$. Next, the triangular moment diagram due to the couple M_1 applied at the left end of a simply supported beam was studied in Problem 13, Chapter 9. The composite bending moment diagram is then found by subtracting the triangular diagram from the parabolic one. This diagram has the appearance shown in the adjacent figure. From this diagram it is evident that the maximum bending moment occurs at the clamped end of the beam.

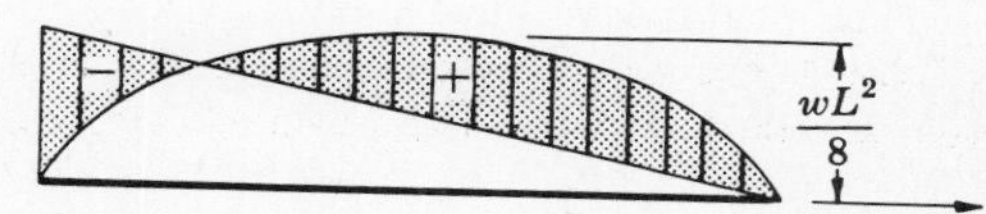

5. The beam discussed in Problem 4 is a 10 WF 21 section. The length of the beam is 20 ft and the maximum permissible bending stress is 20,000 lb/in². Determine the allowable uniform load.

In Problem 4 the maximum bending moment was found to occur at the clamped end *A*. The value of the bending moment there was found to be $M_1 = wL^2/8$.

But we also know that the maximum bending stress at this cross-section occurs at the outer fibers and is given by $s = M_1c/I$, where $I = 106.3 \text{ in}^4$ from the table at the end of Chapter 8. Substituting,

$$20{,}000 = \frac{w(240)^2}{8} \cdot \frac{5}{106.3} \quad \text{and} \quad w = 59.2 \text{ lb/in} = 710 \text{ lb/ft}$$

This is the total uniform load including the weight of the beam. Since the weight of the beam is 21 lb per ft of length, the allowable uniform load is (710 − 21) = 689 lb/ft.

6. Discuss an alternate method for determining the reactions of the uniformly loaded beam described in Problem 4.

The problem may be solved by superposition. Let us temporarily remove the reaction R_1 acting at the right end of the bar. The beam then acts as a cantilever subject to a unifomly distributed load. In Problem 6 of Chapter 9 we found the deflection of the free end of such a beam to be $\Delta = \frac{wL^4}{8EI}$. This deflection is illustrated in the adjoining figure.

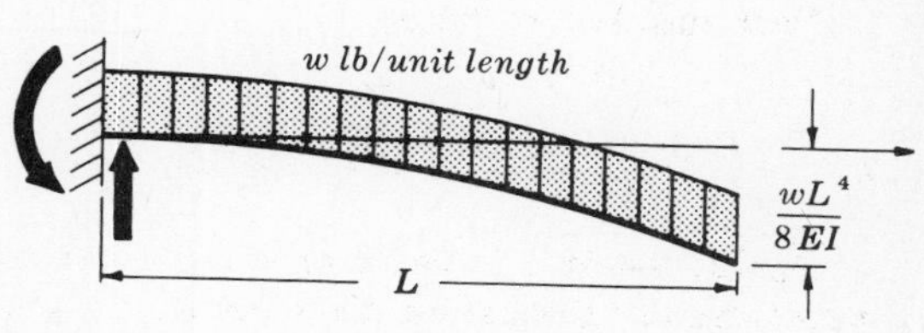

Next, we consider the beam to be subject only to a concentrated upward force R_1 applied at the right end of the beam. In Problem 2 of Chapter 9 we found the deflection of the free end of such a beam to be $\Delta = \frac{R_1L^3}{3EI}$, as illustrated in the adjacent figure.

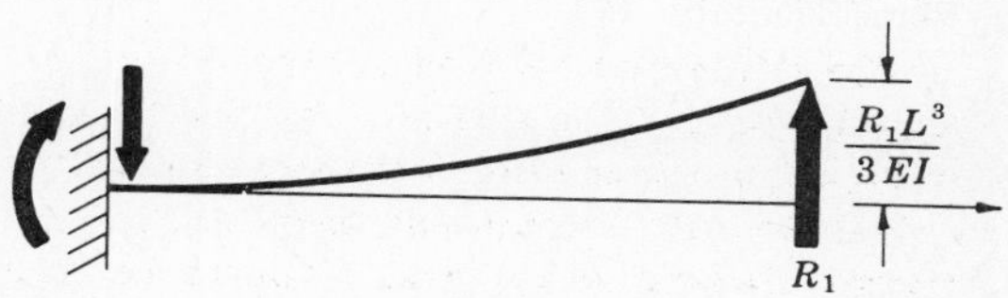

In reality both of these loadings act simultaneously on the beam and the value of R_1 is such that the net or resultant vertical deflection of the right end of the beam is zero. Thus

$$\frac{R_1 L^3}{3EI} = \frac{wL^4}{8EI} \quad \text{or} \quad R_1 = \frac{3}{8}wL$$

This agrees with the value found in Problem 4. Knowing R_1, the values of R_2 and M_1 are found from the statics equations given in Equations (*1*) and (*2*) of Problem 4. The values found of course agree with those given in Equations (*4*) and (*5*) of Problem 4.

7. Consider the over-hanging beam shown in the adjacent diagram. Determine the value of the various reactions.

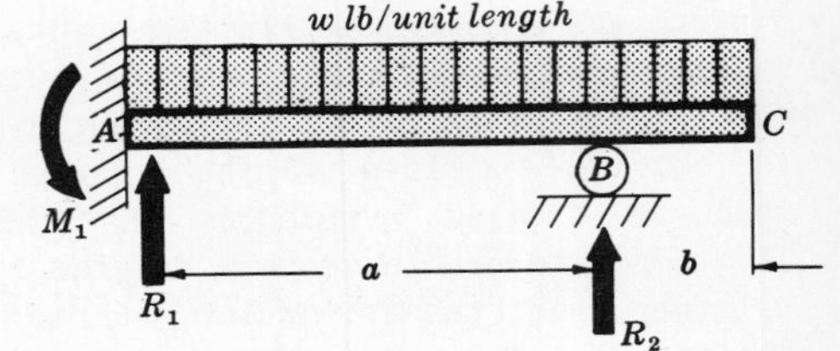

From statics we have

(*1*) $\Sigma M_A = M_1 + R_2 a - w(a+b)^2/2 = 0$

(*2*) $\Sigma F_v = R_1 + R_2 - w(a+b) = 0$

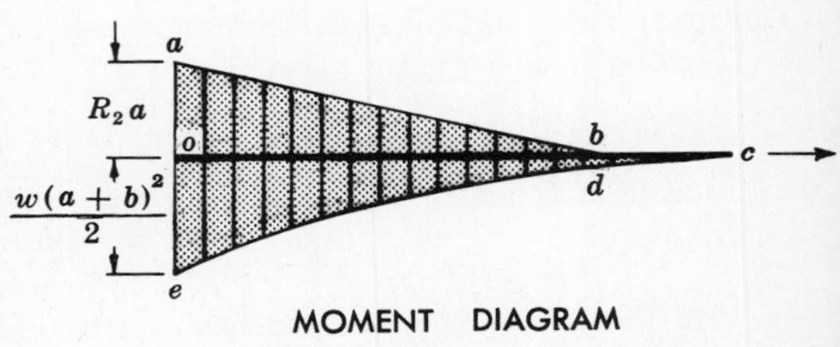

MOMENT DIAGRAM

Again, there are only two equations of equilibrium for this force system yet these equations contain three unknowns, R_1, R_2, and M_1. Thus the system is statically indeterminate and we must look to the deformations to furnish an additional equation. This is possible since the deflection of point *B* is known to be zero. A tangent drawn to the deflected beam at the point *A* is horizontal and coincides with the original position of the beam. The moment diagram, when drawn by parts (working from right to left) is shown in the adjoining diagram.

According to the Second Moment-Area Theorem the deflection of *B* from the tangent at *A* is equal to the moment of the area under the moment diagram between *A* and *B* about a vertical line through *B*, divided by *EI*. This area consists of the triangle *oab* together with the portion of the parabolic area *obde* shown in the above diagram. The moment of the triangular area about a vertical line through *B* is

$$(1/2)(a)(R_2 a)(2a/3)$$

For the parabolic area it is convenient to subtract the moment of the area *bdc* (with respect to the vertical line through *B*) from the moment of the area *obcde* (with respect to the same vertical line). Each of these last two areas has a negative altitude (since the downward uniform load gives rise to a negative bending moment). However it is to be carefully noted that the moment arm of the area *obcde* is positive whereas the moment arm of area *bdc* is negative. The moment of the area *obde* is thus

$$\frac{1}{3}(a+b)\left[-\frac{w(a+b)^2}{2}\right]\left[a - \left(\frac{a+b}{4}\right)\right] - \frac{1}{3}(b)\left[-\frac{wb^2}{2}\right]\left[-\frac{b}{4}\right]$$

Consequently, the Second Moment-Area Theorem becomes

$$\frac{1}{2}(a)(R_2 a)\left(\frac{2}{3}a\right) + \frac{1}{3}(a+b)\left[-\frac{w(a+b)^2}{2}\right]\left[a - \left(\frac{a+b}{4}\right)\right] - \frac{1}{3}(b)\left[-\frac{wb^2}{2}\right]\left[-\frac{b}{4}\right] = 0$$

Solving, $$R_2 = \frac{w(a+b)^3}{2a^2} - \frac{w(a+b)^4}{8a^3} + \frac{wb^4}{8a^3} \qquad (3)$$

From Equation (*1*) $$M_1 = \frac{w(a+b)^2}{2} - \frac{w(a+b)^3}{2a} + \frac{w(a+b)^4}{8a^2} - \frac{wb^4}{8a^2} \qquad (4)$$

Finally, from Equation (*2*) $$R_1 = w(a+b) - \frac{w(a+b)^3}{2a^2} + \frac{w(a+b)^4}{8a^3} - \frac{wb^4}{8a^3} \qquad (5)$$

8. Determine the deflection of the right end *C* of the overhanging beam discussed in Problem 7.

Under the action of the uniform load the deflection curve of the beam has the appearance as shown in the adjoining figure. It is to be noted that although the vertical deflection of point B is zero there is no reason to assume that the tangent to the bent beam is horizontal at that point.

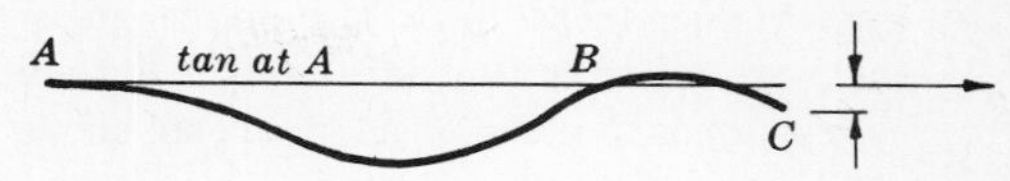

It is still convenient to work with the tangent drawn at the clamped end A, since that tangent is horizontal. According to the Second Moment-Area Theorem the desired deflection of C from the tangent drawn at A is equal to the moment about the vertical line through C of the area under the entire bending moment diagram between A and C, divided by EI. It is to be observed that this theorem is valid even though the beam is statically indeterminate. Thus, referring to the moment diagram shown in Problem 7, we have

$$EI(\Delta_C) = +\frac{1}{2}(a)(R_2 a)(b+\frac{2}{3}a) + \frac{1}{3}(a+b)[-\frac{w}{2}(a+b)^2](\frac{3}{4})(a+b)$$

Substituting the value of R_2 found in Problem 7 and simplifying, we have

$$\Delta_C = \frac{wa^3b}{48EI} - \frac{wa^2b^2}{2EI} - \frac{wab^3}{8EI} - \frac{wb^4}{6EI}$$

9. The uniformly loaded beam is clamped at both ends as shown in Fig. (a). Determine the reactions.

From symmetry, the force reactions at each end must be equal and each is denoted by R_1. Also, the moment reactions must be equal and each of these is represented by M_1. From statics we have

(1) $\quad \Sigma F_v = 2R_1 - wL = 0 \quad$ or $\quad R_1 = wL/2$

Although there were initially two equations of static equilibrium available for such a parallel force system, we have essentially utilized one of these equations when we made use of the symmetry considerations. Thus to determine M_1 we must now examine the deformations of the system. This procedure will furnish the additional equation necessary to complete the analysis of this statically indeterminate system.

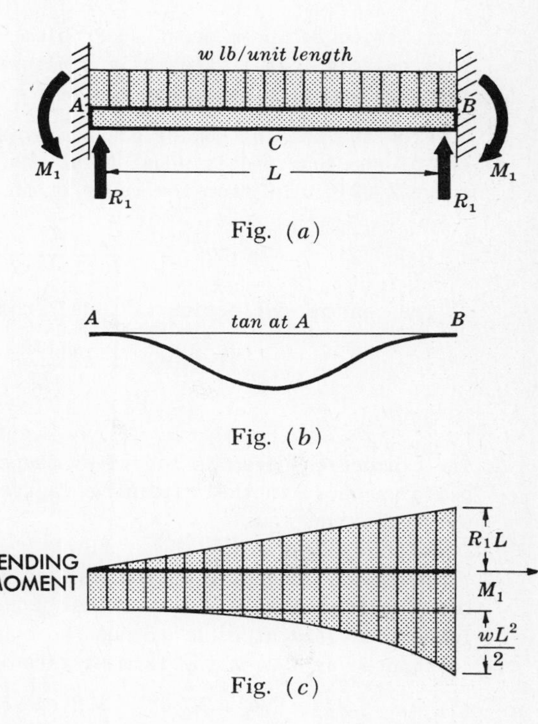

Fig. (a)

Fig. (b)

Fig. (c)

The deflected beam has the symmetrical appearance shown in Fig. (b). Since both ends are clamped, the tangent drawn to the deflected beam at the left end A coincides with that drawn at the right end B, and both are horizontal. We may thus use the Second Moment-Area Theorem which states that the deflection of B from the tangent drawn at A is equal to the moment about the vertical line through B of the area under the bending moment diagram between A and B, divided by EI. The bending moment diagram, drawn by parts working from left to right, appears as in Fig. (c).

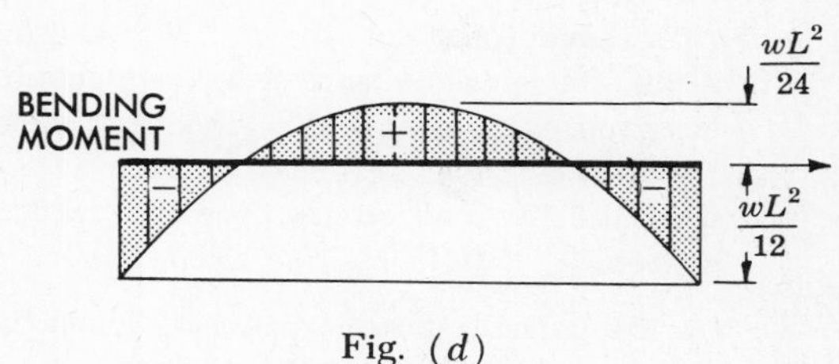

Fig. (d)

From the Second Moment-Area Theorem we thus have

$$\frac{1}{2}(L)(R_1 L)(\frac{L}{3}) + L(-M_1)(\frac{L}{2}) + \frac{1}{3}(L)(-\frac{wL^2}{2})(\frac{L}{4}) = 0$$

Substituting R_1 from Equation (1) and solving, we have

(2) $\quad M_1 = wL^2/12$

The bending moment at the center of the span is now readily found to be $wL^2/24$. It is frequently convenient to present the bending moment diagram in a composite form rather than by parts as shown above. This may be done by superposing on the parabolic diagram due to a uniform load acting on a simply supported beam the rectangular moment diagram corresponding to the end moments. According to the sign convention of Chapter 6 these end moments are negative. The resultant moment diagram then corresponds to the shaded areas shown in Fig. (*d*) above.

10. Determine the central deflection of the clamped end beam of Problem 9.

This deflection is found by applying the Second Moment-Area Theorem between the left end A and the midpoint C of the beam. The desired deflection is given by the deflection of C from the tangent at A. This is found as the moment about the vertical line through C of the area under the moment diagram between A and C, divided by EI. Referring to the moment diagram drawn by parts in Problem 9, we have for the desired deflection

$$EI(\Delta_C) = \frac{1}{2}\left(\frac{L}{2}\right)\left(\frac{1}{2}\right)(R_1 L)\left(\frac{L}{6}\right) + \frac{L}{2}\left(-\frac{wL^2}{12}\right)\left(\frac{L}{4}\right) + \frac{1}{3}\left(\frac{L}{2}\right)\left(-\frac{wL^2}{8}\right)\left(\frac{L}{8}\right)$$

Substituting the value of R_1 found in Problem 9, we have $\Delta_C = \dfrac{-wL^4}{384EI}$. The negative sign indicates that the final position of C lies below the tangent at A.

11. The clamped end beam shown in Problem 9 is a 12 WF 32 section, 20 ft in length. Determine the allowable uniform load if the maximum allowable bending stress is 20,000 lb/in^2. What is the central deflection under this loading ?

From the bending moment diagram presented in Problem 9 the maximum bending moment occurs at either end of the beam and is equal to $wL^2/12$. The bending stress at the outer fibers is given by $s = Mc/I$, where $I = 246.8$ in^4 from the table at the end of Chapter 8. Substituting,

$$20{,}000 = \frac{w(240)^2}{12}\cdot\frac{6}{246.8} \qquad \text{or} \qquad w = 172 \text{ lb/in} = 2060 \text{ lb/ft}$$

The central deflection is found from Problem 10 to be

$$\Delta_C = -\frac{wL^4}{384EI} = -\frac{172(240)^4}{384(30\cdot 10^6)(246.8)} = -0.203 \text{ in.}$$

12. The clamped-end beam is loaded by the couple M_0 applied as shown in the adjoining figure. Determine all reactions.

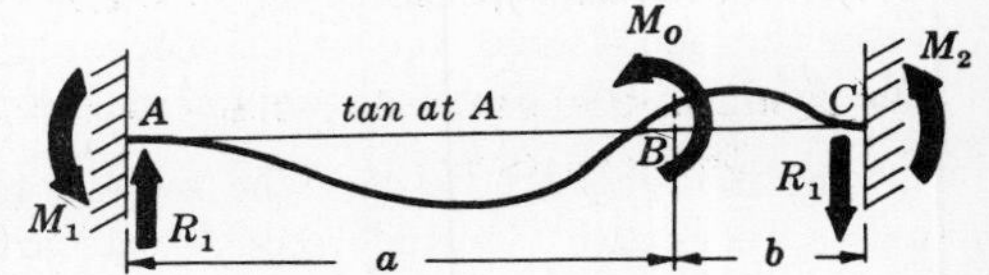

The deformed configuration is indicated by the heavy curved line. For vertical equilibrium the force reactions at each end must be equal and each is denoted by R_1. Also, from statics we have

(*1*) $\quad \Sigma M_A = M_1 + M_2 + M_0 - R_1(a+b) = 0$

This equation contains R_1, M_1, and M_2 as unknowns. Since there are no more statics equations available, the problem is statically indeterminate. Consequently we must supplement Equation (*1*) with two additional equations coming from the deformations of the system.

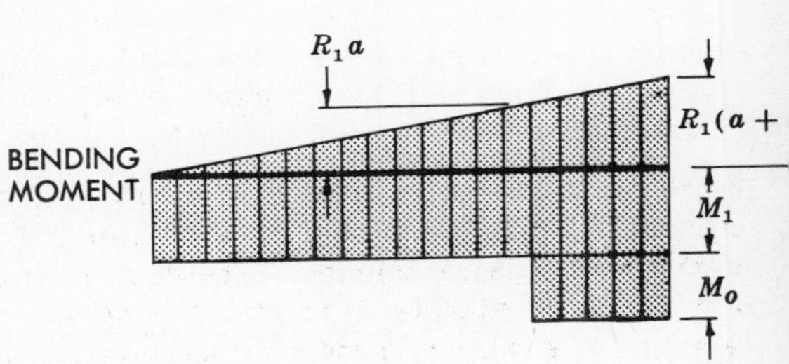

The bending moment diagram drawn by parts (working from left to right) is shown at the right.

The first equation is found from the relation that the tangent drawn to the deformed beam at A remains horizontal. Hence the deflection of the right end of the beam, C, from the tangent at A is zero Employing the Second Moment-Area Theorem between points A and C we have

(2) $$\frac{1}{2}(a+b)[R_1(a+b)](\frac{a+b}{3}) + (a+b)(-M_1)(\frac{a+b}{2}) + b(-M_0)(\frac{b}{2}) = 0$$

The second equation is perhaps found most simply by realizing that since both ends of the beam are clamped the angle between the tangents at A and C is zero. Actually, the tangents to the ends of the deformed beam coincide with the original straight configuration of the beam. The First Moment-Area Theorem may now be employed between points A and C. This theorem states that the angle between the tangents at A and C is equal to the area of the bending moment diagram between A and C, divided by EI. Thus

(3) $$\frac{1}{2}(a+b)[R_1(a+b)] + (a+b)(-M_1) + b(-M_0) = 0$$

Solving Equations (1), (2) and (3), we find

$$(4)\quad R_1 = \frac{6M_0ab}{(a+b)^3}, \qquad (5)\quad M_1 = \frac{M_0(2ab-b^2)}{(a+b)^2}, \qquad (6)\quad M_2 = \frac{M_0(2ab-a^2)}{(a+b)^2}.$$

It is to be carefully noted that there is no reason to assume that the vertical deflection of point B, the point of application of the couple M_0, is zero.

Determine the deflection of point B, the point of application of the couple M_0, in Problem 12.

This is readily determined since the desired deflection is equal to the deflection of point B from the tangent drawn at A. This tangent remains horizontal during deformation of the beam. From the Second Moment-Area Theorem the deflection of B from the tangent at A is given by the moment about the vertical line through B of the area under the bending moment diagram between A and B, divided by EI. Referring to the moment diagram shown in Problem 12 we have

$$EI(\Delta_B) = \tfrac{1}{2}(a)(R_1 a)(a/3) + a(-M_1)(a/2)$$

Substituting the values of R_1 and M_1 found in Problem 12, we have $\Delta_B = \dfrac{M_0a^2b^2(b-a)}{2(a+b)^3EI}$.

The clamped end beam shown in Problem 12 is an 8 WF 40 section. The couple is applied at a point B located such that $a = 4$ ft, $b = 10$ ft. If the maximum allowable bending stress is 18,000 lb/in², determine the allowable value of M_0 together with the deflection of point B when this loading is applied.

We shall first determine the reactions as functions of M_0 from Equations (4), (5), and (6) of Problem 12. These become

$$R_1 = \frac{6M_0(4)(10)}{(4+10)^3} = 0.0875M_0$$

$$M_1 = \frac{M_0[(2)(4)(10)-(10)^2]}{(4+10)^2} = -0.102M_0$$

$$M_2 = \frac{M_0[(2)(4)(10)-(4)^2]}{(4+10)^2} = 0.326M_0$$

We shall express M_0 in units of lb-ft. Although it is not too lengthy a procedure to determine the composite bending moment diagram, it is perhaps simpler to realize that, since there are no distributed loads, the variations of bending moments over the length of the bar must be linear functions, i.e. the composite bending moment diagram will consist of a number of straight line segments. Evidently it is only necessary to calculate the bending moment at four critical points: (a) The left end of the span, (b) The right end of the span, (c) A point immediately to the left of B, and (d) A point immediately to the right of B.

The bending moments at the left and right ends of the span are given by $-0.102M_0$ and $0.326M_0$ respectively. At a point immediately to the left of B the bending moment is

$$0.102M_0 + (0.0875M_0)(4) = 0.452M_0$$

At a point immediately to the right of B the moment is

$$0.326 M_0 - (0.0875 M_0)(10) = -0.548 M_0$$

This last value is obviously the maximum bending moment along the length of the beam. The peak ben ing stresses occur at the outer fibers and are given by $s = Mc/I$. Substituting, and using $I = 146.3$ i. from the table at the end of Chapter 8, we have

$$18{,}000 = \frac{(0.548\,M_0)(12)(4)}{146.3} \qquad \text{or} \qquad M_0 = 100{,}000 \text{ lb-ft}$$

The deflection under the point of application of this load was determined in Problem 13. It is

$$\Delta_B = \frac{M_0 a^2 b^2 (b-a)}{2(a+b)^3 EI} = \frac{(100{,}000)(12)(48)^2(120)^2(72)}{2(168)^3(30 \cdot 10^6)(146.3)} = 0.0688 \text{ in.}$$

15. The clamped-end beam is subject to the centrally applied concentrated force shown in Fig. (a). Determine the reactions.

Fig. (a)

From symmetry the end reactions are equal and are designated by the symbols R_1 and M_1. From statics we have

(1) $\quad \Sigma F_v = 2R_1 - P = 0 \qquad \text{or} \qquad R_1 = P/2$

Also we have essentially utilized one equilibrium equation by making use of the symmetry considerations. Hence for such a parallel force system no other statics equations are available. The problem is thus statically indeterminate and we must examine the deformations of the system to determine M_1.

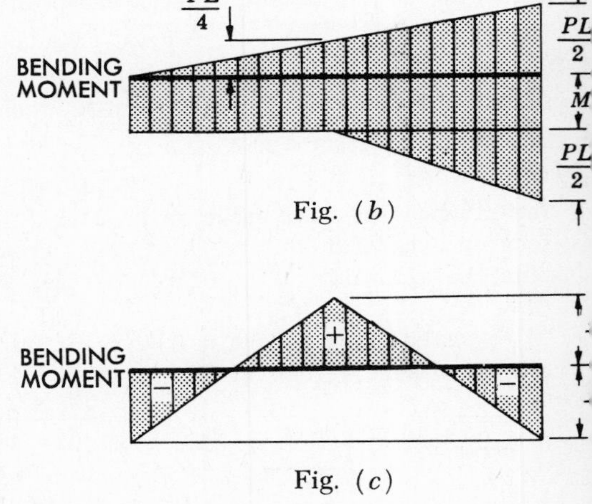

Fig. (b)

Fig. (c)

The deflected beam has the symmetrical appearance shown by the heavy line in Fig. (a). Tangents drawn to the deflected beam at each of the clamped ends A and C remain horizontal and both coincide with the original unbent position of the beam. Thus the angle between these tangents is zero. The bending moment diagram drawn by parts (working from left to right) is shown in Fig. (b).

According to the First Moment-Area Theorem the angle between the tangents at A and C is equal to the area under the bending moment diagram between A and C, divided by EI. Thus

(2) $\quad \tfrac{1}{2}(L)(PL/2) + L(-M_1) + \tfrac{1}{2}(L/2)(-PL/2) = 0 \qquad \text{or} \qquad M_1 = PL/8$

The bending moment at the midpoint of the span is now found to be $PL/8$. Again, it is sometimes sirable to present the bending moment diagram in a composite form rather than by parts as shown abo This may be done by superposing on the triangular diagram due to a concentrated load acting on a s ply supported beam the rectangular moment diagram corresponding to the end moments. According to sign convention of Chapter 6 these end moments are negative. The resultant moment diagram then c responds to the shaded area shown in Fig. (c) above.

16. Determine the central deflection of the beam described in Problem 15.

This deflection may be found by applying the Second Moment-Area Theorem between the left end A the midpoint B of the beam. The desired deflection is given by the deflection of B from the tang drawn at A. This is found as the moment about the vertical line through B of the area under the moment diagram between A and B, divided by EI. Referring to the moment diagram of Problem 15 we have

$$EI(\Delta_B) = \frac{1}{2}(\frac{L}{2})(\frac{PL}{4})(\frac{1}{3}\cdot\frac{L}{2}) + \frac{L}{2}(-M_1)(\frac{L}{4})$$

Substituting the value of M_1 found in Problem 15, we find $\Delta_B = -\dfrac{PL^3}{192EI}$. The negative sign indicates that the final position of B lies below the tangent at A.

17. The span of the clamped-end beam shown in Problem 15 is 22 ft. The centrally applied load is 38,000 lb. Select a WF section to carry this load so as not to exceed a working stress of 18,000 lb/in². Also, determine the deflection under the point of application of the load.

As illustrated in the composite bending moment diagram in Problem 15, the bending moment assumes maximum values at either end of the beam as well as at the midpoint. This value is $PL/8$. At the outer fibers the bending stress is $s = Mc/I = MZ$, where Z denotes the section modulus of the beam. Substituting,

$$18{,}000 = \frac{(38{,}000)(22)(12)}{8Z} \qquad \text{or} \qquad Z = 70.0 \text{ in}^3$$

From the table at the end of Chapter 8 it is seen that a 12 WF 58 section is adequate to support this load. It has $Z = 78.1$ in³. For this section, $I = 476.1$ in⁴.

The central deflection was shown in Problem 16 to be

$$\Delta_B = -\frac{PL^3}{192EI} = -\frac{(38{,}000)(22)^3(12)^3}{(192)(30\cdot 10^6)(476.1)} = -0.255 \text{ in.}$$

18. The horizontal beam shown in Fig. (a) is simply supported at the ends and is connected to a composite elastic vertical rod at its midpoint. The supports of the beam and the top of the copper rod are originally at the same elevation, at which time the beam is horizontal. The temperature of both vertical rods is then decreased 100°F. Find the stress in each of the vertical rods. Neglect the weight of the beam and of the rods. The cross-sectional area of the copper rod is 1 in², $E_{cu} = 15\cdot 10^6$ lb/in², and $\alpha_{cu} = 9.3\cdot 10^{-6}/°$F. The cross-sectional area of the aluminun rod is 2 in², $E_{al} = 10\cdot 10^6$ lb/in², and $\alpha_{al} = 12.8\cdot 10^{-6}/°$F. For the beam, $E = 1.5\cdot 10^6$ lb/in² and $I = 1000$ in⁴.

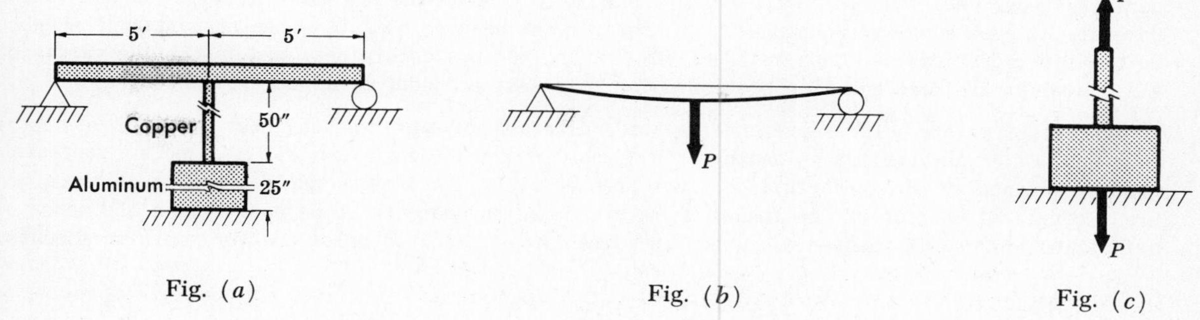

Fig. (a) Fig. (b) Fig. (c)

A free-body diagram of the horizontal beam appears as in Fig. (b) above. Here, P denotes the force exerted upon the beam by the copper rod. Since this force is initially unknown there are three forces acting upon the beam, but only two equations of equilibrium for a parallel force system; hence the problem is statically indeterminate. It will thus be necessary to consider the deformations of the system.

A free-body diagram of the two vertical rods appears as in Fig. (c) above.

The simplest procedure is to temporarily cut the connection between the beam and the copper rod, and then allow the vertical rods to contract freely because of the decrease in temperature. If the horizontal beam offers no restraint, the copper rod will contract an amount

$$\Delta_{cu} = (9.3\cdot10^{-6})(50)(100) = 0.0465 \text{ in.}$$

and the aluminum rod will contract by an amount

$$\Delta_{al} = (12.8\cdot10^{-6})(25)(100) = 0.0320 \text{ in.}$$

However, the beam exerts a tensile force P upon the copper rod and the same force acts in the aluminum rod as shown in Fig. (*c*) above. These axial forces elongate the vertical rods and this elongation (see Problem 1, Chapter 1) is $\dfrac{P(50)}{(1)(15\cdot10^6)} + \dfrac{P(25)}{(2)(10\cdot10^6)}$.

The downward force P exerted by the copper rod upon the horizontal beam causes a vertical deflection of the beam. In Problem 9 of Chapter 10 this central deflection was found to be $\Delta = PL^3/48EI$.

Actually, of course, the connection between the copper rod and the horizontal beam is not cut in the true problem and we realize that the resultant shortening of the vertical rods is exactly equal to the downward vertical deflection of the midpoint of the beam. This change of length of the vertical rods is caused partially by the decrease in temperature and partially by the axial force acting in the rods. For the shortening of the rods to be equal to the deflection of the beam we must have

$$[0.0465 + 0.0320] - \left[\frac{P(50)}{(1)(15\cdot10^6)} + \frac{P(25)}{(2)(10\cdot10^6)}\right] = \frac{P(120)^3}{(48)(1.5\cdot10^6)(1000)}$$

Solving, $P = 2740$ lb. Then $s_{cu} = 2740/1 = 2740 \text{ lb/in}^2$ and $s_{al} = 2740/2 = 1370 \text{ lb/in}^2$.

19. Derive the Three-Moment Theorem for continuous beams.

A continuous beam is one that rests on more than two supports. An example of a continuous beam is shown in the adjacent figure. This particular beam consists of two spans and is subject to a series of uniform and concentrated loads as shown. It will be assumed that the nature of the supports is such that no horizontal reactions are present.

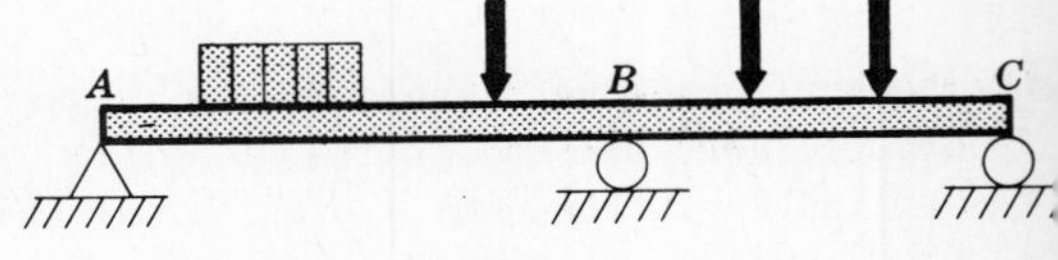

Continuous beams are statically indeterminate and thus it is necessary to supplement the availab[le] statics equations with equations arising from the deformations of the system. One possible way [of] deriving these equations is to treat the vertical forces at the various supports as the unknown[s]. However, an even simpler technique is to consider the bending moments at the supports as the unknown[s]. Deformation equations are then written, the bending moments determined, and lastly the values of t[he] various vertical force reactions determined. This last procedure will be employed here.

The illustrations below represent free-body diagrams of any two adjacent spans of a continuo[us] beam subject to any arbitrary loading. As shown in these diagrams M_A, M_B, and M_C represent t[he] bending moments at the supports A, B, and C respectively. Although the directions of these momen[ts] are of course a function of the loads, we have assumed the moments to be positive in the sense of t[he] definition offered in Chapter 6. Hence the directions indicated below are for positive moments.

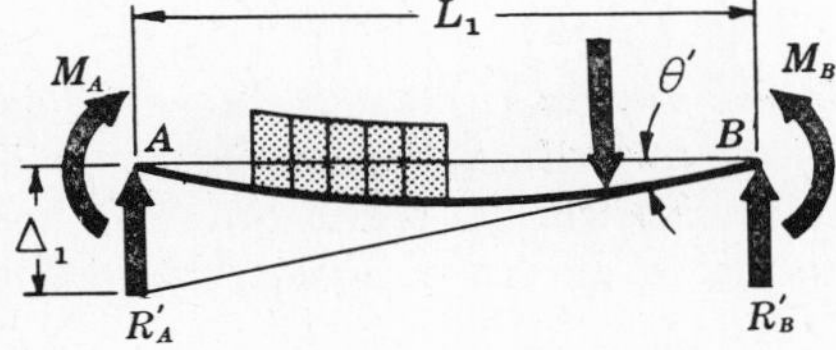

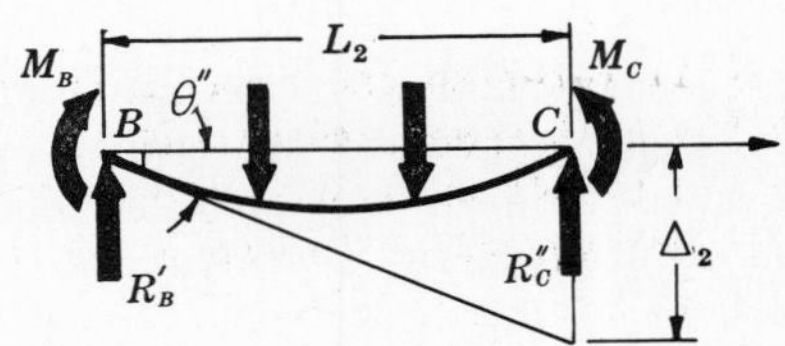

The slope of the deflection curve must be continuous over the central support; hence

$$\theta' = -\theta''$$

The values of these angles will now be found by the Moment-Area method.

We shall first consider the various known loadings to act on corresponding simply supported beams, i.e. the moments M_A, M_B, and M_C are temporarily omitted. The bending moment diagrams for each of the spans L_1 and L_2 may then be determined by the techniques of Chapter 6 and are represented symbolically as follows:

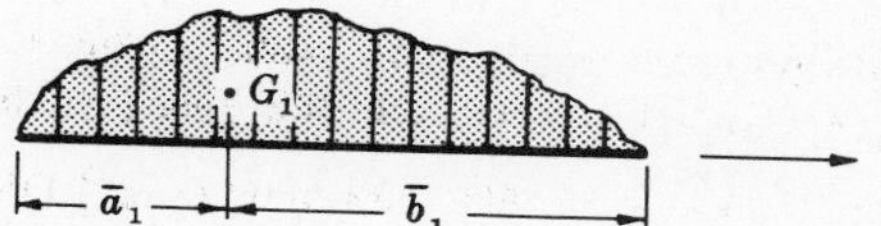

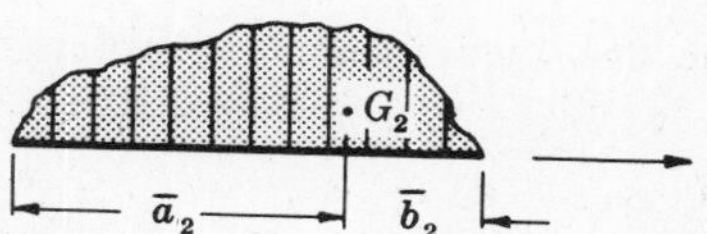

In these diagrams G_1 and G_2 denote the centroids of the areas of the moment diagrams; and $\bar{a}_1$, $\bar{b}_1$, $\bar{a}_2$, and $\bar{b}_2$ have the meanings shown. It is to be carefully noted that these moment diagrams are determined on the assumption that each of the spans is simply supported. We shall let A_1 and A_2 denote the areas of these moment diagrams for the left and right spans respectively.

The deflection of A from the tangent at B is found from the Second Moment-Area Theorem to be $\Delta_1 = \dfrac{A_1\bar{a}_1}{EI}$ and consequently the slope θ' is $\theta' = \dfrac{A_1\bar{a}_1}{L_1EI}$.

It is next necessary to consider the effects of M_A and M_B on the slope θ' in the left span. From a consideration of Problem 13 of Chapter 9, the rotation of the left span at point B produced by these moments is

$$\frac{M_BL_1}{3EI} + \frac{M_AL_1}{6EI}$$

The total angle of rotation is thus the sum of these, or

$$\theta' = \frac{A_1\bar{a}_1}{L_1EI} + \frac{M_BL_1}{3EI} + \frac{M_AL_1}{6EI}$$

Similarly, for the right span we have

$$\theta'' = \frac{A_2\bar{b}_2}{L_2EI} + \frac{M_BL_2}{3EI} + \frac{M_CL_2}{6EI}$$

Substituting in the relationship $\theta' = -\theta''$, we have

$$M_AL_1 + 2M_B(L_1+L_2) + M_CL_2 = -\frac{6A_1\bar{a}_1}{L_1} - \frac{6A_2\bar{b}_2}{L_2}$$

This is the Three-Moment Theorem and in this general form is applicable to any type of loading. Application of this equation to a continuous beam, together with the statics equations, enables one to solve for the various reactions. The above equation is sometimes called Clapeyron's Equation.

). Discuss the special form of the Three-Moment Theorem for uniformly distributed loads acting on two adjacent spans.

We shall refer to the diagrams of Problem 19. Let w_1 represent the intensity of uniform load acting on the left span, and w_2 the intensity acting on the right span. If each of these loadings is considered to act on simply supported spans of lengths L_1 and L_2 respectively, then the bending moment diagrams are parabolic as shown below. In Problem 6 of Chapter 6 the maximum ordinates were found to have the values indicated.

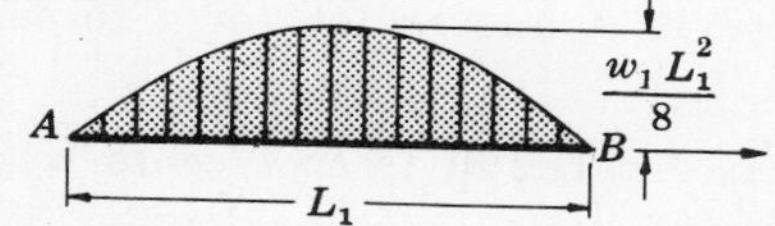

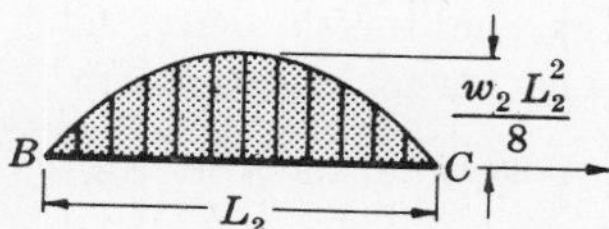

In Problem 3 of Chapter 10 it was shown that the area under such a parabola is 2/3 the area of th surrounding rectangle. Hence, since A_1 represents the area of the left bending moment diagram, we hav

$$A_1 = \frac{2}{3}(L_1)(\frac{w_1L_1^2}{8}) \qquad \text{Similarly,} \quad A_2 = \frac{2}{3}(L_2)(\frac{w_2L_2^2}{8})$$

The centroids of these areas are located at the midpoints of each of these spans.

Substituting these values in the right-hand side of the general form of the Three-Moment Theore presented in Problem 19, we find

$$M_AL_1 + 2M_B(L_1+L_2) + M_CL_2 = -\frac{w_1L_1^3}{4} - \frac{w_2L_2^3}{4}$$

This is the Three-Moment Equation for uniformly distributed loads.

21. Determine the special form of the Three-Moment Theorem for a single concentrated force acting on eac of two adjacent spans.

Again, we shall refer to the diagrams of Problem 19. Let P_1 denote the concentrated force actin on the left span, and a_1 the distance of this load from the left support A. Also, P_2 denotes the con centrated force acting on the right span at a distance b_2 from the right support C. If each of thes loadings is considered to act on simply supported spans of length L_1 and L_2 respectively, then th bending moment diagrams are triangular as shown below.

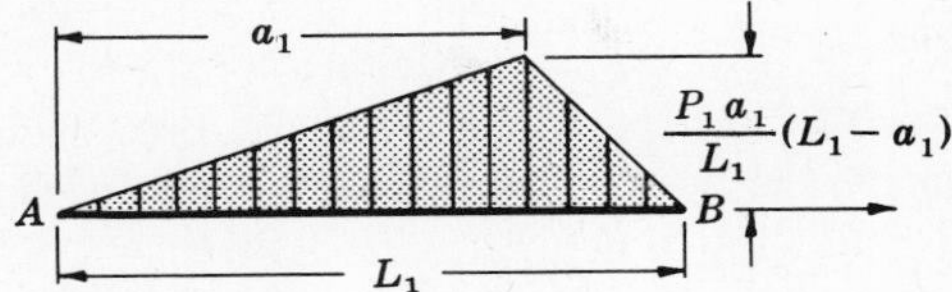

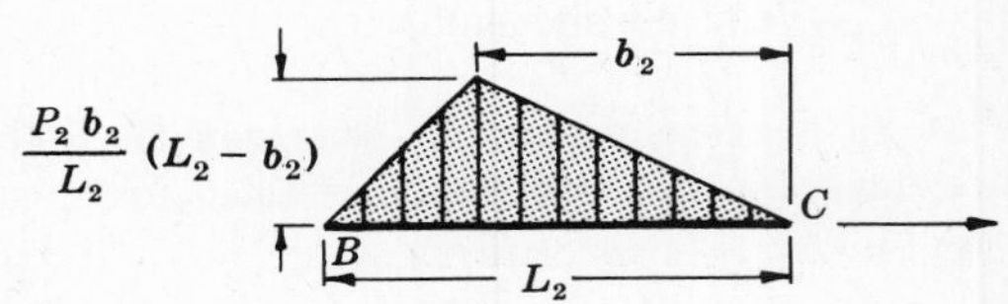

In the general form of the Three-Moment Theorem presented in Problem 19 the quantity $A_1\bar{a}_1$ appear For the left span this designates the moment of the area under the above bending moment diagram abou a vertical line through the left end A. Referring to the above moment diagram for the left span L_1 we may calculate this most easily by dividing the triangle into two triangles, one having a base equa to a_1, the other having a base equal to (L_1-a_1). The product of the area of each of these triangle times the distance from its centroid to the vertical line through A yields the required product $A_1\bar{a}_1$ Thus

$$A_1\bar{a}_1 = \frac{1}{2}(a_1)(\frac{P_1a_1}{L_1})(L_1-a_1)(\frac{2}{3}a_1) + \frac{1}{2}(L_1-a_1)[(\frac{P_1a_1}{L_1})(L_1-a_1)][a_1+\frac{1}{3}(L_1-a_1)]$$

From this, after simplification, we obtain $\dfrac{6A_1\bar{a}_1}{L_1} = \dfrac{P_1a_1}{L_1}(L_1^2-a_1^2)$.

Similarly, for the right span L_2 we find $\dfrac{6A_2\bar{b}_2}{L_2} = \dfrac{P_2b_2}{L_2}(L_2^2-b_2^2)$.

For a number of concentrated loads the Three-Moment Theorem thus becomes

$$M_AL_1 + 2M_B(L_1+L_2) + M_CL_2 = -\sum\frac{P_1a_1}{L_1}(L_1^2-a_1^2) - \sum\frac{P_2b_2}{L_2}(L_2^2-b_2^2)$$

where the summation sign is used so as to include the effects of all concentrated loads.

22. The two-span continuous beam shown in Fig. (a) below supports a uniform load of w lb per unit lengt Determine the various reactions.

The Three-Moment Theorem for uniform loads as given in Problem 20 is applicable. Here the load constant across the entire beam and we have $w_1 = w_2 = w$. Also, $L_1 = L_2 = L$. The ends A and C are simply supported and hence we have $M_A = M_C = 0$. Substituting in the Three-Moment Theorem,

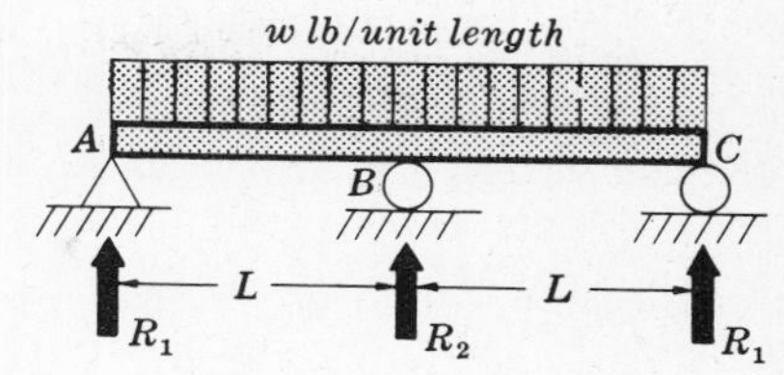

Fig. (a)

$$M_A L_1 + 2M_B(L_1 + L_2) + M_C L_2 = -\frac{w_1 L_1^3}{4} - \frac{w_2 L_2^3}{4}$$

we obtain
$$0 + 2M_B(2L) + 0 = -\frac{wL^3}{4} - \frac{wL^3}{4}$$

or
$$M_B = -\frac{wL^2}{8}$$

Perhaps the simplest method for determining the reactions is to write the expression for the bending moment at B (which has just been determined) in terms of the moments of the forces to the left of B. Thus

$$R_1 L - wL(\frac{L}{2}) = -\frac{wL^2}{8} \qquad \text{or} \qquad R_1 = \frac{3}{8}wL$$

The end forces designated as R_1 are equal by symmetry. From statics,

$$2(\frac{3}{8}wL) + R_2 - 2wL = 0 \qquad \text{or} \qquad R_2 = \frac{5}{4}wL$$

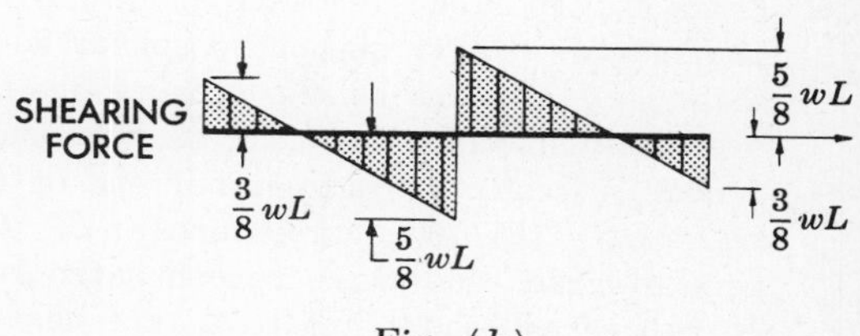

Fig. (b)

The shear diagram may now be constructed by the usual methods of Chapter 6 and has the appearance shown in Fig. (b).

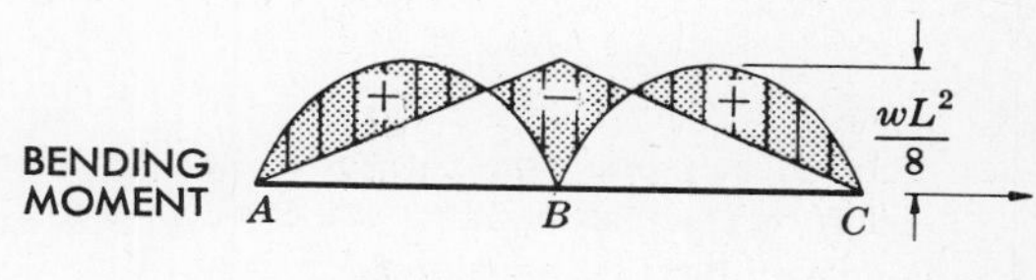

Fig. (c)

The bending moment diagram may also be plotted by the techniques of Chapter 6. However, a somewhat simpler procedure for continuous beams is to plot the moment diagram for the load acting on each span on the basis that the span is simply supported. For this loading this diagram is of course parabolic for the uniform load. Then, the moment diagram due to the moments at the supports is constructed. This moment is zero at either end of the beam and equal to $-wL^2/8$ at the point B. From a consideration of Problem 13 of Chapter 9 it is evident that the variation of moment from either end of the beam to the midpoint (due to the moment $-wL^2/8$ *only*) is linear, i.e. straight lines must connect the value of the moment at the midpoint with the values at the ends, which are zero. Also, the moment diagrams due to the uniform load are positive, whereas those due to M_B are negative. By superposition of these diagrams the final form appears as shown by the shaded areas in Fig. (c) above.

23. The two-span continuous beam shown in Fig. (a) supports the centrally applied forces. Determine the various reactions.

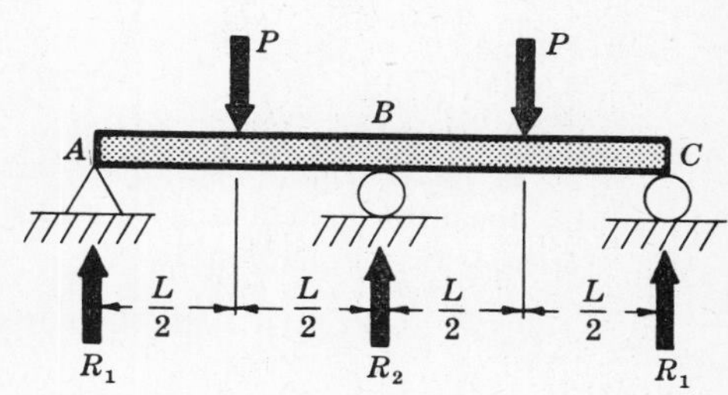

Fig. (a)

The Three-Moment Theorem for concentrated loads as given in Problem 21 is applicable. Here we have $P_1 = P_2 = P$, $L_1 = L_2 = L$, and $a_1 = b_2 = L/2$. The ends A and C are simply supported and hence we have $M_A = M_C = 0$. Substituting in the Three-Moment Theorem,

$$M_A L_1 + 2M_B(L_1 + L_2) + M_C L_2 = -\sum \frac{P_1 a_1}{L_1}(L_1^2 - a_1^2) - \sum \frac{P_2 b_2}{L_2}(L_2^2 - b_2^2)$$

we obtain
$$0 + 2M_B(2L) + 0 = -\frac{2P(L/2)}{L}(L^2 - L^2/4) \qquad \text{or} \qquad M_B = -\frac{3}{16}PL.$$

We may now express this bending moment at B in terms of moments of the forces to the left of B as follows:

$$R_1 L - \frac{PL}{2} = -\frac{3PL}{16} \qquad \text{or} \qquad R_1 = \frac{5}{16}P$$

The end forces designated as R_1 are equal by symmetry. From statics,

$$2(\frac{5P}{16}) + R_2 - 2P = 0 \qquad \text{or} \qquad R_2 = \frac{11}{8}P$$

The shear diagram thus appears as in Fig. (*b*).

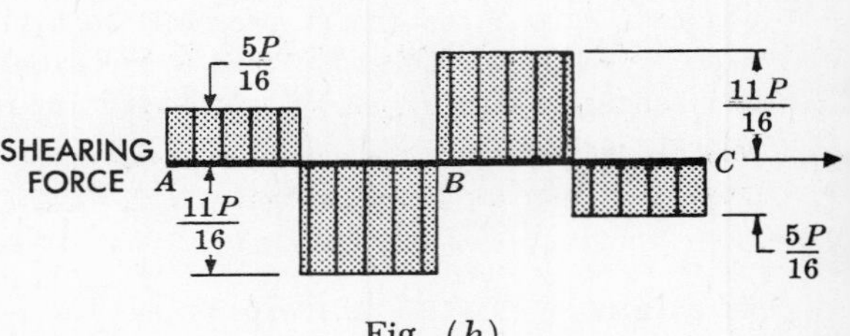

Fig. (*b*)

The bending moment diagram may be plotted by the technique mentioned in Problem 22. The moment diagrams for the loads on each span on the basis that the span is simply supported appear as triangles of altitude $PL/4$. The moment diagram due to the moments over the supports is next constructed. It varies linearly from zero at either end of the beam to a value of $-3PL/16$ at the midpoint B. Again, the moment diagrams due to the concentrated loads are positive whereas those due to M_B are negative. Superposition of these diagrams yields the final form of the moment diagram as shown by the shaded areas in Fig. (*c*).

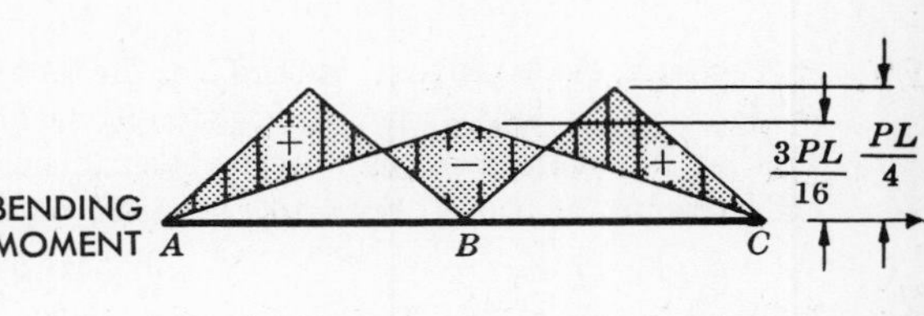

Fig. (*c*)

24. The three-span continuous beam is subject to the uniform load as well as the two concentrated loads shown. Determine the four reactions.

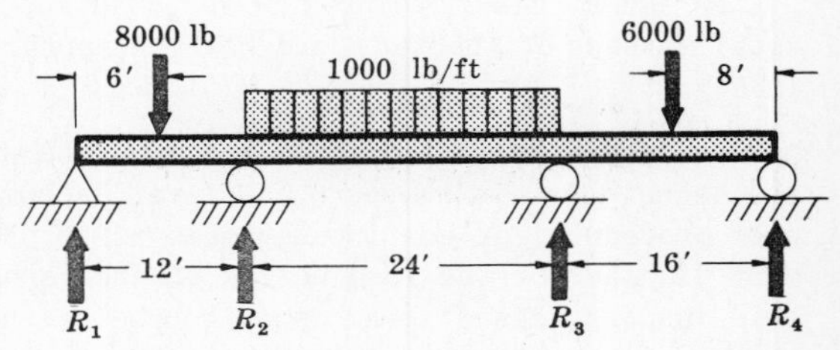

We shall denote the moments acting at the supports from left to right by M_1, M_2, M_3, and M_4 respectively. We immediately have $M_1 = M_4 = 0$, since these ends are simply supported.

We shall first apply the Three-Moment Theorem to the left and central spans. This will evidently give rise to an equation containing M_2 and M_3 as unknowns. Since these two spans are subject to a concentrated force and a uniform load respectively, the special forms of the Three-Moment Theorem developed in Problems 20 and 21 are applicable. We must take $L_1 = 12$ ft, $L_2 = 24$ ft. We thus have

$$0 + 2M_2(12+24) + M_3(24) = -\frac{8000(6)}{12}[(12)^2 - (6)^2] - \frac{1000(24)^3}{4}$$

Simplifying,

(*a*) $$3M_2 + M_3 = -18{,}000 - 144{,}000$$

Next we shall apply the Three-Moment Theorem to the central and right spans. This gives rise to another equation containing M_2 and M_3. It is to be carefully noted that we must now take $L_1 = 24$ ft, $L_2 = 16$ ft. This equation is

$$M_2(24) + 2M_3(24+16) + 0 = -\frac{1000(24)^3}{4} - \frac{6000(8)}{16}[(16)^2 - (8)^2]$$

Simplifying,

(*b*) $$M_2 + 3.33M_3 = -144{,}000 - 24{,}000$$

Solving equations (*a*) and (*b*) simultaneously, we find $M_2 = -41{,}300$ lb-ft and $M_3 = -38{,}000$ lb-ft.

We may now express the bending moment M_2 at support 2 in terms of moments of the forces to the left of this reaction as follows:

$$12R_1 - 8000(6) = -41{,}300 \qquad \text{or} \qquad R_1 = 560 \text{ lb}$$

Similarly for the bending moment M_3 at support 3,

$$36(560) + 24R_2 - 8000(30) - 1000(24)(12) = -38{,}000 \qquad \text{or} \qquad R_2 = 19{,}600 \text{ lb}$$

Working from the right end, $16R_4 - 6000(8) = -38,000$ or $R_4 = 640$ lb.

Lastly, $40(640) + 24R_3 - 1000(24)(12) - 6000(32) = -41,300$ or $R_3 = 17,200$ lb.

It is to be observed that two equations of static equilibrium could have been used to determine R_3 and R_4 rather than the last two equations if desired. However, the above procedure has the advantage that the statics equations are still available for checking the accuracy of the above numerical work.

SUPPLEMENTARY PROBLEMS

25. The clamped-end beam is supported at the right end, clamped at the left and carries the two concentrated forces shown in Fig. (*a*) below. Determine the reaction at the wall and the reaction at the right end of the beam. *Ans.* $4P/3$ acting upward at left end, $PL/3$ acting counterclockwise at left end, $2P/3$ acting upward at right end.

26. Determine the deflection under the point of application of the force P located a distance $L/3$ from the right end of the beam described in Problem 25. *Ans.* $7PL^3/486EI$

27. The beam discussed in Problem 25 is a 10 WF 54 section. The length is 15 ft and the maximum permissible bending stress is 18,000 lb/in^2. Determine the maximum allowable value of each load P. *Ans.* 18,400 lb.

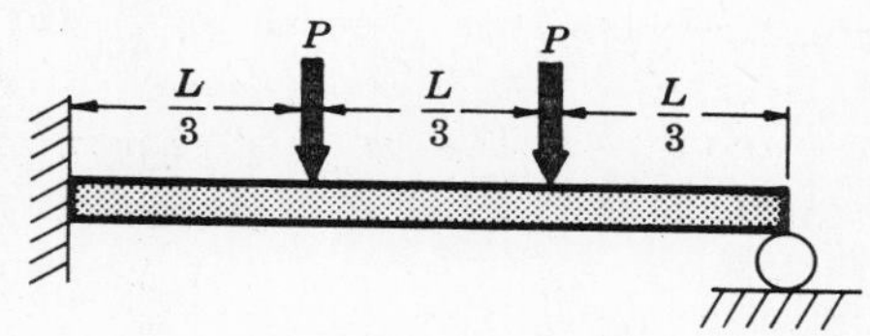

Fig. (*a*) Prob. 25

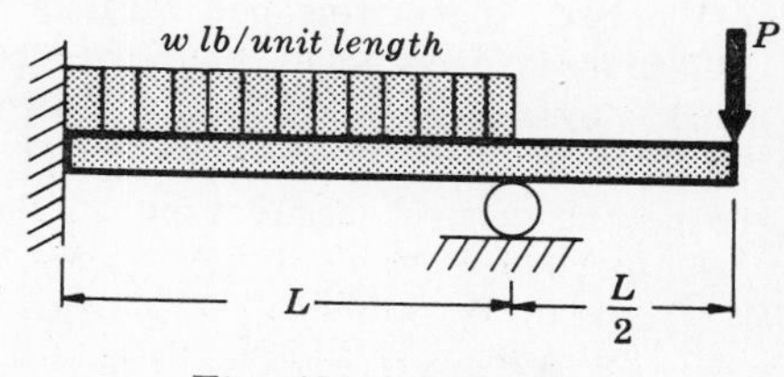

Fig. (*b*) Prob. 28

28. The clamped-end beam is supported at an intermediate point and loaded as shown in Fig. (*b*) above. Determine the various reactions.

Ans. $(\frac{5}{8}wL - \frac{3}{4}P)$ upward at left end, $(\frac{1}{8}wL^2 - \frac{1}{4}PL)$ counterclock. at left end, $(\frac{3}{8}wL + \frac{7}{4}P)$ upward at support.

29. For the beam shown in Problem 28, determine the deflection of the right end of the beam (under the point of application of the force P). *Ans.* $\dfrac{wL^4}{96EI} - \dfrac{5PL^3}{48EI}$

30. The clamped-end beam is supported at an intermediate point and loaded as shown in the adjacent figure. Determine the various reactions.
Ans. $3P/10$ acting downward at left end
$PL/10$ acting clockwise at left end
$13P/10$ acting upward at support

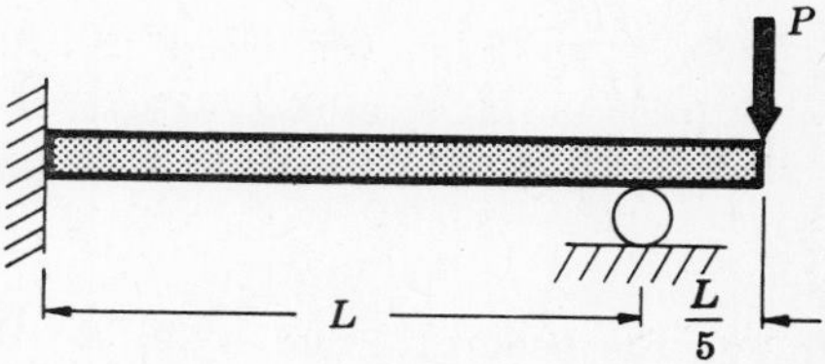

31. For the beam shown in Problem 30 determine the deflection under the point of application of force P.
Ans. $\dfrac{19PL^3}{1500EI}$

32. For the beam shown in Problem 30 take $L = 20$ ft and $P = 20,000$ lb. Select a wide-flange section adequate to carry this load without exceeding a bending stress of 18,000 lb/in^2. *Ans.* 12 WF 45

33. The beam shown in Fig. (*c*) below, supports a concentrated load as well as a partial uniform load. Determine the reaction at the right end of the beam. *Ans.* $(\frac{81}{128}P + \frac{7}{128}wL)$

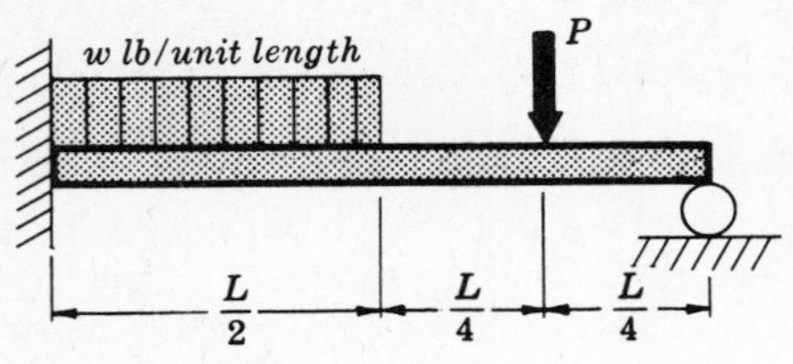

Fig. (c) Prob. 33

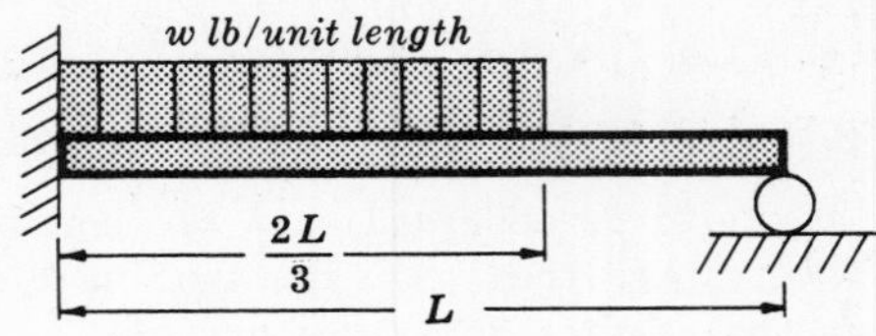

Fig. (d) Prob. 34

34. The beam shown in Fig. (d) above, supports the uniform load extending over two-thirds of its length. Determine the reaction at the right end of the beam. *Ans.* $10\,wL/81$

35. A beam is clamped at both ends and supports a uniform load over its right half as shown in the adjacent diagram. Determine all reactions.

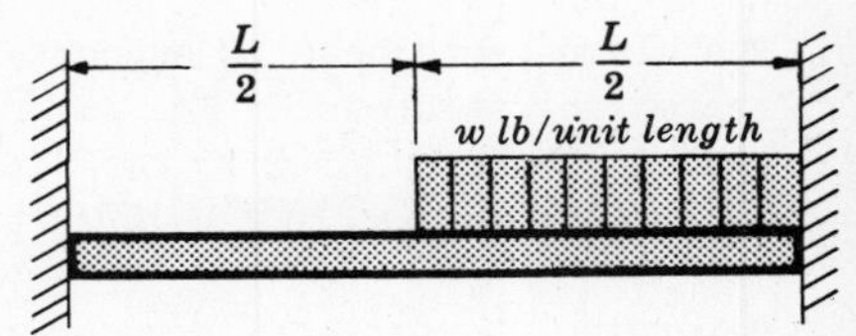

Ans. $3\,wL/32$ acting upward at left end
$5\,wL^2/192$ acting counterclockwise at left end
$13\,wL/32$ acting upward at right end
$11\,wL^2/192$ acting clockwise at right end

36. Determine the central deflection of the beam described in Problem 35. *Ans.* $wL^4/768\,EI$

37. A beam is clamped at both ends and supports two symmetrically located concentrated forces as shown in Fig. (e) below. Determine the various reactions.
Ans. An upward force equal to P together with a counterclockwise moment equal to $\frac{51}{400}PL$ at the left end. Symmetric reactions at the right end.

38. Determine the central deflection of the beam described in Problem 37. *Ans.* $\dfrac{9\,PL^3}{4000\,EI}$

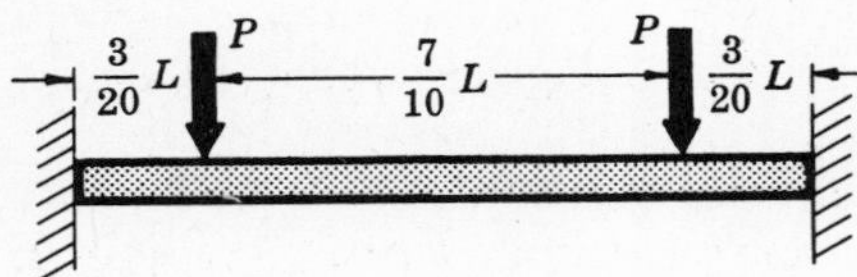

Fig. (e) Prob. 37

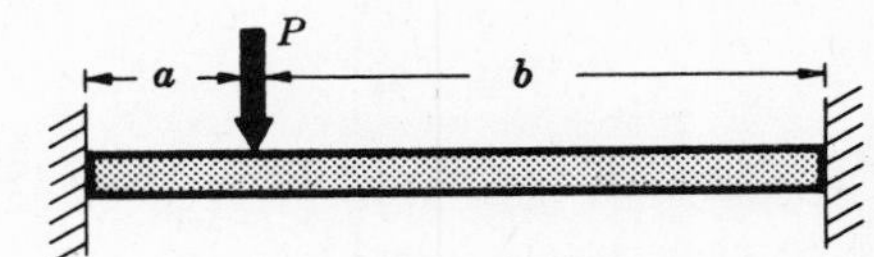

Fig. (f) Prob. 39

39. A beam is clamped at both ends and loaded by the single concentrated force shown in Fig. (f) above. Determine the various reactions.

Ans. $\dfrac{Pb^2}{L^3}(3a+b)$ acting upward at left end, $\dfrac{Pab^2}{L^2}$ acting counterclockwise at left end,

$\dfrac{Pa^2}{L^3}(a+3b)$ acting upward at right end, $\dfrac{Pa^2b}{L^2}$ acting clockwise at right end.

40. For the beam of Problem 39, $P = 10{,}000$ lb, $a = 3$ ft, and $b = 12$ ft. Determine the end reactions.
Ans. An upward force of 8960 lb and a counterclockwise moment of 19,200 lb-ft acting at the left end. An upward force of 1040 lb and a clockwise moment of 4800 lb-ft acting at the right end.

41. Select a wide-flange section adequate to carry the load for the beam discussed in Problem 40. The permissible bending stress is 20,000 lb/in^2. *Ans.* 8 WF 19

42. The beam shown in Fig. (g) below is clamped at the left end, supported at the right, and is loaded by a couple M_0 as shown. Determine the reaction at the right support. *Ans.* $\dfrac{3M_0\,a(a+2b)}{2(a+b)^3}$

43. For the beam shown in Problem 42, determine the deflection under the point of application of the applied moment M_0. *Ans.* $\dfrac{M_0 a^2 b(a^2 - 2b^2)}{4(a+b)^3 EI}$

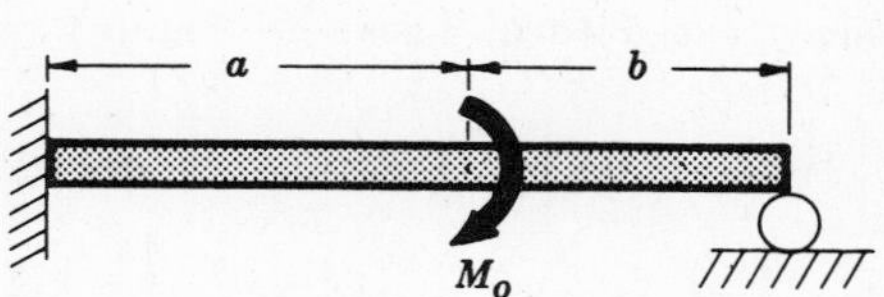

Fig. (*g*) Prob. 42

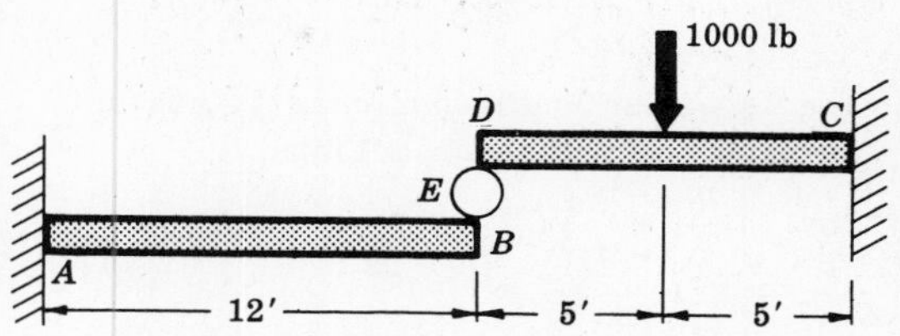

Fig. (*h*) Prob. 44

44. *AB* and *CD* are cantilever beams with a roller *E* between their end points. A load of 1000 lb is applied as shown in Fig. (*h*) above. Both beams are made of steel for which $E = 30 \cdot 10^6$ lb/in². For beam *AB*, $I = 50$ in⁴ ; for *CD*, $I = 80$ in⁴ . Find the reaction at *E*. *Ans.* 83 lb.

45. The two-span beam is supported at *B* and *C* and clamped at *A*. Each span carries the uniformly distributed load shown in Fig. (*i*) below. Determine the reactions at *B* and *C*. *Ans.* R_B = 2260 lb, R_C = 970 lb

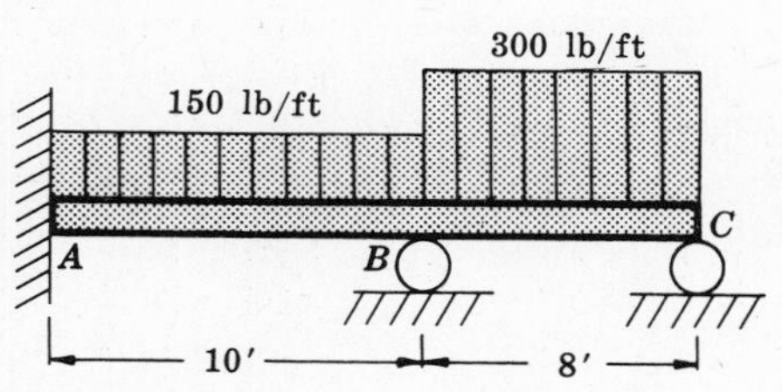

Fig. (*i*) Prob. 45

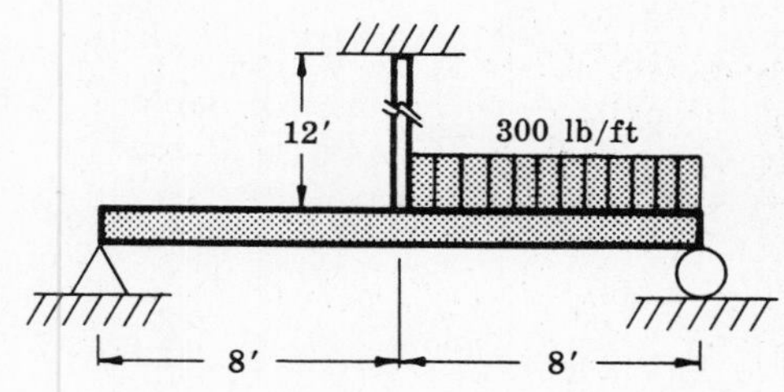

Fig. (*j*) Prob. 46

46. A 16 ft beam carries a uniform load over the right half of its span and is supported at the center of the span by a vertical rod as shown in Fig. (*j*) above. The rod is steel, 12 ft in length, 0.5 in² in cross-sectional area, and $E_s = 30 \cdot 10^6$ lb/in². The beam is wood 4 in. by 8 in. in cross-section and $E_w = 1.5 \cdot 10^6$ lb/in². Determine the stress in the vertical steel rod. *Ans.* 2960 lb/in².

47. The three-span continuous beam is subject to the uniform load shown in Fig. (*k*) below. Determine the various reactions as well as the maximum bending moment in the beam.

Ans. Reactions: $\frac{4}{10}wL$, $\frac{11}{10}wL$, $\frac{11}{10}wL$, $\frac{4}{10}wL$ Maximum moment: $\frac{1}{10}wL^2$

48. The continuous beam shown in Problem 47 is a 12 WF 32 section, and L = 12 ft. Determine the allowable load per unit length so as not to exceed an allowable bending stress of 18,000 lb/in².
Ans. 428 lb/ft

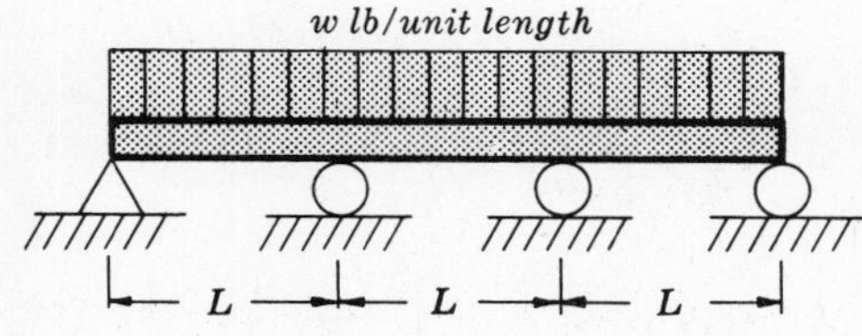

Fig. (*k*) Prob. 47

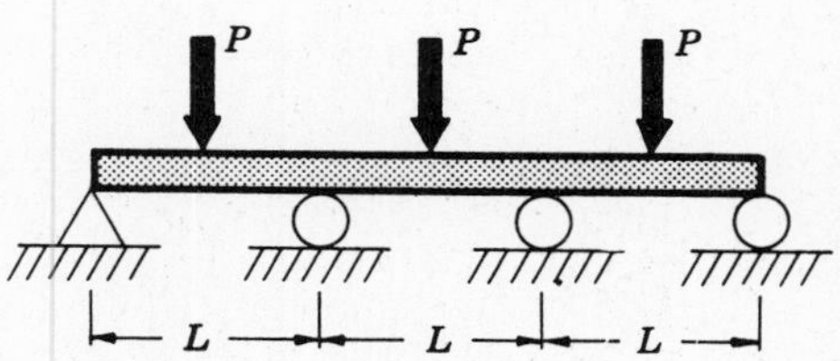

Fig. (*l*) Prob. 49

49. The three-span continuous beam is subject to the three centrally placed concentrated loads shown Fig. (l) above. Determine the various reactions as well as the maximum bending moment in the beam.

Ans. Reactions; $\frac{7}{20}P$, $\frac{23}{20}P$, $\frac{23}{20}P$, $\frac{7}{20}P$ Maximum moment: $\frac{7}{40}PL$

50. The two-span continuous beam is subject to the single concentrated force shown in Fig. (m) below. Determine the various reactions.

Ans. $\left[\frac{Pb}{L} - \frac{Pa}{4L^3}(L^2 - a^2)\right]$ upward, $\left[\frac{Pa}{L} + \frac{Pa}{2L^3}(L^2 - a^2)\right]$ upward, $\left[\frac{Pa}{4L^3}(L^2 - a^2)\right]$ downward.

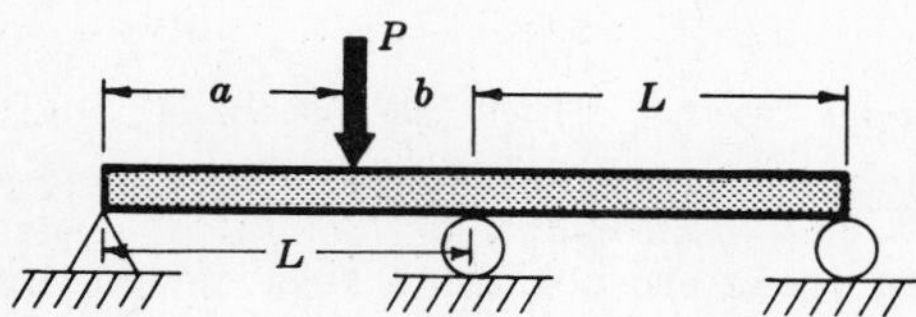

Fig. (m) Prob. 50

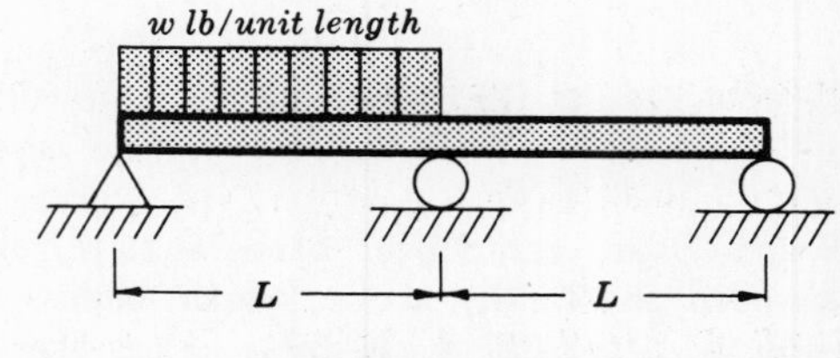

Fig. (n) Prob. 51

51. The two span continuous beam is subject to the uniform load shown in Fig. (n) above. Determine the various reactions. Also, determine the maximum bending moment in the beam.

Ans. $\frac{7}{16}wL$ acting upward, $\frac{5}{8}wL$ acting upward, $\frac{1}{16}wL$ acting downward

Maximum moment $= \frac{49}{512}wL^2$

CHAPTER 12

Columns

DEFINITION OF A COLUMN. A long slender bar subject to axial compression is called a *column*. The term *column* is frequently used to describe a vertical member, whereas the word *strut* is occasionally used in regard to inclined bars.

TYPE OF FAILURE OF A COLUMN. Failure of a column occurs by buckling, i.e. by lateral deflection of the bar. In comparison it is to be noted that failure of a short compression member occurs by yielding of the material. Buckling, and hence failure, of a column may occur even though the maximum stress in the bar is less than the yield point of the material.

EXAMPLES OF COLUMNS. Many aircraft structural components, certain members in roof and bridge trusses, locomotive connecting rods, and vertical floor supports in buildings are common examples of columns.

DEFINITION OF THE CRITICAL LOAD OF A COLUMN. The critical load of a long slender bar subject to axial compression is that value of the axial force that is just sufficient to keep the bar in a slightly deflected configuration. The adjacent figure shows a pin-ended bar in a buckled configuration due to the critical load P.

SLENDERNESS RATIO OF A COLUMN. The ratio of the length of the column to the minimum radius of gyration of the cross-sectional area is termed the slenderness ratio of the bar. This ratio is of course dimensionless. The method of determining the radius of gyration of an area was discussed in Chapter 7.

If the column is free to rotate at each end, then buckling takes place about that axis for which the radius of gyration is a minimum.

CRITICAL LOAD OF A LONG SLENDER COLUMN. If a long slender bar of constant cross-section is pinned at each end and subject to axial compression, the load P_{cr} that will cause buckling is given by

$$P_{cr} = \frac{\pi^2 EI}{L^2}$$

where E denotes the modulus of elasticity, I the minimum moment of inertia of the cross-sectional area about an axis through the centroid, and L the length of the bar. The derivation of this formula is presented in Problem 1.

This formula was first derived by a Swiss mathematician, Leonhard Euler (1707-1783) and the load P_{cr} is frequently called the Euler buckling load. As discussed in Problem 2, this expression is not valid if the corresponding axial stress, found from the expression $s_{cr} = P_{cr}/A$, where A represents the cross-sectional area of the bar, exceeds the proportional limit of the material. For example, for a steel bar having a proportional limit of 30,000 lb/in^2, the above formula is valid only for columns whose slenderness ratio exceeds 100. The value of P_{cr} represented by this formula is a failure load; consequently, a safety factor must be introduced to obtain a design load. Applications of this expression may be found in Problems 7-11.

DESIGN FORMULAS FOR COLUMNS HAVING INTERMEDIATE SLENDERNESS RATIOS. The design of compression members having large values of the slenderness ratio proceeds according to the Euler formula presented above, together with an appropriate safety factor. For the design of shorter compression members it is customary to employ any one of the many empirical formulas giving a relationship between the critical stress and the slenderness ratio of the bar. Actually, the formulas usually present an expression for the working stress as a function of the slenderness ratio, i.e. a safety factor has already been incorporated into the expression. Only two of the many existing empirical relations will be considered in this book.

The first, a so-called straight-line formula, originated in the Chicago Building Code and states that the allowable working stress in a column is given by

$$s_w = 16,000 - 70(L/r)$$

where L/r represents the slenderness ratio of the bar. These specifications state that this expression is to be used only in the range $30 < L/r < 120$ for so-called main members and as high as $L/r = 150$ for so-called secondary members, i.e. bars used as lateral bracing between roof trusses, or bars used to reduce the slenderness ratio of a long column by bracing it at some intermediate point. This formula is discussed in detail in Problem 13. For an application, see Problem 15.

The second relation, to be found in the specifications of the American Institute of Steel Construction (A.I.S.C.) is a so-called parabolic formula and states that the allowable working stress in a column is given by

$$s_w = 17,000 - 0.485(L/r)^2$$

provided L/r is less than 120. This expression is discussed in detail in Problem 14. For applications, see Problems 16, 17, 18, 20, 21.

The effect of each of these two expressions is to reduce the working stress in a column for increasing values of the slenderness ratio.

DESIGN OF ECCENTRICALLY LOADED COLUMNS. Several methods exist for the rational analysis and design of an eccentrically loaded column. Only one of these will be presented here. For a bar subject to a compressive force P_o acting through the centroid of the cross-section together with an additional force P applied with an eccentricity e (measured from the centroid) the maximum stress is

$$s = \frac{P + P_o}{A} + \frac{Pec}{I}$$

where A represents the cross-sectional area of the bar and I is the moment of inertia of the cross-sectional area with respect to the axis about which bending occurs. As in Chapter 8, c denotes the distance from the neutral axis to the extreme fibers of the bar. It is necessary to use either the A.I.S.C. specifications or the Chicago Building Code formula to obtain a safe value of the allowable compressive stress for use in conjunction with this expression. The use of this formula for eccentrically loaded columns is discussed in Problem 19. For applications, see Problems 20 and 21.

SOLVED PROBLEMS

1. Determine the critical load for a long slender pin-ended bar loaded by an axial compressive force at each end. The line of action of the forces passes through the centroid of the cross-section of the bar.

The critical load is defined to be that axial force that is just sufficient to hold the bar in a slightly deformed configuration. Under the action of the load P the bar has the deflected shape shown in the adjacent figure.

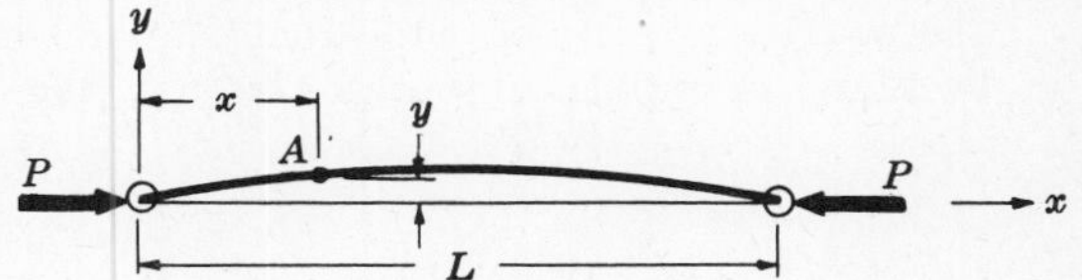

It is of course necessary that one end of the bar be able to move axially with respect to the other end in order that the lateral deflection may take place. The differential equation of the deflection curve is the same as that presented in Chapter 9, namely

$$EI \frac{d^2y}{dx^2} = M \tag{1}$$

Here the bending moment at the point A having coordinates (x,y) is merely the moment of the force P applied at the left end of the bar about an axis through the point A and perpendicular to the plane of the page. It is to be carefully noted that this force produces curvature of the bar that is concave downward, which, according to the sign convention of Chapter 6, constitutes negative bending. Hence the bending moment is $M = -Py$. Thus we have

$$EI \frac{d^2y}{dx^2} = -Py \tag{2}$$

If we set

$$\frac{P}{EI} = k^2 \tag{3}$$

this last equation becomes

$$\frac{d^2y}{dx^2} + k^2y = 0 \tag{4}$$

This equation is readily solved by any one of several standard techniques discussed in works on differential equations. However, the solution is almost immediately apparent. We need merely find a function which when differentiated twice and added to itself (times a constant) is equal to zero. Evidently either $\sin kx$ or $\cos kx$ possesses this property. In fact, a combination of these terms in the form

$$y = C \sin kx + D \cos kx \tag{5}$$

may also be taken to be a solution of equation (4). This may be readily checked by substitution of y as given by equation (5) into (4).

Having obtained y in the form given in equation (5), it is next necessary to determine C and D. At the left end of the bar, $y = 0$ when $x = 0$. Substituting these values in equation (5) we obtain

$$0 = 0 + D \quad \text{or} \quad D = 0$$

At the right end of the bar, $y = 0$ when $x = L$. Substituting these values in (5) with $D = 0$ we obtain

$$0 = C \sin kL$$

Evidently either $C = 0$ or $\sin kL = 0$. But if $C = 0$ then y is everywhere zero and we have only the trivial case of a straight bar which is the configuration prior to the occurrence of buckling.

Since we are not interested in this solution, then we must take

6) $$\sin kL = 0$$

For this to be true, we must have

7) $$kL = n\pi \text{ radians} \quad (n = 1, 2, 3, \ldots)$$

Substituting $k^2 = P/EI$ in equation (7) we find

8) $$\sqrt{\frac{P}{EI}}\,L = n\pi \quad \text{or} \quad P = \frac{n^2\pi^2 EI}{L^2}$$

The smallest value of this load P evidently occurs when $n = 1$. Then we have the so-called first mode of buckling where the critical load is given by

9) $$P_{cr} = \frac{\pi^2 EI}{L^2}$$

This is called Euler's buckling load for a pin-ended column. The deflection shape corresponding to this load is

10) $$y = C \sin\sqrt{\frac{P}{EI}}\,x$$

Substituting in this equation from equation (9) we obtain

11) $$y = C \sin\frac{\pi x}{L}$$

Thus the deflected shape is a sine curve. Because of the approximations introduced in the derivation of equation (1) it is not possible to obtain the amplitude of the buckled shape, denoted by C in equation (11).

As may be seen from equation (9), buckling of the bar will take place about that axis in the cross-section for which I assumes a minimum value.

2. Determine the axial stress in the column considered in Problem 1.

In the derivation of the equation $EI(d^2y/dx^2) = M$ used to determine the critical load in Problem 1, it was assumed that there is a linear relationship between stress and strain (see Problem 1, Chapter 9.) Thus the critical load indicated by equation (9) of Problem 1 is correct only if the proportional limit of the material has not been exceeded.

The axial stress in the bar immediately prior to the instant when the bar assumes its buckled configuration is given by

1) $$s_{cr} = \frac{P_{cr}}{A}$$

where A represents the cross-sectional area of the bar. Substituting for P_{cr} its value as given by equation (9) of Problem 1, we find

2) $$s_{cr} = \frac{\pi^2 EI}{AL^2}$$

But from Chapter 7 we know that we may write

3) $$I = Ar^2$$

where r represents the so-called radius of gyration of the cross-sectional area. Substituting this value in equation (2), we find

4) $$s_{cr} = \frac{\pi^2 EAr^2}{AL^2} = \pi^2 E\left(\frac{r}{L}\right)^2$$

$$s_{cr} = \frac{\pi^2 E}{(L/r)^2} \tag{5}$$

The ratio L/r is called the *slenderness ratio* of the column.

Let us consider a steel column having a proportional limit of 30,000 lb/in^2 and $E = 30\cdot 10^6$ lb/in^2. The stress of 30,000 lb/in^2 marks the upper limit of stress for which equation (5) may be used. To find the value of L/r corresponding to these constants, we substitute in equation (5) and obtain

$$30{,}000 = \frac{\pi^2(30\cdot 10^6)}{(L/r)^2} \qquad \text{or} \qquad L/r \approx 100$$

Thus for this material the buckling load as given by equation (9), Problem 1 and the axial stress as given by equation (5) are valid only for those columns having $L/r \geqslant 100$. For those columns having $L/r < 100$, the compressive stress exceeds the proportional limit before elastic buckling takes place and the above equations are not valid.

Equation (5) may be plotted as shown in Fig. (a) below. For the particular values of proportional limit and modulus of elasticity assumed above, the portion of the curve to the left of $L/r = 100$ is not valid. Thus for this material, point A marks the upper limit of applicability of the curve.

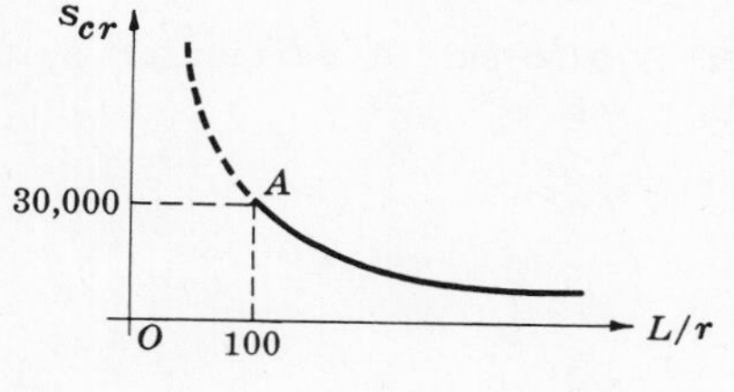

Fig. (a) Prob. 2

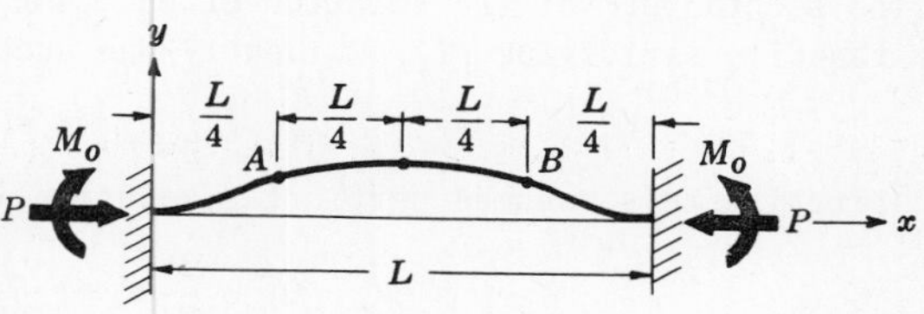

Fig. (b) Prob. 3

3. Determine the critical load for a long slender bar clamped at each end and loaded by an axial compressive force at each end.

The critical load is that axial compressive force P that is just sufficient to keep the bar in a slightly deformed configuration, as shown in Fig. (b) above. The moments M_o at each end of the bar represent the actions of the supports on the column; these moments prevent any angular rotation of the bar at either end.

Inspection of the above deflection curve for the buckled column indicates that the central portion of the bar between points A and B corresponds to the deflection curve for the pin-ended column discussed in Problem 1. Thus for the fixed-end column, the length $L/2$ corresponds to the entire length L for the pin-ended bar. Hence the critical load for a clamped-end bar may be found from equation (9), Problem 1 by replacing L by $L/2$. This yields

$$P_{cr} = \frac{\pi^2 EI}{(L/2)^2} = \frac{4\pi^2 EI}{L^2}$$

Again, it is assumed that the maximum stress in the column does not exceed the proportional limit of the material.

The above formula, derived here on an intuitive basis, could be derived more rigorously by a solution of the usual differential equation for a bent bar. This is carried out in detail in Problem 4.

4. Determine the critical load for the long slender clamped end bar described in Problem 3 by direct solution of the governing differential equation.

Let us introduce the x-y coordinate system shown there and let (x, y) represent the coordinates of an arbitrary point on the bar. The bending moment at this point is found as the sum of the moments of the forces to the left of this section about an axis through this point and perpendicular to the plane of the page. Hence at this point we have

$$M = -Py + M_o$$

Using the usual differential equation for the bending of a bar, we have

1) $$EI\frac{d^2y}{dx^2} = -Py + M_o \quad \text{or} \quad \frac{d^2y}{dx^2} + \frac{P}{EI}y = \frac{M_o}{EI}$$

As discussed in texts on differential equations, the solution to this equation consists of two parts. The first part is merely the solution of the so-called homogeneous equation obtained by setting the right hand side of equation *(1)* equal to zero. We must then solve the equation

2) $$\frac{d^2y}{dx^2} + \frac{P}{EI}y = 0$$

But the solution to this equation has already been found in Problem 1 to be

3) $$y = A_1 \cos\sqrt{\frac{P}{EI}}\,x + B_1 \sin\sqrt{\frac{P}{EI}}\,x$$

The second part of the solution of equation *(1)* is given by a so-called particular solution, i.e. any function satisfying *(1)*. Evidently one such function is given by

$$y = c_1 \quad (= \text{constant})$$

Substituting this assumed particular solution in *(1)* we find

$$0 + \frac{P}{EI}c_1 = \frac{M_o}{EI} \quad \text{or} \quad c_1 = \frac{M_o}{P}$$

Thus, a particular solution is

4) $$y = M_o/P$$

The general solution of equation *(1)* is given by the sum of the solutions represented by equations *(3)* and *(4)*, or

5) $$y = A_1 \cos\sqrt{\frac{P}{EI}}\,x + B_1 \sin\sqrt{\frac{P}{EI}}\,x + \frac{M_o}{P}$$

Consequently

6) $$\frac{dy}{dx} = -A_1\sqrt{\frac{P}{EI}}\sin\sqrt{\frac{P}{EI}}\,x + B_1\sqrt{\frac{P}{EI}}\cos\sqrt{\frac{P}{EI}}\,x$$

At the left end of the bar we have $y = 0$ when $x = 0$. Substituting these values in equation *(5)* we find $0 = A_1 + M_o/P$. Also, at the left end of the bar we have $dy/dx = 0$ when $x = 0$; substituting in *(6)* we obtain $0 = 0 + B_1\sqrt{P/EI}$ or $B_1 = 0$.

At the right end of the bar we have $dy/dx = 0$ when $x = L$; substituting in *(6)*, with $B_1 = 0$ we find

$$0 = -A_1\sqrt{\frac{P}{EI}}\sin\sqrt{\frac{P}{EI}}\,L$$

But $A_1 = -M_o/P$ and since this ratio is not zero, then $\sin\sqrt{P/EI}\,L = 0$. This occurs only when $\sqrt{P/EI}\,L = n\pi$ where $n = 1, 2, 3, \ldots$ Consequently

7) $$P_{cr} = \frac{n^2\pi^2EI}{L^2}$$

For the so-called first mode of buckling illustrated in Problem 3 the deflection curve of the bent bar has a horizontal tangent at $x = L/2$, i.e. $dy/dx = 0$ there. Equation *(6)* may now be written in the form

6') $$\frac{dy}{dx} = \frac{M_o}{P}\cdot\frac{n\pi}{L}\cdot\sin\frac{n\pi x}{L}$$

And since $dy/dx = 0$ at $x = L/2$, we find

$$0 = \frac{M_o}{P}\cdot\frac{n\pi}{L}\cdot\sin\frac{n\pi}{2}$$

The only manner in which this equation may be satisfied is for n to assume even values, i.e. $n = 2,4,6,\ldots$

Thus for the smallest possible value of $n = 2$, equation *(7)* becomes

$$P_{cr} = \frac{4\pi^2 EI}{L^2}$$

This is the critical load for a clamped end bar subject to axial compression. The result found by the less rigorous treatment of Problem 3 is thus confirmed.

5. Determine the critical load for a long slender bar clamped at one end, free at the other, and loaded by an axial compressive force applied at the free end.

The critical load is that axial compressive force P that is just sufficient to keep the bar in a slightly deformed configuration, as shown in the adjoining figure. The moment M_o represents the effect of the support in preventing any angular rotation of the left end of the bar.

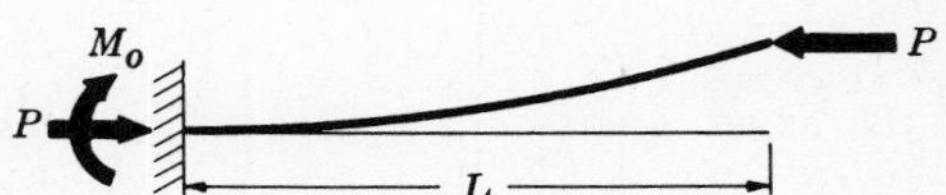

Inspection of the above deflection curve for the buckled column indicates that the entire bar corresponds to one-half of the deflected pin-ended bar discussed in Problem 1. Thus for the column under consideration, the length L corresponds to $L/2$ for the pin-ended column. Hence the critical load for the present column may be found from equation *(9)*, Problem 1, by replacing L by $2L$. This yields

$$P_{cr} = \frac{\pi^2 EI}{(2L)^2} = \frac{\pi^2 EI}{4L^2}$$

6. Determine the slenderness ratio for a timber column 8 × 10 in. in cross-section and 25 ft long.

As mentioned in Problem 1, buckling of such a bar will take place about that axis in the cross-section for which the moment of inertia assumes a minimum value. This moment of inertia for a rectangular area about an axis through its centroid is

$$I = bh^3/12 = 10(8^3)/12 = 426 \text{ in}^4$$

The cross-sectional area is 80 in^2; hence the minimum radius of gyration is

$$r = \sqrt{I/A} = \sqrt{426/80} = 2.31 \text{ in.}$$

The slenderness ratio $\frac{L}{r}$ is thus $\frac{L}{r} = \frac{25(12)}{2.31} = 130.$

7. A steel bar of rectangular cross-section 1.5 in. by 2 in. and pinned at each end is subject to axial compression. If the proportional limit of the material is 33,000 lb/in^2, and $E = 30\cdot10^6$ lb/in^2, determine the minimum length for which Euler's equation may be used to determine the buckling load.

The minimum moment of inertia is $I = \frac{1}{12}bh^3 = \frac{1}{12}(2)(1.5)^3 = 0.562 \text{ in}^4$.

Hence the least radius of gyration is $r = \sqrt{\frac{I}{A}} = \sqrt{\frac{0.562}{(1.5)(2)}} = 0.434$ in.

The axial stress for such an axially loaded bar was found in Problem 2 to be

$$s_{cr} = \frac{\pi^2 E}{(L/r)^2}$$

The minimum length for which Euler's equation may be applied is found by placing the critical stress in the above formula equal to 33,000 lb/in^2. Doing this, we obtain

$$33{,}000 = \frac{\pi^2(30\cdot 10^6)}{(L/0.434)^2} \quad \text{or} \quad L = 41.0 \text{ in.}$$

8. Consider again a rectangular steel bar 1.5 in. by 2 in. in cross-section, pinned at each end and subject to axial compression. The bar is 70 in. long and $E = 30\cdot 10^6$ lb/in^2. Determine the buckling load using Euler's formula.

The minimum moment of inertia of this cross-section was found in Problem 7 to be 0.562 in^4. Applying the expression for buckling load given in equation 9, Problem 1, we find

$$P_{cr} = \frac{\pi^2 EI}{L^2} = \frac{\pi^2(30\cdot 10^6)(0.562)}{(70)^2} = 34{,}000 \text{ lb}$$

The axial stress corresponding to this load is $s_{cr} = \dfrac{P_{cr}}{A} = \dfrac{34{,}000}{(1.5)(2)} = 11{,}300$ lb/in^2.

9. Compare the buckling strengths of two long slender pin-end bars, one of solid circular cross-section 2 in. in diameter, the other of solid square cross-section and having the same cross-sectional area as the circular bar. The columns are the same lengths and are made of the same material. Use the Euler column theory.

For the bar of circular cross-section, $I = \pi D^4/64 = \pi(2^4)/64 = 0.785$ in^4. The buckling load is consequently $P_{cr} = \pi^2 E(0.785)/L^2$.

The area of the circular bar is $\pi(1)^2 = 3.14$ in^2; hence the square bar is $\sqrt{3.14} = 1.78$ in. on a side. The moment of inertia of the square bar about an axis through the centroid of the cross-section is $I = bh^3/12 = 1.78(1.78)^3/12 = 0.820$ in^4. The buckling load is thus $P'_{cr} = \pi^2 E(0.820)/L^2$.

We may now form the ratio $P_{cr}/P'_{cr} = 0.785/0.820 = 0.958$. Thus the buckling load of the circular bar is 95.8% that of the square bar.

10. Consider a long slender bar of circular cross-section 2 in. in diameter and clamped at each end. The material is steel for which $E = 30\cdot 10^6$ lb/in^2. Determine the minimum length for which Euler's equation may be used to determine the buckling load if the proportional limit of the material is 35,000 lb/in^2.

From Problem 3 the buckling load is known to be $P_{cr} = 4\pi^2 EI/L^2$.

The axial stress prior to buckling is given by $s_{cr} = P_{cr}/A = 4\pi^2 EI/AL^2$. But $I = Ar^2$ where r denotes the minimum radius of gyration of the cross-section. (Actually, all radii of gyration are equal because of symmetry). Substituting

$$s_{cr} = \frac{4\pi^2 E(Ar^2)}{AL^2} = \frac{4\pi^2 E}{(L/r)^2} \tag{1}$$

From Problem 9 the moment of inertia of the circular cross-section about a diametral axis is 0.785 in^4. Then the radius of gyration is $r = \sqrt{I/A} = \sqrt{0.785/\pi(1^2)} = 0.50$ in.

The minimum length for which Euler's equation may be applied is found by setting the critical stress in equation *(1)* equal to 35,000 lb/in^2. Then

$$35{,}000 = \frac{4\pi^2(30\cdot 10^6)}{(L/0.50)^2} \quad \text{or} \quad L = 92.0 \text{ in.}$$

1. Determine the critical load for a 10 WF 21 section acting as a pin-ended column. The bar is 12 ft long and $E = 30\cdot10^6$ lb/in^2. Use Euler's theory.

From the table at the end of Chapter 8 we find the minimum moment of inertia to be 9.7 in^4. This is the value that must be used in the expression for the buckling load. Thus

$$P_{cr} \quad = \quad \pi^2 EI/L^2 \quad = \quad \pi^2(30\cdot10^6)(9.7)/(144)^2 \quad = \quad 138{,}000 \text{ lb}$$

2. In Problem 2 it was found that the limit of applicability of Euler's formula for the buckling load of a column corresponds to the proportional limit of the material. Discuss the design of compression members having slenderness ratios less than the value corresponding to the proportional limit of the material.

In Problem 2 the axial stress in the pin-ended bar immediately prior to buckling was found to be

$$s_{cr} \quad = \quad \frac{\pi^2 E}{(L/r)^2}$$

where L/r denotes the slenderness ratio of the column. For a steel column having a proportional limit of 30,000 lb/in^2 and $E = 30\cdot10^6$ lb/in^2 the above expression for the stress, and consequently the corresponding expression for the buckling load, was found to be valid only for values of L/r greater than 100. The design procedure for columns having L/r greater than 100 is thus established.

Several design procedures are available for columns having slenderness ratios less than 100 (or the corresponding value for values of proportional limit and modulus different from those used above). The first procedure seeks to extend Euler's formula to shorter columns where the critical stress is above the proportional limit of the material by employing a so-called reduced modulus E' instead of the constant value of the modulus of elasticity. This reduced modulus is not a constant but instead varies with the slenderness ratio of the column. With this technique the upper limit of applicability of the above expression for the axial stress is raised approximately to the yield point of the material. If the yield point of the material is 40,000 lb/in^2, the minimum value of slenderness ratio for which the above expression for axial stress is valid is approximately 60. This reduced modulus concept may be represented approximately by the straight line BC in the adjoining diagram.

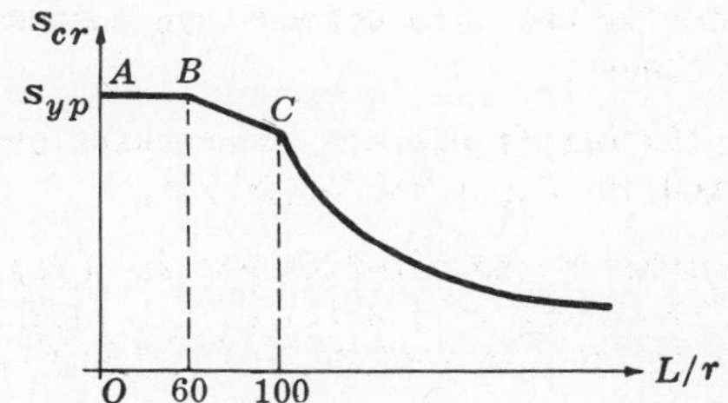

For values of the slenderness ratio less than 60, the critical stress may be considered to be equal to the yield point of the material. This is represented by the straight line AB in the diagram above. Thus, in this diagram, the broken line ABC together with Euler's curve determines the critical stress in a bar for all values of the slenderness ratio. It is to be noted that no safety factor has been introduced in presenting these values. Columns having values of the slenderness ratio corresponding to BC are frequently said to be of "intermediate length."

To obtain working stresses the ordinates to the above diagram must be divided by some number representing a factor of safety. Experimental evidence indicates that both the eccentricity of loading and the initial imperfections always present in the column tend to increase with increasing values of L/r. Hence a variable safety factor is frequently employed, ranging from 2.0 for very short bars to 3.5 for very long slender columns. Formulas for working stresses in columns are presented in Problems 13 and 14.

For design purposes it is customary to represent the above relations between s_{cr} and L/r by empirical formulas. Certain of these will be presented in the following two problems.

3. Discuss the various formulas relating to design of columns based upon straight-line relationships between working stress and slenderness ratio.

These linear or straight-line relations assume that the critical stress, when it exceeds the proportional limit of the material, may be represented by an equation of the form

$$s_{cr} = a - b(L/r)$$

where a and b are constants depending upon the physical properties of the material. Such an expressio may give the critical stress, or the values of the constants a and b may be adjusted so that a saf working stress is indicated by the above formula.

Usually the latter is the case. One of the most commonly used straight line formulas is that give by the Chicago Building Code. This expression gives the safe working stress s_w in the following form

$$s_w = 16,000 - 70(L/r)$$

This expression is to be used for $30 < L/r < 120$ for main members and $30 < L/r < 150$ in so-called sec ondary members, such as lateral bracing in roof and bridge trusses. The same building code specifie a working stress of 14,000 lb/in^2 for bars having a slenderness ratio less than 30.

It is evident that the purpose of the above formula for the working stress is to reduce the critica compressive stress (based upon very short columns) for increasing values of the slenderness ratio.

14. Discuss the various formulas relating to design of columns based upon parabolic relationships betwee working stress and slenderness ratio.

These parabolic relations assume that the critical stress, when it exceeds the proportional limi of the material, may be represented by an equation of the form

$$s_{cr} = a - b(L/r)^2$$

where again a and b are constants depending upon the physical properties of the material. Usually, th constants a and b are selected so as to make the parabola represented by the above equation tangent t Euler's curve and also to make the critical stress equal to the yield point of the material for ver short bars. Again, as in Problem 13, such an expression may give the critical stress, or the values c a and b may be adjusted so that a safe working stress is indicated by the formula.

The latter case is exemplified by a formula suggested by the American Institute of Steel Construc tion (A.I.S.C.) wherein the working stress is given by

$$s_w = 17,000 - 0.485(L/r)^2$$

for the design of main members having L/r less than 120. For secondary members having $120 < L/r < 20$ these same specifications indicate the following formula:

$$s_w = \frac{18,000}{1 + [\frac{1}{18,000}(\frac{L}{r})^2]}$$

These expressions, as well as those offered in Problem 13 assume hinged end conditions. They may used for other end conditions by using the modified length concepts mentioned in Problems 3 and 5.

15. A bar of circular cross-section, 2 in. in diameter and 4.5 ft long, is loaded by axial compressive force The bar is pinned at each end. Determine the maximum safe load the bar may carry using the Chica Building Code formula.

The moment of inertia is $I = \pi(2^4)/64 = 0.785$ in^4.

The radius of gyration is $r = \sqrt{I/A} = \sqrt{0.785/\pi(1^2)} = 0.50$ in.

The slenderness ratio is $L/r = 4.5(12)/(0.5) = 108$.

The working stress is $s_w = 16,000 - 70L/r = 16,000 - 70(108) = 8420$ lb/in^2.

Then $P = A \cdot s_w = \pi(1^2)(8420) = 26,500$ lb.

16. Determine the maximum safe load for the column discussed in Problem 15 by use of the A.I.S.C. formul

According to this formula the working stress is given by

$$s_w = 17{,}000 - 0.485(L/r)^2 = 17{,}000 - 0.485(108)^2 = 11{,}350 \text{ lb/in}^2$$

Then $P = A \cdot s_w = \pi(1^2)(11{,}350) = 35{,}800$ lb.

The difference in the results obtained by use of the straight-line and parabolic formulas reflects the differences in the safety factor involved in the Chicago Building Code and the A.I.S.C. specifications.

. What is the maximum safe axial compressive load that an 8 WF 40 section 18 ft long may carry if the bar is pinned at each end? Use the A.I.S.C. formula.

From the table at the end of Chapter 8 we find the minimum moment of inertia of this section to be 49.0 in^4 and the cross-sectional area 11.76 in^2. Then the least $r = \sqrt{I/A} = \sqrt{49.0/11.76} = 2.04$ in.

The slenderness ratio is $L/r = 18(12)/2.04 = 106$

The working stress is $s_w = 17{,}000 - 0.485(L/r)^2 = 11{,}540 \text{ lb/in}^2$.

Hence $P = A \cdot s_w = 11.76(11{,}540) = 135{,}800$ lb.

. Select a WF section to carry an axial compressive load of 160,000 lb. The bar is 16 ft 8 in. long. Use the A.I.S.C. specifications. The bar is pinned at each end.

The working stress is given by

1) $$s_w = 17{,}000 - 0.485(L/r)^2$$

Substituting the given values of P and L (= 200 in.) in this expression, we obtain

2) $$160{,}000/A = 17{,}000 - 0.485(200/r)^2$$

The solution of this equation may be obtained by a trial-and-error method. As a first approximation, let us find the minimum area by setting the axial stress equal to 17,000 lb/in^2, even though this is, of course, greater than the allowable working stress. Doing this we find

$$160{,}000/A = 17{,}000 \quad \text{or} \quad A = 9.5 \text{ in}^2$$

Hence we need not consider any section having an area less than 9.5 in^2.

Let us first investigate a 10 WF 37 section. From the table at the end of Chapter 8 the minimum moment of inertia is 42.2 in^4 and the cross-sectional area is 10.88 in^2. The least radius of gyration is consequently $r = \sqrt{42.2/10.88} = 1.97$ in. The slenderness ratio is thus $L/r = 200/1.97 = 102$. From equation *(1)* the allowable working stress in this bar is

$$s_w = 17{,}000 - 0.485(102)^2 = 11{,}950 \text{ lb/in}^2$$

The maximum safe load is thus

$$P = A \cdot s_w = 10.88(11{,}950) = 130{,}000 \text{ lb}$$

Since this is less than the design load of 160,000 lb, the section is too light.

Let us next investigate a 10 WF 45 section. According to the table in Chapter 8 this section has a minimum moment of inertia of 53.2 in^4 and a cross-sectional area of 13.24 in^2. The least radius of gyration is $r = \sqrt{53.2/13.24} = 2.00$. The slenderness ratio is $L/r = 200/2.00 = 100$. The allowable working stress is

$$s_w = 17{,}000 - 0.485(100)^2 = 12{,}150 \text{ lb/in}^2$$

The maximum safe load is thus

$$P = A \cdot s_w = 13.24(12{,}150) = 161{,}000 \text{ lb}$$

Since this value slightly exceeds the design load, the 10 WF 45 section is adequate.

. All of the expressions for allowable axial loads on columns presented thus far in this chapter have assumed that the load is applied through the centroid of the cross-section of the bar. Discuss one method for investigating and designing the column in the case where the load is eccentrically applied.

Frequently, columns are loaded in such a manner that the line of action of the axial loads lies at a distance e from the centroid of the cross-section of the bar as shown in the adjacent diagram.

One of the simplest as well as one of the most conservative methods of analysis is to neglect th effect of the lateral deflections of the column on the moment arm of the axial load. The maxim stress in the column occurs at those fibers most remote from the neutral axis of the cross-sectio The stress at these outer fibers is the sum of the normal stress due to the axial loading plus th bending stress arising from the eccentricity of loading. For a column with an axial load P_o and load P applied with eccentricity e, the maximum stress is

$$s = \frac{P + P_o}{A} + \frac{Pec}{I}$$

where A denotes the cross-sectional area of the bar, I the moment of inertia, and c the distance fr the neutral axis to the outer fibers.

The actual stress calculated from this equation should not exceed the design stress, which is de termined on the basis of one of the formulas for axially loaded columns presented previously in th chapter. In calculating the design stress from any of these formulas (such as the A.I.S.C. specifica tions) the least radius of gyration should be used, regardless of the axis about which the bendi will occur.

20. An 8 WF 40 section 12 ft in length is subject to an axial compressive force of 5000 lb applied throu the centroid of the cross-section. What additional compressive force may be applied simultaneous with an eccentricity of 6 ft? Use the A.I.S.C. specifications.

The working stress is given by $s_w = 17{,}000 - 0.485(L/r)^2$. Hence from Problem 19, we immediate have the following equation for determining the eccentric load P:

1) $$17{,}000 - 0.485(\frac{L}{r})^2 = \frac{P + P_o}{A} + \frac{Pec}{I}$$

The least radius of gyration for an 8 WF 40 section was found in Problem 17 to be 2.04 in. Als from the table at the end of Chapter 8 we have $A = 11.76\ \text{in}^2$, $I = 49.0\ \text{in}^4$, and $c = 4$ in. Substitu ing these values, together with $P_o = 5000$ lb in equation *(1)* we obtain

$$17{,}000 - 0.485(\frac{144}{2.04})^2 = \frac{P + 5000}{11.76} + \frac{P(72)(4)}{49.0} \quad \text{or} \quad P = 2360 \text{ lb}$$

It is to be noted that L/r is 71, which falls within the range of applicability of the A.I.S. formula.

21. Select a WF section adequate to support a concentric axial compressive load of 80,000 lb (applied at A) together with an eccentric load of 60,000 lb applied 3 in. from the center of the section at the point B shown in the adjacent diagram. Use the A.I.S.C. specifications. The bar is 20 ft long and the ends are pinned.

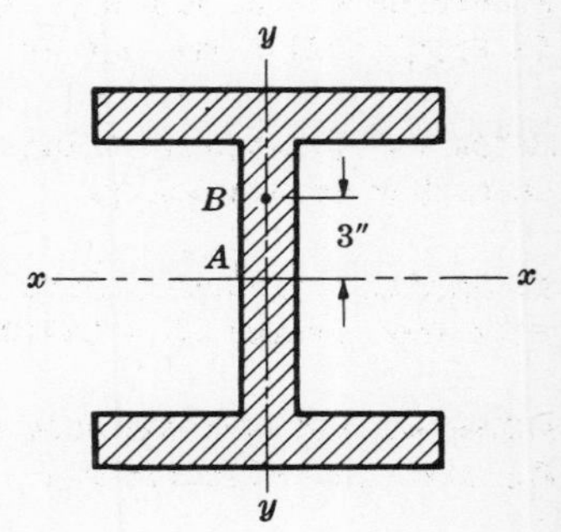

Using the A.I.S.C. formula together with the equation presented in Problem 19, we have

1) $$17{,}000 - 0.485(\frac{L}{r})^2 = \frac{P + P_o}{A} + \frac{Pec}{I}$$

Note that the least radius of gyration is to be used in determining the slenderness ratio L/r regardless of the axis about which bending occurs. Equation *(1)* must be solved by trial-and-error. Although the allowable stress is evidently less than 17,000 lb/in^2, that value may be used to determine the minimum area of the column. The total compressive force is 140,000 lb and this requires a minimum area of 140,000/17,000 = 8.20 in^2. Hence there is no need to investigate any section having an area smaller than this value.

As an initial trial let us investigate a 10 WF 29 section. From the table at the end of Chapter 8 the minimum moment of inertia is 15.2 in^4 and the area is 8.53 in^2. The slenderness ratio is thus $240/\sqrt{15.2/8.53} = 178$, which exceeds the allowable value of 120 for which the A.I.S.C. formula is valid.

Next, let us investigate a 10 WF 49 section. The slenderness ratio is $240/\sqrt{93.0/14.40} = 94.5$. From the table at the end of Chapter 8 the moment of inertia about the bending axis (the x-x axis) is 272.9 in^4. Substituting these values in equation *(1)* we have

$$17{,}000 - 0.485(94.5)^2 \quad = \quad \frac{60{,}000 + 80{,}000}{14.40} + \frac{60{,}000(3)(5)}{272.9}$$

Simplifying,
$$12{,}680 \neq 13{,}020$$

This relation tells us that the allowable stress is 12,680 lb/in^2, whereas the actual stress in this column is 13,020 lb/in^2. Consequently a heavier section is required.

Let us investigate a 10 WF 54 section. The slenderness ratio is $240/\sqrt{103.9/15.88} = 94.0$, and the moment of inertia about the bending axis is 305.7 in^4. Substituting these values in equation *(1)* we find

$$17{,}000 - 0.485(94.0)^2 \quad = \quad \frac{60{,}000 + 80{,}000}{15.88} + \frac{60{,}000(3)(5)}{305.7}$$

Simplifying,
$$12{,}710 \neq 11{,}780$$

The allowable stress in this column is thus 12,710 lb/in^2, whereas the actual stress is 11,780 lb/in^2. This section is consequently satisfactory.

SUPPLEMENTARY PROBLEMS

A steel bar of solid circular cross-section is 2 in. in diameter. The bar is pinned at each end and subject to axial compression. If the proportional limit of the material is 36,000 lb/in^2 and $E = 30\cdot 10^6$ lb/in^2, determine the minimum length for which Euler's formula is valid. Also, determine the value of the Euler buckling load if the column has this minimum length. *Ans.* 45.5 in., 112,800 lb

If the length of the column in Problem 22 is increased to 80 in., determine the Euler buckling load. *Ans.* 36,400 lb

Determine the slenderness ratio for a steel column of solid circular cross-section 4 in. in diameter and 9 ft long. *Ans.* 108

According to the A.I.S.C. specifications, what is the load-carrying capacity of the column described in Problem 24? The bar is pin-ended. *Ans.* 143,000 lb

Using the Chicago Building Code specifications, determine the load-carrying capacity of the column described in Problem 24. *Ans.* 106,500 lb

Using Euler's theory determine the critical load for a 10 WF 54 column 25 ft long. The bar is pinned at each end. Assume $E = 30\cdot 10^6$ lb/in^2. *Ans.* 342,000 lb

Determine the maximum safe axial compressive load for a 10 WF 54 pin-end column 18 ft long. Use the A.I.S.C. specifications. *Ans.* 214,000 lb

Select a WF section to carry an axial compressive load of 120,000 lb. The bar is pin-ended and is 14 ft long. Use the A.I.S.C. specifications. *Ans.* 8 WF 31

30. A column is made from a 22 ft length of standard welded steel pipe. The outside diameter of the pip is 8.625 in., the inside diameter is 8.071 in. The cross-sectional area of the pipe is 7.265 in^2 an the moment of inertia about a diametral axis is 63.3 in^4. A compressive force with an eccentricity c 1 in. is applied to the pipe. Determine the maximum such eccentric load that may be applied. Use th A.I.S.C. specification. *Ans.* 62,800 lb

31. Select a WF section adequate to support a concentric load of 52,000 lb, together with an eccentri load of 40,000 lb applied 4 in. from the center of the section at a point on the axis of symmetry bi secting the width of the web of the beam. The bar is 18 ft long and has pinned ends. Use the A.I.S.C specifications. *Ans.* 12 WF 40

32. The column shown below is pinned at both ends and is free to expand into the opening at the upper enc The bar is steel, 1 in. in diameter and occupies the position shown at 60°F. Determine the highes temperature to which the column may be heated before it will buckle. Take $\alpha = 6 \cdot 10^{-6}/°F$ and $E = 30 \cdot 1C$ lb/in^2. Neglect the weight of the column. *Ans.* 95.2°F

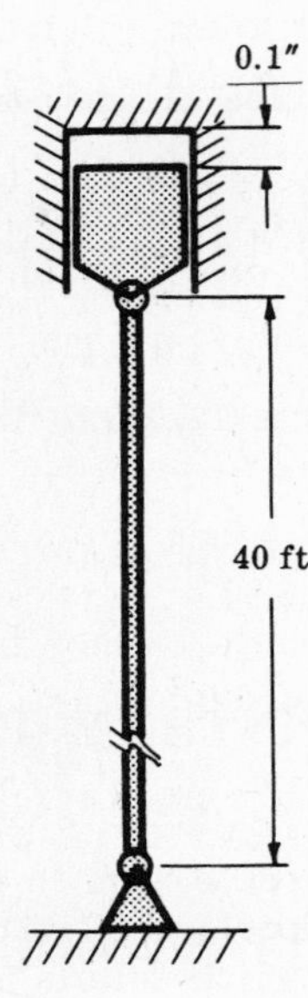

CHAPTER 13

Riveted Joints

INTRODUCTION. Structural members are usually joined together by either rivets or welds. Common applications of riveted joints are to be found in aircraft, pressure vessels, steam boilers, tanks, penstocks, plate girders, trusses, and ship structures.

TYPES OF RIVETED JOINTS. Two common types of riveted joint for joining plates are to be found in practice. They are known as lap joints and butt joints.

LAP JOINTS have the two plates lapped over each other and connected with one or more rows of rivets. For applications see Problems 2-6.

BUTT JOINTS have the two plates butted together and are connected by two cover plates, each main plate and the cover plates being fastened together by one or more rows of rivets. For applications see Problems 7-9.

TYPES OF LAP JOINT. Common types of lap joint are shown in Figure 1 below. They are designated as single riveted lap joints and double riveted lap joints respectively.

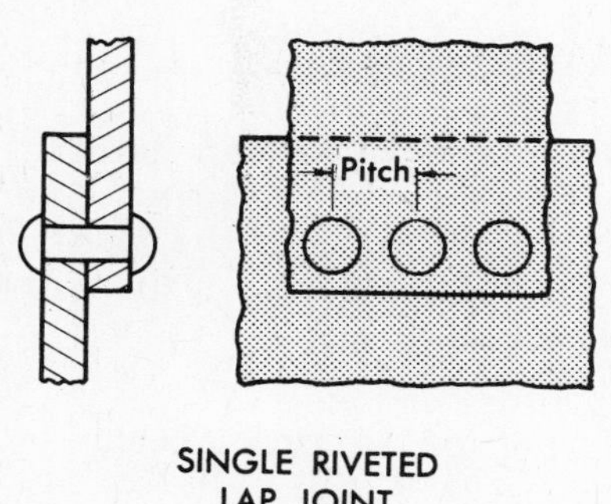

SINGLE RIVETED LAP JOINT

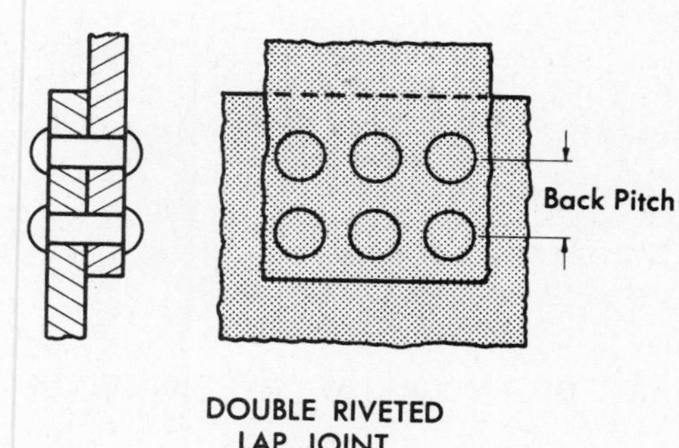

DOUBLE RIVETED LAP JOINT

Fig. 1

TYPES OF BUTT JOINT. Common types of butt joint are shown in Figure 2 below. They are designated as single riveted butt joints and double riveted butt joints respectively. In butt joints, particularly those used on boilers, the cover plates are occasionally of different widths. The short cover plate is always placed on the outside of the boiler or pressure vessel to permit caulking of the joint, hence insuring maximum tightness.

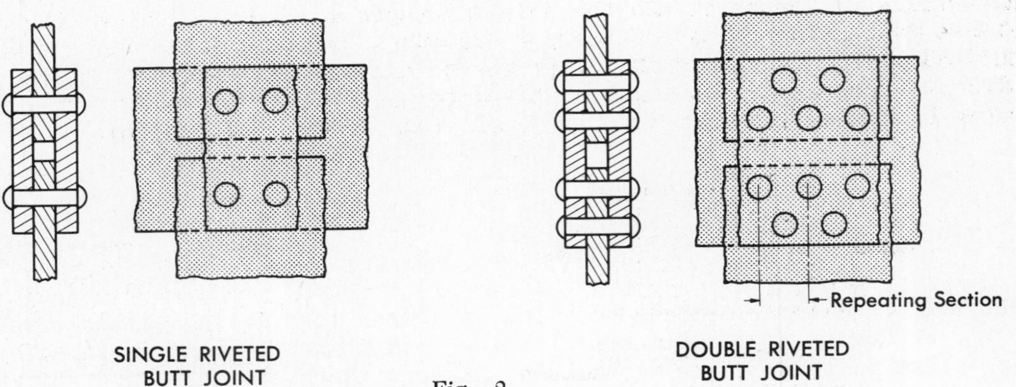

SINGLE RIVETED BUTT JOINT

DOUBLE RIVETED BUTT JOINT

Fig. 2

PITCH. The distance from center to center of rivets lying in a row is termed the *pitch*. The distance is illustrated in Figure 1 for the single riveted lap joint. The pitch may of course vary from one row to another of a joint. The greatest pitch is called the *long pitch*, the smallest distance the *short pitch*.

BACK PITCH. The distance between the center lines of two rows of rivets is called the *back pitch*. This term is illustrated in Figure 1 for the double riveted lap joint. The back pitch varies from $2\frac{1}{2}$ to $3\frac{1}{2}$ times the diameter of the rivets.

REPEATING SECTION. A *repeating section* consists of a group of rivets whose pattern repeats itself along the length of the joint. A typical repeating section is illustrated in Figure 2 for the double riveted butt joint. It is usually more convenient to base all strength calculations upon the strength of a repeating section, rather than upon the strength of the entire length of the joint. For other illustrations see Problems 2 through 9 inclusive.

EFFICIENCY. The ratio of strength of the joint to the strength of an unriveted solid plate of the same length is termed the *efficiency* of the joint. The strength as used in this ratio is usually based upon the ultimate strength of the joint. The working strength differs from this by a factor of safety of 5 or more. For examples see Problems 4, 5, 8, 9.

METHODS OF FAILURE OF RIVETED JOINTS. The principal types of failure are:

a) Shearing of the rivets in either single shear or double shear, as shown in Figure 3. The tendency is to cut through the rivet across the section lying in the plane between the plates it connects.

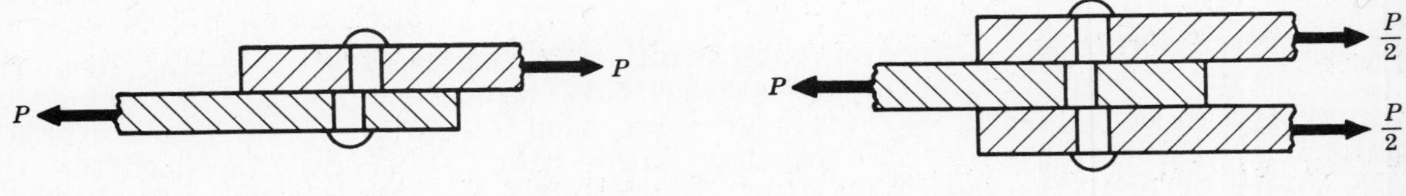

RIVET SUBJECT TO SINGLE SHEAR

RIVET SUBJECT TO DOUBLE SHEAR

Fig. 3

b) Bearing, or crushing of the plate or rivet caused by the pressure between the cylindrical surface of the rivet and the hole, as shown in Figure 4 below. In calculating the resistance to bearing it is customary to use the product of the projected area of the cylindrical hole, i.e. the diameter of the hole times the thickness of the plate, times the ultimate compressive strength of the material. The two traces of this projected area are shown by dotted lines in the diametral plane of the rivet below.

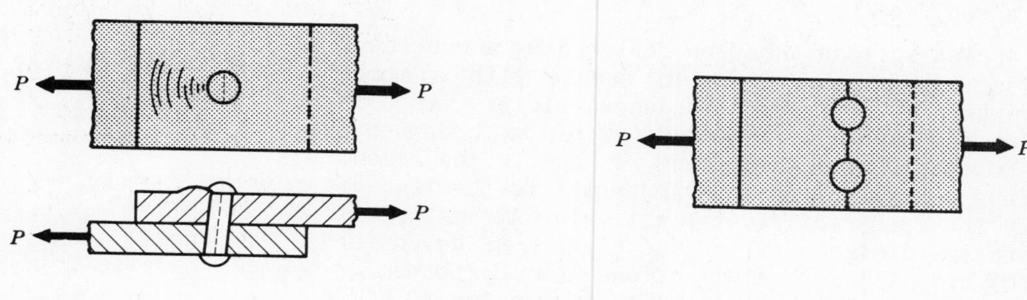

Fig. 4

Fig. 5

c) Tearing of the plate between rivet holes due to lack of tensile strength on a section along a row of rivets. This type of failure is illustrated in Figure 5 above.

d) Tearing along a diagonal. This is illustrated in Figure 6 below. However this type of failure is unlikely to happen if the back pitch is at least $1\frac{3}{4}$ times the rivet diameter. In this book it will be assumed that the back pitch bears at least this ratio to the rivet diameter, and hence we will not investigate this type of failure.

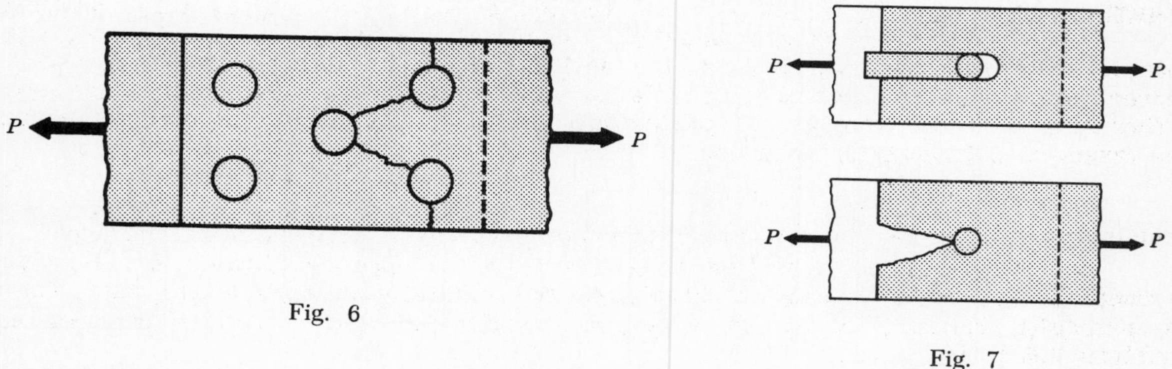

Fig. 6

Fig. 7

e) Shearing of the plate or possibly *tearing of the plate* between a rivet hole and the edge of the plate as shown in Figure 7 above. These types of failure are unlikely to occur if the distance between the center of the rivet hole and the edge of the plate is approximately twice the diameter of the rivet. In this book it will be assumed that this is true, and hence we will not investigate this type of failure.

Consequently, it is necessary in any riveted joint discussed in this book to investigate only three modes of failure: *(a)* shearing, *(b)* bearing, and *(c)* tearing. This is illustrated in Problems 4-9.

ECCENTRIC RIVETED JOINTS. Frequently in structural work riveted joints are required to withstand eccentric loadings. When the line of action of the applied force P does not go through the center of gravity O of the rivet group, as shown in Figure 8 below, it is evident that the plate shown will tend to rotate about some centerpoint in the direction of the moment $M = P \cdot L$.

The tendency of this kind of connection to rotate is often so great that the stresses on the rivets due to direct load are much less than the stresses due to moment. Therefore it is important to design an eccentric joint in such a way that it will be able to resist both the shear due to the vertical load and the moment due to the eccentricity.

The eccentric load P may be replaced by a concentric force P acting through O and a couple with a moment $M = P \cdot L$, where L represents the eccentricity. Then the stress on the rivets will be made up of two components, s_a due to the direct shear, and s_m due to the moment. The stress s_a will be uniform and equal to the load divided by the number of rivets, while the stress s_m will vary with the distance (r) of the rivet from the centerpoint (O) and will act in directions normal to the r-lines, that connect the rivets with the center (O). For an application, see Problem 10.

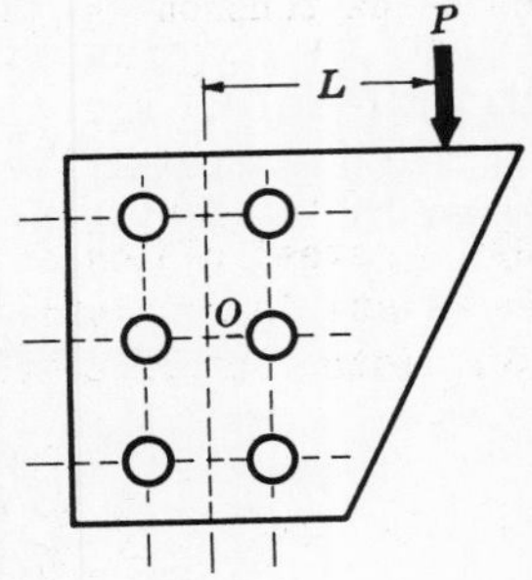

Fig. 8

SOLVED PROBLEMS

1. A load is applied to a steel plate supported by a single pin as shown in Fig. (a) below. Determine the maximum shearing stress in the pin, the maximum bearing stress and the tensile stress acting on the net section of the plate.

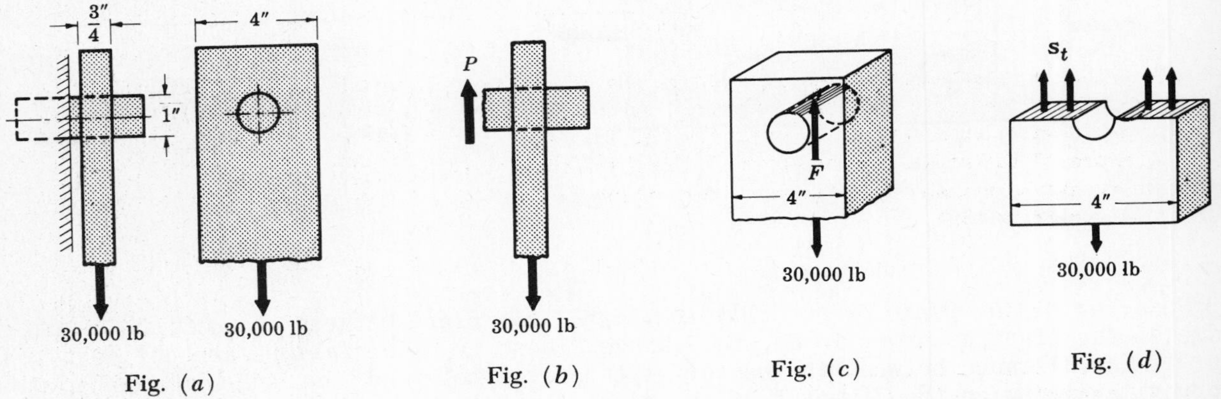

Fig. (a) Fig. (b) Fig. (c) Fig. (d)

The pin is stressed by the shear force shown in Fig. (b) above and denoted by P. The area over which this shear force acts is the cross-sectional area of the pin, i.e. $\frac{1}{4}\pi(1)^2 = 0.785$ in^2. The shearing stress is thus

$$s_s = P/A_s = 30,000/0.785 = 38,200 \text{ lb/in}^2$$

To determine the bearing stress, we determine the force that must be resisted in bearing between the pin and the plate. This is illustrated by the force F shown in Fig. (c) above. For equilibrium $F = 30,000$ lb. The bearing area is given by the projection of the above shaded area on a horizontal plane through the diameter of the pin, i.e. $A_b = (1)(3/4) = 0.75$ in^2. The bearing stress is thus

$$s_b = F/A_b = 30,000/0.75 = 40,000 \text{ lb/in}^2$$

The net section is subject to the tensile stresses s_t shown in Fig. (*d*) above. The area effective in resisting tension is $A_t = (4-1)(3/4) = 2.25\ \text{in}^2$. The tensile stress on the shaded area is thus

$$s_t = 30{,}000/2.25 = 13{,}300\ \text{lb/in}^2$$

2. Two 5/8 in. steel plates are joined by a single-riveted lap joint as shown in Fig. (*a*) below. The pitch of the rivets is 3 in. and the rivet diameter is 7/8 in. The load carried by the plates is 4500 lb per inch of width. Determine the maximum shearing, bearing and tensile stresses in the joint.

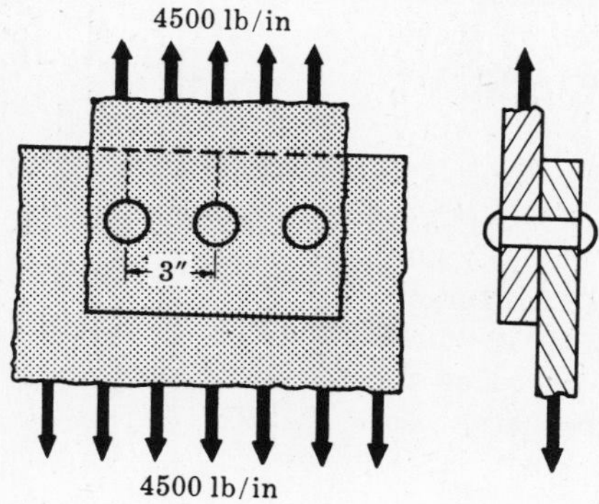

Fig. (*a*)

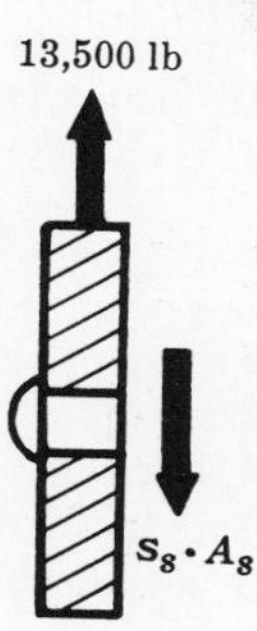

Fig. (*b*)

A typical repeating section lies between the dotted lines shown in Fig. (*a*). The load on such a section is 3(4500) = 13,500 lb. The free-body diagram in Fig. (*b*) illustrates the shear acting in each rivet.

In attempting to pull the joint apart, the cross-section of each rivet is consequently stressed by the shear force shown. Rivet holes are usually 1/16 in. larger in diameter than the rivet and it is customary to assume that the rivet fills the hole completely. Hence the shearing stress is given by

$$s_s = \frac{P}{A_s} = \frac{13{,}500}{\frac{1}{4}\pi(7/8 + 1/16)^2} = 19{,}600\ \text{lb/in}^2$$

To determine the bearing stress, we shall consider a section as shown in Fig. (*c*) below. Bearing action takes place along the curved surfaces indicated by the heavy lines. The bearing area for each rivet is found by considering the projection of this semi-cylindrical curved area on a vertical diametral plane of the rivet. For one rivet this projected area is $A_b = (5/8)(7/8 + 1/16) = 0.585\ \text{in}^2$. Again, the rivet has been assumed to fill the hole, which is 1/16 in. larger than the rivet diameter. The bearing stress is thus

$$s_b = P/A_b = 13{,}500/0.585 = 21{,}200\ \text{lb/in}^2$$

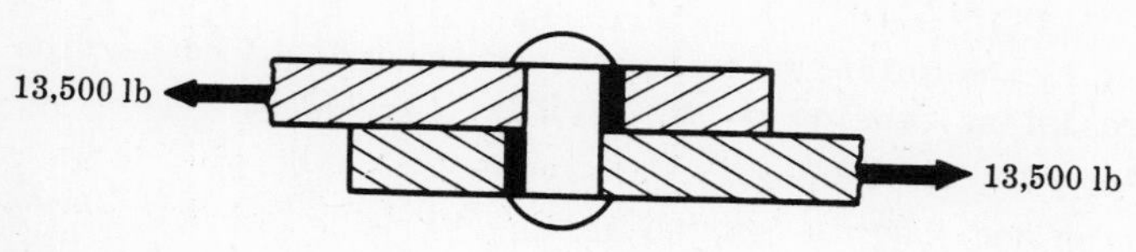

Fig. (*c*)

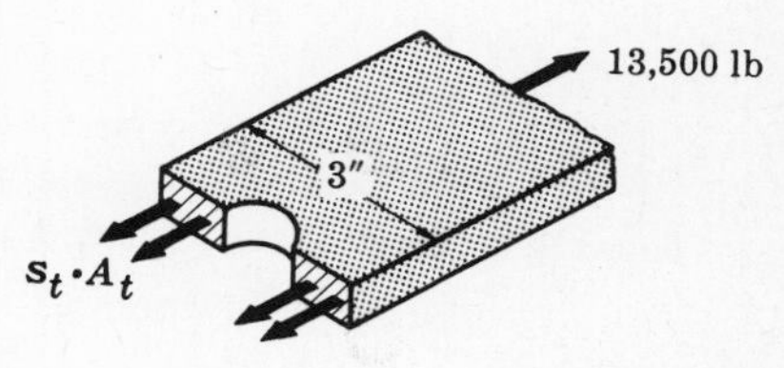

Fig. (*d*)

To determine the maximum tensile stress in the joint, we shall consider the net shaded section shown in Fig. (*d*) above. The effective area in resisting tension is $A_t = (5/8)(3 - 15/16) = 1.29\ \text{in}^2$. The tensile stress on the shaded area is thus

$$s_t = P/A_t = 13{,}500/1.29 = 10{,}500\ \text{lb/in}^2$$

3. Consider the double-riveted lap joint shown in the adjoining diagram where the pitch of both rows of rivets is $3\frac{1}{2}$ in. The rivets are staggered and of 1 in. diameter. Each plate is 5/8 in. thick. The force acting on a repeating section is 20,000 lb. Find the maximum shearing, bearing and tensile stresses in the joint.

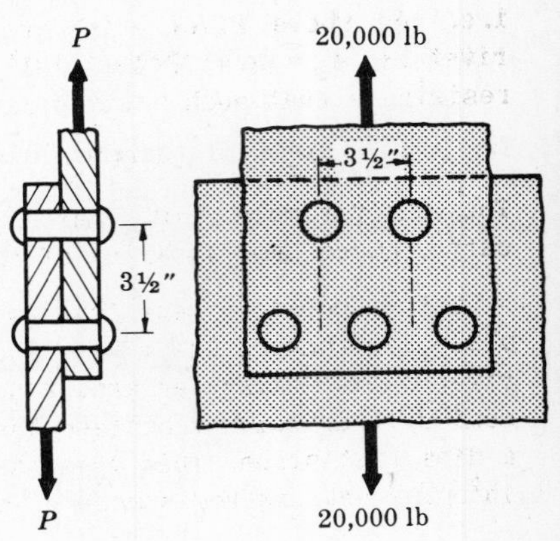

In a repeating section such as that shown between the dotted lines in the adjacent diagram, we have two half rivets and one whole rivet, or two complete rivets. In attempting to pull the joint apart, one cross-section of each of these rivets is loaded in shear (so-called single shear) and the shearing stress is given by

$$s_s = \frac{P}{A_s} = \frac{20,000}{2(\pi/4)(1+1/16)^2} = 11,300 \text{ lb/in}^2$$

It is to be noted that rivet holes are usually 1/16 in. larger in diameter than the rivet and the rivet is assumed to fill the hole completely. Hence the factor of 1/16 in. is added to the denominator of the above fraction used for computing the shearing stress.

To determine the bearing stress we shall consider a section such as that shown in Fig. (*a*) below. The upper and lower halves of each rivet are equally loaded in bearing along the surfaces indicated by heavy lines. As before, there are two complete rivets included in the repeating section. The bearing area of each rivet is found by considering the projection of the curved area on a vertical diametral plane of the rivet. For one rivet this projected area is given by the half-height of the rivet (5/8 in.) times the diameter of the rivet hole (17/16 in.) or $A_b = (5/8)(17/16) = 0.67 \text{ in}^2$, where again the 1/16 in. has been added to the rivet diameter since it is customary to assume that the rivet fills the hole. The bearing stress is consequently

$$s_b = P/A_b = 20,000/2(0.67) = 15,000 \text{ lb/in}^2$$

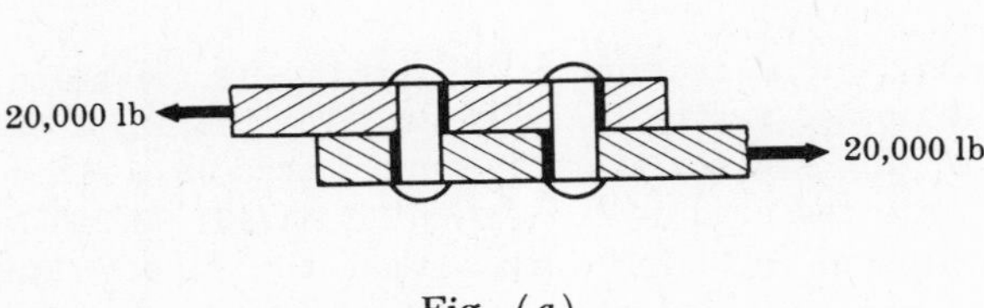

Fig. (*a*)

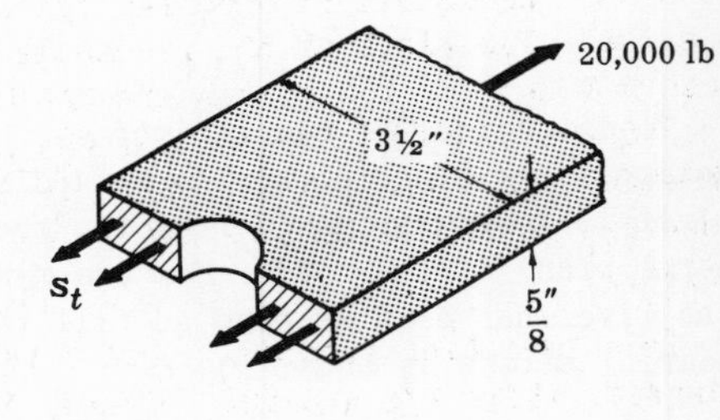

Fig. (*b*)

To determine the maximum tensile stress in the joint, we must consider the net shaded section shown in Fig. (*b*) above. The area effective in resisting tension is $A_t = (5/8)(7/2 - 17/16) = 1.52 \text{ in}^2$. The tensile stress on the shaded area is thus

$$s_t = P/A_t = 20,000/1.52 = 13,100 \text{ lb/in}^2$$

The same result could have been obtained by considering a section through the other row of rivets. In this case we would have deducted two half rivets, which would of course have led to the above result.

4. Consider the single-riveted lap joint shown below. The pitch of the rivets is $2\frac{1}{2}$ in., the plate thickness is $\frac{1}{2}$ in., and the rivets are $\frac{3}{4}$ in. diameter. The ultimate stresses recommended by the A.S.M.E. Boiler Code are: tension 55,000 lb/in², shear 44,000 lb/in², compression 95,000 lb/in². Determine the allowable load on a repeating section and also the efficiency of the joint.

A repeating section is shown between the dotted lines in the diagram. Such a section contains two half rivets, i.e. one whole rivet. The cross-sectional area of this rivet is $A_s = (\pi/4)(3/4 + 1/16)^2 = 0.518$ in^2. The shearing resistance that such a rivet can offer is

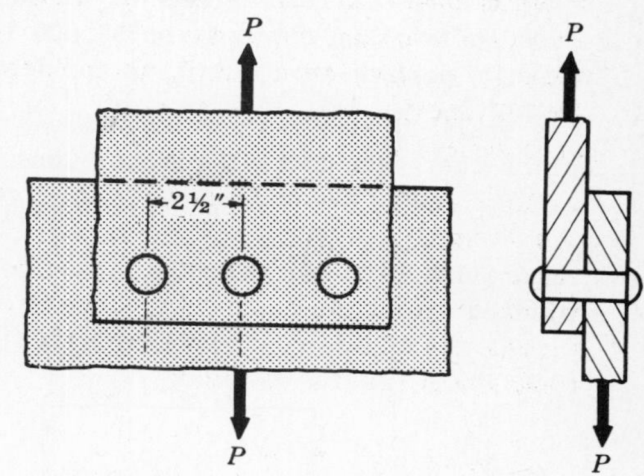

$$s_s A_s = 44{,}000(0.518) = 22{,}800 \text{ lb}$$

Thus the ultimate load that the repeating section can carry, based upon shearing of the rivet, is 22,800 lb.

Let us consider next the resistance to crushing of the plate in front of one rivet, i.e. a bearing failure. The rivet bears against an area equal to the projection of the semi-cylindrical area between the rivet and the hole upon a diametral plane. This area is $A_b = \frac{1}{2}(3/4 + 1/16) = 0.406$ in^2. The bearing resistance of one rivet is thus

$$s_b A_b = 95{,}000(0.406) = 38{,}600 \text{ lb}$$

Thus the ultimate load that the repeating section can carry, based upon bearing strength, is 38,600 lb.

Lastly, we shall determine the load that will tear the plate between rivet holes. The net area resisting such tension is $A_t = \frac{1}{2}(5/2 - 13/16) = 0.845$ in^2. The resistance to tearing is thus

$$s_t A_t = 55{,}000(0.845) = 46{,}500 \text{ lb}$$

The mode of failure that offers the least resistance is thus shearing of the rivet, at a load of 22,800 lb acting on the repeating section. Consequently this is the allowable load on a repeating section. In an actual design this value should of course be divided by a safety factor, ranging from 3 to 5.

The tensile resistance of a solid plate $2\frac{1}{2}$ in. wide and $\frac{1}{2}$ in. thick is $(2\frac{1}{2})(\frac{1}{2})(55{,}000) = 68{,}600$ lb. The efficiency of the joint is thus $(22{,}800/68{,}600)(100) = 33.2\%$.

. Consider the double-riveted lap joint shown in the adjacent diagram. The pitch of the rivets is 3 in., the plate thickness 5/8 in., and the rivets are 3/4 in. in diameter. According to the A.S.M.E. Boiler Code the ultimate stresses recommended are: tension 55,000 lb/in^2, shear 44,000 lb/in^2, compression 95,000 lb/in^2. Determine the efficiency of this joint.

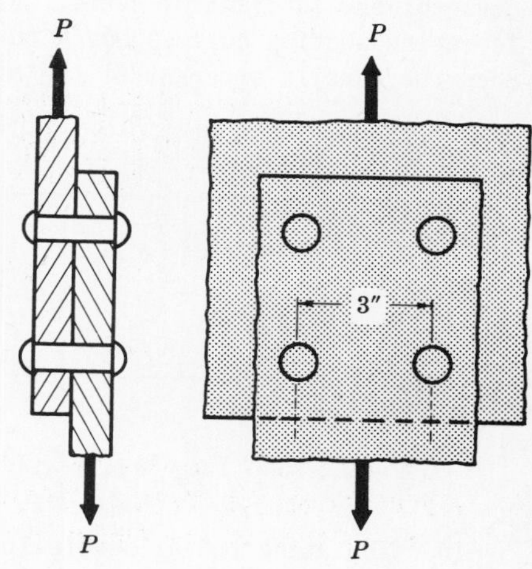

In a repeating section, shown between dotted lines in the diagram, there are two full rivets. The shearing resistance of a repeating section is

$$P_s = 2[(\pi/4)(3/4 + 1/16)^2](44{,}000) = 46{,}000 \text{ lb}$$

The bearing resistance of a repeating section is

$$P_b = 2[(3/4 + 1/16)(5/8)](95{,}000) = 96{,}500 \text{ lb}$$

The load that will tear the plate between rivet holes is

$$P_t = [(5/8)(3 - 13/16)](55{,}000) = 75{,}400 \text{ lb}$$

The minimum failure load is thus 46,000 lb, corresponding to a shearing type failure.

The tensile resistance of a solid plate 3 in. wide and $\frac{1}{2}$ in. thick is $(3)(\frac{1}{2})55{,}000 = 82{,}500$ lb. The efficiency of the joint is thus $(46{,}000/82{,}500)(100) = 55.8\%$.

Consider the design of a boiler having a double-riveted longitudinal lap-type seam. The diameter of the boiler is 85 in., the plate thickness 3/4 in., the width of a repeating section is 3 in., and 3/4 in.

rivets are used. Determine the allowable internal pressure that may exist within the boiler. Use the ultimate stresses recommended in the A.S.M.E. Boiler Code as follows: tension 55,000 lb/in^2, shear 44,000 lb/in^2, and compression 95,000 lb/in^2, together with a safety factor of 5 for all types of stress. A portion of the boiler and its longitudinal seam appears in Fig. (*a*) below.

We shall first determine the allowable force P that may act on a repeating section. In a repeating section there are two full rivets. Consequently the allowable load based upon the shear resistance is

$$P_s = 2[(\pi/4)(3/4 + 1/16)^2](44,000/5) = 9150 \text{ lb}$$

The allowable load based upon the bearing resistance, again considering two full rivets and their projected areas, is

$$P_b = 2[(3/4)(3/4 + 1/16)](95,000/5) = 23,200 \text{ lb}$$

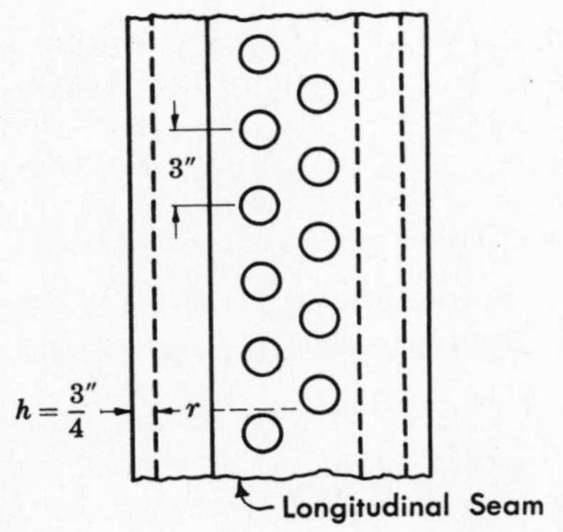

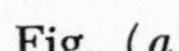

Fig. (*a*)

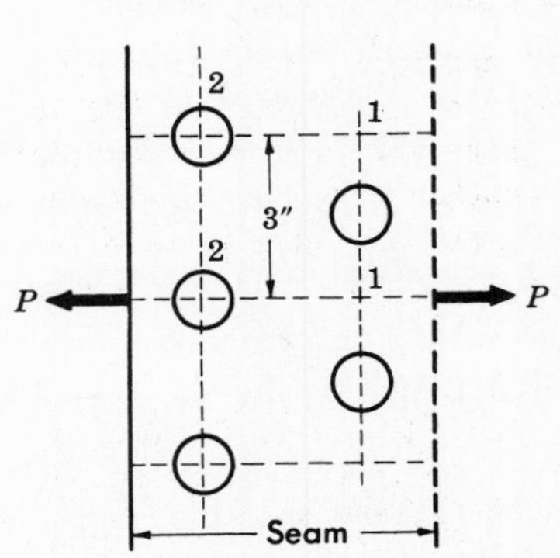

Fig. (*b*)

The load that will cause tensile failure between holes is determined as follows. Examination of a repeating section indicates that two net sections should be investigated. These are designated as 1-1 and 2-2 in Fig. (*b*) above. The forces P represent the circumferential force arising in the boiler because of the internal pressure. These forces were discussed in detail in Problem 1, Chapter 3. The repeating section corresponding to section 1-1 appears as in Fig. (*c*) below. The allowable load based upon the tensile strength of the section 1-1 is

$$P_t' = [(3/4)(3 - 13/16)](55,000/5) = 18,000 \text{ lb}$$

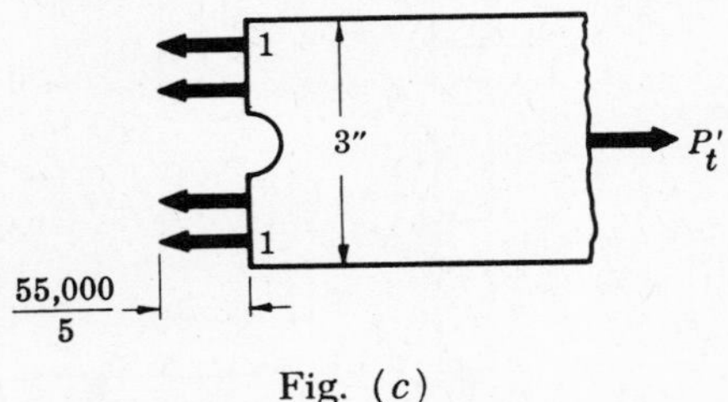

Fig. (*c*)

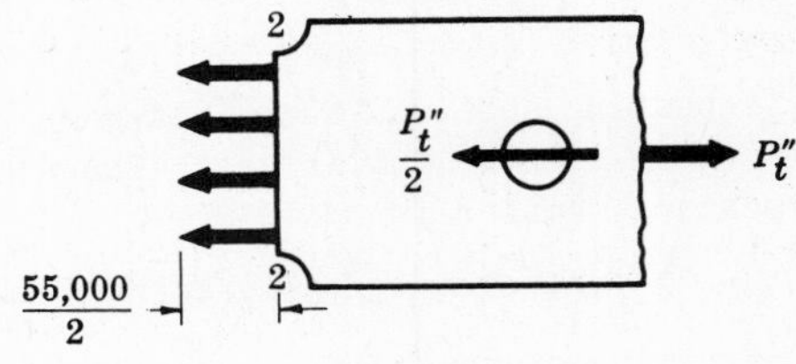

Fig. (*d*)

In calculating the allowable load based upon the tensile strength of section 2-2, shown in Fig. (*d*) above, it must be remembered that the rivet in the back row withstands a force $P_t''/2$ in shear. Consequently the free-body diagram of the repeating section corresponding to section 2-2 appears as in Fig. (*d*) above. For equilibrium

$$\Sigma F_h = P_t'' - \frac{P_t''}{2} - \frac{55,000}{2}(\frac{3}{4})(3 - \frac{13}{16}) = 0 \qquad \text{or} \qquad P_t'' = 36,000 \text{ lb}$$

Thus the governing load on the repeating section is 9150 lb, as determined by the shearing resistance. This load of 9150 lb, which is in the circumferential direction, corresponds to a circumferential stress of $9150/[3(3/4)] = 4060$ lb/in^2. From Problem 1, Chapter 3 the circumferential stress

s_c in the boiler is given by

$$s_c = pr/h$$

where p represents the internal pressure, r the radius of the boiler and h the wall thickness. The circumferential stress of 4060 lb/in^2 corresponds to a pressure p that may be found by substituting in the last equation. Then $4060 = p(85/2)/(3/4)$ or $p = 72$ lb/in^2.

7. Consider the single-riveted butt joint shown in Fig. (a) below, where the rivet pitch is 3 in., the main plates are each 1/2 in. thick and the cover plates are each 3/8 in. thick. The rivet diameter is 3/4 in. According to the A.S.M.E. Boiler Code the ultimate stresses recommended are: tension 55,000 lb/in^2, shear 44,000 lb/in^2, compression 95,000 lb/in^2. Determine the allowable load on a repeating section, based upon a safety factor of 5.

The load applied to the upper plate is transmitted through that plate to the upper rivets, then to the two outer or cover plates, and then through the lower rivets and into the lower plate. We shall determine three values of the load-carrying capacity of a repeating section, one based upon shearing strength, another upon bearing strength, and a third upon tensile resistance. The allowable load is then the minimum of these three values.

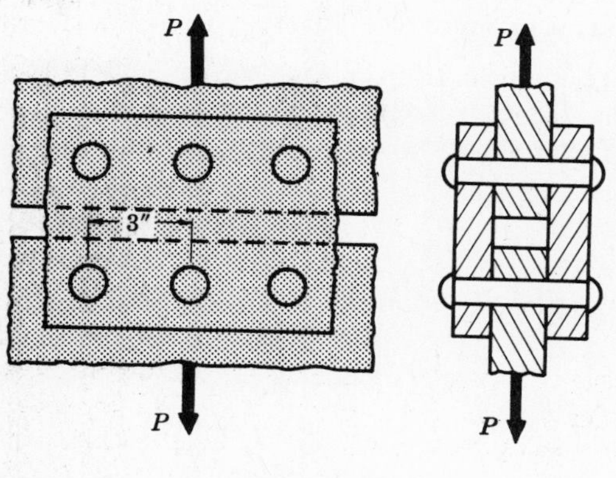

Fig. (a)

Fig. (b)

The allowable load based upon the shearing strength of the rivets is found by considering the free-body diagram of the upper main plate shown in Fig. (b) above. It is to be noted that this rivet is in a state of *double shear*, i.e. shearing action occurs along each of two surfaces. Using the ultimate shearing strength of 44,000 lb/in^2 together with a safety factor of 5, we find

$$\tfrac{1}{2}P_s = \tfrac{1}{4}\pi(3/4 + 1/16)^2(44,000/5) \quad \text{or} \quad P_s = 9140 \text{ lb}$$

The allowable load based upon the bearing strength is found by considering the free-body diagram of the upper main-plate shown in Fig. (c) below. The load P_b causing the allowable bearing stress is

$$P_b = \tfrac{1}{2}(3/4 + 1/16)(95,000/5) = 7700 \text{ lb}$$

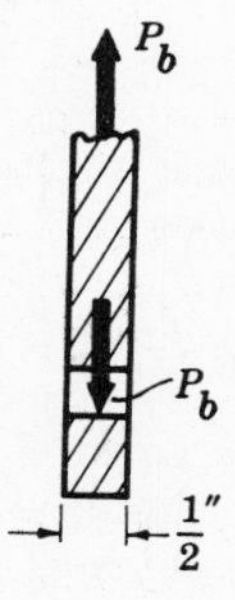

Fig. (c)

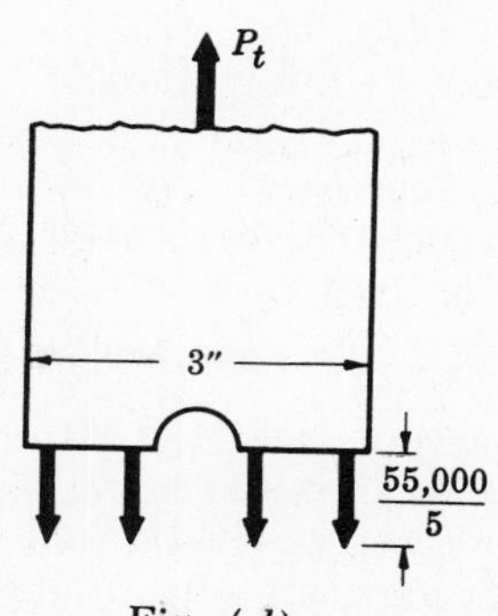

Fig. (d)

Note that it is unnecessary to consider the possibility of a bearing failure in the cover plates because the combined thickness of these two plates (3/4 in.) provides a greater bearing area than exists in the center or main plate.

The allowable load based upon the tensile resistance may be found by considering the free-body diagram of the upper main plate shown in Fig. (*d*) above. The load P_t causing tearing between rivets in the main plate is thus

$$P_t \;=\; \tfrac{1}{2}(3-13/16)(55{,}000/5) \;=\; 12{,}000 \text{ lb}$$

It is to be noted that it is unnecessary to consider tearing of the cover plates since the combined thickness of these two plates provides a greater area to resist tension than exists in the main plate.

The allowable load on a repeating section is thus 7700 lb, as determined by the bearing capacity.

8. Consider the double-riveted butt joint shown in Fig. (*a*) below, where the rivet pitch is 3 in., the main plates are each 1/2 in. thick, and the cover plates are each 3/8 in. thick. The rivet diameter is 3/4 in. According to the A.I.S.C. (American Institute of Steel Construction) Code for Buildings the allowable stresses are: tension 20,000 lb/in^2, shear 15,000 lb/in^2, compression 40,000 lb/in^2. Determine the allowable load on a repeating section and the efficiency of the joint.

A repeating section is shown between the dotted lines in Fig. (*a*). We shall determine one value of the allowable load based upon shearing strength, another based upon bearing strength, and a third based upon tensile resistance. The minimum of these three values is then the allowable load.

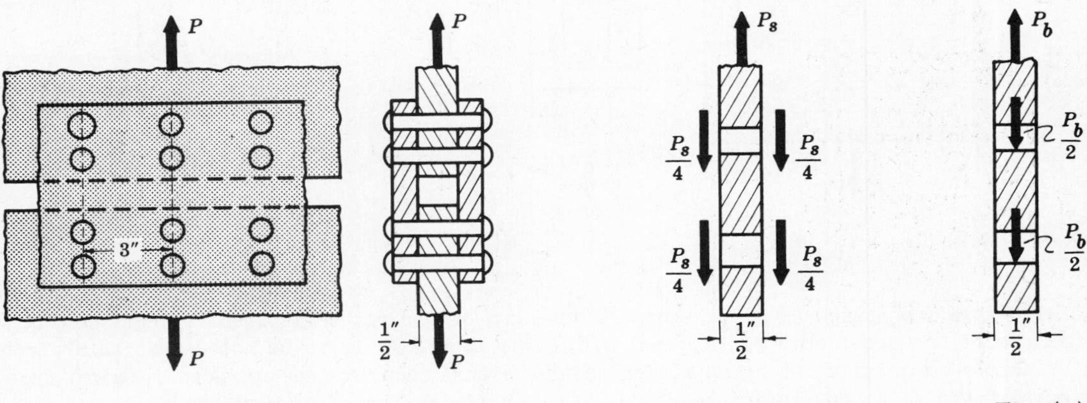

Fig. (*a*) Fig. (*b*) Fig. (*c*)

The allowable load based upon the shearing strength of the rivets is found by considering the free-body diagram of the upper main plate shown in Fig. (*b*) above. Again, as in Problem 7, each rivet is subject to *double shear*. Since the allowable shear stress is 15,000 lb/in^2 we have

$$\tfrac{1}{4}P_s \;=\; \tfrac{1}{4}\pi(3/4+1/16)^2(15{,}000) \qquad \text{or} \qquad P_s = 31{,}200 \text{ lb}$$

The allowable load based upon the bearing strength is found by considering the free-body diagram of the upper main plate shown in Fig. (*c*) above. The load P_b causing the allowable bearing stress is

$$\tfrac{1}{2}P_b \;=\; \tfrac{1}{2}(3/4+1/16)(40{,}000) \qquad \text{or} \qquad P_b = 32{,}500 \text{ lb}$$

The load necessary to cause the allowable bearing stress in the cover plates is of course in excess of 32,500 lb because the combined thickness of the two cover plates (3/4 in.) is greater than the $\tfrac{1}{2}$ in. thickness of the main plate.

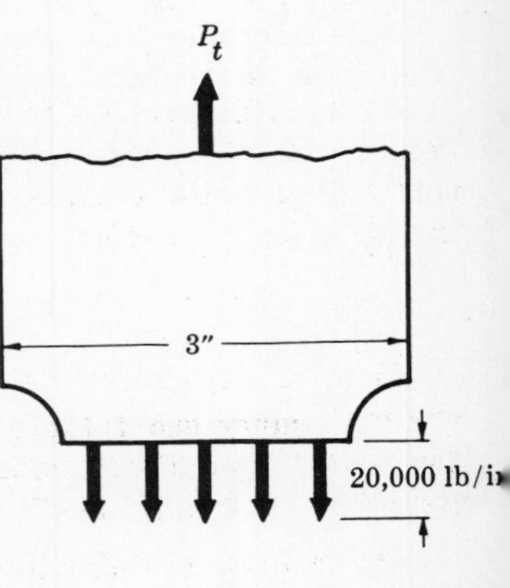

Fig. (*d*)

The allowable load based upon the tensile resistance may be found by considering the free-body diagram of the upper main plate shown in Fig. (d) above. The load P_t causing tearing between rivets in the main plate is thus

$$P_t = \tfrac{1}{2}(3 - 13/16)(20{,}000) = 21{,}200 \text{ lb}$$

It is unnecessary to investigate the load that would cause tearing of the cover plates since it would be in excess of 21,200 lb. This is because the combined thickness of the two cover plates is greater than that of the main plate.

The allowable load on a repeating section is thus 21,200 lb, as determined by the tensile strength.

If there were no rivet holes in the main plate its allowable tensile strength would be

$$\tfrac{1}{2}(3)(20{,}000) = 30{,}000 \text{ lb}$$

The efficiency of the joint is defined to be the ratio of the allowable load to the allowable tensile strength of the solid plate of the same width and thickness, or (21,200/30,000)(100) = 70.8 %.

9. Consider the double-riveted butt joint shown in Fig. (a) below, in which one cover plate is w[illegible]er than the other. This construction is frequently employed in fabricating tight joints, and calking is then applied to the edges of the narrow plate. The rivet pitch is 3 in. in the inner row and 6 in. in the outer row. The main plates are 1/2 in. thick, the cover plates 5/16 in. thick, and the rivet diameter is 3/4 in. According to the A.I.S.C. Code for Buildings the allowable stresses are: tension 20,000 lb/in^2, shear 15,000 lb/in^2, compression 40,000 lb/in^2. Determine the allowable load on a repeating section and the efficiency of the joint.

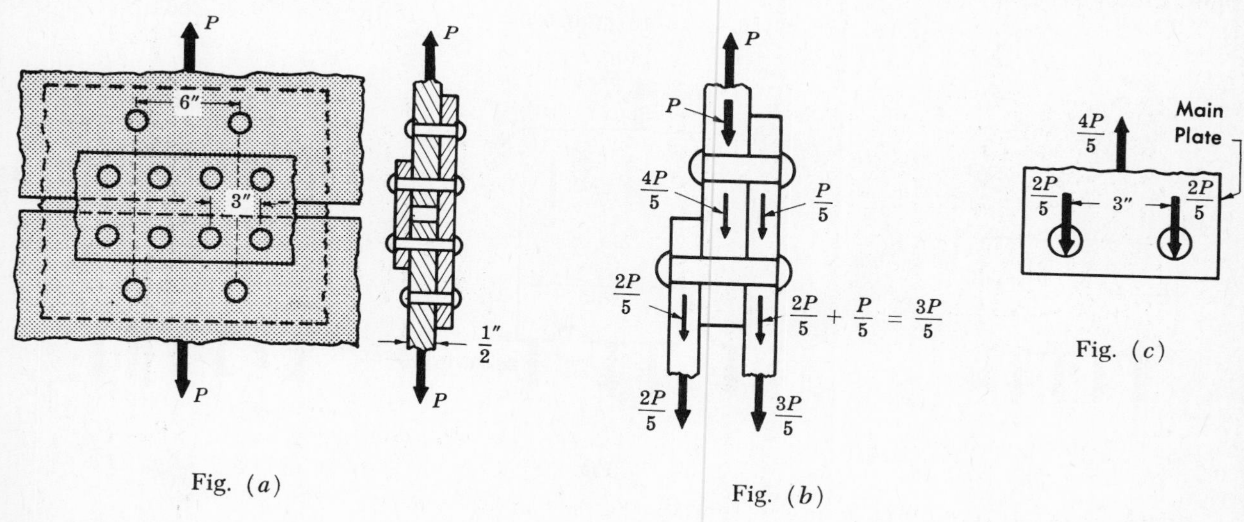

Fig. (a)

Fig. (b)

Fig. (c)

A repeating section is shown between the dotted lines in Fig. (a) above. For each main plate we have two rivets in double shear and one in single shear, i.e. five areas to resist shearing. We shall assume that the shearing resistances of all of these areas are equal and thus the allowable load based upon the shearing resistance is

$$P_s = 5[\tfrac{1}{4}\pi(3/4 + 1/16)^2(15{,}000)] = 39{,}000 \text{ lb}$$

Since the shearing resistances of the rivets were assumed to be equal, the single rivet in the outer row will carry one-fifth of the total load P and each of the two inner rivets will carry two-fifths of the total load. The distribution of the load P through the upper half of the joint may then be represented as in Fig. (b) above.

Let us next consider bearing resistance. Evidently the critical surface exists in the inner rivet row of the main plate, since the combined thicknesses of the cover plates are greater than that of the main plate. The bearing force acting on each rivet of the inner row is seen from the free-body diagram of the main plate, Fig. (c) above, to have the value $2P/5$. Note that the main plate shown in this diagram has been cut *below* the outer row of rivets, hence the force in the plate is $4P/5$, and not P. The allowable bearing load is thus

$$2P_b/5 \;=\; \tfrac{1}{2}(3/4+1/16)(40{,}000) \qquad \text{or} \qquad P_b \;=\; 40{,}600 \text{ lb}$$

Next, we shall investigate several allowable loads, each based upon the tensile resistance of the various elements of a repeating section. First, the tension across the net section of the main plate at the outer row of rivets is to be considered. This may be evaluated by considering the free-body diagram in Fig. (d) below.

$$P_t' \;=\; \tfrac{1}{2}(6-13/16)(20{,}000) \;=\; 58{,}000 \text{ lb}$$

Secondly, the tension across the net section of the main plate at the inner row of rivets is to be considered. This tension may be shown as in Fig. (e) below. It is to be noted that each of the forces $P_t''/10$ shown represents the bearing force exerted by a half rivet (of the outer row) upon the main plate. For equilibrium,

$$\Sigma F_v \;=\; P_t'' - P_t''/5 - \tfrac{1}{2}[6-2(13/16)](20{,}000) \;=\; 0 \qquad \text{or} \qquad P_t'' = 54{,}600 \text{ lb}$$

Lastly, the tension across the cover plate is to be investigated. Since the longer cover plate has been found to carry 3/5 of the load, the stresses in it will be greater than in the shorter cover plate. The critical section of the longer cover plate occurs at the inner row of rivets, since that section has the minimum area. A free-body diagram appears as in Fig. (f) below. Each of the forces $P_t'''/5$ shown represents the bearing action of a rivet on this cover plate. For equilibrium,

$$\Sigma F_v \;=\; 3P_t'''/5 - (5/16)[6-2(13/16)](20{,}000) \;=\; 0 \qquad \text{or} \qquad P_t''' = 45{,}600 \text{ lb}$$

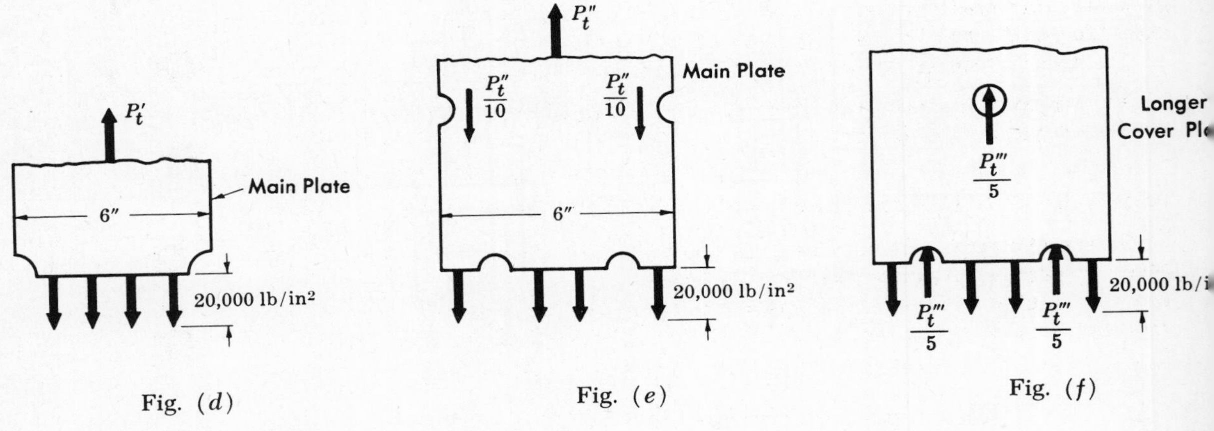

Fig. (d) Fig. (e) Fig. (f)

Thus the allowable load on a repeating section is 39,000 lb, as determined by the shearing resistance. If the main plate were solid across a 6 in. width, its load carrying capacity would be

$$\tfrac{1}{2}(6)(20{,}000) \;=\; 60{,}000 \text{ lb}$$

The joint efficiency is thus $(39{,}000/60{,}000)(100) = 65.0\%$.

10. An eccentrically loaded riveted joint is shown in the adjoining diagram. The applied load is 18,000 lb, acting with an eccentricity of 8 in. from the geometric center of the group of six 3/4 in. rivets. Determine the maximum shearing stress in the rivets.

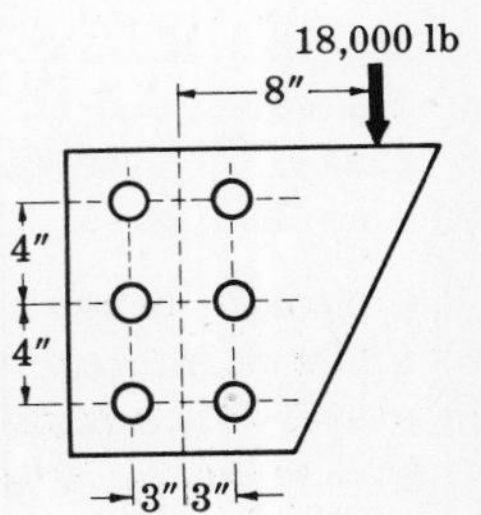

In Fig. (a) below, the free-body diagram shows the resistance offered by the rivets in contact with the plate to the applied force. This resistance consists of a force of 18,000 lb acting through the centroid of the rivet group together with a couple of magnitude (18,000)(8) = 144,000 lb-in. The resisting force of 18,000 lb arises because of the vertical shearing effect of the load; the couple arises because of the twisting effect of the eccentric load.

The upward resisting force of 18,000 lb is the resultant of six vertical shearing forces, all assumed to be of equal value (3000 lb), and distributed over the six rivets. These six shears are shown in Fig. (*b*) below.

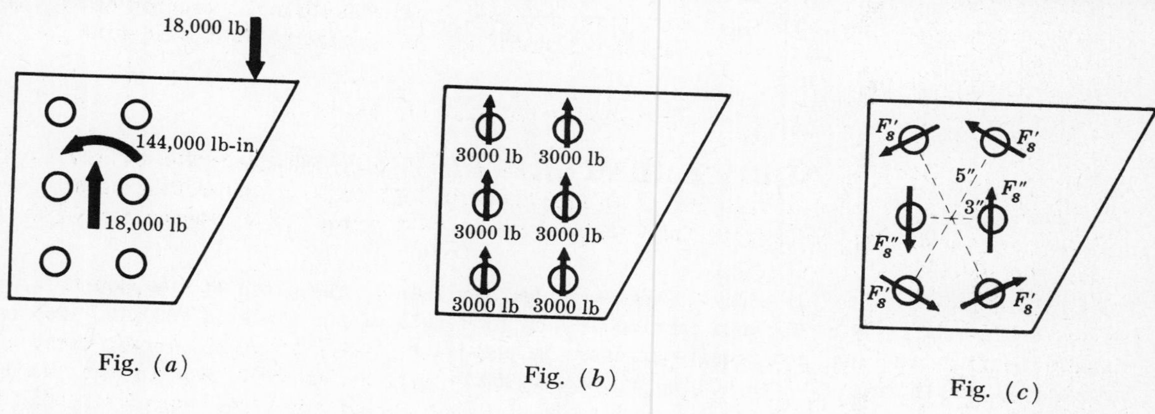

Fig. (*a*) Fig. (*b*) Fig. (*c*)

The resisting moment of 144,000 lb-in is the moment resultant of additional shearing forces acting on the rivets, each being in a direction perpendicular to the radial line from the centroid of the rivet group to the rivet under consideration. These shearing forces are denoted by F_s' and F_s'' and appear as in Fig. (*c*) above. Because of symmetry all four corner rivets are subject to the same force F_s' as shown. The action of a group of rivets in resisting an eccentric load is analogous to the action of a circular shaft resisting torsion (see Chapter 5). Hence it is reasonable to assume that the shear forces F_s' and F_s'' are proportional to the distances of the respective rivets from the centroid of the group. Then we may write $F_s'/5 = F_s''/3$ or $F_s'' = (3/5)F_s'$.

The sum of the moments of these six shear forces must be equal to the moment of the resultant couple of 144,000 lb-in. Then

$$4F_s'(5) + 2F_s''(3) = 144,000$$

Substitute $F_s'' = (3/5)F_s'$ and solve to obtain $F_s' = 6100$ lb, $F_s'' = 3660$ lb.

The resulting resisting forces in the rivets are found by superposition of these forces with the 3000 lb vertical shearing forces found earlier. These shearing forces are of course vector quantities and the resultant shear force on each rivet must be determined by the usual parallelogram law for vector addition. These resultants R_1 through R_4 are shown in Fig. (*d*) below.

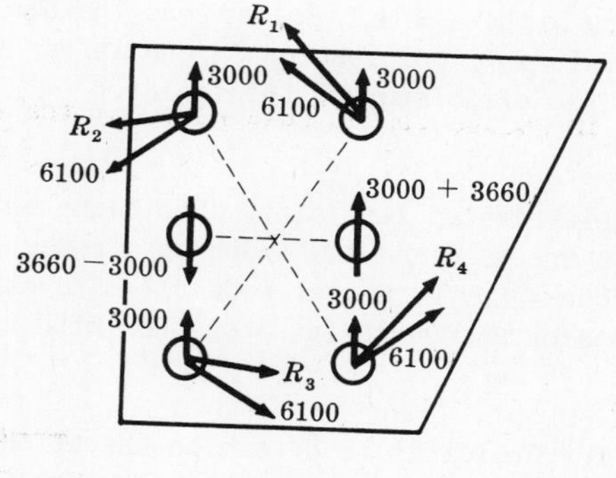

Fig. (*d*)

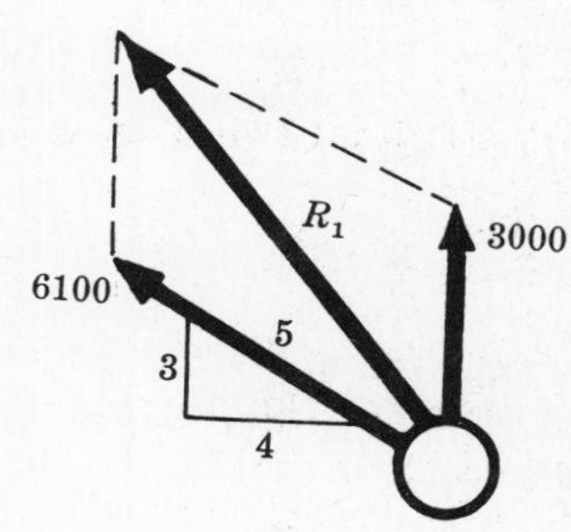

Fig. (*e*)

By inspection of the above vector additions it is evident that R_1 and R_4 are equal in magnitude. Also, this magnitude exceeds that of R_2 or R_3. An enlarged sketch of the forces acting on the upper right-hand rivet appears in Fig. (*e*) above. R_1 is found either analytically or graphically to be 8250 lb. This value exceeds the vertical shear force of 6660 lb acting on the central rivet in the right-hand column and hence is the maximum shearing force acting on any rivet.

The maximum shearing stress is thus given by $(s_s)_{max} = R_1/A_s$ where A_s denotes the rivet area, the rivet as usual being assumed to fill the hole, which is 1/16 in. larger than the rivet itself. Thus we have for the maximum shearing stress

$$(s_s)_{max} = \frac{8250}{\frac{1}{4}\pi(3/4 + 1/16)^2} = 15,900 \text{ lb/in}^2$$

SUPPLEMENTARY PROBLEMS

11. Two $\frac{1}{2}$ in. steel plates are joined by a single-riveted lap joint. The pitch of the rivets is $2\frac{1}{2}$ in. and the rivet diameter is $\frac{3}{4}$ in. The load carried by a $2\frac{1}{2}$ in. length of the plate is 7500 lb. Determine the maximum shearing, bearing, and tensile stresses in the joint.
Ans. s_s = 14,400 lb/in^2, s_b = 18,400 lb/in^2, s_t = 8850 lb/in^2

12. A boiler 3 ft in diameter is made of 1/4 in. steel plate. The longitudinal seam is joined by a single row of 5/8 in. rivets, spaced 2 in. center to center. The allowable unit stresses are: in tension 16,000 lb/in^2, in shear 12,000 lb/in^2, in compression 24,000 lb/in^2. What is the efficiency of the joint and what maximum pressure should be permitted in the boiler? Assume the rivet holes to be 1/8 in. larger than the rivets. *Ans.* Efficiency = 46.0 %, Internal pressure = 51 lb/in^2

13. A cylindrical tank 25 in. in diameter is subject to an internal pressure of 280 lb/in^2. The longitudinal seam of the tank is a double-riveted lap joint in which the pitch of the rivets in both rows is 3 in. The rivets are staggered and of diameter 3/4 in. The wall thickness of the tank is 5/8 in. Determine the maximum shearing, bearing, and tensile stresses in the joint.
Ans. s_s = 10,100 lb/in^2, s_b = 10,300 lb/in^2, s_t = 7700 lb/in^2

14. Two 5/8 in. plates are joined by a single-riveted lap joint. The pitch of the rivets is 3 in. and the rivet diameter is 7/8 in. The ultimate stresses suggested by the A.S.M.E. Boiler Code are: tension 55,000 lb/in^2, shear 44,000 lb/in^2, compression 95,000 lb/in^2. Determine the allowable load on a repeating section using a safety factor of 5 and also the efficiency of the joint.
Ans. 6100 lb, 29.6 %

15. Two plates 5/8 in. thick are joined by a double-riveted lap joint. The pitch of both rows of rivets is $3\frac{1}{2}$ in., and the rivets are 1 in. in diameter. According to the A.S.M.E. Boiler Code the ultimate stresses recommended are: tension 55,000 lb/in^2, shear 44,000 lb/in^2, compression 95,000 lb/in^2. Determine the ultimate strength of a repeating section as well as the efficiency of this joint.
Ans. 78,000 lb, 65.0 %

16. The longitudinal seam of a boiler consists of a double-riveted lap joint. The diameter of the boiler is 120 in., the plate thickness is 7/8 in., 1 in. rivets are used and the width of a repeating section is 3 in. Determine the allowable internal pressure if the A.S.M.E. Boiler Code ultimate stresses of tension 55,000 lb/in^2, shear 44,000 lb/in^2, and compression 95,000 lb/in^2, together with a safety factor of 5 are used in the design. *Ans.* 87 lb/in^2

17. Re-examine Problem 7 but use now the A.I.S.C. Code for Buildings to determine the allowable load on a repeating section. According to this code the allowable stresses are: tension 20,000 lb/in^2, shear 15,000 lb/in^2, compression 40,000 lb/in^2. *Ans.* 15,600 lb

18. The longitudinal seam of a boiler consists of a double-riveted butt joint. The diameter of the boiler is 120 in., the plate thickness is 7/8 in., 1 in. rivets are used and the width of a repeating section is 3 in. The cover plates are each 1/2 in. thick and both are of the same width. Determine the allowable

internal pressure if the A.S.M.E. Boiler Code ultimate stresses of tension 55,000 lb/in^2, shear 44,000 lb/in^2, and compression 95,000 lb/in^2, together with a safety factor of 5 are used. Compare this result with that obtained for the same boiler in Problem 16, where a double-riveted lap joint was considered. *Ans.* 104 lb/in^2

19. Two 15 in. by 3/8 in. steel plates are to be joined by a lap joint using 7/8 in. rivets. The allowable stresses are: tension 8000 lb/in^2, shear 13,500 lb/in^2, compression 27,000 lb/in^2. Design a joint of maximum efficiency. *Ans.* A joint utilizing 12 rivets arranged in six rows as 1:2:3:3:2:1. The joint will withstand 94,500 lb at an efficiency of 93.3%.

20. Consider the eccentrically loaded riveted joint shown in Fig. (*a*) below. The rivets are each 3/4 in. diameter. Determine the maximum shearing stress in the rivets. *Ans.* 8350 lb/in^2

21. Consider the eccentrically loaded riveted joint shown in Fig. (*b*) below. The rivets are each 5/8 in. diameter. Determine the maximum shearing stress in the rivets. *Ans.* 11,000 lb/in^2

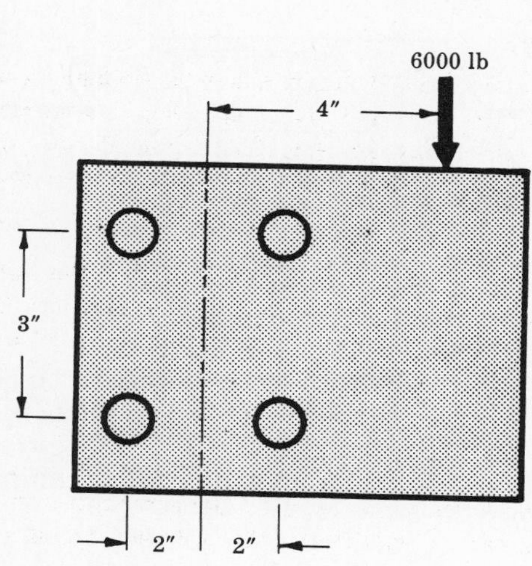

Fig. (*a*) Prob. 20

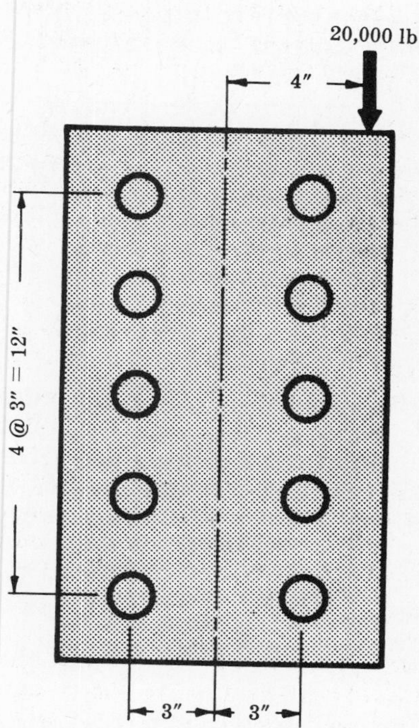

Fig. (*b*) Prob. 21

CHAPTER 14

Welded Connections

TYPES OF WELDS. Two common types of welded joints for joining plates are to be found in structural practice. They are known as butt welds and fillet welds. These are illustrated in Figure 1 below. Butt welds are also sometimes called Vee-welds. Butt welds may act only in tension or compression, whereas fillet welds undergo shear as well as tension or compression and frequently bending in addition. Welding is done either by an electric arc process or by use of gas. Electric arc welding is by far the more commonly used of these two methods.

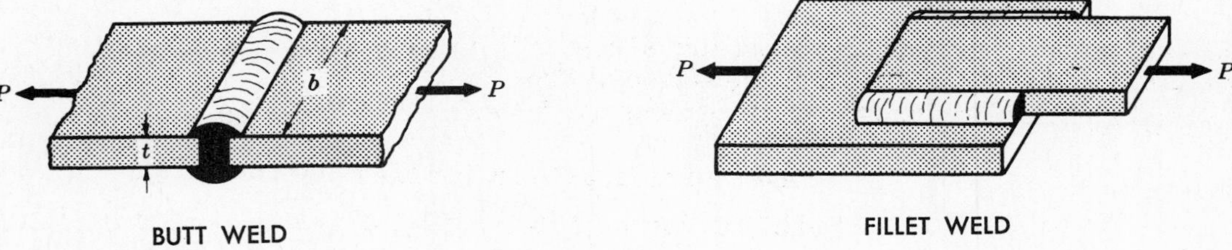

Fig. 1

STRENGTHS OF BUTT WELDS. The strength of the butt weld shown above is assumed to be equal to the net cross-sectional area through the weld multiplied by the allowable working stress in either tension or compression for the weld material. The net area is taken to be the length of the weld times the thickness of the thinner of the two plates being joined. Consequently we have

$$P = s_w\, b\, t$$

where s_w denotes the working stress, t the plate thickness, and b the plate width. This is illustrated in Problem 1.

STRENGTHS OF FILLET WELDS. Before we can calculate the strength of the fillet weld shown above, it is first necessary to define several characteristic dimensions of the cross-section of the weld. For the cross-section appearing in the adjacent diagram, the weld has equal length *legs* and the minimum cross-sectional dimension of the weld is termed the *throat* distance. Evidently the throat distance is equal to the product of the leg length and sin 45°. It is customary to assume that only the shearing strength of such a weld need be considered because failure usually occurs by shear at 45° along the throat. Thus the strength of the fillet weld is taken to be equal to the net area at the throat section times the allowable working stress in shear of the weld material. This is illustrated in Problems 2 and 3.

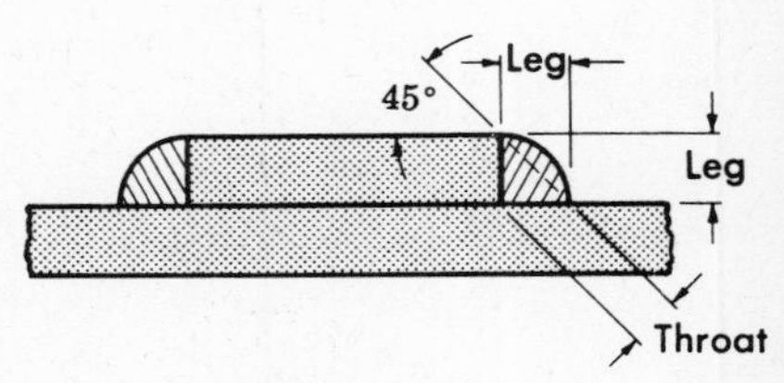

In the fillet weld shown in Figure 1 above, the weld runs in the direction of loading.

Occasionally a fillet weld perpendicular to the direction of loading is added across the end of the narrower plate, in addition to the welds shown there. According to studies carried out by the Welding Research Council, fillet welds placed perpendicular to the direction of applied load are somewhat stronger (per unit length of weld) than fillet welds in the direction of loading. Nevertheless it is common practice to treat both of these types of fillet welds as being of equal strength.

WORKING STRESSES IN WELDS. The Code for Fusion Welding of the American Welding Society specifies the following working stresses for structural welds:

Shear stress..........11,300 lb/in²

Tensile stress........13,000 lb/in²

Compressive stress....18,000 lb/in²

For fillet welds the working stress of 11,300 lb/in² may be expressed in another form that is frequently convenient for design purposes. For a 1/8 in. fillet (i.e. either leg is 1/8 in.) this working stress gives rise to an allowable shear per inch length of weld of

$$0.125(0.707)(11,300) = 1000 \text{ lb}$$

Thus we see that the allowable shearing stress of 11,300 lb/in² was selected so as to lead to simplicity in design considerations. From this result we readily have the following allowable shearing loads per inch length of weld:

2500 lb for a 5/16 in. fillet

3000 lb for a 3/8 in. fillet

3500 lb for a 7/16 in. fillet

4000 lb for a 1/2 in. fillet

SPECIAL CASES OF FILLET WELDS.

a) TORSION OF A CIRCULAR FILLET WELD. This is shown in Figure 2 below. The torque T acts on the circular shaft of diameter d. The leg of the fillet weld is designated as a. In Problem 4 it is shown that the maximum shearing stress in the fillet weld is given approximately by

$$(s_s)_{max} = \frac{2.83T}{\pi a d^2}$$

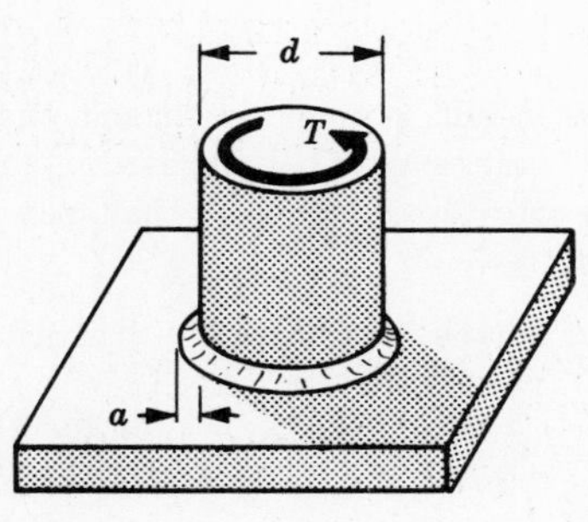

Fig. 2

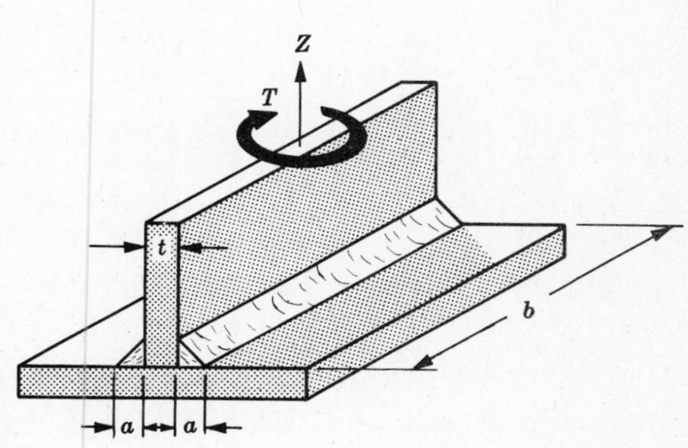

Fig. 3

b) TORSION RESISTED BY LONG ADJACENT FILLET WELDS. This is shown in Figure 3 above. The torque T acts on the vertical plate which is attached to the horizontal plate by two identical fillet welds of length b and having legs designated as a. In Problem 6 it is shown that the maximum shearing stress in the fillet welds is given approximately by

$$(s_s)_{max} = \frac{4.24T}{ab^2}$$

c) TORSION RESISTED BY WIDELY SEPARATED FILLET WELDS. This is the case if the vertical member shown in Figure 3 above is very thick, so that t is of the same order of magnitude as b. In this case a rational analysis becomes very difficult. The use of the torsion formula as in case *(a)* above (see Problem 4) may not be justified because in Chapter 5 it was emphasized that the torsion formula is valid only for members of circular cross-section, and not for ones that are rectangular in shape. Although use of this formula may give some rough estimate of the maximum shearing stress in the fillet weld, it should not be thought that the result obtained is anywhere near exact. This is true because the portion of the weld at the greatest distance from the center of the vertical member may not be the most highly stressed. However, if both the vertical and horizontal members are quite heavy and rigid, the shearing stresses in the fillet welds may be approximately proportional to their distances from the centroid of the weld areas and in this case it may be possible to obtain a reasonable estimate of the shearing stresses in the fillet welds by application of the torsion formula.

SOLVED PROBLEMS

1. A so-called plain butt weld joining two plates is shown in the adjoining diagram. The plates are each $\frac{1}{2}$ in. thick and 8 in. wide. The Code for Fusion Welding of the American Welding Society specifies an allowable working stress of 13,000 lb/in^2 for a tensile loading of such a welded joint. Determine the allowable tensile load P that may be applied to the plates.

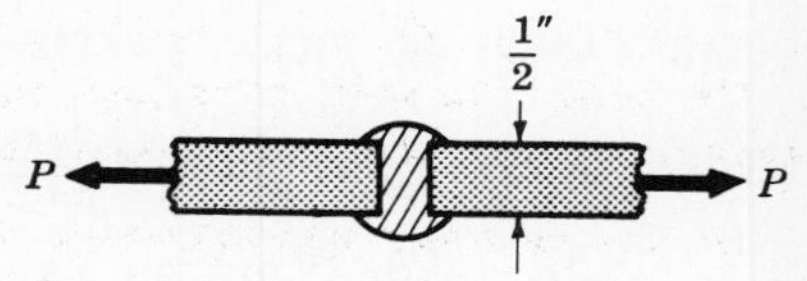

The strength of such a weld is usually taken to be equal to the product of the tensile working stress of the weld material by the net cross-sectional area of the weld. The tensile working stress is given by the specification to be 13,000 lb/in^2 and the net cross-sectional area is the length of the weld (8 in.) times the plate thickness ($\frac{1}{2}$ in.). Thus the allowable tensile load P is

$$P = 13{,}000(8)(\tfrac{1}{2}) = 52{,}000 \text{ lb}$$

2. A fillet weld joining two plates is shown below. The plates are each $\frac{1}{2}$ inch thick and the welds are each 7 inches long. The Code for Fusion Welding specifies an allowable working stress of 11,300 lb/in^2 for a shear loading of such a welded joint. Determine the allowable tensile load P that may be applied to the plates. The load is applied midway between the two welds.

Such fillet welds have equal *legs*, as shown in the cross-section below, and the minimum dimension of the cross-section is termed *throat*.

The throat dimension is $\frac{1}{2}\sin 45° = 0.353$ in.

The effective weld area that resists shearing is given by the length of the weld times the throat dimension, or weld area = 7(0.353) = 2.47 in^2 for each of the two welds.

Thus the allowable tensile load P is given by the product of the working stress in shear times the

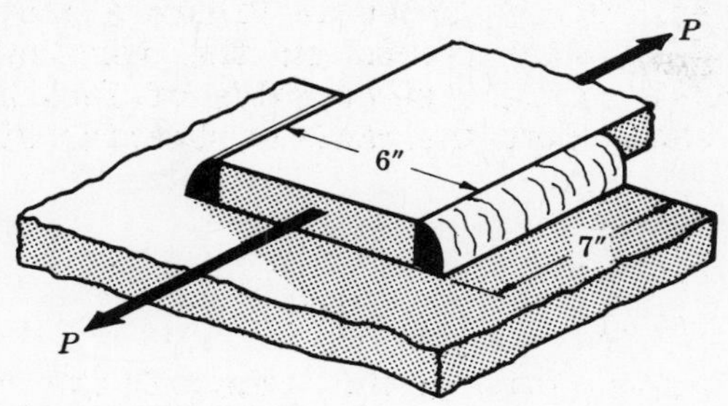

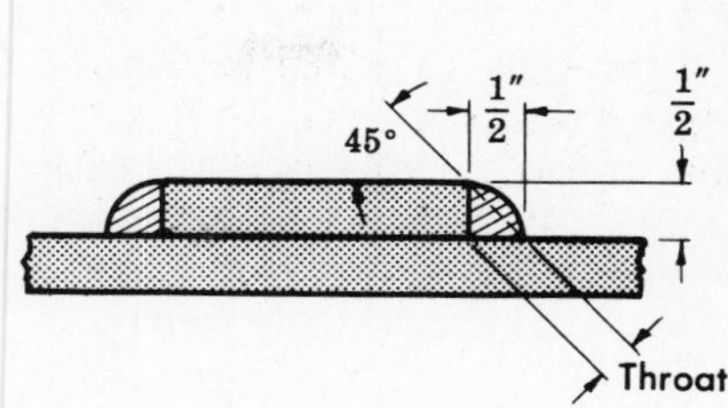

area resisting shear or

$$P = 11,300(2)(2.47) = 56,000 \text{ lb}$$

Using a shear value of 4000 lb/in of weld, we obtain the same value: $P = 4000(14) = 56,000$ lb.

Such fillet welds are obviously subject to bending stresses as well as tensile stresses in addition to the above shearing stresses. These stresses are usually not considered because failure of fillet welds occurs by shear along the throat of the weld.

3. In the adjoining illustration the 5/8 in. plates are loaded by the forces of 30,000 lb applied eccentrically as shown. Determine the required lengths L_1 and L_2 of the fillet welds so that they will be equally stressed in shear. Use the Code for Fusion Welding specifications.

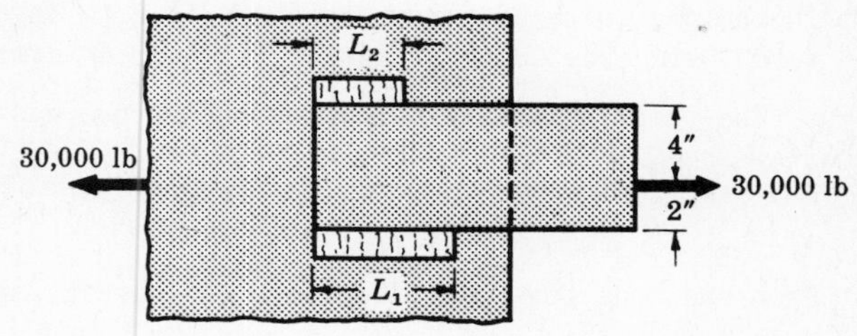

The legs of the welds are equal to the plate thickness, 5/8 in.; hence the throat of the weld is

$$(5/8) \sin 45^\circ = 0.442$$

The effective weld area that resists shear on these throat sections is thus $(L_1 + L_2)(0.442)$ in^2, where L_1 and L_2 denote the weld lengths as shown above.

It is customary to proportion the lengths L_1 and L_2 so that the welds are equally stressed in shear. This means that the resultant of these two shearing forces in the welds must coincide with the line of action of the 30,000 lb load. In other words, the moments of the two shear forces in the welds about any point on this line of action must be equal. Thus for an allowable shearing stress of 11,300 lb/in^2 we have

$$11,300(0.442)(L_1)(2) = 11,300(0.442)(L_2)(4) \qquad \text{or} \qquad L_1 = 2L_2$$

The required weld length to resist the load of 30,000 lb is given by

$$(L_1 + L_2)(0.442)(11,300) = 30,000$$

But $L_1 = 2L_2$; hence $(3L_2)(0.442)(11,300) = 30,000$ and $L_2 = 2.0$ in., $L_1 = 4.0$ in.

4. Determine the maximum shearing stress in the fillet weld joining a circular shaft to a rigid plate as shown below. The shaft is subject to an applied torque T. Each leg of the fillet weld is designated as a.

If the shaft is considered to be oriented vertically, then the plate must lie in a horizontal plane. The shearing stress in the fillet weld in a horizontal plane coinciding with the upper face of the horizontal plate is readily found from the simple torsion formula presented in Chapter 5. The intensity of this stress is given by

$$s_s = T\rho/J = T(\tfrac{1}{2}d)/J$$

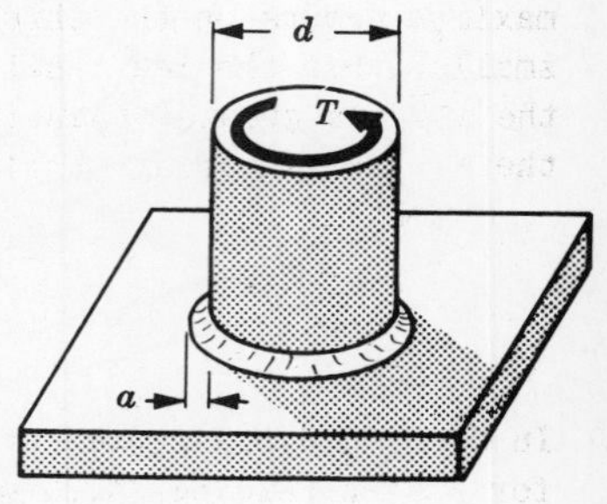

But $J = \int \rho^2 da = (\tfrac{1}{2}d)^2 \pi ad$, since the leg of the fillet weld is small compared to the dimension d and hence ρ may be taken to be constant. Thus

$$s_s = \frac{T(\tfrac{1}{2}d)}{(\tfrac{1}{2}d)^2 \pi ad} = \frac{2T}{\pi ad^2}$$

This shearing stress occurs in a horizontal plane, along a leg of the fillet weld. This is not the maximum shearing stress. The maximum shearing stress occurs on the throat of the weld or at 45° to this horizontal plane. As may be seen from the third diagram in this chapter, the length of the throat is given by the product of the leg length and sin 45°. Since the throat distance is smaller than the leg length, the shearing stress across the throat is greater than in a plane coinciding with a leg. Thus along the 45° throat section we have the maximum shearing stress

$$(s_s)_{max} = \frac{2T}{\pi ad^2(0.707)} = \frac{2.83T}{\pi ad^2}$$

5. In the preceding problem, the shaft is 2 in. in diameter and is joined to the plate by a 1/4 in. fillet weld. Using the Code for Fusion Welding, determine the maximum torque the welded joint can sustain.

The relationship to determine the torque was found in Problem 4 to be

$$(s_s)_{max} = \frac{2.83T}{\pi ad^2}$$

From the Code for Fusion Welding, the working stress in shear may not exceed 11,300 lb/in². Hence

$$11{,}300 = \frac{2.83T}{\pi(0.25)(2)^2} \quad \text{or} \quad T = 12{,}600 \text{ lb-in}$$

6. Determine the maximum shearing stress in the fillet welds connecting the two rigid plates shown in the adjacent figure. The vertical plate is subject to a torque T acting in a horizontal plane as shown. The fillet welds are identical and the legs of each are of length a.

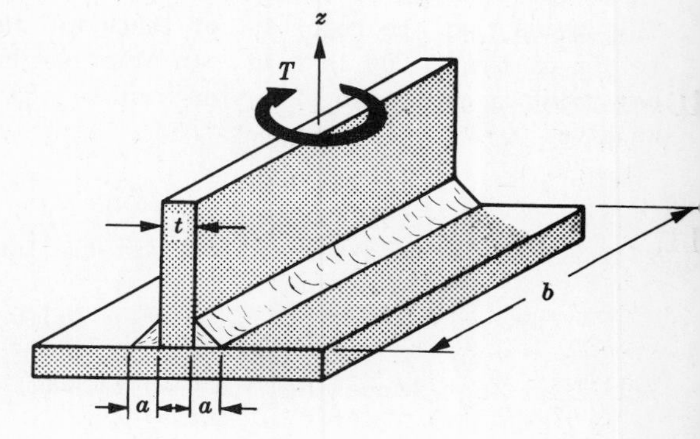

The effect of the applied torque T is to tend to rotate the vertical plate about the z-axis through its midpoint. This rotation is resisted by shearing stresses developed between the two fillet welds and the horizontal plate. If the plates are each quite rigid it is reasonable to assume that the intensities of these horizontal shearing stresses vary from zero at the z-axis to a minimum at the ends of the plate, i.e. at $\pm b/2$. Let us denote the shearing stress in a horizontal plane at the ends of the plate by s_s. This variation of shearing stress is exactly analogous to the variation of normal stress over the depth b of a beam subject to pure bending. Hence by analogy the stress s_s at the outer fibers is

$$s_s = \frac{Mc}{I} = \frac{T(b/2)}{(2a)(b^3)/12} = \frac{3T}{ab^2}$$

Again, as in Problem 4 this is not the maximum shearing stress in the fillet weld but rather the maximum occurs on the throat of the weld or at 45° to this horizontal plane. The throat distance is smaller than the leg length by a factor of sin 45°; hence the shearing stress across the throat of the weld is greater than in a plane coinciding with a leg. Thus along the 45° throat section we have the maximum shearing stress

$$(s_s)_{max} = \frac{3T}{ab^2(0.707)} = \frac{4.24T}{ab^2}$$

7. In Problem 6 the plates are each 20 in. long and are joined by two 3/8 in. fillet welds. Using the Code for Fusion Welding, determine the maximum torque the welded joint can resist.

The relationship between maximum shearing stress in the welds and the torque was shown in Problem 6 to be

$$(s_s)_{max} = \frac{4.24T}{ab^2}$$

From the Code for Fusion Welding, the allowable shearing stress is 11,300 lb/in². Then

$$11{,}300 = \frac{4.24T}{(3/8)(20)^2} \qquad \text{or} \qquad T = 400{,}000 \text{ lb-in}$$

SUPPLEMENTARY PROBLEMS

8. Two 5 in. wide by ½ in. thick steel plates are butt welded at their ends. The Code for Fusion Welding of the American Welding Society specifies an allowable working stress of 13,000 lb/in² for a tensile loading of such a joint. Determine the allowable tensile load that may be applied to the plates. *Ans.* 32,400 lb

9. A spherical gas tank is made of 5/8 in. steel plate hemispheres butt welded together. The Code for Fusion Welding specifies an allowable working stress of 13,000 lb/in² for a tensile loading of such a joint. The tank is 40 ft in diameter. Determine the allowable internal pressure to which the tank may be subject. *Ans.* 68 lb/in²

10. Two plates are joined by fillet welds as shown in Fig. (*a*) below and subject to a tensile loading of 80,000 lb. What length L of 7/16 in. weld is required to resist this load? The allowable working stress in shear for the weld material is 11,300 lb/in². *Ans.* L = 11.5 in.

11. A 5 × 5 × ½ in. structural angle having the cross-section shown in Fig. (*b*) below, is fillet welded with ½ in. welds to a flat steel plate. A tensile force of 65,000 lb is applied to the joint as shown. This force acts through the centroid of the angle cross-section, which is located 1.43 in. from the outer faces of the angle. Determine the required lengths L_1 and L_2 of the fillet welds so that they will be equally stressed in shear. The allowable shearing stress in the welds is 11,300 lb/in².
Ans. L_1 = 11.6 in., L_2 = 4.7 in.

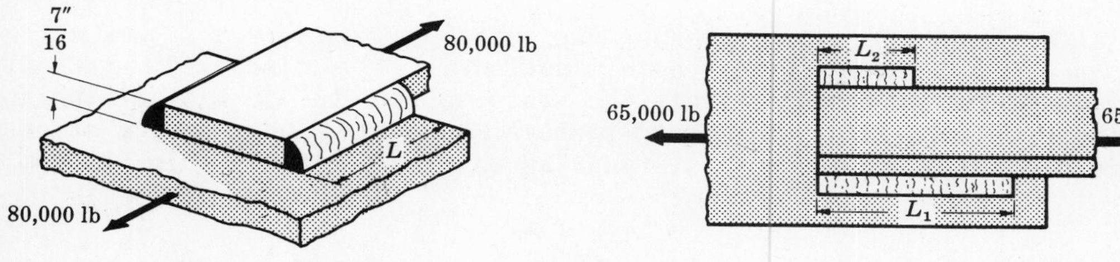

Fig. (*a*) Prob. 10

Fig. (*b*) Prob. 11

CHAPTER 15

Combined Stresses

INTRODUCTION. Thus far in this book we have considered stresses arising in bars subject to axial loading, shafts subject to torsion, and beams subject to bending as well as several cases involving thin-walled pressure vessels and riveted joints. It is to be noted that we have considered a bar, for example, to be subject to only *one* loading at a time, such as bending. Frequently such bars are simultaneously subject to several of the previously mentioned loadings, such as bending and torsion for example, and it is required to determine the state of stress under these conditions. Since normal and shearing stress are vector quantities, considerable care must be exercised in combining the stresses given by the expressions for single loadings as derived in previous chapters. It is the purpose of this chapter to investigate the state of stress on an arbitrary plane through an element in a body subject to several simultaneous loadings.

GENERAL CASE OF TWO-DIMENSIONAL STRESS. In general if a plane element is removed from a body it will be subject to the normal stresses s_x and s_y together with the shearing stress s_{xy} as shown in Figure 1 below.

SIGN CONVENTION. For normal stresses, tensile stresses are considered to be positive, compressive stress negative. For shearing stresses, the positive sense is that illustrated in Figure 1.

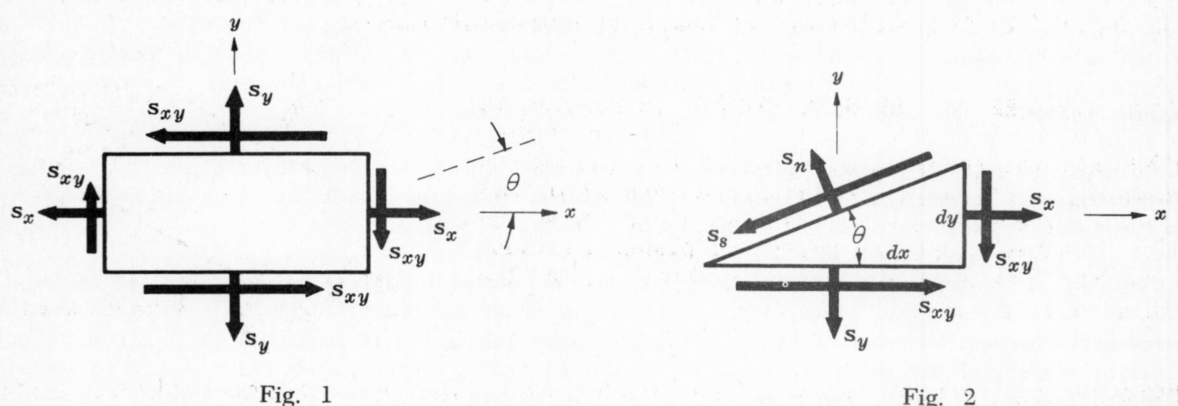

Fig. 1

Fig. 2

STRESSES ON AN INCLINED PLANE. We shall assume that the stresses s_x, s_y, and s_{xy} are known. (Their determination will be discussed in Chapter 16.) Frequently it is desirable to investigate the state of stress on a plane inclined at an angle θ to the x-axis, as shown in Figure 1 above. The normal and shearing stresses on such a plane are denoted by s_n and s_s and appear as in Figure 2 above. In Problem 13 it is shown that

$$s_n = \left(\frac{s_x + s_y}{2}\right) - \left(\frac{s_x - s_y}{2}\right)\cos 2\theta + s_{xy}\sin 2\theta$$

and

$$s_s = \frac{1}{2}(s_x - s_y)\sin 2\theta + s_{xy}\cos 2\theta$$

Thus, for any value of θ, s_n and s_s may be obtained from these expressions. For applications see Problems 2, 9, 11.

PRINCIPAL STRESSES. There are certain values of the angle θ that lead to maximum and minimum values of s_n for a given set of stresses s_x, s_y, and s_{xy}. These maximum and minimum values that s_n may assume are termed *principal stress* and are given by

$$(s_n)_{max} = \left(\frac{s_x + s_y}{2}\right) + \sqrt{\left(\frac{s_x - s_y}{2}\right)^2 + (s_{xy})^2}$$

and

$$(s_n)_{min} = \left(\frac{s_x + s_y}{2}\right) - \sqrt{\left(\frac{s_x - s_y}{2}\right)^2 + (s_{xy})^2}$$

These expressions are derived in Problem 13. For applications see Problems 9, 11, 15, 18.

DIRECTIONS OF PRINCIPAL STRESSES; PRINCIPAL PLANES. The angles designated as θ_p between the x-axis and the planes on which the principal stresses occur are given by the equation

$$\tan 2\theta_p = \frac{-s_{xy}}{\left(\frac{s_x - s_y}{2}\right)}$$

This expression also is derived in Problem 13. For applications see Problems 9, 11, 15, 18. As shown there we always have two values of θ_p satisfying this equation. The stress $(s_n)_{max}$ occurs on one of these planes, and the stress $(s_n)_{min}$ occurs on the other. The planes defined by the angles θ_p are known as *principal planes*.

SHEARING STRESSES ON PRINCIPAL PLANES. In Problem 13 it is demonstrated that the shearing stresses on the planes on which $(s_n)_{max}$ and $(s_n)_{min}$ occur are always zero, regardless of the values of s_x, s_y, and s_{xy}. Thus, an element oriented along the principal planes and subject to the principal stresses appears as in the adjoining diagram.

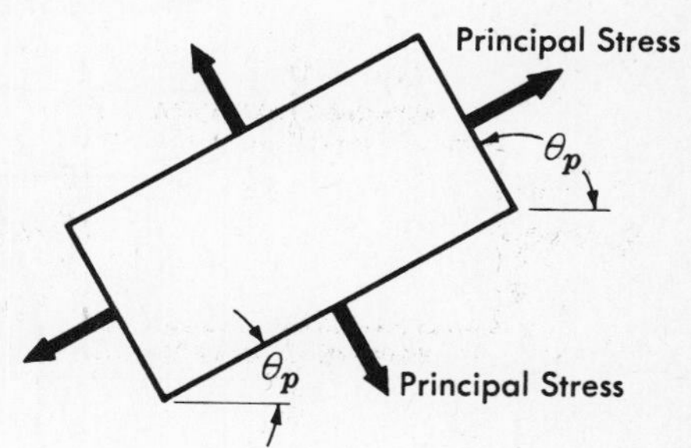

MAXIMUM SHEARING STRESS. There are certain values of the angle θ that lead to a maximum value of s_s for a given set of stresses s_x, s_y, and s_{xy}. The maximum value of the shearing stress is given by

$$s_s = \pm\sqrt{\left(\frac{s_x - s_y}{2}\right)^2 + (s_{xy})^2}$$

This expression is derived in Problem 13. For applications see Problems 3, 9, 11, 15, 18.

DIRECTIONS OF MAXIMUM SHEARING STRESS. The angles θ_s between the x-axis and the planes on which the maximum shearing stresses occur are given by the equation

$$\tan 2\theta_s = \frac{\left(\frac{s_x - s_y}{2}\right)}{s_{xy}}$$

This expression also is derived in Problem 13. For applications see Problems 3, 9, 11, 1 18. There are always two values of θ_s satisfying this equation. The shearing stress co responding to the positive square root given above occurs on one of the planes desi nated by θ_s, the shearing stress corresponding to the negative square root occurs on t other plane.

NORMAL STRESSES ON PLANES OF MAXIMUM SHEARING STRESS. In Problem 13 it is demonstrated that the normal stress on each of the planes of maximum shearing stress (which are of course 90° apart) is given by

$$s_n' = \frac{s_x + s_y}{2}$$

Thus an element oriented along the planes of maximum shearing stress appears as in the adjoining figure. This is illustrated in Problems 9, 11, 15, 18.

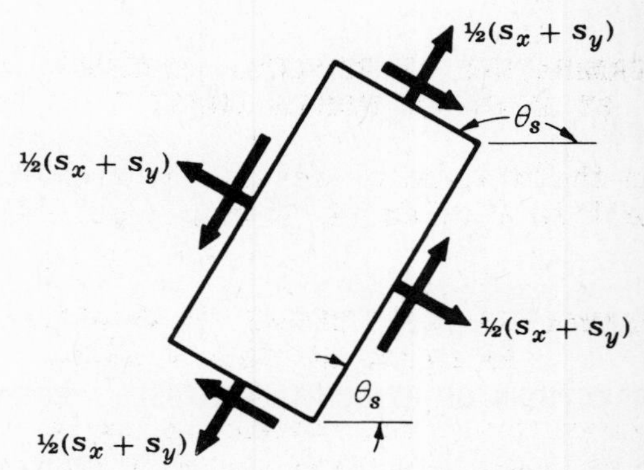

MOHR'S CIRCLE. All of the information contained in the above equations may be present in a convenient graphical form known as Mohr's circle. In this represe tation normal stresses are plotted along the horizontal axis and shearing stresses alo the vertical axis. The stresses s_x, s_y, and s_{xy} are plotted to scale and a circle drawn through these points having its center on the horizontal axis. Mohr's circle f an element subject to the general case of plane stress appears as follows:

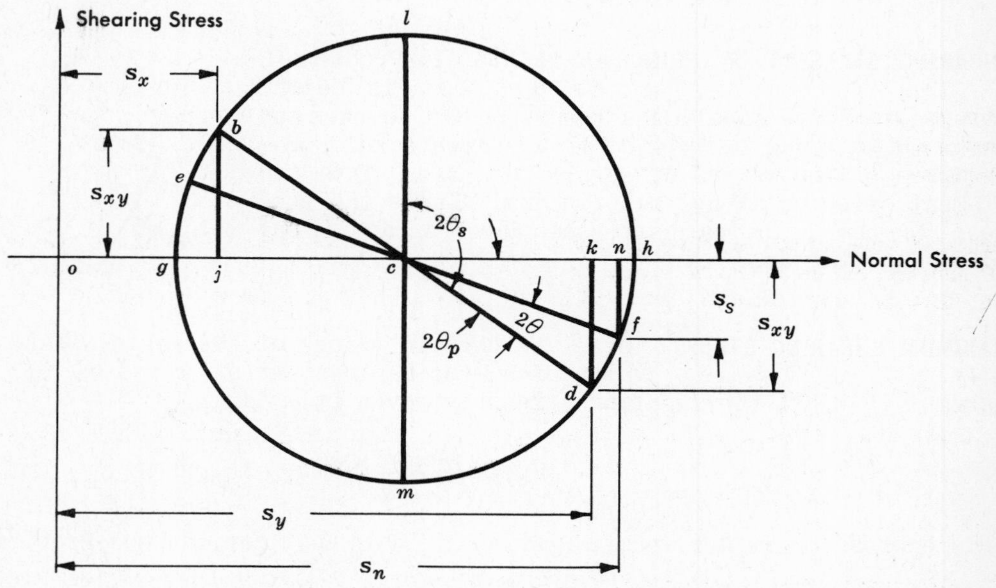

For applications see Problems 4, 5, 6, 8, 10, 12, 14, 16, 17, 19.

ɜN CONVENTIONS USED WITH MOHR'S CIRCLE. Tensile stresses are considered to be positive and compressive stresses negative. Thus ten-le stresses are plotted to the right of the origin in the above figure and compressive resses to the left. With regard to shearing stresses it is to be carefully noted that different sign convention exists than is used in connection with the above-mentioned uations. We shall refer to a plane element subject to ıearing stresses and appearing as in the adjoining dia-ram. We shall say that shearing stresses are positive they tend to rotate the element clockwise, negative if ıey tend to rotate it counterclockwise. Thus for the ove element the shearing stresses on the vertical faces e positive, those on the horizontal faces are negative.

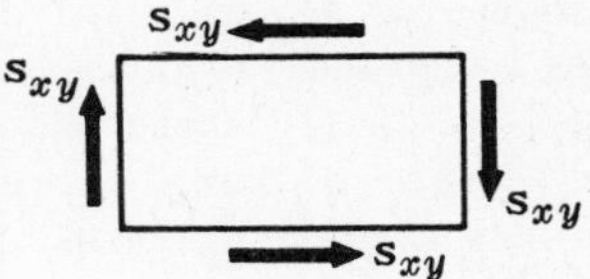

:TERMINATION OF PRINCIPAL STRESSES BY MEANS OF MOHR'S CIRCLE. When Mohr's circle has been drawn, the principal stresses are represented by the line segments *og* and *oh* respectively. These may either be scaled om the diagram or determined from the geometry of the figure. This is explained in ɜtail in Problem 14. For application see Problems 8, 10, 12, 16, 17, 19.

ETERMINATION OF STRESSES ON AN ARBITRARY PLANE BY MEANS OF MOHR'S CIRCLE. To determine the normal and shearing stresses on a plane inclined at a counterclockwise angle θ with the *x*-axis ɜ measure a counterclockwise angle equal to 2θ from the diameter *bd* of Mohr's circle. ıe end points of this diameter *bd* represent the stress conditions in the original *x-y* irections, i.e. they represent the stresses s_x, s_y, and s_{xy}. The angle 2θ corresponds ɔ the diameter *ef*. The coordinates of point *f* represent the normal and shearing stresses ı the plane at an angle θ to the *x*-axis. That is, the normal stress s_n is represented y the abscissa *on* and the shearing stress is represented by the ordinate *nf*. This is iscussed in detail in Problem 14. For applications see Problems 4, 5, 6, 8, 14, 17.

SOLVED PROBLEMS

. Let us consider a straight bar of uniform cross-section loaded in axial tension. Determine the normal and shearing stress intensities on a plane inclined at an angle θ to the axis of the bar. Also, determine the magnitude and direction of the maximum shearing stress in the bar.

This is the same elastic body that was considered in Chapter 1, but there the stresses that were studied were normal stresses in the direction of the axial force acting on the bar. In Fig. (*a*) below, *P* denotes the axial force acting on the bar, *A* the area of the cross-section perpendicular to the axis of the bar, and from Chapter 1 the normal stress *s* is given by $s = P/A$.

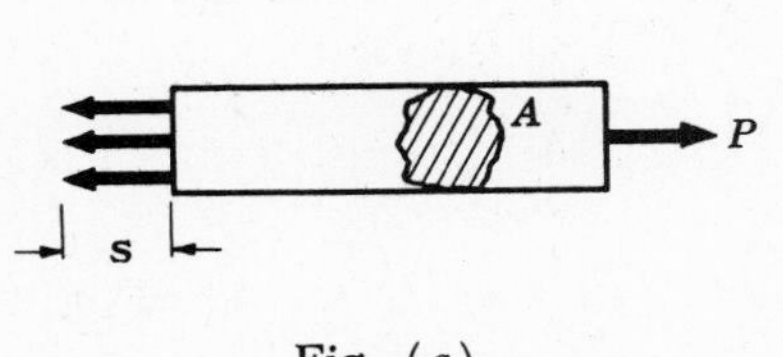

Fig. (*a*)

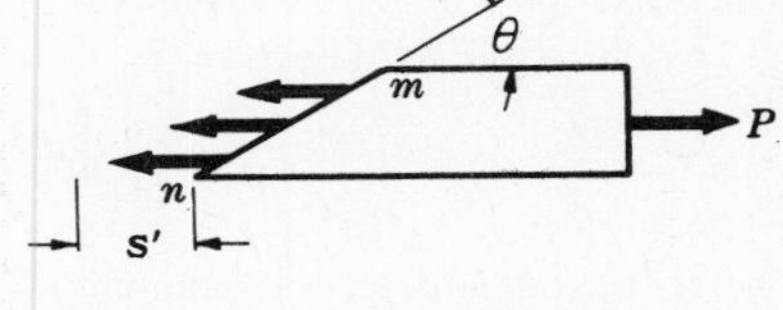

Fig. (*b*)

Suppose now that instead of using the above cutting plane which is perpendicular to the axis of the bar, that we instead pass a plane through the bar at an angle θ with the axis of the bar. Such a plane *m-n* is shown in Fig. (*b*) above. Since we must still have equilibrium of the bar in the horizontal direction there must evidently be distributed horizontal stresses acting over this inclined plane as shown. Let us designate the magnitude of these stresses by s'. Evidently the area of the inclined cross-section is $A/\sin\theta$ and for equilibrium of forces in the horizontal direction we have

$$s'(A/\sin\theta) = P \qquad \text{or} \qquad s' = (P\sin\theta)/A$$

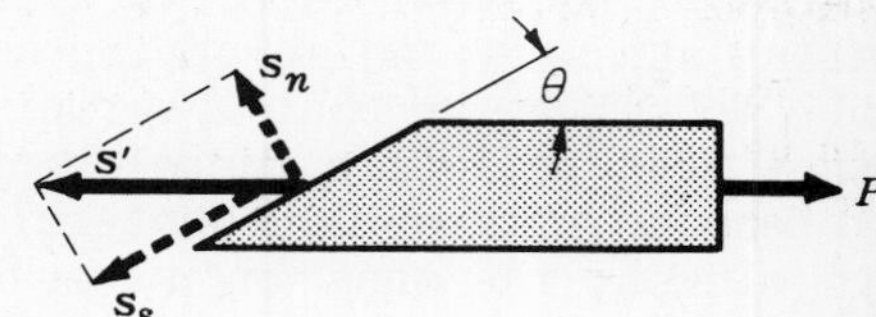

In the adjacent diagram let us consider only a single stress vector s' and resolve it into two components, one normal to the inclined plane mn and one tangential to this plane. We shall denote the first of these components by s_n to denote a normal stress, and the second by s_s to represent a shearing stress.

Since the angle between s' and s_s is θ we immediately have the relations

$$s_s = s'\cos\theta \qquad \text{and} \qquad s_n = s'\sin\theta$$

But $s' = (P\sin\theta)/A$. Substituting this value in the above equations, we obtain

$$s_s = P\sin\theta\cos\theta/A \qquad \text{and} \qquad s_n = P\sin^2\theta/A$$

But $s = P/A$. Hence we may write these in the form

$$s_s = s\sin\theta\cos\theta \qquad \text{and} \qquad s_n = s\sin^2\theta$$

Now, employing the trigonometric identities

$$\sin 2\theta = 2\sin\theta\cos\theta \qquad \text{and} \qquad \sin^2\theta = (1 - \cos 2\theta)/2$$

we may write

1) $$s_s = \tfrac{1}{2}s\sin 2\theta$$

2) $$s_n = \tfrac{1}{2}s(1 - \cos 2\theta)$$

These expressions give the normal and shearing stresses on a plane inclined at an angle θ to the axis of the bar.

From these equations it is evident that the shearing stress is maximum when $\sin 2\theta$ assumes its maximum value of unity, i.e. when $2\theta = 90°$ or $\theta = 45°$. The value of this maximum shearing stress is evidently $s_s = \frac{1}{2}s$. Also, the normal stress is maximum when $\cos 2\theta$ assumes its minimum value of -1, i.e. when $2\theta = 180°$ or $\theta = 90°$. For this value of θ the normal stress has the value $s_n = s$. Consequently the maximum normal stress acts over cross-sections perpendicular to the axis of the bar.

Thus we have the very interesting result that the maximum shearing stress in an axially loaded bar occurs on the planes at 45° to the direction of loading, and further on these planes the value of this maximum shearing stress is $s_s = \frac{1}{2}s$, i.e. the maximum shearing stress is one-half the maximum normal stress.

2. A bar of cross-section 1.3 in^2 is acted upon by axial tensile forces of 15,000 lb applied at each end of the bar. Determine the normal and shearing stresses on a plane inclined at 30° to the direction of loading.

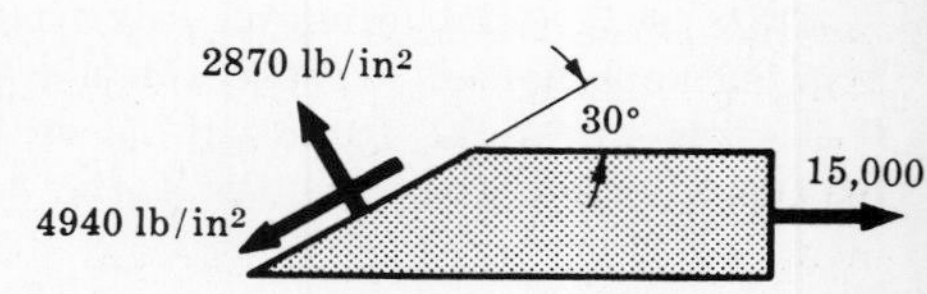

From Problem 1 the normal stress on a cross-section perpendicular to the axis of the bar is

$$s = \frac{P}{A} = \frac{15,000}{1.3} = 11,500\ \text{lb/in}^2$$

The normal stress on a plane at an angle θ with the direction of loading was found in Problem 1 to be $s_n = \frac{1}{2}s(1 - \cos 2\theta)$. For $\theta = 30°$ this becomes

$$s_n = \tfrac{1}{2}(11,500)(1 - \cos 60°) = 2870\ \text{lb/in}^2$$

The shearing stress on a plane at an angle θ with the direction of loading was found in Prob. 1 to be $s_s = \frac{1}{2}s\sin 2\theta$. For $\theta = 30°$ this becomes

$$s_s = \tfrac{1}{2}(11,500)\sin 60° = 4940 \text{ lb/in}^2$$

These stresses together with the axial load of 15,000 lb may be represented as shown in the diagram above.

3. Determine the maximum shearing stress in the axially loaded bar described in Problem 2.

The shearing stress on a plane at an angle θ with the direction of the load was shown in Problem 1 to be $s_s = \frac{1}{2}s \sin 2\theta$. This is maximum when $2\theta = 90°$, i.e. when $\theta = 45°$. For this loading we have $s = 11,500$ lb/in^2 and when $\theta = 45°$ the shear stress is

$$s_s = \tfrac{1}{2}(11,500)\sin 90° = 5750 \text{ lb/in}^2$$

That is, the maximum shearing stress is equal to one-half of the maximum normal stress.

The normal stress on this 45° plane may be found from the expression

$$s_n = \tfrac{1}{2}s(1 - \cos 2\theta) = \tfrac{1}{2}(11,500)(1 - \cos 90°) = 5750 \text{ lb/in}^2$$

4. Discuss a graphical representation of equations *(1)* and *(2)* of Problem 1.

According to these equations the normal and shearing stresses on a plane inclined at an angle θ to the direction of loading are given by

$$s_n = \tfrac{1}{2}s(1 - \cos 2\theta) \quad \text{and} \quad s_s = \tfrac{1}{2}s \sin 2\theta$$

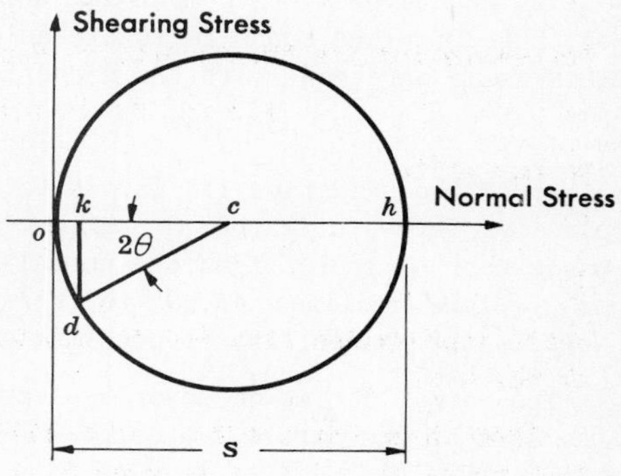

To represent these relations graphically it is customary to introduce a rectangular cartesian coordinate system, plotting normal stresses as abscissas and shearing stresses as ordinates.

Let us proceed by first laying off to some convenient scale the normal stress s (taken to be tensile) along the positive horizontal axis. The midpoint of this line segment, point c in the adjacent diagram, serves as the center of a circle whose diameter is s as shown. The radius of this circle, denoted by oc, ch and cd, is $\frac{1}{2}s$. The angle 2θ is measured positive in a counterclockwise direction from the radial line oc. From the above figure we immediately have the relations

$$kd = s_s = \tfrac{1}{2}s \sin 2\theta, \qquad ok = oc - kc = \tfrac{1}{2}s - \tfrac{1}{2}s\cos 2\theta = s_n = \tfrac{1}{2}s(1 - \cos 2\theta)$$

It is to be noted that the scales used in the horizontal and vertical directions are equal.

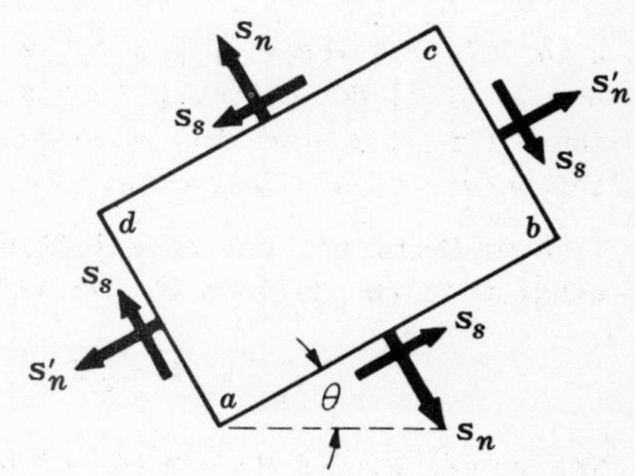

Thus the abscissa of point d represents the normal stress and the ordinate the shearing stress acting on a plane at an angle θ with the axis of the bar subject to tension. In plotting this diagram tensile stresses are regarded as positive in algebraic sign and compressive stresses are taken to be negative. Let us return to Problem 1 and examine a free-body diagram of an element taken from the surface of the inclined section on which the stresses s_n and s_s act. Such a diagram appears at the right. We shall consider shearing stresses to be positive if they tend to rotate the element clockwise, negative if they tend to rotate the element counterclockwise. This sign convention is used only in this graphical representation, not in the analytical treatment of Problem 1. Since the shearing stresses found in Problem 1 were actually those acting on face dc of the above element, they should be regarded as negative. Hence in the above circular diagram representing normal and shearing stresses the shearing stress on plane dc appears as an ordinate kd plotted in the negative sense.

This diagram, termed Mohr's Circle was first presented by O. Mohr in 1882. It represents the variation of normal and shearing stresses on all inclined planes passing through a given point in the body It is a convenient graphical representation of equations *(1)* and *(2)* of Problem 1.

5. Consider again the axially loaded bar discussed in Problem 2. Use Mohr's circle to determine the normal and shearing stresses on the 30° plane.

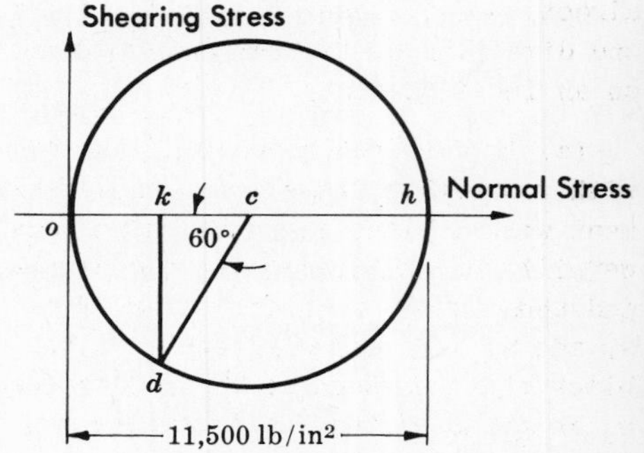

The normal stress of 11,500 lb/in^2 is laid off along the horizontal axis to some convenient scale and a circle is drawn with this line as a diameter. The angle $2\theta = 2(30°) = 60°$ is measured counterclockwise from *oc* as shown in the adjacent diagram.

The coordinates of the point *d* are

$$kd = s_s = -\tfrac{1}{2}(11{,}500)\sin 60° = -4940 \text{ lb/in}^2$$

and

$$ok = s_n = oc - kc = \tfrac{1}{2}(11{,}500) - \tfrac{1}{2}(11{,}500)\cos 60° = 2870 \text{ lb/in}^2$$

The negative sign accompanying the value of the shearing stress indicates that the shearing stress on this 30° plane tends to rotate an element bounded by this plane in a counterclockwise direction. This is in agreement with the direction of the shearing stress illustrated in Problem 2.

6. A bar of cross-section 1.3 in^2 is acted upon by axial compressive forces of 15,000 lb applied to each end of the bar. Using Mohr's circle, find the normal and shearing stresses on a plane inclined at 30° to the direction of loading. Neglect the possibility of buckling of the bar.

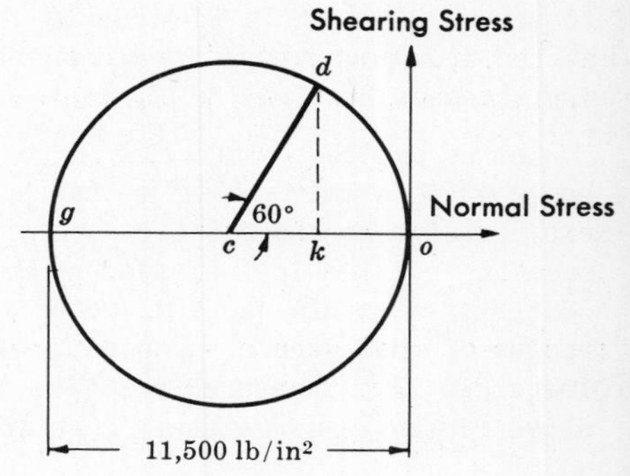

The normal stress on a cross-section perpendicular to the axis of the bar is

$$s = P/A = -15{,}000/1.3 = -11{,}500 \text{ lb/in}^2$$

We shall first lay off this compressive normal stress to some convenient scale along the negative end of the horizontal axis. The midpoint of this line segment, point *c* in the adjoining diagram, serves as the center of a circle whose diameter is 11,500 lb/in^2 to the scale shown.

The angle $2\theta = 2(30°) = 60°$ with the vertex at *c* is measured counterclockwise from *co* as shown. Th The abscissa of point *d* represents the normal stress and the ordinate the shearing stress on the desire 30° plane. The coordinates of point *d* are

$$kd = s_s = \tfrac{1}{2}(11{,}500)\sin 60° = 4940 \text{ lb/in}^2$$

and

$$ok = s_n = oc - ck = \tfrac{1}{2}(11{,}500) - \tfrac{1}{2}(11{,}500)\cos 60° = 2870 \text{ lb/in}^2$$

It is to be noted that line segment *ok* lies to the left of the origin of coordinates, hence this norma stress is compressive.

The positive algebraic sign accompanying the shearing stress indicates that the shearing stress on the 30° plane tends to rotate an element (denoted by dotted lines) bounded by this plane in a clockwise direction. The directions of the normal and shearing stresses together with the axial load of 15,000 lb are shown in the adjoining diagram.

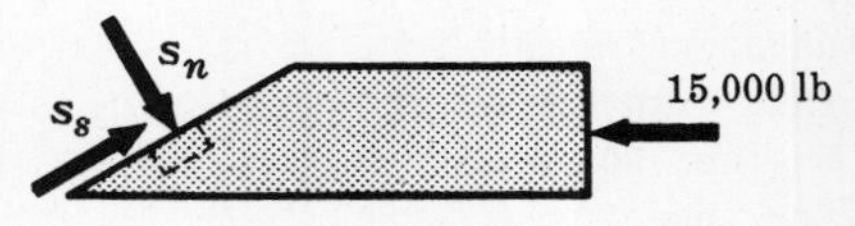

7. Consider a plane element removed from a stressed elastic body and subject to the normal and shearin

stresses s_x and s_{xy} respectively as shown in the adjoining diagram. (*a*) Determine the normal and shearing stress intensities on a plane inclined at an angle θ to the normal stress s_x. (*b*) Determine the maximum and minimum values of the normal stress that may exist on inclined planes and the directions of these stresses. (*c*) Determine the magnitude and direction of the maximum shearing stress that may exist on an inclined plane.

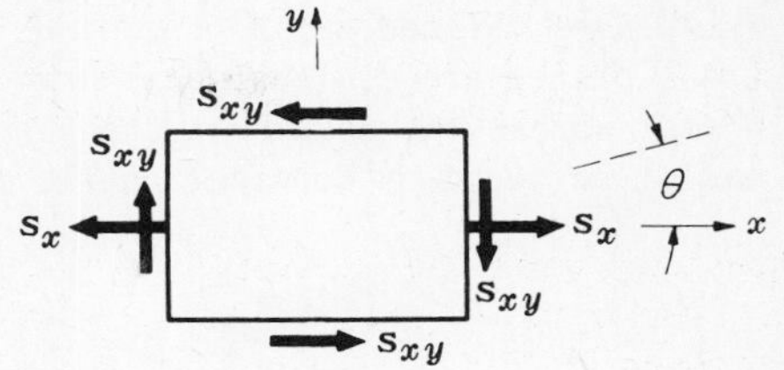

(*a*) The desired normal and shearing stresses acting on an inclined plane are internal quantities with respect to the element shown above. We shall follow the customary procedure of cutting this element with a plane in such a manner as to render the desired stresses external to the new body; that is, we will cut the originally rectangular element along the plane inclined at an angle θ with the x-axis and thus obtain a triangular element as shown below. The normal and shearing stresses designated as s_n and s_s respectively represent the effect of the remaining portion of the originally rectangular block that has been removed. Consequently, the problem reduces to finding the unknown stresses s_n and s_s in terms of the known stresses s_x and s_{xy}. It is to be observed that in the free-body diagram of the triangular element, the vectors indicate stresses acting on the various faces of the element and not forces. Each of these stresses is assumed to be uniformly distributed over the area upon which it acts. The thickness of the element perpendicular to the plane of the paper is denoted by t.

Let us introduce the N and T axes normal and tangent to the inclined plane as shown in the adjoining figure. First, we shall sum forces in the N-direction. For equilibrium we have

$$\Sigma F_N = s_n t\, ds - s_x t\, dy \sin\theta - s_{xy} t\, dy \cos\theta - s_{xy} t\, dx \sin\theta = 0$$

But from trigonometry $dy = ds \sin\theta$, $dx = ds\cos\theta$. Substituting these relations in the equilibrium equation above, we find

$$s_n(ds) = s_x(ds)\sin^2\theta + 2s_{xy}(ds)\sin\theta\cos\theta$$

Next, employing the identities $\sin^2\theta = \frac{1}{2}(1-\cos 2\theta)$ and $\sin 2\theta = 2\sin\theta\cos\theta$, we obtain

(1) $$s_n = \tfrac{1}{2}s_x(1-\cos 2\theta) + s_{xy}\sin 2\theta = \tfrac{1}{2}s_x - \tfrac{1}{2}s_x\cos 2\theta + s_{xy}\sin 2\theta$$

Thus the normal stress s_n on any plane inclined at an angle θ with the x-axis is known as a function of s_x, s_{xy}, and θ.

Next we shall consider the equilibrium of the forces acting on the triangular element in the T-direction. This leads to the equation

$$\Sigma F_T = s_s\, t\, ds - s_x\, t\, dy\cos\theta + s_{xy}\, t\, dy \sin\theta - s_{xy}\, t\, dx\cos\theta = 0$$

Substituting $dy = ds\sin\theta$ and $dx = ds\cos\theta$, we obtain

$$s_s(ds) = +\, s_x(ds)\sin\theta\cos\theta - s_{xy}(ds)\sin^2\theta + s_{xy}(ds)\cos^2\theta$$

Employing the identities $\cos 2\theta = \cos^2\theta - \sin^2\theta$ and $\sin 2\theta = 2\sin\theta\cos\theta$, this becomes

(2) $$s_s = \tfrac{1}{2}s_x\sin 2\theta + s_{xy}\cos 2\theta$$

Thus the shearing stress s_s on any plane inclined at an angle θ with the x-axis is known as a function of s_x, s_{xy}, and θ.

(*b*) To determine the maximum value that the normal stress s_n may assume as the angle θ varies, we shall differentiate equation (1) with respect to θ and set this derivative equal to zero. Thus

$$\frac{d(s_n)}{d\theta} = +\, s_x\sin 2\theta + 2\, s_{xy}\cos 2\theta = 0$$

The values of θ leading to maximum and minimum values of the normal stress are consequently

(3) $$\tan 2\theta_p = \frac{-\, s_{xy}}{\frac{1}{2}s_x}$$

The planes defined by the angles θ_p are called *principal planes*. The normal stresses that exist c these planes are designated as *principal stresses*. They are the maximum and minimum values that th normal stress may assume in the element under consideration. The values of the principal stresses ma easily be found by considering the following graphical interpretation of equation *(3)*:

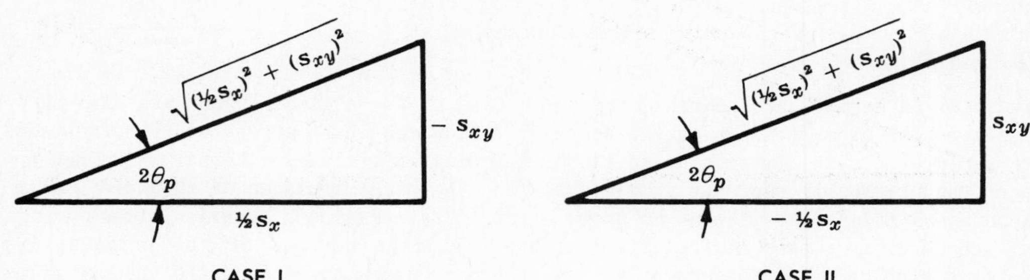

Evidently the tangent of either of these angles designated as $2\theta_p$ has the value given in equation *(3)* Thus there are two solutions to *(3)*, consequently two values of $2\theta_p$ (differing by 180°) and also tw values of θ_p. These values of θ_p differ by 90°. It is to be noted that the above two diagrams bear r direct relationship to the triangular element whose free-body diagram was considered earlier.

The values of $\sin 2\theta_p$ and $\cos 2\theta_p$ as found from the above two diagrams may now be substituted i equation *(1)* to yield the maximum and minimum values of the normal stresses. Observing that

$$\sin 2\theta_p = \frac{\mp s_{xy}}{\sqrt{(\frac{1}{2}s_x)^2 + (s_{xy})^2}}, \qquad \cos 2\theta_p = \frac{\pm \frac{1}{2}s_x}{\sqrt{(\frac{1}{2}s_x)^2 + (s_{xy})^2}}$$

where the upper signs pertain to Case I and the lower signs to Case II, we obtain from equation *(1)*

4) $$s_n = \tfrac{1}{2}s_x \mp (\tfrac{1}{2}s_x)\frac{\frac{1}{2}s_x}{\sqrt{(\frac{1}{2}s_x)^2 + (s_{xy})^2}} \mp \frac{(s_{xy})^2}{\sqrt{(\frac{1}{2}s_x)^2 + (s_{xy})^2}} = \tfrac{1}{2}s_x \pm \sqrt{(\tfrac{1}{2}s_x)^2 + (s_{xy})^2}$$

The maximum normal stress is

5) $$(s_n)_{\max} = \tfrac{1}{2}s_x + \sqrt{(\tfrac{1}{2}s_x)^2 + (s_{xy})^2}$$

The minimum normal stress is

6) $$(s_n)_{\min} = \tfrac{1}{2}s_x - \sqrt{(\tfrac{1}{2}s_x)^2 + (s_{xy})^2}$$

The stresses given by equations *(5)* and *(6)* are the principal stresses and they occur on the principa planes defined by equation (3). By substituting one of the values of θ_p from equation (3) into (1) one may readily determine which of the two principal stresses is acting on that plane. The other prin cipal stress naturally acts on the other principal plane.

By substituting the values of the angles $2\theta_p$ as given by equation (3) and the above two diagram into equation (2), it is readily seen that the shearing stresses s_s on the principal planes are zero

(c) To determine the maximum value the shearing stress s_s may assume as the angle θ varies, w shall differentiate equation (2) with respect to θ and set this derivative equal to zero. Thus

$$\frac{d(s_s)}{d\theta} = s_x \cos 2\theta - 2s_{xy} \sin 2\theta = 0$$

The values of θ leading to maximum values of the shearing stress are consequently

7) $$\tan 2\theta_s = \frac{\frac{1}{2}s_x}{s_{xy}}$$

The planes defined by the two solutions to this equation are the planes of maximum shearing stress.

Again, a graphical interpretation of equation (7) is convenient. The two values of the angle $2\theta_s$ satisfying this equation may be represented as follows:

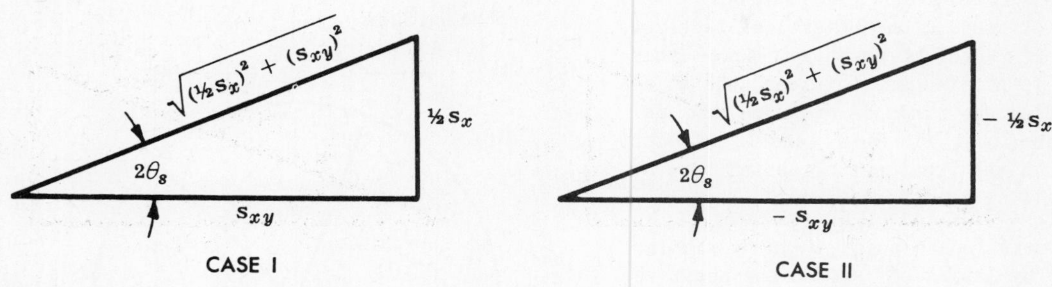

From these diagrams we have

$$\sin 2\theta_s = \frac{\pm\frac{1}{2}s_x}{\sqrt{(\frac{1}{2}s_x)^2 + (s_{xy})^2}}, \qquad \cos 2\theta_s = \frac{\pm s_{xy}}{\sqrt{(\frac{1}{2}s_x)^2 + (s_{xy})^2}}$$

where the upper (positive) signs pertain to Case I and the lower (negative) signs apply to Case II. Substituting these values in equation (2) we obtain

$$8)\qquad s_s = (\tfrac{1}{2}s_x)\frac{\pm\frac{1}{2}s_x}{\sqrt{(\frac{1}{2}s_x)^2 + (s_{xy})^2}} + (s_{xy})\frac{\pm s_{xy}}{\sqrt{(\frac{1}{2}s_x)^2 + (s_{xy})^2}} = \pm\sqrt{(\tfrac{1}{2}s_x)^2 + (s_{xy})^2}$$

Here the positive sign represents the maximum shearing stress, the negative sign the minimum shearing stress.

If we compare equations (3) and (7) it is evident that the angles $2\theta_p$ and $2\theta_s$ differ by 90°, since the tangents of these angles are the negative reciprocals of one another. Hence the planes defined by the angles θ_p and θ_s differ from one another by 45°; that is, the planes of maximum shearing stress are oriented 45° from the planes of maximum normal stress.

It is also of interest to determine the normal stresses on the planes of maximum shearing stress. These planes are defined by equation (7). If we now substitute these values of $\sin 2\theta_s$ and $\cos 2\theta_s$ in equation (1) for the normal stress we find

$$9)\qquad s_n' = \tfrac{1}{2}s_x - (\tfrac{1}{2}s_x)\frac{\pm s_{xy}}{\sqrt{(\frac{1}{2}s_x)^2 + (s_{xy})^2}} + (s_{xy})\frac{\pm\frac{1}{2}s_x}{\sqrt{(\frac{1}{2}s_x)^2 + (s_{xy})^2}} = \tfrac{1}{2}s_x$$

Thus on each of the planes of maximum shearing stress we have a normal stress of magnitude $\frac{1}{2}s_x$.

. Discuss a graphical representation of the analysis presented in Problem 7.

For given values of s_x and s_{xy} we proceed as follows:

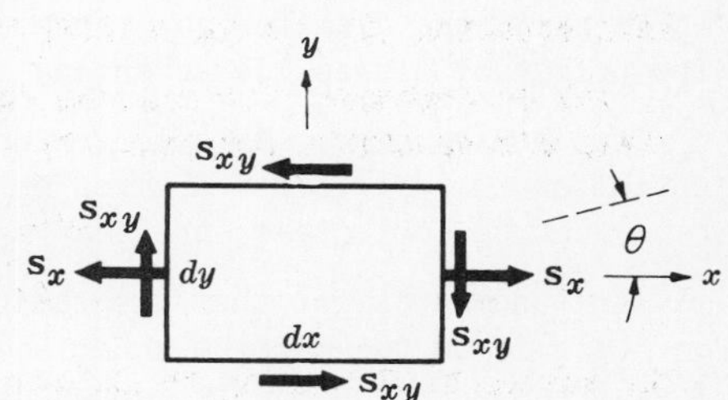

a) Introduce a rectangular coordinate system in which normal stresses are represented along the horizontal axis and shearing stresses along the vertical axis. The scales used on these two axes must be equal.

b) With reference to the original rectangular element considered in Problem 7 and reproduced here, we shall introduce the sign convention that shearing stresses are positive if they tend to rotate the element clockwise, negative if they tend to rotate it counterclockwise. Here the shearing stresses on the vertical faces are positive, those on the horizontal faces are negative. Also, tensile stresses are considered to be positive and compressive stresses negative.

c) We first locate point b by laying out s_x and s_{xy} to their given values. The shear stress s_{xy} on the vertical faces on which s_x acts is positive, hence this value is plotted as positive in the adjacent diagram. This is drawn on the assumption that s_x is a tensile stress, although the treatment presented here is valid if s_x is compressive.

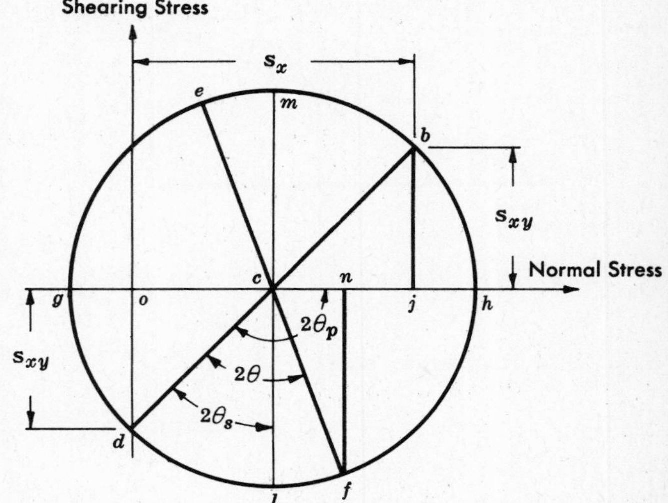

d) We next locate point d in a similar manner by laying off s_{xy} on the negative side of the vertical axis. Actually, this point d corresponds to the negative shearing stresses s_{xy} existing on the horizontal faces of the element together with a zero value normal stress acting on those same faces.

e) Next, we draw line bd, locate the midpoint c, and draw a circle having its center at c and radius equal to cb. This is known as Mohr's circle.

We shall first show that the points g and h along the horizontal diameter of the circle represent the principal stresses. To do this we note that the point c lies at a distance $\frac{1}{2}s_x$ from the origin of the coordinate system. From the right triangle relationship we have

$$(cd)^2 = (oc)^2 + (od)^2 \qquad \text{or} \qquad cd = \sqrt{(\tfrac{1}{2}s_x)^2 + (s_{xy})^2}$$

Also, we have $cd = ch = cg$. Hence, the x-coordinate of point h is $(oc + ch)$ or

$$\tfrac{1}{2}s_x + \sqrt{(\tfrac{1}{2}s_x)^2 + (s_{xy})^2}$$

But this expression is exactly the maximum principal stress, as found in equation *(5)* of Problem 7. Likewise the x-coordinate of point g is $(oc - cg)$. But this quantity is negative; hence og lies to the left of the origin, and point g symbolizes a compressive stress. This stress becomes

$$\tfrac{1}{2}s_x - \sqrt{(\tfrac{1}{2}s_x)^2 + (s_{xy})^2}$$

But this expression is exactly the minimum principal stress, as found in equation *(6)* of Problem 7. Consequently the points g and h represent the principal stresses existing in the original element. We see that the tangent of $\angle ocd = s_{xy}/(\frac{1}{2}s_x)$. But from equation *(3)* of Problem 7, $\tan 2\theta_p = -s_{xy}/(\frac{1}{2}s_x)$; and by comparison of these two relations we see that $\angle hcd = 2\theta_p$, since $\tan(180° - \theta) = -\tan\theta$. Thus a counterclockwise rotation from the diameter bd (corresponding to the stresses in the x-y directions) leads us to the diameter gh, representing the principal planes, on which the principal stresses occur. The principal planes lie at an angle θ_p from the x-direction.

Thus Mohr's circle is a convenient device for finding the principal stresses, since one can merely establish the circle for a given set of stresses s_x and s_{xy} then measure og and oh. These abscissas represent the principal stresses to the same scale used in plotting s_x and s_{xy}.

It is now apparent that the radius of Mohr's circle, represented by cd, where $cd = \sqrt{(\frac{1}{2}s_x)^2 + (s_{xy})^2}$, corresponds to the maximum shearing stress, as found in equation *(8)* of Problem 7. Actually, the shearing stress on any plane is represented by the ordinate to Mohr's circle; hence we should consider the radial lines cl and cm as representing the maximum shearing stresses. The angle dcl is evidently $2\theta_s$ and hence it is apparent that the double angle between the planes of maximum normal stress and the planes of maximum shearing stress ($\angle lch$) is 90°; thus the planes of maximum shearing stress are oriented 45° from the planes of maximum normal stress.

Evidently the end points of the diameter bd represent the stresses acting in the original x and y directions. We shall now demonstrate that the end points of any other diameter, such as ef (at any angle 2θ with bd), represent the stresses on a plane inclined at an angle θ to the x-axis. To do this we note that the abscissa of point f is given by

$$\begin{aligned} s_n &= oc + cn = \tfrac{1}{2}s_x + (cf)\cos(2\theta_p - 2\theta) \\ &= \tfrac{1}{2}s_x + (cf)(\cos 2\theta_p \cos 2\theta + \sin 2\theta_p \sin 2\theta) \\ &= \tfrac{1}{2}s_x + \sqrt{(\tfrac{1}{2}s_x)^2 + (s_{xy})^2}\,(\cos 2\theta_p \cos 2\theta + \sin 2\theta_p \sin 2\theta) \end{aligned}$$

But from inspection of triangle cod appearing in Mohr's circle it is evident that

1) $$\sin 2\theta_p = \frac{s_{xy}}{\sqrt{(\tfrac{1}{2}s_x)^2 + (s_{xy})^2}} \quad \text{and} \quad \cos 2\theta_p = \frac{-\tfrac{1}{2}s_x}{\sqrt{(\tfrac{1}{2}s_x)^2 + (s_{xy})^2}}$$

Substituting the values of s_{xy} and $\tfrac{1}{2}s_x$ from these two equations into the previous equation, we find

$$s_n = \tfrac{1}{2}s_x - \tfrac{1}{2}s_x \cos 2\theta + s_{xy} \sin 2\theta$$

But this is exactly the normal stress on a plane inclined at an angle θ to the x-axis as derived in equation *(1)* of Problem 7.

Next we observe that the ordinate of point f is given by

$$\begin{aligned} s_s &= nf = (cf)\sin(2\theta_p - 2\theta) \\ &= \sqrt{(\tfrac{1}{2}s_x)^2 + (s_{xy})^2}\,(\sin 2\theta_p \cos 2\theta - \cos 2\theta_p \sin 2\theta) \end{aligned}$$

Again, substituting the values of s_{xy} and $\tfrac{1}{2}s_x$ from equations *(1)* into this equation, we find

$$s_s = \tfrac{1}{2}s_x \sin 2\theta + s_{xy} \cos 2\theta$$

But this is exactly the shearing stress on a plane inclined at an angle θ to the x-axis as derived in equation *(2)* of Problem 7.

Hence the coordinates of point f on Mohr's circle represent the normal and shearing stresses on a plane inclined at an angle θ to the x-axis.

A plane element in a body is subjected to a normal stress in the x-direction of 12,000 lb/in² as well as a shearing stress of 4000 lb/in², as shown below. *(a)* Determine the normal and shearing stress intensities on a plane inclined at an angle of 30° to the normal stress. *(b)* Determine the maximum and minimum values of the normal stress that may exist on inclined planes and the directions of these stresses. *(c)* Determine the magnitude and direction of the maximum shearing stress that may exist on an inclined plane.

a) In accordance with the notation of Problem 7 we have $s_x = 12{,}000$ lb/in² and $s_{xy} = 4000$ lb/in². From equation *(1)* of Problem 7, the normal stress on a plane inclined at an angle θ to the x-axis is given by

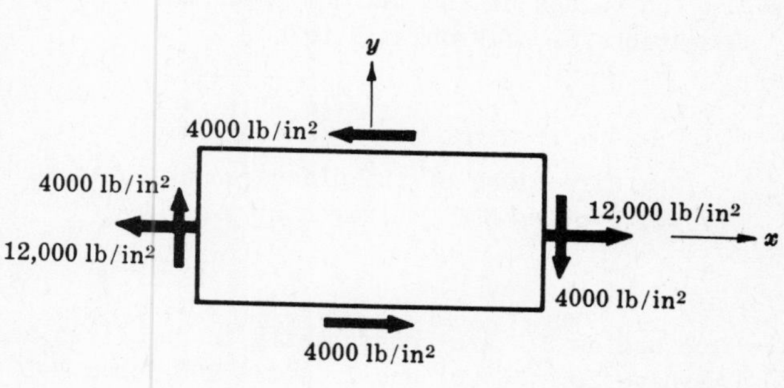

$$s_n = \tfrac{1}{2}s_x - \tfrac{1}{2}s_x \cos 2\theta + s_{xy} \sin 2\theta$$

Substituting the above values of s_x and s_{xy}, when $\theta = 30°$ this becomes

$$\begin{aligned} s_n &= \tfrac{1}{2}(12{,}000) - \tfrac{1}{2}(12{,}000)\cos 60° + 4000 \sin 60° \\ &= 6470 \text{ lb/in}^2 \end{aligned}$$

From equation *(2)* of Problem 7 the shearing stress on any plane inclined at an angle θ to the x-axi is given by

$$s_s = \tfrac{1}{2}s_x \sin 2\theta + s_{xy} \cos 2\theta$$

Substituting the above values of s_x and s_{xy}, when $\theta = 30°$ this becomes

$$s_s = \tfrac{1}{2}(12,000) \sin 60° + 4000 \cos 60° = 5200 + 2000 = 7200 \text{ lb/in}^2$$

The positive directions of the normal and shearing stresses on an inclined plane were illustrated in the second diagram of Problem 7. In accordance with this sign convention the stresses on the 30° plane thus appear as in the adjoining diagram.

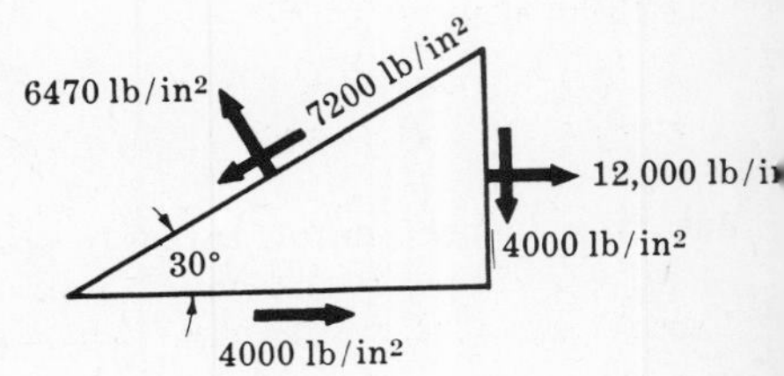

b) The values of the principal stresses, that is, the maximum and minimum values of the normal stresses existing in this element, were given by equations *(5)* and *(6)* of Problem 7. From equation *(5)* for the maximum normal stress we have

$$(s_n)_{\max} = \tfrac{1}{2}s_x + \sqrt{(\tfrac{1}{2}s_x)^2 + (s_{xy})^2} = 6000 + \sqrt{(6000)^2 + (4000)^2} = 13,220 \text{ lb/in}^2$$

From equation *(6)* for the minimum normal stress we have

$$(s_n)_{\min} = \tfrac{1}{2}s_x - \sqrt{(\tfrac{1}{2}s_x)^2 + (s_{xy})^2} = 6000 - \sqrt{(6000)^2 + (4000)^2} = -1220 \text{ lb/in}^2$$

The directions of the planes on which these principal stresses occur were found in equation *(3)* Problem 7 to be

$$\tan 2\theta_p = -\frac{s_{xy}}{\tfrac{1}{2}s_x} = -\frac{4000}{6000} = -\frac{2}{3}$$

Since the tangent of the angle $2\theta_p$ is negative, the two values of $2\theta_p$ lie in the second and fou quadrants. In the second quadrant, $2\theta_p = 146°20'$; in the fourth quadrant, $2\theta_p' = 326°20'$. Consequen we have the principal planes defined by $\theta_p = 73°10'$ and $\theta_p' = 163°10'$. If $\theta_p = 73°10'$, together with given values of s_x and s_{xy}, is now substituted in equation *(1)* of Problem 7, we find

$$\begin{aligned} s_n &= \tfrac{1}{2}s_x - \tfrac{1}{2}s_x \cos 2\theta + s_{xy} \sin 2\theta \\ &= 6000 - 6000 \cos 146°20' + 4000 \sin 146°20' \\ &= 6000 - 6000(-0.833) + 4000(0.554) = 13,220 \text{ lb/in}^2 \end{aligned}$$

Thus the principal stress of 13,220 lb/in^2 occurs on the principal plane oriented at 73°10' to the x-axis. The principal stresses thus appear as in the adjoining diagram. As stated in Problem 7 the shearing stresses on these principal planes are zero.

1220 lb/in²
13,220 lb/in²
13,220 lb
73°10′
1220 lb/in²

c) The values of the maximum shearing stresses were found in equation *(8)* of Problem 7 to be

$$s_s = \pm\sqrt{(\tfrac{1}{2}s_x)^2 + (s_{xy})^2} = \pm\sqrt{(6000)^2 + (4000)^2} = \pm 7220 \text{ lb/in}^2$$

The directions of the planes on which these maximum shearing stresses occur were found in equa *(7)* of Problem 7 to be given by

$$\tan 2\theta_s = \frac{\tfrac{1}{2}s_x}{s_{xy}} = \frac{6000}{4000} = \frac{3}{2}$$

The angles $2\theta_s$ are consequently in the first and third quadrants, since the tangent is positive. we have $2\theta_s = 56°20'$ and $2\theta_s' = 236°20'$, or $\theta_s = 28°10'$ and $\theta_s' = 118°10'$. The shearing stress on any p inclined at an angle θ with the x-axis was found in equation *(2)* of Problem 7 to be

$$s_s = \tfrac{1}{2}s_x \sin 2\theta + s_{xy} \cos 2\theta$$

Substituting $s_x = 12{,}000$ lb/in^2, $s_{xy} = 4000$ lb/in^2 and $\theta = 28°10'$, we find

$$s_s = \tfrac{1}{2}(12{,}000) \sin 56°20' + 4000 \cos 56°20' = +7220 \text{ lb/in}^2$$

Thus the shearing stress on the 28°10' plane is positive. The positive sense of shearing stress was shown in the second diagram of Problem 7.

The normal stresses on the planes of maximum shearing stress are found from equation *(9)* of Problem 7 to be

$$s_n' = \tfrac{1}{2}s_x = \tfrac{1}{2}(12{,}000) = 6000 \text{ lb/in}^2$$

This normal stress acts on each of the planes of maximum shearing stress as shown in the adjacent diagram.

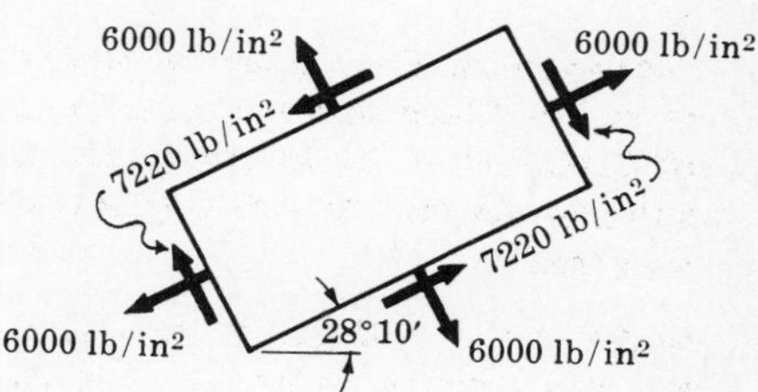

10. A plane element is subject to the stresses shown in Fig. *(a)* below. Using Mohr's circle, determine *a)* the principal stresses and their directions, *b)* the maximum shearing stresses and the directions of the planes on which they occur.

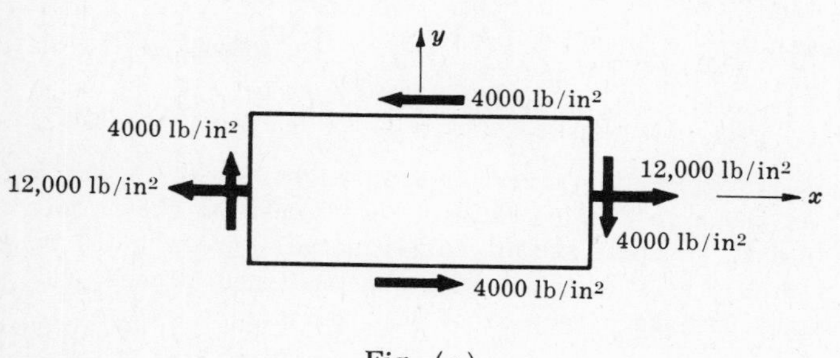

Fig. *(a)*

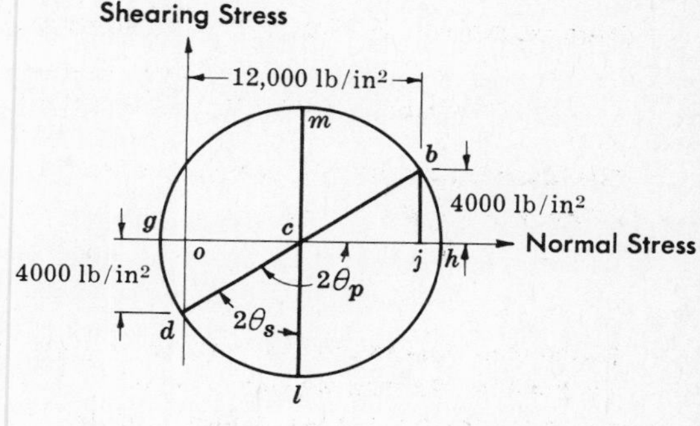

Fig. *(b)*

The procedure for the construction of Mohr's circle was outlined in Problem 8. Proceeding in accordance with the instructions there we realize that the shearing stress on the vertical faces of the above element are positive, whereas those on the horizontal faces are negative. Thus the stress condition of $s_x = 12{,}000$ lb/in^2, $s_{xy} = 4000$ lb/in^2 existing on the vertical faces of the element plots as point b in Fig. *(b)* above. The stress condition of $s_{xy} = -4000$ lb/in^2 together with a zero normal stress on the horizontal faces plots as point d. Line bd is drawn, its midpoint c is located, and a circle of radius $cb = cd$ is drawn with c as a center. This is Mohr's circle. The end points of the diameter bd represent the stress conditions existing in the element if it has the original orientation shown above.

a) The principal stresses are represented by points g and h, as shown in Problem 8. The principal stresses may be determined either by direct measurement from the above diagram or by realizing that the coordinate of c is 6000, and that $cd = \sqrt{(6000)^2 + (4000)^2} = 7220$. Therefore the minimum principal stress is

$$(s_n)_{\min} = og = (oc - cg) = 6000 - 7220 = -1220 \text{ lb/in}^2$$

Also, the maximum principal stress is

$$(s_n)_{\max} = oh = (oc + ch) = 6000 + 7220 = 13{,}220 \text{ lb/in}^2$$

The angle $2\theta_p$ designated above is given by

$$\tan 2\theta_p = -\frac{4000}{6000} = -\frac{2}{3} \qquad \text{or} \qquad \theta_p = 73°10'$$

This value could also be obtained by measurement of $\angle dch$ in Mohr's circle. From this it is readily seen that the principal stress represented by point h acts on a plane oriented 73°10' from the original x-axis. The principal stresses thus appear as in the adjacent diagram. It is evident from Mohr's circle that the shearing stresses on these planes are zero, since points g and h lie on the horizontal axis of Mohr's circle.

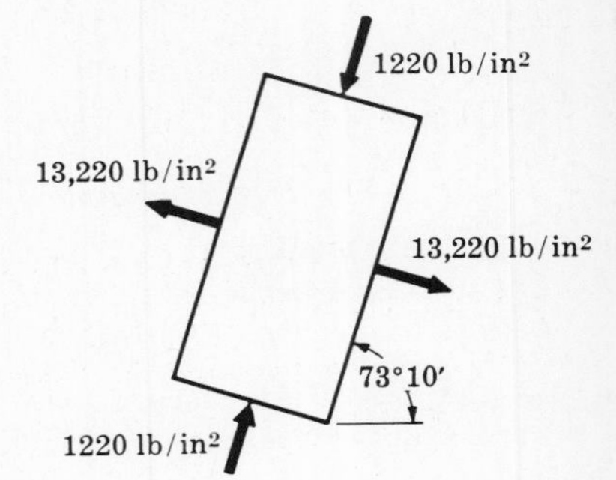

b) The maximum shearing stress is represented by cl in Mohr's circle. This radius has already been found to be equal to 7220 lb/in^2. The angle $2\theta_s$ may be found either by direct measurement from the above plot or simply by subtracting 90° from the angle $2\theta_p$, which has already been determined. This leads to $2\theta_s = 56°20'$ and $\theta_s = 28°10'$. The shearing stress represented by point l is negative, hence on this 28°10' plane the shearing stress tends to rotate the element in a counterclockwise direction. Also, from Mohr's circle the abscissa of point l is 6000 lb/in^2 and this represents the normal stress occurring on the planes of maximum shearing stress. The maximum shearing stresses thus appear as in the adjoining figure.

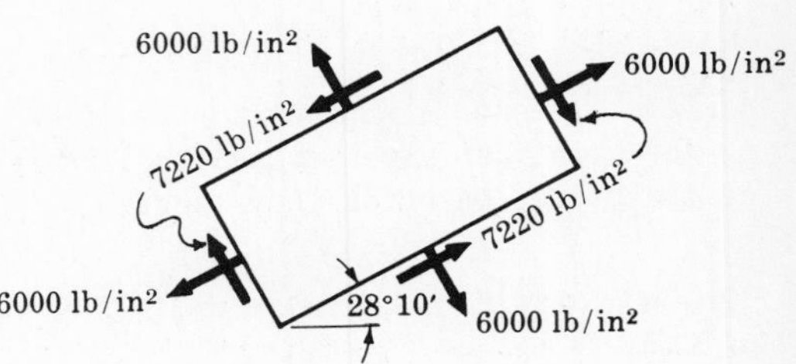

11. A plane element in a body is subject to a normal compressive stress in the x-direction of 12,000 lb/in^2 as well as a shearing stress of 4000 lb/in^2 as shown in Fig. (a) below. (a) Determine the normal and shearing stress intensities on a plane inclined at an angle of 30° to the normal stress. (b) Determine the maximum and minimum values of the normal stress that may exist on inclined planes and the direction of these stresses. (c) Find the magnitude and direction of the maximum shearing stress that may exist on an inclined plane.

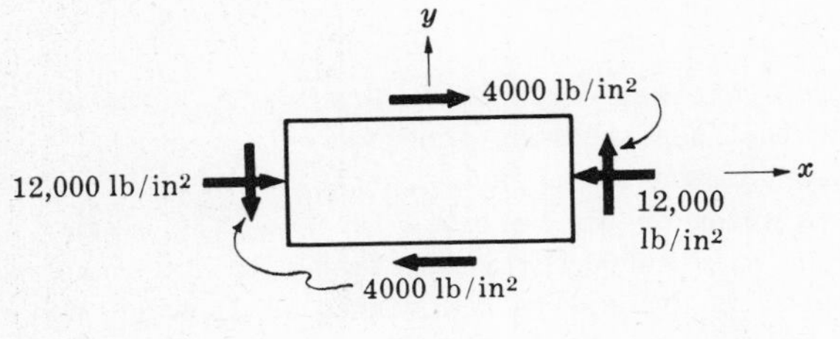

Fig. (a)

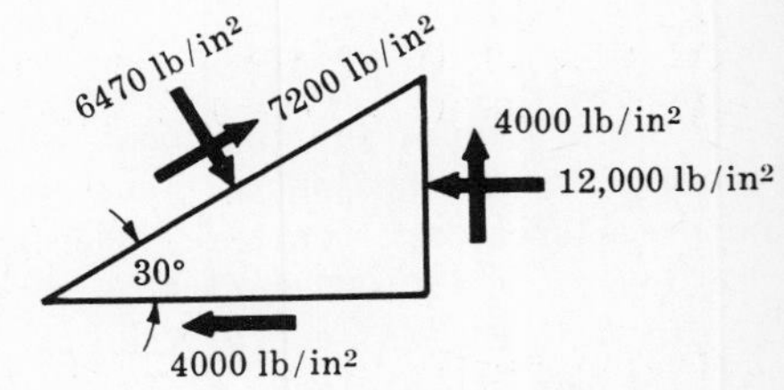

Fig. (b)

a) By the sign convention for normal and shearing stresses adopted in Problem 7 we have here s_x = $-12,000$ lb/in^2, $s_{xy} = -4000$ lb/in^2. From equation (1) of Problem 7 the normal stress on the 30° plane is

$$s_n = -12,000/2 - (-12,000/2)\cos 60° - 4000 \sin 60° = -6470 \text{ lb/in}^2$$

From equation (2) of Problem 7 the shearing stress on the 30° plane is

$$s_s = \tfrac{1}{2}(-12,000)\sin 60° - 4000\cos 60° = -7200 \text{ lb/in}^2$$

The positive directions of the normal and shearing stresses on an inclined plane were illustrated in the second diagram of Problem 7. By this sign convention the stresses on the 30° plane appear as in Fig. (b) above.

b) The values of the principal stresses were given by equations (5) and (6) of Problem 7. From equation (5), we have

$$(s_n)_{max} = -12,000/2 + \sqrt{(-12,000/2)^2 + (-4000)^2} = 1220 \text{ lb/in}^2$$

From equation *(6)*,

$$(s_n)_{min} = -12,000/2 - \sqrt{(-12,000/2)^2 + (-4000)^2} = -13,220 \text{ lb/in}^2$$

The tensile principal stress is usually referred to as the maximum, even though its absolute value is smaller than that of the compressive stress.

The directions of the planes on which these principal stresses occur are given by equation *(3)* of Problem 7 to be

$$\tan 2\theta_p = -\frac{s_{xy}}{\frac{1}{2}s_x} = -\frac{-4000}{-12,000/2} = -2/3$$

The angles defined by $2\theta_p$ lie in the second and fourth quadrants since the tangent is negative. Hence $2\theta_p = 146°20'$ and $2\theta_p' = 326°20'$. Thus the principal planes are defined by $\theta_p = 73°10'$ and $\theta_p' = 163°10'$. If $\theta_p = 73°10'$, together with the given values of s_x and s_{xy}, is now substituted in equation *(1)* of Problem 7 we find

$$\begin{aligned} s_n &= \tfrac{1}{2}s_x - \tfrac{1}{2}s_x \cos 2\theta + s_{xy} \sin 2\theta \\ &= -12,000/2 - (-12,000/2)\cos 146°20' - 4000 \sin 146°20' = -13,220 \text{ lb/in}^2 \end{aligned}$$

Thus the principal stress of $-13,220$ lb/in^2 occurs on the principal plane oriented at 73°10' to the x-axis. The principal stresses thus appear as in the adjacent diagram. The shearing stresses on these principal planes are zero.

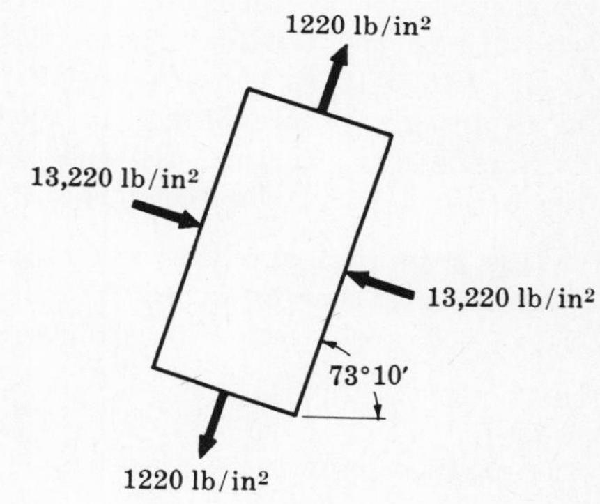

c) The value of the maximum shearing stress is found from equation *(8)* of Problem 7 to be

$$\begin{aligned} s_s &= \pm\sqrt{(\tfrac{1}{2}s_x)^2 + (s_{xy})^2} \\ &= \pm\sqrt{(-12,000/2)^2 + (-4000)^2} = \pm 7220 \text{ lb/in}^2 \end{aligned}$$

The directions of the planes on which these shearing stresses occur was found in equation *(7)* of Problem 7 to be

$$\tan 2\theta_s = \frac{\frac{1}{2}s_x}{s_{xy}} = \frac{-12,000/2}{-4000} = \frac{3}{2}$$

Thus $2\theta_s = 56°20'$ and $2\theta_s' = 236°20'$; then $\theta_s = 28°10'$ and $\theta_s' = 118°10'$. From equation *(2)* of Problem 7, the shearing stress on any plane inclined at an angle θ with the x-axis is

$$\begin{aligned} s_s &= \tfrac{1}{2}s_x \sin 2\theta + s_{xy} \cos 2\theta \\ &= \tfrac{1}{2}(-12,000)\sin 56°20' - 4000 \cos 56°20' = -7220 \text{ lb/in}^2 \end{aligned}$$

Thus the shearing stress on the 28°10' plane is negative. The positive sense of shearing stress was shown in the second diagram of Problem 7.

The normal stresses on the planes of maximum shearing stress were found in equation *(9)* of Problem 7 to be

$$s_n' = \tfrac{1}{2}s_x = -12,000/2 = -6000 \text{ lb/in}^2$$

This normal stress acts on each of the planes of maximum shearing stress, as shown in the adjoining diagram.

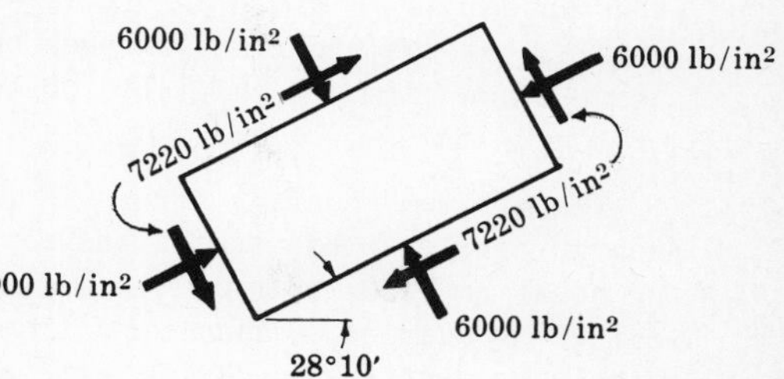

12. A plane element is subject to the stresses shown in Fig. (*a*) below. Using Mohr's circle, determine (*a* the principal stresses and their directions, (*b*) the maximum shearing stresses and the directions o the planes on which they occur.

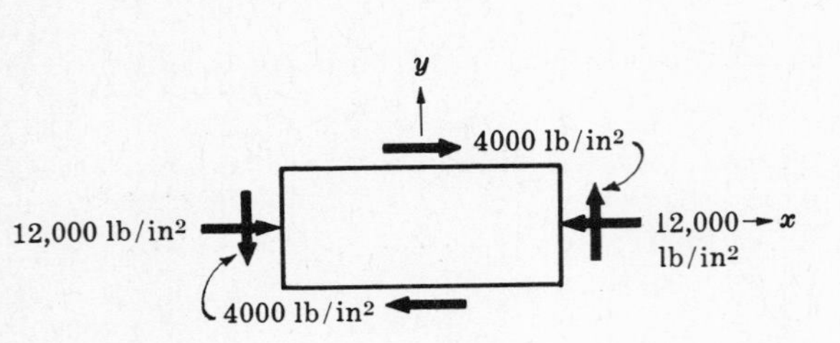

Fig. (*a*)

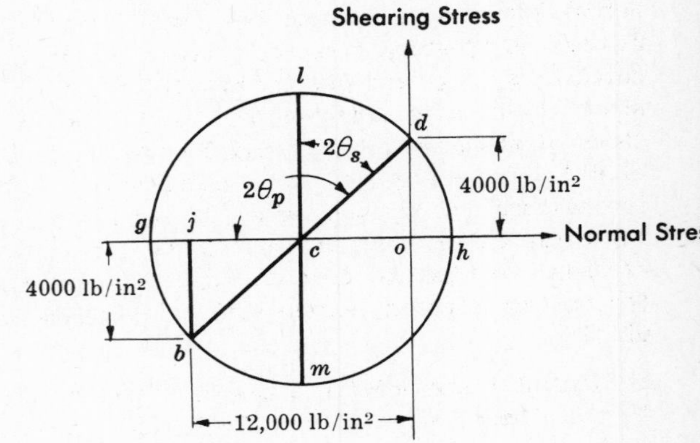

Fig. (*b*)

The procedure for the construction of Mohr's circle was outlined in Problem 8. Following the i structions there, the shearing stresses on the vertical faces of the above element are negative, thos on the horizontal faces are positive. Thus the stress condition of $s_x = -12,000$ lb/in², $s_{xy} = -4000$ lb/i existing on the vertical faces of the element plots as point *b* in Fig. (*b*) above. The stress conditi of $s_{xy} = 4000$ lb/in² together with a zero normal stress on the horizontal faces plots as point *d*. Li *bd* is drawn, its midpoint *c* is located, and a circle of radius $cb = cd$ is drawn with *c* as a cente This is Mohr's circle. The end points of the diameter *bd* represent the stress conditions existing the element if it has the original orientation shown above.

The principal stresses are represented by points *g* and *h*, as demonstrated in Problem 8. They may determined either by direct measurement from the above diagram or by realizing that the coordinate *c* is -6000, and that $cd = \sqrt{(6000)^2 + (4000)^2} = 7220$. Thus the minimum principal stress is

$$(s_n)_{\min} = og = +(oc + cg) = -6000 - 7220 = -13,220 \text{ lb/in}^2$$

The maximum principal stress is

$$(s_n)_{\max} = oh = ch - co = 7220 - 6000 = 1220 \text{ lb/in}^2$$

The angle $2\theta_p$ designated above is given by $\tan 2\theta_p = -4000/6000 = -2/3$ since $\tan(180° - \theta) = -\tan\theta$. Hence $2\theta_p = 146°20'$, $\theta_p = 73°10'$. This value could of course have been obtained by direct measurement of angle *dcg* in Mohr's circle. Thus the principal stress of $-13,220$ lb/in² represented by point *g* acts on a plane oriented 73°10' from the original *x*-axis. The principal stresses thus appear as in Fig. (*c*). It is evident from Mohr's circle that the shearing stresses on these planes are zero, since points *g* and *h* lie on the horizontal axis of Mohr's circle.

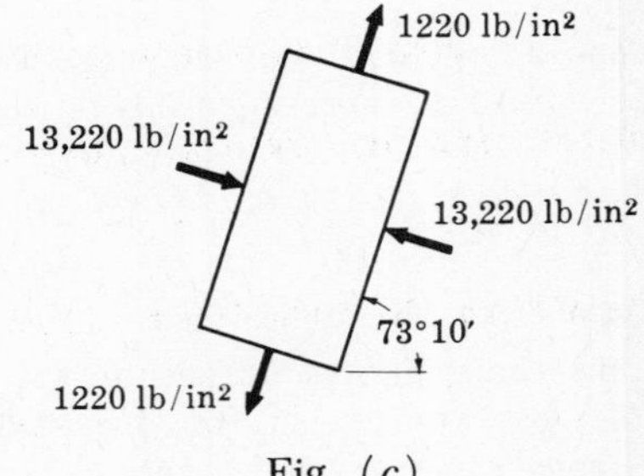

Fig. (*c*)

The maximum shearing stress is represented by *cl* in Mohr's circle. This radius has already been found to be equal to 7220 lb/in². The angle $2\theta_s$ may be found either by direct measurement from Mohr's circle or simply by subtracting 90° from the above value of $2\theta_p$. This leads to $\theta_s = 28°10'$. The shearing stress represented by point *l* is positive, hence on this 28°10' plane the shearing stress tends to rotate the element in a clockwise direction. Also, from Mohr's circle the abscissa of point *l* is -6000 lb/in² and this represents

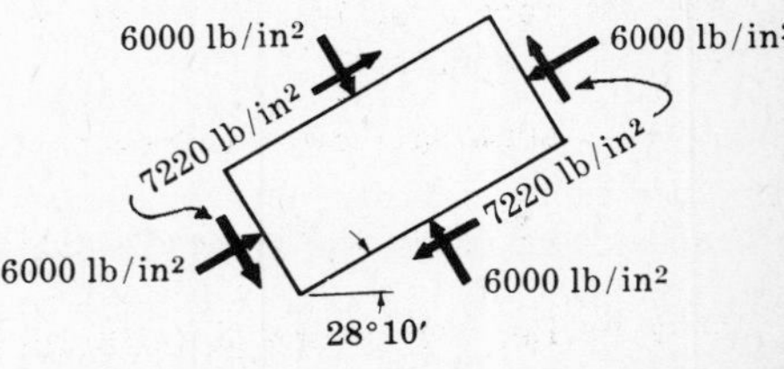

Fig. (*d*)

the normal stress occurring on the planes of maximum shearing stresses, as shown in Fig. (*d*) above.

13. Consider a plane element removed from a stressed elastic member. In general such an element will be subject to normal stresses in each of two perpendicular directions, as well as shearing stresses. Let these stresses be denoted by s_x, s_y, and s_{xy} and have the positive directions shown in the adjoining diagram. (*a*) Determine the magnitudes of the normal and shearing stresses on a plane inclined at an angle θ to the x-axis. (*b*) Also, determine the maximum and minimum values of the normal stress that may exist on inclined planes and the directions of these stresses. (*c*) Lastly, find the magnitude and direction of the maximum shearing stress that may exist on an inclined plane.

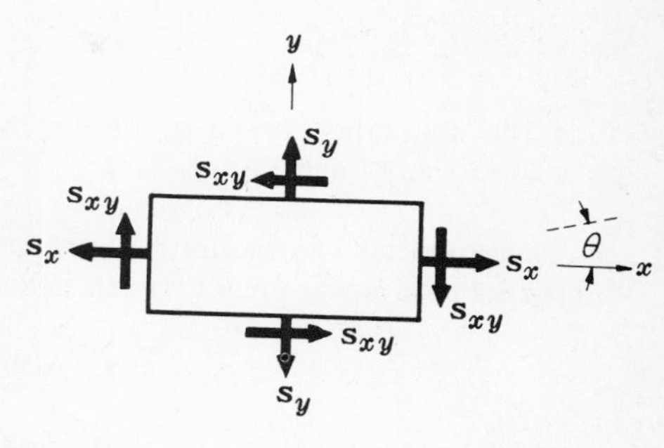

a) Evidently the desired stresses acting on the inclined planes are internal quantities with respect to the element shown above. Following the usual procedure of introducing a cutting plane so as to render the desired quantities external to the new section, we cut the originally rectangular element along the plane inclined at the angle θ to the x-axis and thus obtain the triangular element shown in the adjacent diagram. Since we have removed half of the material in the rectangular element, we must replace it by the effect that it exerted upon the remaining lower triangle shown and this effect in general consists of both normal and shearing forces acting along the inclined plane. We shall designate the magnitudes of the stresses corresponding to these forces by s_n and s_s respectively. Thus our problem reduces to finding the unknown stresses s_n and s_s in terms of the known stresses s_x, s_y, and s_{xy}. Chapter 16 illustrates the manner of determination of the stresses s_x, s_y, and s_{xy}. It is to be carefully noted that the free-body diagram at the right indicates stresses acting on the various faces of the element, and not forces. Each of these stresses is assumed to be uniformly distributed over the area on which it acts.

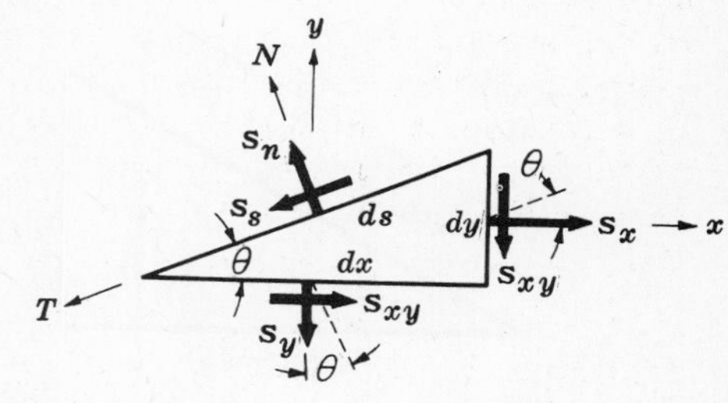

We shall introduce the N and T axes normal and tangential to the inclined plane as shown. Let t denote the thickness of the element perpendicular to the plane of the page. Let us begin by summing forces in the N direction. For equilibrium we have

$$\Sigma F_N = s_n\,t\,ds - s_x\,t\,dy\,\sin\theta - s_{xy}\,t\,dy\,\cos\theta - s_y\,t\,dx\,\cos\theta - s_{xy}\,t\,dx\,\sin\theta = 0$$

Substituting $dy = ds\,\sin\theta$, $dx = ds\,\cos\theta$ in the equilibrium equation,

$$s_n\,ds = s_x\,ds\,\sin^2\theta + s_y\,ds\,\cos^2\theta + 2s_{xy}\,ds\,\sin\theta\,\cos\theta$$

Introducing the identities $\sin^2\theta = \frac{1}{2}(1-\cos 2\theta)$, $\cos^2\theta = \frac{1}{2}(1+\cos 2\theta)$, $\sin 2\theta = 2\sin\theta\,\cos\theta$, we find

$$s_n = \tfrac{1}{2}s_x(1-\cos 2\theta) + \tfrac{1}{2}s_y(1+\cos 2\theta) + s_{xy}\sin 2\theta \quad \text{or}$$

1) $$s_n = \tfrac{1}{2}(s_x+s_y) - \tfrac{1}{2}(s_x-s_y)\cos 2\theta + s_{xy}\sin 2\theta$$

Thus the normal stress s_n on any plane inclined at an angle θ with the x-axis is known as a function of s_x, s_y, s_{xy}, and θ.

Next, summing forces acting on the element in the T-direction, we find

$$\Sigma F_T = s_s\,t\,ds - s_x\,t\,dy\,\cos\theta + s_{xy}\,t\,dy\,\sin\theta - s_{xy}\,t\,dx\,\cos\theta + s_y\,t\,dx\,\sin\theta = 0$$

Substituting for dx and dy as before we get

$$s_s\,ds = s_x\,ds\,\sin\theta\cos\theta - s_{xy}\,ds\,\sin^2\theta + s_{xy}\,ds\,\cos^2\theta - s_y\,ds\,\sin\theta\cos\theta$$

Introducing the previous identities and the relation $\cos 2\theta = \cos^2\theta - \sin^2\theta$, this last equation becomes

2) $$s_s = \tfrac{1}{2}(s_x - s_y)\sin 2\theta + s_{xy}\cos 2\theta$$

Thus the shearing stress s_s on any plane inclined at an angle θ with the x-axis is known as a function of s_x, s_y, s_{xy}, and θ.

b) To determine the maximum value that the normal stress s_n may assume as the angle θ varies, we shall differentiate equation *(1)* with respect to θ and set this derivative equal to zero. Thus

$$d(s_n)/d\theta = (s_x - s_y)\sin 2\theta + 2s_{xy}\cos 2\theta = 0$$

Hence the values of θ leading to maximum and minimum values of the normal stress are given by

3) $$\tan 2\theta_p = -\frac{s_{xy}}{\tfrac{1}{2}(s_x - s_y)}$$

The planes defined by the angles θ_p are called *principal planes*. The normal stresses that exist on these planes are designated as *principal stresses*. They are the maximum and minimum values that the normal stress may assume in the element under consideration. The values of the principal stresses may easily be found by considering the following graphical interpretation of equation *(3)*:

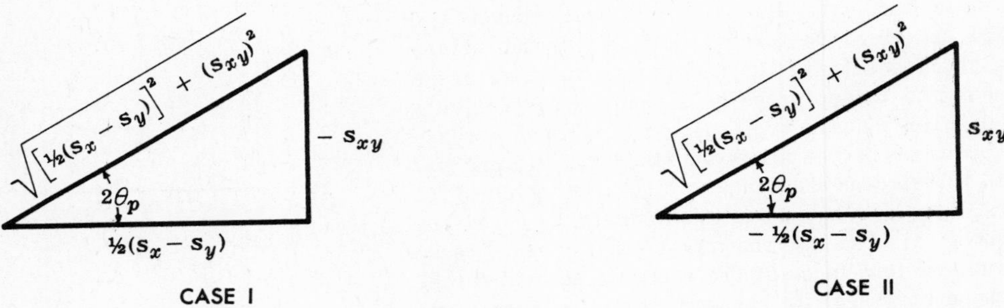

Evidently the tangent of either of these angles designated as $2\theta_p$ has the value given in equation *(3)*. Thus there are two solutions of equation *(3)*, consequently two values of $2\theta_p$ (differing by 180°) and also two values of θ_p (differing by 90°). It is to be noted that the above two diagrams bear no direct relationship to the triangular element whose free-body diagram was considered earlier.

The values of $\sin 2\theta_p$ and $\cos 2\theta_p$ as found from the above two diagrams may now be substituted in equation *(1)* to yield the maximum and minimum values of the normal stresses. Observing that

$$\sin 2\theta_p = \frac{\mp s_{xy}}{\sqrt{[\tfrac{1}{2}(s_x - s_y)]^2 + (s_{xy})^2}}, \qquad \cos 2\theta_p = \frac{\pm\tfrac{1}{2}(s_x - s_y)}{\sqrt{[\tfrac{1}{2}(s_x - s_y)]^2 + (s_{xy})^2}}$$

where the upper signs pertain to Case I and the lower to Case II, we obtain from *(1)*

4) $$s_n = \tfrac{1}{2}(s_x + s_y) \pm \sqrt{[\tfrac{1}{2}(s_x - s_y)]^2 + (s_{xy})^2}$$

The maximum normal stress is

5) $$(s_n)_{max} = \tfrac{1}{2}(s_x + s_y) + \sqrt{[\tfrac{1}{2}(s_x - s_y)]^2 + (s_{xy})^2}$$

The minimum normal stress is

6) $$(s_n)_{min} = \tfrac{1}{2}(s_x + s_y) - \sqrt{[\tfrac{1}{2}(s_x - s_y)]^2 + (s_{xy})^2}$$

The stresses given by equations *(5)* and *(6)* are the principal stresses and they occur on the principal planes defined by equation *(3)*. By substituting one of the values of θ_p from equation *(3)* into equation *(1)*, one may readily determine which of the two principal stresses is acting on that plane. The other principal stress naturally acts on the other principal plane.

By substituting the values of the angle $2\theta_p$ as given by equation *(3)* and the above two diagrams representing the sine and cosine functions into equation *(2)*, it is readily seen that the shearing stresses s_s on the principal planes are zero.

c) To determine the maximum value that the shearing stress s_s may assume as the angle θ varies, we shall differentiate equation *(2)* with respect to θ and set this derivative equal to zero. Thus

$$d(s_s)/d\theta \;=\; (s_x - s_y)\cos 2\theta \;-\; 2s_{xy}\sin 2\theta \;=\; 0$$

The values of θ leading to the maximum values of the shearing stress are thus

7) $$\tan 2\theta_s \;=\; \tfrac{1}{2}(s_x - s_y)/s_{xy}$$

The planes defined by the two solutions to this equation are the planes of maximum shearing stress.

Again, a graphical interpretation of equation *(7)* is convenient. The two values of the angle $2\theta_s$ satisfying this equation may be represented as follows:

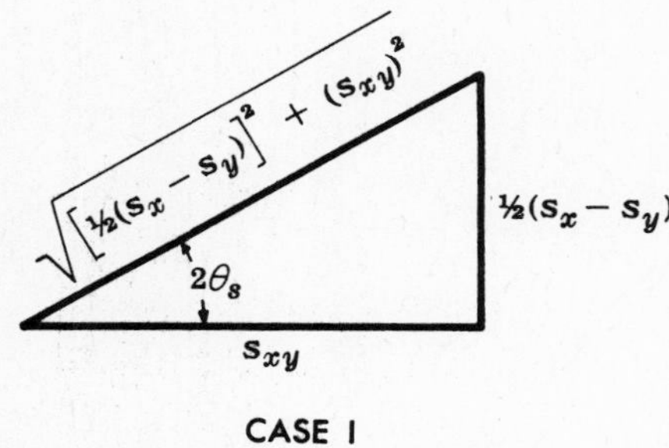

CASE I

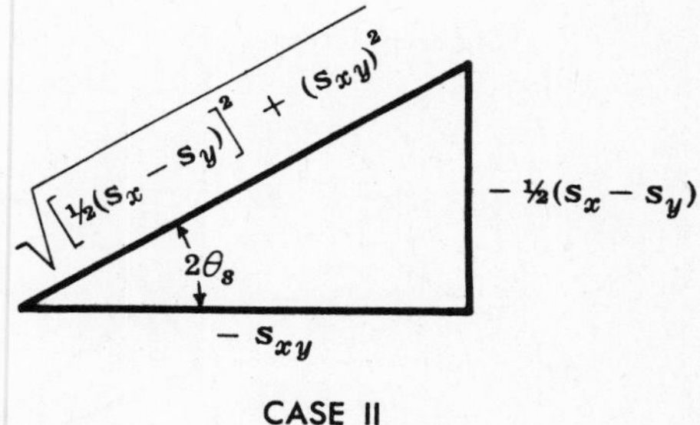

CASE II

From these diagrams we have

$$\sin 2\theta_s \;=\; \frac{\pm\,\tfrac{1}{2}(s_x - s_y)}{\sqrt{[\tfrac{1}{2}(s_x - s_y)]^2 + (s_{xy})^2}} \qquad \cos 2\theta_s \;=\; \frac{\pm\, s_{xy}}{\sqrt{[\tfrac{1}{2}(s_x - s_y)]^2 + (s_{xy})^2}}$$

where the upper (positive) sign refers to Case I and the lower (negative) sign applies to Case II. Substituting these values in equation *(2)* we find

8) $$s_s \;=\; \pm\sqrt{[\tfrac{1}{2}(s_x - s_y)]^2 + (s_{xy})^2}$$

Here the positive sign represents the maximum shearing stress, the negative sign the minimum shearing stress.

If we compare equations *(3)* and *(7)*, it is evident that the angles $2\theta_p$ and $2\theta_s$ differ by 90° since the tangents of these angles are the negative reciprocals of one another. Hence the planes defined by the angles θ_p and θ_s differ by 45°, i.e. the planes of maximum shearing stress are oriented 45° from the planes of maximum normal stress.

It is also of interest to determine the normal stresses on the planes of maximum shearing stress. These planes are defined by equation *(7)*. If we now substitute these values of $\sin 2\theta_s$ and $\cos 2\theta_s$ in equation *(1)* for normal stress we find

9) $$s_n' \;=\; \tfrac{1}{2}(s_x + s_y)$$

Thus on each of the planes of maximum shearing stress is a normal stress of magnitude $\tfrac{1}{2}(s_x + s_y)$.

14. Discuss a graphical representation of the analysis presented in Problem 12.

For given values of s_x, s_y, and s_{xy} we proceed as follows:

a) Introduce a rectangular coordinate system in which normal stresses are represented along the horizontal axis and shearing stresses along the vertical axis. The scales used on these two axes must be equal.

b) With reference to the original rectangular element considered in Problem 13 and reproduced in the adjacent diagram we shall introduce the sign convention that shearing stresses are positive if they tend to rotate the element clockwise, negative if they tend to rotate it counterclockwise. Here the shearing stresses on the vertical faces are positive, those on the horizontal faces are negative. Also, tensile normal stresses are considered to be positive, compressive stresses negative.

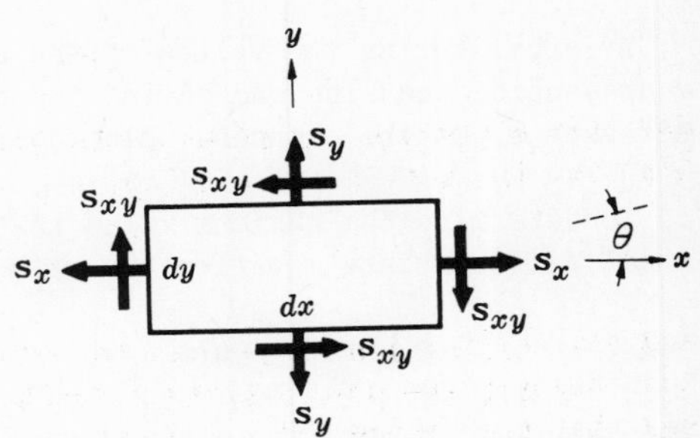

c) We first locate point *b* by laying out s_x and s_{xy} to their given values. The shear stress s_{xy} on the vertical faces on which s_x acts is positive, hence this value is plotted as positive in the diagram below.

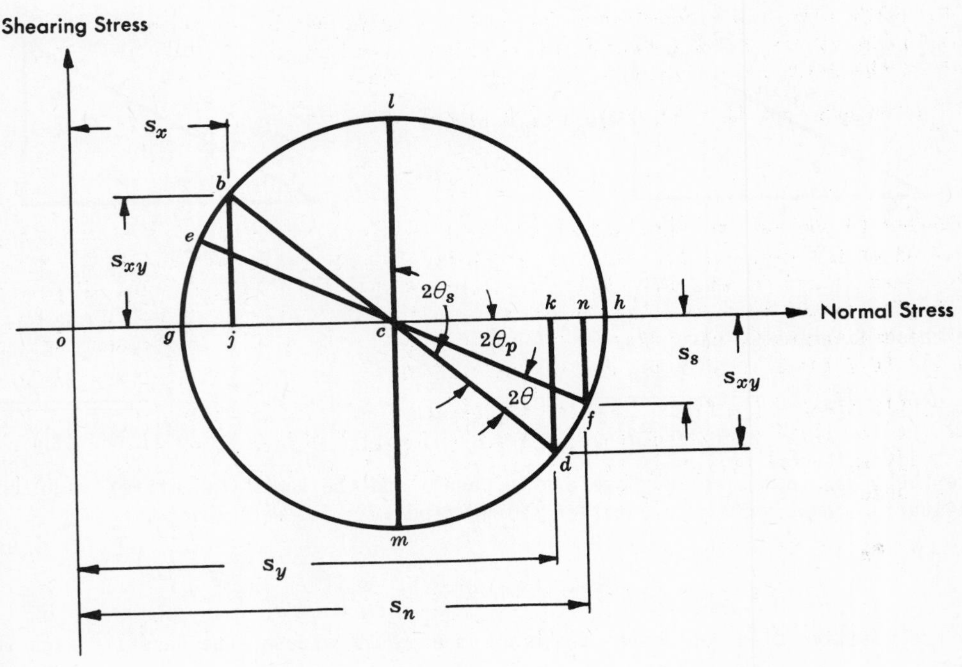

d) We next locate point *d* in a similar manner by laying off s_y and s_{xy} to their given values. The diagram above is drawn on the assumption that $s_y > s_x$, although the treatment presented here holds if $s_y < s_x$. The shear stress s_{xy} on the horizontal faces on which s_y acts is negative, hence this value is plotted below the reference axis.

e) Next, we draw line *bd*, locate midpoint *c*, and draw a circle having its center at *c* and radius equal to *cb*. This is known as Mohr's circle.

We shall first show that the points *g* and *h* along the horizontal diameter of the circle represent the principal stresses. To do this we note that the point *c* lies at a distance $\frac{1}{2}(s_x + s_y)$ from the origin of the coordinate system. Also, the line segment *jk* is of length $(s_y - s_x)$, hence *ck* is of length

$\frac{1}{2}(s_y - s_x)$. From the right triangle relationship we have

$$(cd)^2 = (ck)^2 + (kd)^2 \quad \text{or} \quad cd = \sqrt{[\tfrac{1}{2}(s_x - s_y)]^2 + (s_{xy})^2}$$

Also, $cg = ch = cd$. Hence the x-coordinate of point h is $(oc + ch)$ or

$$\tfrac{1}{2}(s_x + s_y) + \sqrt{[\tfrac{1}{2}(s_x - s_y)]^2 + (s_{xy})^2}$$

But this expression is exactly the maximum principal stress, as found in equation *(5)* of Problem 13. Likewise the x-coordinate of point g is $(oc - gc)$ or

$$\tfrac{1}{2}(s_x + s_y) - \sqrt{[\tfrac{1}{2}(s_x - s_y)]^2 + (s_{xy})^2}$$

and this expression is exactly the minimum principal stress, as found in equation *(6)* of Problem 13. Consequently the points g and h represent the principal stresses existing in the original element. We see that the tangent of $\angle kcd = dk/ck = s_{xy}/\frac{1}{2}(s_y - s_x)$. But from equation *(3)* of Problem 13 we had

$$\tan 2\theta_p = -\frac{s_{xy}}{\frac{1}{2}(s_x - s_y)}$$

and by comparison of these two relations we see that $\angle kcd = 2\theta_p$, i.e. a counterclockwise rotation from the diameter bd (corresponding to the stresses in the x-y directions) leads us to the diameter gh, representing the principal planes, on which the principal stresses occur. The principal planes lie at an angle θ_p from the x-direction.

Thus Mohr's circle is a convenient device for finding the principal stresses, since one can merely establish the circle for a given set of stresses s_x, s_y, s_{xy}, then measure og and oh. These abscissas represent the principal stresses to the same scale used in plotting s_x, s_y, s_{xy}.

It is now apparent that the radius of Mohr's circle, represented by cd, where

$$cd = \sqrt{[\tfrac{1}{2}(s_x - s_y)]^2 + (s_{xy})^2}$$

corresponds to the maximum shearing stress as found in equation *(8)* of Problem 13. Actually, the shearing stress on any plane is represented by the ordinate to Mohr's circle; hence we should consider the radial lines cl and cm as representing the maximum shearing stresses. The angle dcl is evidently $2\theta_s$ and hence it is apparent that the double angle between the planes of maximum normal stress and the planes of maximum shearing stress ($\angle kcl$) is $90°$; hence the planes of maximum shearing stress are oriented $45°$ from the planes of maximum normal stress.

Evidently the end points of the diameter bd represent the stresses acting in the original x and y directions. We shall now demonstrate that the end points of any other diameter such as ef (at an angle 2θ with bd) represent the stresses on a plane inclined at an angle θ to the x-axis. To do this we note that the abscissa of point f is given by

$$\begin{aligned} s_n &= oc + cn \\ &= \tfrac{1}{2}(s_x + s_y) + (cf)\cos(2\theta_p - 2\theta) \\ &= \tfrac{1}{2}(s_x + s_y) + (cf)(\cos 2\theta_p \cos 2\theta + \sin 2\theta_p \sin 2\theta) \\ &= \tfrac{1}{2}(s_x + s_y) + \sqrt{[\tfrac{1}{2}(s_x - s_y)]^2 + (s_{xy})^2}\,(\cos 2\theta_p \cos 2\theta + \sin 2\theta_p \sin 2\theta) \end{aligned}$$

But from an inspection of triangle ckd appearing in Mohr's circle it is evident that

1) $$\sin 2\theta_p = \frac{s_{xy}}{\sqrt{[\frac{1}{2}(s_x - s_y)]^2 + (s_{xy})^2}}, \qquad \cos 2\theta_p = \frac{\frac{1}{2}(s_y - s_x)}{\sqrt{[\frac{1}{2}(s_x - s_y)]^2 + (s_{xy})^2}}$$

Substituting the values of s_{xy} and $\frac{1}{2}(s_y - s_x)$ from these last two equations into the previous equation, we find

$$s_n = \tfrac{1}{2}(s_x + s_y) - \tfrac{1}{2}(s_x - s_y)\cos 2\theta + s_{xy}\sin 2\theta$$

But this is exactly the normal stress on a plane inclined at an angle θ to the x-axis as derived in equation *(1)* of Problem 13.

Next we observe that the ordinate of point f is given by

$$\begin{aligned} s_s = nf &= (cf)\sin(2\theta_p - 2\theta) \\ &= (cf)(\sin 2\theta_p \cos 2\theta - \cos 2\theta_p \sin 2\theta) \\ &= \sqrt{[\tfrac{1}{2}(s_x - s_y)]^2 + (s_{xy})^2}\,(\sin 2\theta_p \cos 2\theta - \cos 2\theta_p \sin 2\theta) \end{aligned}$$

Again, substituting the values of s_{xy} and $\frac{1}{2}(s_y - s_x)$ from equation *(1)* into this equation we find

$$s_s = s_{xy}\cos 2\theta + \tfrac{1}{2}(s_x - s_y)\sin 2\theta$$

But this is exactly the shearing stress on a plane inclined at an angle θ to the x-axis as derived in equation *(2)* of Problem 13.

Hence the coordinates of point f on Mohr's circle represent the normal and shearing stresses on a plane inclined at an angle θ to the x-axis.

15. A plane element is subject to the stresses shown in the adjacent diagram. Determine *(a)* the principal stresses and their directions, *(b)* the maximum shearing stresses and the directions of the planes on which they occur.

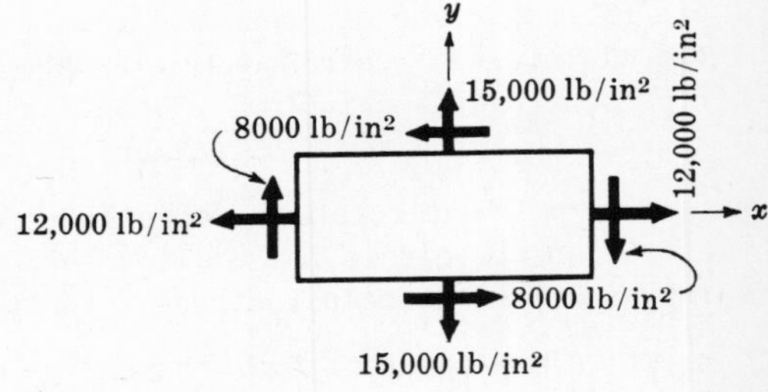

a) In accordance with the notation of Problem 13 we have $s_x = 12,000$ lb/in², $s_y = 15,000$ lb/in², and $s_{xy} = 8000$ lb/in². The maximum normal stress is, by equation *(5)* of Problem 13,

$$\begin{aligned} (s_n)_{max} &= \tfrac{1}{2}(s_x + s_y) + \sqrt{[\tfrac{1}{2}(s_x - s_y)]^2 + (s_{xy})^2} \\ &= \tfrac{1}{2}(12,000 + 15,000) + \sqrt{[\tfrac{1}{2}(12,000 - 15,000)]^2 + (8000)^2} \\ &= 13,500 + 8150 = 21,650 \text{ lb/in}^2 \end{aligned}$$

The minimum normal stress is given by equation *(6)* of Problem 13 to be

$$\begin{aligned} (s_n)_{min} &= \tfrac{1}{2}(s_x + s_y) - \sqrt{[\tfrac{1}{2}(s_x - s_y)]^2 + (s_{xy})^2} \\ &= 13,500 - 8150 = 5350 \text{ lb/in}^2 \end{aligned}$$

From equation *(3)* of Problem 13 the directions of the principal planes on which these stresses of 21,650 lb/in² and 5350 lb/in² occur are given by

$$\tan 2\theta_p = -\frac{s_{xy}}{\frac{1}{2}(s_x - s_y)} = -\frac{8000}{\frac{1}{2}(12,000 - 15,000)} = 5.33$$

Then $2\theta_p = 79°24'$, $259°24'$ and $\theta_p = 39°42'$, $129°42'$.

To determine which of the above principal stresses occurs on each of these planes we return to equation *(1)* of Problem 13, namely

$$s_n = \tfrac{1}{2}(s_x + s_y) - \tfrac{1}{2}(s_x - s_y)\cos 2\theta + s_{xy}\sin 2\theta$$

and substitute $\theta = 39°42'$ together with the given values of s_x, s_y, and s_{xy} to obtain

$$s_n = \tfrac{1}{2}(12,000 + 15,000) - \tfrac{1}{2}(12,000 - 15,000)\cos 79°24' + 8000\sin 79°24' = 21,650 \text{ lb/in}^2$$

Thus an element oriented along the principal planes and subject to the above principal stresses appears as in the following diagram. The shearing stresses on these planes are zero.

The maximum shearing stress was found in equation *(8)* of Problem 13 to be

$$s_s = \pm\sqrt{[\tfrac{1}{2}(s_x - s_y)]^2 + (s_{xy})^2}$$
$$= \pm\sqrt{[\tfrac{1}{2}(12,000 - 15,000)]^2 + (8000)^2}$$
$$= \pm\ 8150\ \text{lb/in}^2$$

From equation *(7)* of Problem 13 the planes on which these maximum shearing stresses occur are defined by the equation

$$\tan 2\theta_s = \frac{\tfrac{1}{2}(s_x - s_y)}{s_{xy}} = -\ 0.188$$

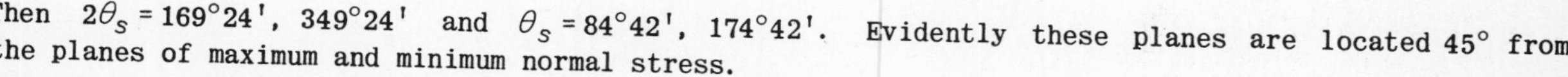

Then $2\theta_s = 169°24'$, $349°24'$ and $\theta_s = 84°42'$, $174°42'$. Evidently these planes are located 45° from the planes of maximum and minimum normal stress.

To determine whether the shearing stress is positive or negative on the 84°42′ plane, we return to equation *(2)* of Problem 13, namely

$$s_s = \tfrac{1}{2}(s_x - s_y)\sin 2\theta + s_{xy}\cos 2\theta$$

and substitute $\theta = 84°42'$ together with the given values of s_x, s_y, and s_{xy} to obtain

$$s_s = \tfrac{1}{2}(12,000 - 15,000)\sin 169°24' + 8000\cos 169°24' = -\ 8150\ \text{lb/in}^2$$

The negative sign indicates that the shearing stress is directed oppositely to the assumed positive direction shown in the first figure in Problem 13. Lastly, the normal stresses on these planes of maximum shearing stress are found from equation *(9)* of Problem 13 to be

$$s_n' = \tfrac{1}{2}(s_x + s_y)$$
$$= \tfrac{1}{2}(12,000 + 15,000) = 13,500\ \text{lb/in}^2$$

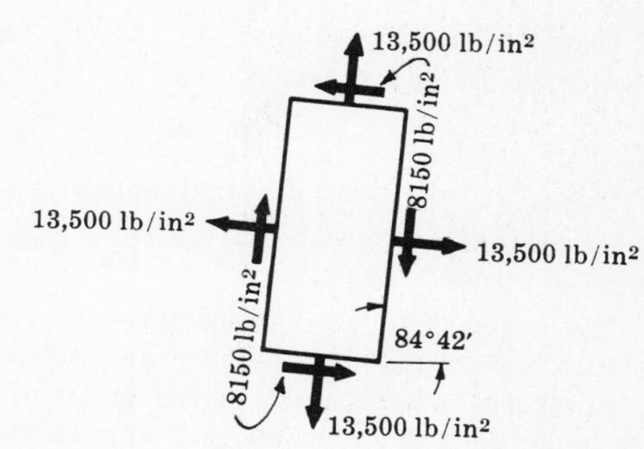

Consequently the orientation of the element for which the shearing stresses are maximum appears as in the adjoining diagram.

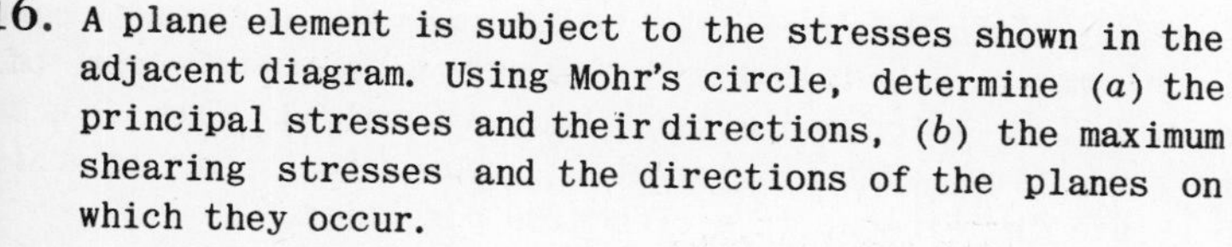

16. A plane element is subject to the stresses shown in the adjacent diagram. Using Mohr's circle, determine *(a)* the principal stresses and their directions, *(b)* the maximum shearing stresses and the directions of the planes on which they occur.

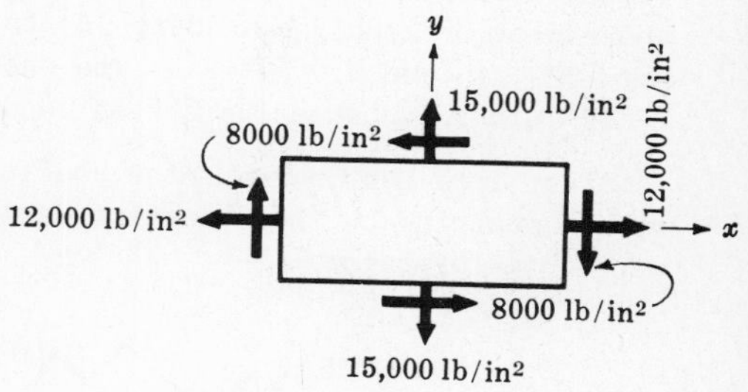

The procedure for the construction of Mohr's circle was outlined in Problem 14. Proceeding in accordance with the instructions there, we realize that the shearing stresses on the vertical faces of the above element are positive, whereas those on the horizontal faces are negative. Thus the stress condition of $s_x = 12,000\ \text{lb/in}^2$, $s_{xy} = 8000\ \text{lb/in}^2$ existing on the vertical faces of the element plots as point *b* in the diagram below. The stress condition of $s_y = 15,000\ \text{lb/in}^2$, $s_{xy} = -\ 8000\ \text{lb/in}^2$ existing on the horizontal faces plots as point *d*. Line *bd* is drawn, its midpoint *c* is located, and a circle of radius *cb* = *cd* is drawn with *c* as a center. This is Mohr's circle. The end points of the diameter *bd* represent the stress conditions existing in the element if it has the original orientation shown above.

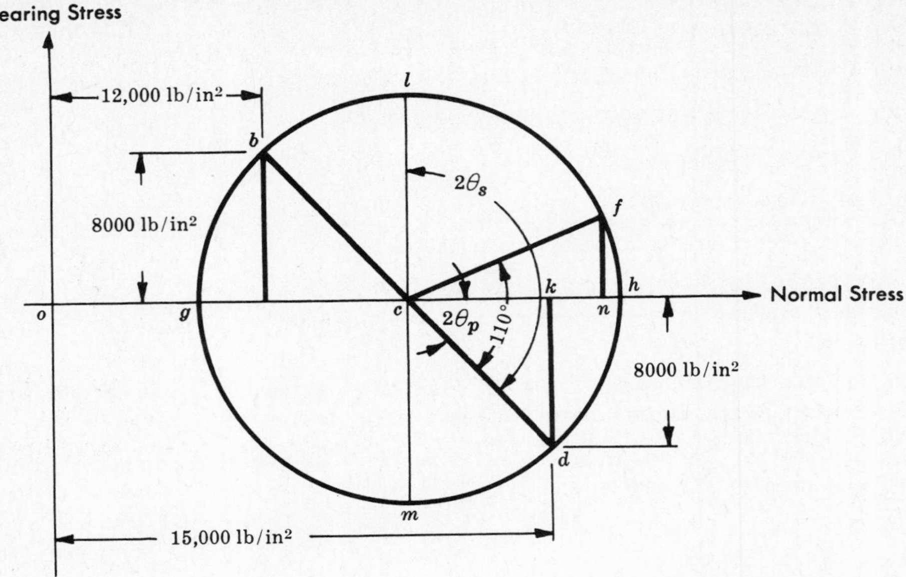

The principal stresses are represented by points g and h as demonstrated in Problem 14. The principal stress may be determined either by direct measurement from the above diagram or by realizing that the coordinate of c is 13,500, that $ck = 1500$ and that $cd = \sqrt{(1500)^2 + (8000)^2} = 8150$. Thus the minimum principal stress is

$$(s_n)_{\min} = og = oc - cg = 13,500 - 8150 = 5350 \text{ lb/in}^2$$

Also, the maximum principal stress is

$$(s_n)_{\max} = oh = oc + ch = 13,500 + 8150 = 21,650 \text{ lb/in}^2$$

The angle $2\theta_p$ is given by $\tan 2\theta_p = 8000/1500 = 5.33$ from which $\theta_p = 39°42'$. This value could also be obtained by measurement of $\angle dck$ in Mohr's circle. From this it is readily seen that the principal stress represented by point h acts on a plane oriented 39°42' from the original x-axis. The principal stresses thus appear as in the Fig. (a) below. It is evident from Mohr's circle that the shearing stresses on these planes are zero, since points g and h lie on the horizontal axis of Mohr's circle.

The maximum shearing stress is represented by cl in Mohr's circle. This radius has already been found to represent 8150 lb/in². The angle $2\theta_s$ may be found either by direct measurement from the above plot or simply by adding 90° to the angle $2\theta_p$, which has already been determined. This leads to $2\theta_s = 169°24'$ and $\theta_s = 84°42'$. The shearing stress represented by point l is positive, hence on this 84°42' plane the shearing stress tends to rotate the element in a clockwise direction.

Also, from Mohr's circle the abscissa of point l is 13,500 lb/in² and this represents the normal stress occuring on the planes of maximum shearing stress. The maximum shearing stresses thus appear as in Fig. (b) below.

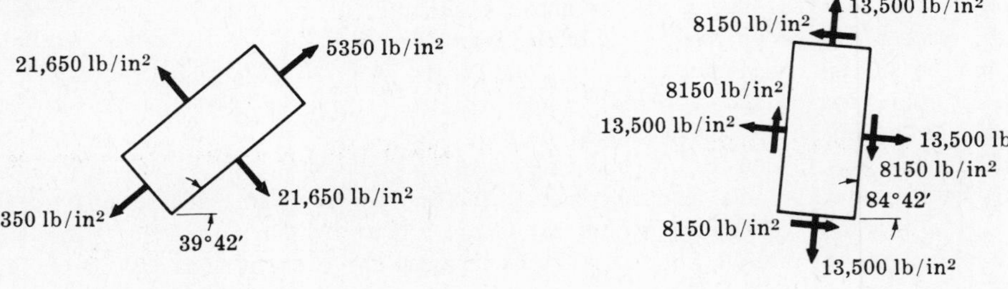

Fig. (a) Fig. (b)

17. For the element discussed in Problem 16, determine the normal and shearing stresses on a plane making an angle of 55° measured counterclockwise from the positive end of the x-axis.

According to the properties of Mohr's circle discussed in Problem 14, we realize that the end points of the diameter bd represent the stress conditions occurring on the original x-y planes. On any plane inclined at an angle θ to the x-axis the stress conditions are represented by the coordinates of a point f, where the radius cf makes an angle of 2θ with the original diameter bd. This angle 2θ appearing in Mohr's circle is measured in the same direction as the angle representing the inclined plane, namely counterclockwise.

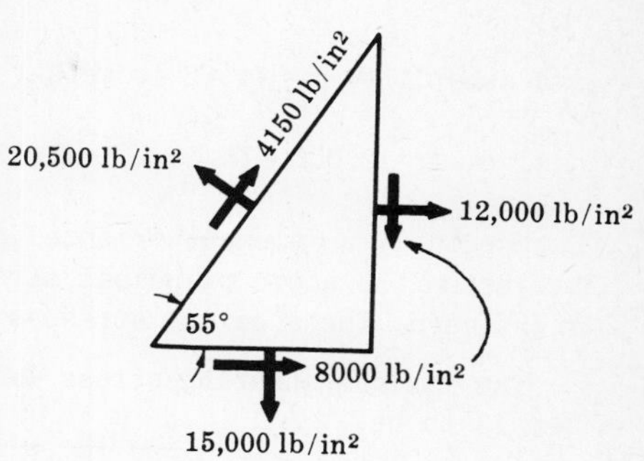

Hence in the Mohr's circle appearing in Problem 16, we merely measure a counterclockwise angle of $2(55°) = 110°$ from line cd. This locates point f. The abscissa of point f represents the normal stress on the desired 55° plane and may be found either by direct measurement or by realizing that

$$on = oc + cn = 13,500 + 8150\cos(110° - 79°24') = 20,500 \text{ lb/in}^2$$

The ordinate of point f represents the shearing stress on the desired 55° plane and may be found from the relation

$$fn = 8150\sin(110° - 79°24') = 4150 \text{ lb/in}^2$$

The stresses acting on the 55° plane may thus be represented as in the above diagram.

18. A plane element is subject to the stresses shown in the adjoining diagram. Determine (a) the principal stresses and their directions, (b) the maximum shearing stresses and the directions of the planes on which they occur.

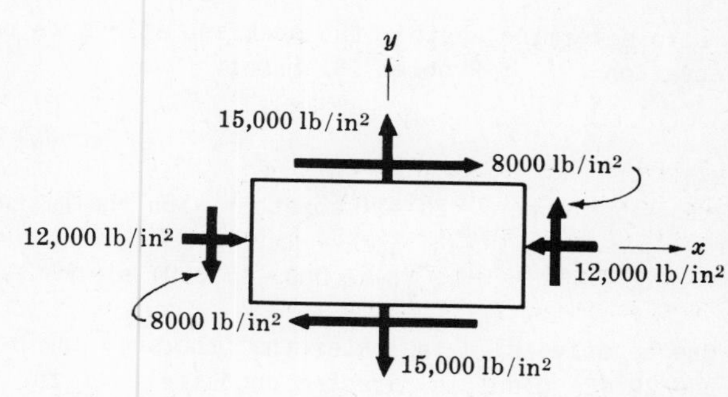

a) In accordance with the notation of Problem 13, $s_x = -12,000$ lb/in², $s_y = 15,000$ lb/in², and $s_{xy} = -8000$ lb/in². The maximum normal stress is given by equation (5) of Problem 13 to be

$$\begin{aligned}(s_n)_{max} &= \tfrac{1}{2}(s_x + s_y) + \sqrt{[\tfrac{1}{2}(s_x - s_y)]^2 + (s_{xy})^2}\\ &= \tfrac{1}{2}(-12,000 + 15,000) + \sqrt{[\tfrac{1}{2}(-12,000 - 15,000)]^2 + (-8000)^2}\\ &= 1500 + 15,700 = 17,200 \text{ lb/in}^2\end{aligned}$$

The minimum normal stress is given by equation (6) of Problem 13 to be

$$\begin{aligned}(s_n)_{min} &= \tfrac{1}{2}(s_x + s_y) - \sqrt{[\tfrac{1}{2}(s_x - s_y)]^2 + (s_{xy})^2}\\ &= 1500 - 15,700 = -14,200 \text{ lb/in}^2\end{aligned}$$

From equation (3) of Problem 13 the directions of the principal planes on which these stresses of 17,200 lb/in² and −14,200 lb/in² occur are given by

$$\tan 2\theta_p = -\frac{s_{xy}}{\tfrac{1}{2}(s_x - s_y)} = -\frac{-8000}{\tfrac{1}{2}(-12,000 - 15,000)} = -0.592$$

Then $2\theta_p = 149°24', 329°24'$ and $\theta_p = 74°42', 164°42'$.

To determine which of the above principal stresses occurs on each of these planes we return to equation *(1)* of Problem 13, namely

$$s_n \;=\; \tfrac{1}{2}(s_x+s_y) \;-\; \tfrac{1}{2}(s_x-s_y)\cos 2\theta \;+\; s_{xy}\sin 2\theta$$

and substitute $\theta = 74°42'$ together with the given values of s_x, s_y, and s_{xy} to obtain

$$s_n \;=\; \tfrac{1}{2}(-12{,}000+15{,}000) \;-\; \tfrac{1}{2}(-12{,}000-15{,}000)\cos 149°24' \;-\; 8000\sin 149°24' \;=\; -14{,}200\ \text{lb/in}^2$$

Consequently an element oriented along the principal planes and subject to the above principal stresses appears as in the adjoining figure. The shearing stresses on these planes are zero.

17,200 lb/in²
14,200 lb/in²
14,200 lb/in
74°42′
17,200 lb/in²

The maximum shearing stress was found in equation *(8)* of Problem 13 to be

$$\begin{aligned} s_s &= \pm\sqrt{[\tfrac{1}{2}(s_x-s_y)]^2 + (s_{xy})^2} \\ &= \pm\sqrt{[\tfrac{1}{2}(-12{,}000-15{,}000)]^2 + (-8000)^2} \\ &= \pm\,15{,}700\ \text{lb/in}^2 \end{aligned}$$

From equation *(7)* of Problem 13 the planes on which these maximum shearing stresses occur are defined by

$$\tan 2\theta_s \;=\; \tfrac{1}{2}(s_x-s_y)/s_{xy} \;=\; 1.69$$

Then $2\theta_s = 59°24'$, $239°24'$ and $\theta_s = 29°42'$, $119°42'$. It is apparent that these planes are located 45 from the planes of maximum and minimum normal stress.

To determine whether the shearing stress is positive or negative on the $29°42'$ plane we return t equation *(2)* of Problem 13, namely

$$s_s \;=\; \tfrac{1}{2}(s_x-s_y)\sin 2\theta \;+\; s_{xy}\cos 2\theta$$

and substitute $\theta = 29°42'$ together with the given values of s_x, s_y, and s_{xy} to obtain

$$s_s \;=\; \tfrac{1}{2}(-12{,}000-15{,}000)\sin 59°24' \;-\; 8000\cos 59°24' \;=\; -15{,}700\ \text{lb/in}^2$$

The negative sign indicates that the shearing stress on the $29°42'$ plane is directed oppositely to the assumed positive direction shown in the first diagram in Problem 13. The normal stresses on these planes of maximum shearing stress were found in equation *(9)* of Problem 13 to be

$$\begin{aligned} s_n' &= \tfrac{1}{2}(s_x+s_y) \\ &= \tfrac{1}{2}(-12{,}000+15{,}000) \;=\; 1500\ \text{lb/in}^2 \end{aligned}$$

Consequently the orientation of the element for which the shearing stresses are a maximum appears as in the adjoining diagram.

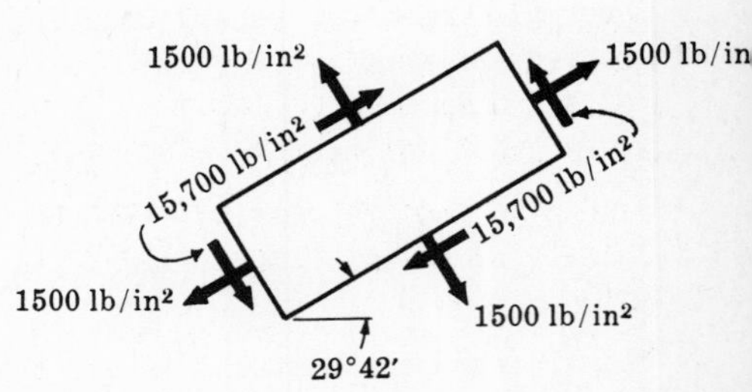

19. A plane element is subject to the stresses shown in Fig. *(a)* below. Using Mohr's circle determin *(a)* The principal stresses and their directions, *(b)* the maximum shearing stresses and the directic of the planes on which they occur.

Again we refer to Problem 14 for the procedure for constructing Mohr's circle. In accordance wi the sign convention outlined there the shearing stresses on the vertical faces of the element a negative, those on the horizontal faces positive. Thus the stress condition of $s_x = -12{,}000$ lb/i

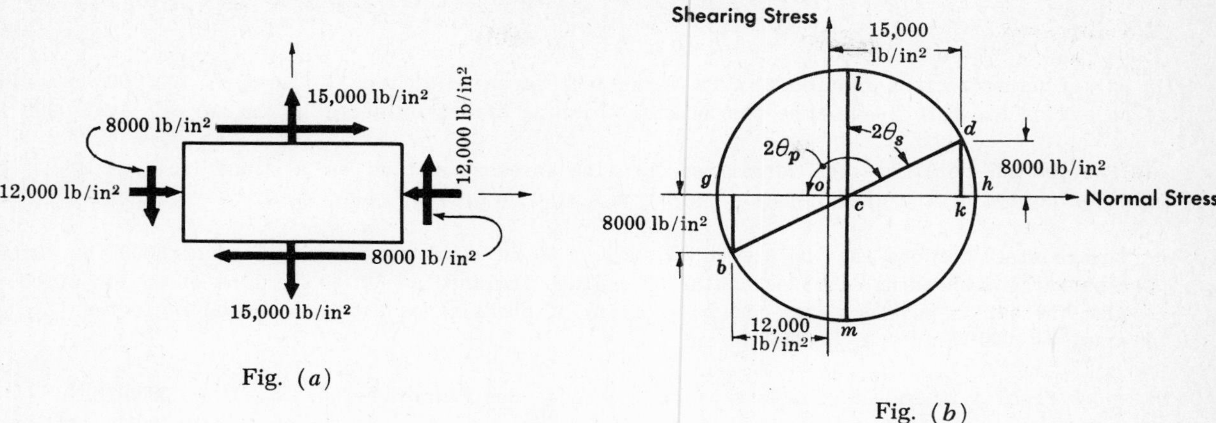

Fig. (a)

Fig. (b)

s_{xy} = − 8000 lb/in^2 existing on the vertical faces of the element plots as point b in Fig. (b) above. The stress condition of s_y = 15,000 lb/in^2, s_{xy} = 8000 lb/in^2 existing on the horizontal faces plots as point d. Line bd is drawn, its midpoint c is located, and a circle of radius $cb = cd$ is drawn with c as a center. This is Mohr's circle. The end points of the diameter bd represent the stress conditions existing in the element if it has the original orientation shown above.

The principal stresses are represented by points g and h as shown in Problem 14. They may be found either by direct measurement from the above diagram or by realizing that the coordinate of c is 1500, that ck = 13,500, and that $cd = \sqrt{(13,500)^2 + (8000)^2}$ = 15,700. Thus the minimum principal stress is

$$(s_n)_{min} = og = oc - cg = 1500 - 15,700 = -14,200 \text{ lb/in}^2$$

Also, the maximum principal stress is

$$(s_n)_{max} = oh = oc + ch = 1500 + 15,700 = 17,200 \text{ lb/in}^2$$

The angle $2\theta_p$ is given by $\tan 2\theta_p = -8000/13,500 = -0.592$, from which $\theta_p = 74°42'$. This value could also be obtained by measurement of $\angle dcg$ in Mohr's circle. From this it is readily seen that the principal stress represented by point g acts on a plane oriented 74°42' from the original x-axis. The principal stresses thus appear as in Fig. (c) below. Since the ordinates of points g and h are each zero the shearing stresses on these planes are zero.

The maximum shearing stress is represented by cl in Mohr's circle. This radius has already been found to represent 15,700 lb/in^2. The angle $2\theta_s$ may be found either by direct measurement from the above plot or simply by subtracting 90° from the angle $2\theta_p$ which has already been determined. This leads to $2\theta_s = 59°24'$ and $\theta_s = 29°42'$. The shearing stress represented by point l is positive, hence on this 29°42' plane the shearing stress tends to rotate the element in a clockwise direction.

Also, from Mohr's circle the abscissa of point l is 1500 lb/in^2 and this represents the normal stress occurring on the planes of maximum shearing stress. The maximum shearing stresses thus appear as in Fig. (d) below.

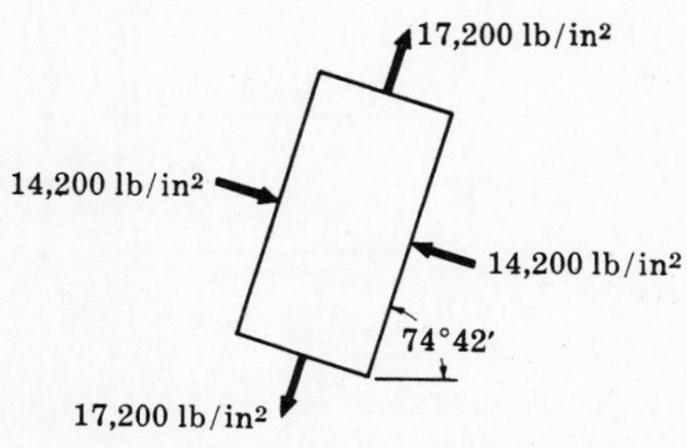

Fig. (c)

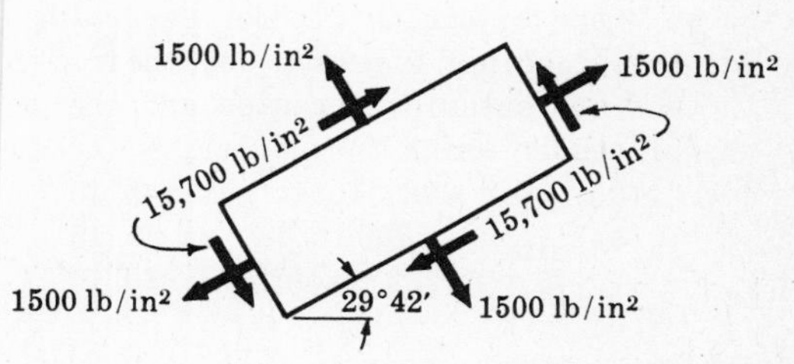

Fig. (d)

SUPPLEMENTARY PROBLEMS

20. A bar of uniform cross-section 2×3 in. is subject to an axial tensile force of 108,000 lb applied at each end of the bar. Determine the maximum shearing stress existing in the bar. *Ans.* 9000 lb/in^2

21. In Problem 20, determine the normal and shearing stresses acting on a plane inclined at 20° to the line of action of the axial loads. *Ans.* s_n = 2100 lb/in^2, s_s = 5800 lb/in^2

22. A square steel bar one inch on a side is subject to an axial compressive load of 8000 lb. Determine the normal and shearing stresses acting on a plane inclined at 30° to the line of action of the axial loads. The bar is so short that the possibility of buckling as a column may be neglected.
Ans. s_n = – 2000 lb/in^2, s_s = – 3460 lb/in^2

23. Re-work Problem 22 by use of Mohr's circle. *Ans.* See Fig. (*a*) below. $s_n = ko$ = – 2000 lb/in^2
$s_s = dk$ = 3460 lb/in^2

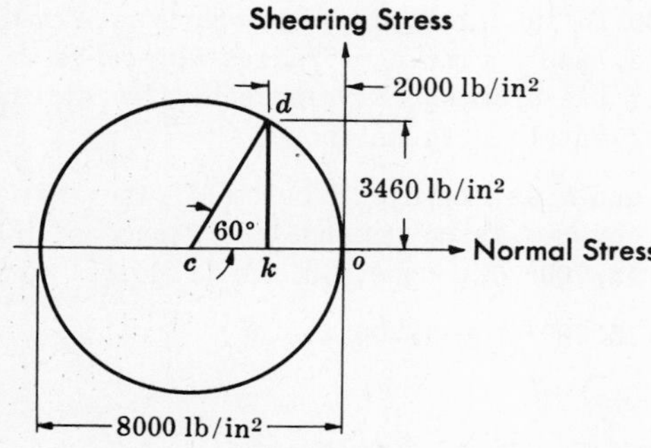

Fig. (*a*) Prob. 23

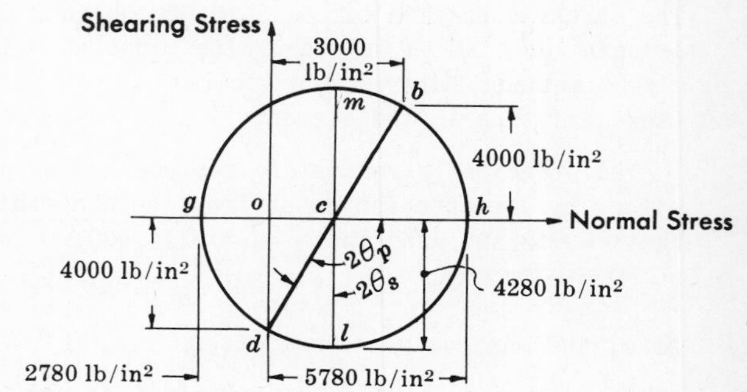

Fig. (*b*) Prob. 26

24. A plane element in a body is subject to the stresses s_x = 3000 lb/in^2, s_y = 0 and s_{xy} = 4000 lb/in^2. Determine analytically the normal and shearing stresses existing on a plane inclined at 45° to the *x*-axis. *Ans.* s_n = 5500 lb/in^2, s_s = 1500 lb/in^2

25. For the element of Problem 24, determine analytically the principal stresses and their directions as well as the maximum shearing stresses and the directions of the planes on which they occur.
Ans. $(s_n)_{max}$ = 5780 lb/in^2 at 55°15', $(s_n)_{min}$ = – 2780 lb/in^2 at 145°15', s_s = 4280 lb/in^2 at 10°15'

26. Re-work Problem 25 by the use of Mohr's circle. *Ans.* See Fig. (*b*) above.

27. A plane element in a body is subject to the stresses shown in the adjoining figure. Determine analytically (*a*) the principal stresses and their directions, (*b*) the maximum shearing stresses and the directions of the planes on which they occur.
Ans. $(s_n)_{max}$ = 2780 lb/in^2 at 145°15'
$(s_n)_{min}$ = – 5780 lb/in^2 at 55°15'
s_s = 4280 lb/in^2 at 10°15'

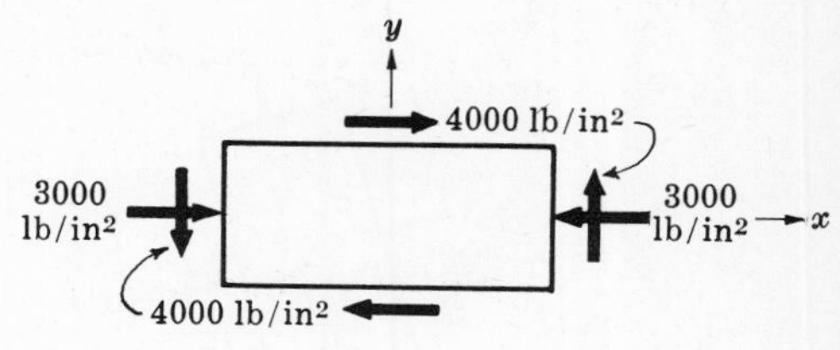

28. For the element shown in Problem 27 determine the normal and shearing stresses acting on a plane inclined at 30° to the *x*-axis. *Ans.* s_n = – 5710 lb/in^2, s_s = – 3300 lb/in^2

29. A plane element is subject to the stresses s_x = 8000 lb/in^2 and s_y = 8000 lb/in^2. Determine analytically the maximum shearing stress existing in the element. *Ans.* Zero

30. What form does Mohr's circle assume for the loading described in Problem 29?
Ans. A point on the horizontal axis, located a distance of 8000 lb/in^2 (to scale) from the origin.

31. A plane element is subject to the stresses s_x = 8000 lb/in^2, and s_y = – 8000 lb/in^2. Determine analytically the maximum shearing stress existing in the element. What is the direction of the planes on which the maximum shearing stresses occur? *Ans.* 8000 lb/in^2 at 45°

32. For the element described in Problem 31, determine analytically the normal and shearing stresses acting on a plane inclined at 30° to the x-axis. *Ans.* s_n = – 4000 lb/in^2, s_s = 6920 lb/in^2

33. Draw Mohr's circle for a plane element subject to the stresses s_x = 8000 lb/in^2 and s_y = – 8000 lb/in^2. From Mohr's circle determine the stresses acting on a plane inclined at 20° to the x-axis.
Ans. See Fig. (*c*) below. s_n = *on* = – 6130 lb/in^2, s_s = *nf* = – 5130 lb/in^2

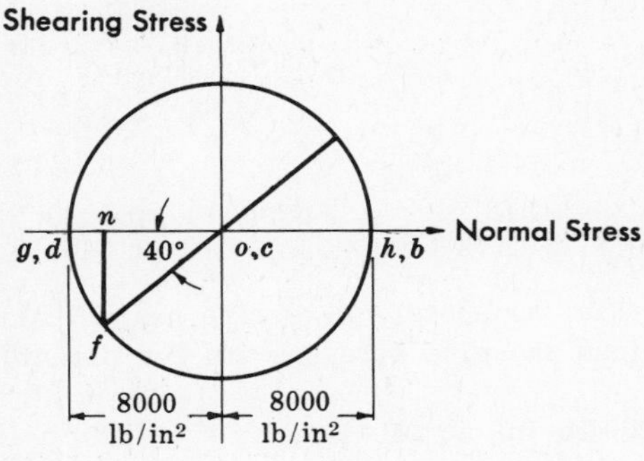

Fig. (*c*) Prob. 33

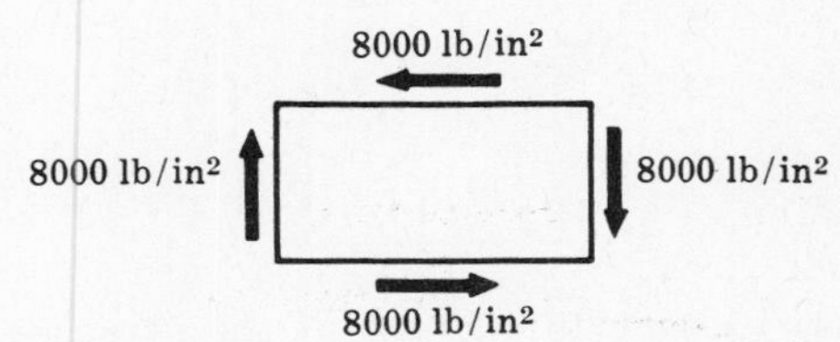

Fig. (*d*) Prob. 34

34. A plane element removed from a thin-walled cylindrical shell loaded in torsion is subject to the shearing stresses shown in Fig. (*d*) above. Determine the principal stresses existing in this element and the directions of the planes on which they occur. *Ans.* 8000 lb/in^2 at 45°

35. A plane element is subject to the stresses shown in Fig. (*e*) below. Determine analytically (*a*) the principal stresses and their directions, (*b*) the maximum shearing stresses and the directions of the planes on which they act.
Ans. $(s_n)_{max}$ = 24,940 lb/in^2 at 121°45', $(s_n)_{min}$ = 7060 lb/in^2 at 31°45', s_s = 8940 lb/in^2 at 76°45'

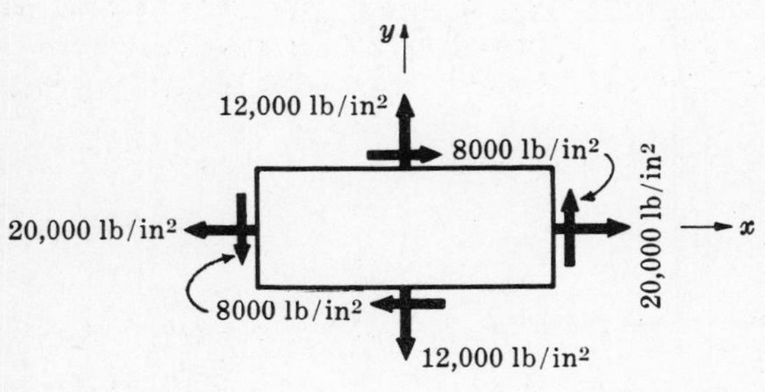

Fig. (*e*) Prob. 35

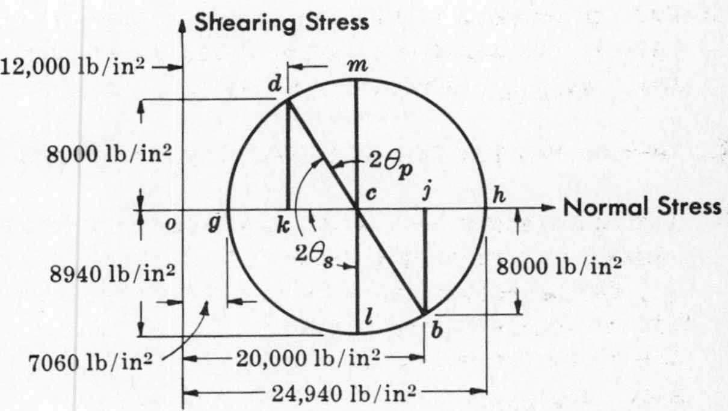

Fig. (*f*) Prob. 36

36. Re-work Problem 35 by the use of Mohr's circle. *Ans.* See Fig. (*f*) above.

37. Consider again the element shown in Problem 35. Determine analytically the normal and shearing stresses on a plane inclined at an angle of 20° to the x-axis. *Ans.* s_n = 7790 lb/in^2, s_s = − 3530 lb/in^2

38. Re-work Problem 37 by the use of Mohr's circle. *Ans.* See Fig. (*g*) below. $s_n = on$ = 7790 lb/in^2
$s_s = nf$ = 3530 lb/in^2

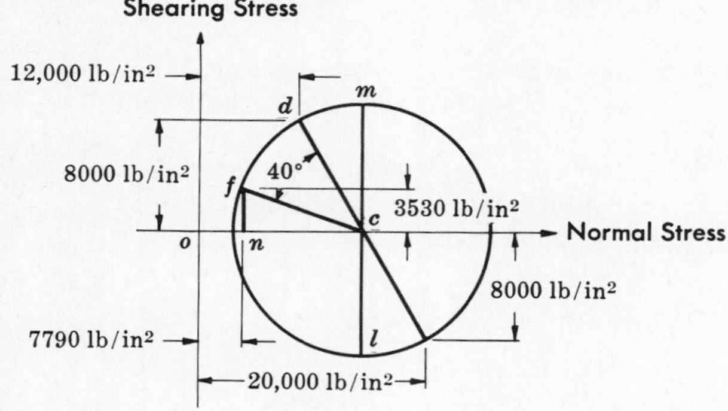

Fig. (*g*) Prob. 38

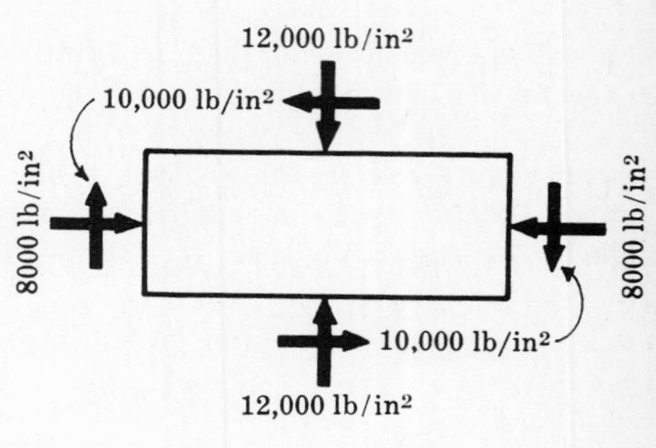

Fig. (*h*) Prob. 39

39. A plane element is subject to the stresses shown in Fig. (*h*) above. Determine analytically (*a*) the principal stresses and their directions, (*b*) the maximum shearing stresses and the directions of the planes on which they act.
Ans. $(s_n)_{max}$ = 200 lb/in^2 at 50°40', $(s_n)_{min}$ = − 20,200 lb/in^2 at 140°40', s_s = 10,200 lb/in^2 at 5°40

40. Re-work Problem 39 by the use of Mohr's circle. *Ans.* See Fig. (*i*) below.

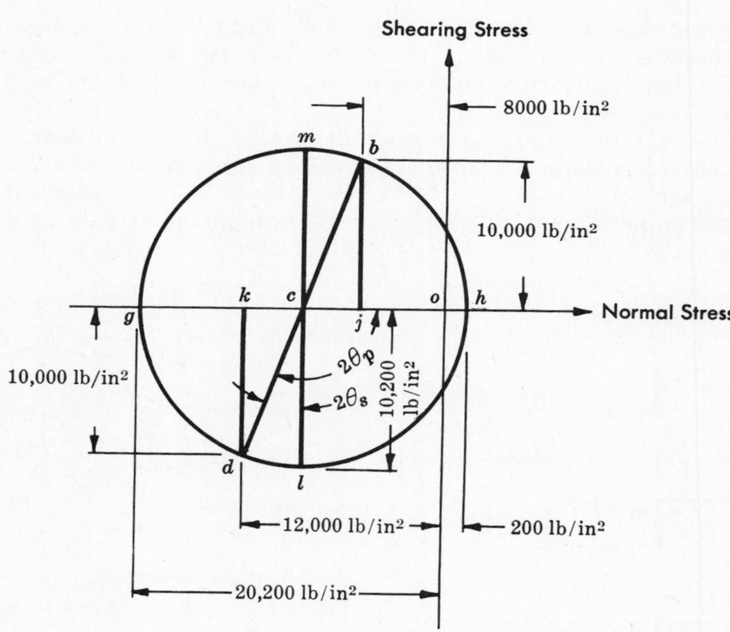

Fig. (*i*) Prob. 40

CHAPTER 16

Eccentrically Loaded Members and Members Subject to Combined Loadings

AXIALLY LOADED MEMBERS SUBJECT TO CONCENTRIC LOADS. In Chapters 1 and 2 we considered numerous cases of straight bars subject to either tensile or compressive loads. In each of these problems it was required that the action line of the applied force pass through the centroid of the cross-section of the member. No problems were considered for which this was not true.

AXIALLY LOADED MEMBERS SUBJECT TO ECCENTRIC LOADS. In this chapter we shall consider those cases where the action line of the applied force acting on a bar in either tension or compression does not pass through the centroid of the cross-section. A typical example of such a loading is shown in the adjoining figure. For those cross-sections of the bar that are perpendicular to the direction of the load, the resultant stress at any point is the sum of the direct stress due to a concentric load of equal magnitude P plus a bending stress due to a couple of moment Pe. This first stress is found from the expression derived in Chapter 1, namely $s = P/A$. The second stress is found from the formula for bending stress presented in Chapter 8, namely $s = My/I$. Applications may be found in Problems 1, 2, and 3.

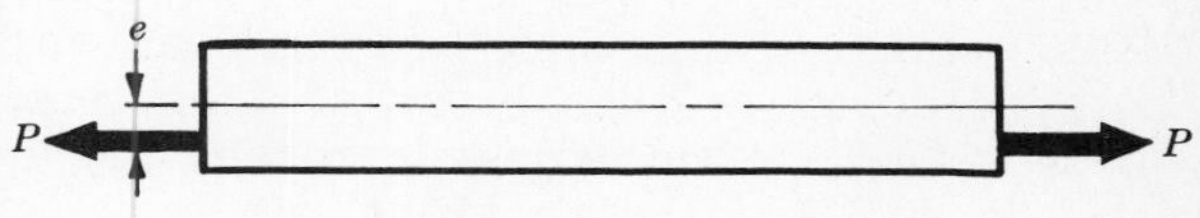

CYLINDRICAL SHELLS SUBJECT TO COMBINED INTERNAL PRESSURE AND AXIAL TENSION. In Chapter 3 we considered the stresses arising in a thin-walled cylindrical shell subject to uniform internal pressure. There it was shown that a longitudinal stress given by $s_L = pr/2t$ as well as a circumferential stress given by $s_c = pr/t$ exist because of the internal pressure p. If in addition an axial tension P is acting simultaneously with the internal pressure, then there arises an additional longitudinal stress given by $s = P/A$ where A denotes the cross-sectional area of the shell. The resultant stress in the longitudinal direction is thus the algebraic sum of these two longitudinal stresses and the resultant stress in the circumferential direction is equal to that due to the internal pressure. Applications may be found in Problem 4.

CYLINDRICAL SHELLS SUBJECT TO COMBINED TORSION AND AXIAL TENSION. In Chapter 5 we considered the stresses arising in a thin-walled cylindrical shell subject to torsion. There it was shown that a shearing stress given by $s_{xy} = T\rho/J$ exists on cross-sections perpendicular to the axis of the cylinder. If in addition an axial tension P is acting simultaneously with the torque, then there arises a longitudinal stress given by $s = P/A$. This loading is illustrated in the adjoining diagram. In this case the stresses due to these two loadings are acting in different directions and use must be made of the results obtained in Chapter 15.

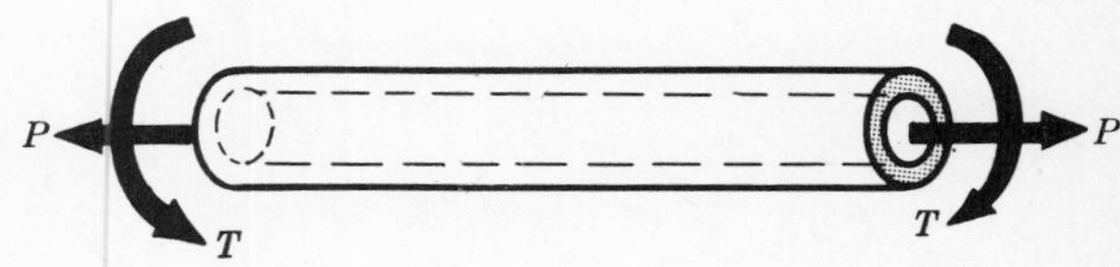

In this manner it will be possible to obtain the principal stresses due to these two loads acting simultaneously. For applications see Problems 5 and 6.

CIRCULAR SHAFT SUBJECT TO COMBINED AXIAL TENSION AND TORSION. This loading is illustrated below. Due to the axial tensile force P there exists a uniform longitudinal tensile stress given by $s = P/A$, where A denotes the cross-sectional area of the bar. From Chapter 5 we know that there exists a torsional shearing stress over any cross-section perpendicular to the axis given by $s_{xy} = T\rho/J$. Again, the stresses due to these two loadings are acting in different directions and the results of Chapter 15 must be employed to obtain the values of the principal stresses at any point or to obtain the state of stress on any plane inclined at some angle to a generator of the shaft. For applications see Problems 7 and 8.

CIRCULAR SHAFT SUBJECT TO COMBINED BENDING AND TORSION. This loading is illustrated below. Again, from Chapter 5 we know that there exists a torsional shearing stress over any cross-section perpendicular to the axis given by $s_{xy} = T\rho/J$. From Chapter 8 we know that there also exists a bending stress perpendicular to this cross-section, i.e. in the direction of the axis of the shaft, given by $s = My/I$. Since these stresses are acting in different directions the results of Chapter 15 must be employed to obtain the values of the principal stresses at any point in the shaft or to obtain the state of stress on any plane inclined to a generator of the shaft. For applications see Problems 9, 10, and 11.

SOLVED PROBLEMS

1. Consider the short block loaded by a compressive force of 10,000 lb. The force is applied with an eccentricity of 1 inch as shown in Fig. (*a*) below. Determine the stresses at the outer fibers of the block.

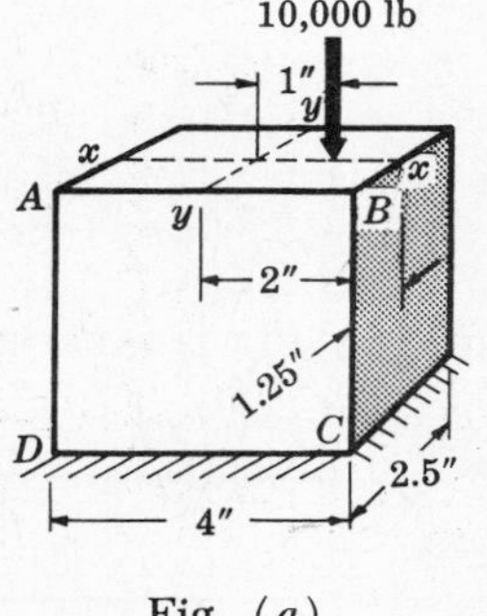

Fig. (*a*)

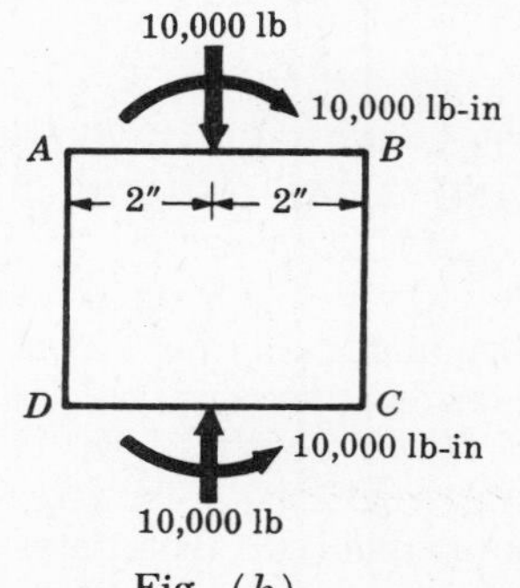

Fig. (*b*)

The force may be replaced by a statically equivalent force system consisting of a force of 10,000 lb acting through the centroid together with a couple of magnitude 10,000 lb-in acting about the y-y axis. The free-body diagram then has the appearance shown in Fig. (*b*) above.

Due to the centrally applied load of 10,000 lb we have a uniformly distributed compressive stress

$$s_1 = \frac{P}{A} = \frac{10,000}{4(2.5)} = 1000 \text{ lb/in}^2$$

distributed over any horizontal cross-section. This appears as in Fig. (c) below.

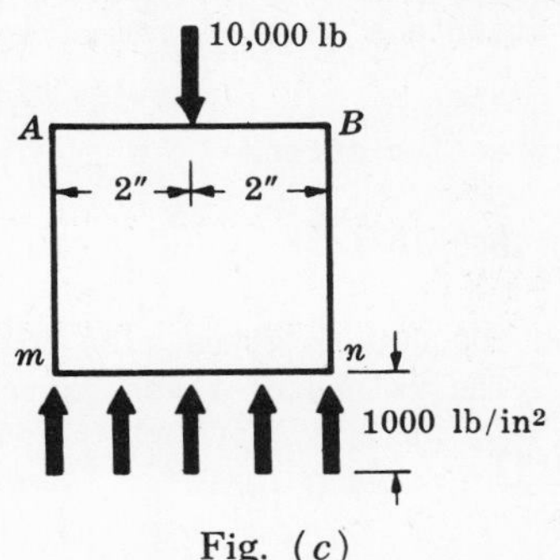

Fig. (c)

10,000 lb-in
A
B
m
n
1500 lb/in^2

Fig. (d)

Due to the couple of 10,000 lb-in we have a linearly varying distribution of normal or bending stress across the cross-section as described in Chapter 8. The stresses at the extreme fibers are given by

$$s_2 = \frac{Mc}{I} = \frac{10,000(2)}{2.5(4^3)/12} = 1500 \text{ lb/in}^2$$

This stress distribution appears as in Fig. (d) above.

The stresses acting along the horizontal cross-section are all directed vertically and hence may be superposed to yield a normal stress at m of

$$s_m = -1000 + 1500 = 500 \text{ lb/in}^2$$

and a normal stress at n of

$$s_n = -1000 - 1500 = -2500 \text{ lb/in}^2$$

In these equations negative signs indicate compression and positive signs tension. It is of course assumed that the resultant stresses do not exceed the proportional limit of the material, otherwise it would not be permissible to apply the above expression for bending stresses.

This method of analysis is valid only if the x and y axes are axes of symmetry of the cross-section.

2. A short block is loaded by a compressive force of 100,000 lb acting 2 in. from one axis and 3 in. from another axis of an 8×8 in. cross-section as shown in Fig. (a) below. Determine the maximum tensile and compressive stresses in the cross-section.

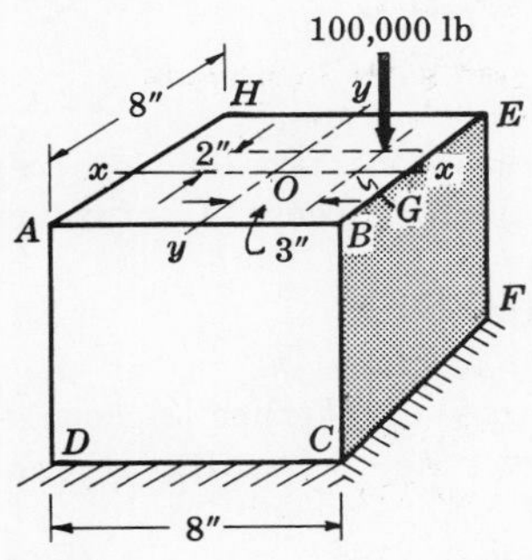

Fig. (a)

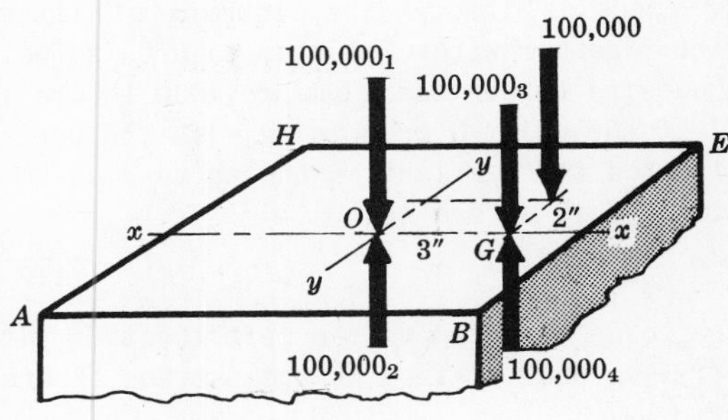

Fig. (b)

Let us consider point O, the geometric center of the cross-section, and point G located on one axis of symmetry and lying 3 inches from O. Through each of these two points let us introduce a pair of equal and opposite forces, each of magnitude 100,000 lb. The upper surface then has the appearance as shown in Fig. (b) above.

These four forces that have been added are designated as $100,000_1$, $100,000_2$, etc., and constitute self-equilibrating system. Thus they do not change the original stressed state of the body but merel provide a simplified medium of calculation.

The force $100,000_1$ lb produces a uniformly distributed compressive stress over any horizontal cross section. The forces $100,000_4$ lb and 100,000 lb constitute a couple giving rise to bending about th x-x axis. The forces $100,000_2$ lb and $100,000_3$ lb constitute a couple giving rise to bending about th y-y axis.

Due to the force $100,000_1$ lb we have a uniform compressive stress

$$s_1 = \frac{100,000}{(8)(8)} = 1560 \text{ lb/in}^2$$

The couple consisting of the forces $100,000_4$ lb and 100,000 lb gives rise to maximum tension alon the line AB and maximum compression along the line HE. The values of these extreme fiber stresses ar

$$s_2 = \frac{M_x c}{I_x} = \frac{100,000(2)(4)}{8(8)^3/12} = 2350 \text{ lb/in}^2$$

The couple consisting of the forces $100,000_2$ lb and $100,000_3$ lb gives rise to maximum tension alon the line AH and maximum compression along the line BE. The values of these extreme fiber stresses ar

$$s_3 = \frac{M_y c}{I_y} = \frac{100,000(3)(4)}{8(8)^3/12} = 3520 \text{ lb/in}^2$$

The maximum compressive stress is thus along line EF and is given by

$$s_4 = -1560 - 2350 - 3520 = -7430 \text{ lb/in}^2$$

The maximum tensile stress occurs along line AD and is equal to

$$s_5 = -1560 + 2350 + 3520 = +4310 \text{ lb/in}^2$$

The stresses s_4 and s_5 are directed vertically.

It is to be observed that this method of analysis is valid only for those cases where the x and axes are axes of symmetry of the cross-section.

3. The bracket shown in the adjoining figure is loaded by a force of 4000 lb applied $1\frac{1}{2}$ in. from the centroid. Determine the stresses at the extreme fibers of a vertical cross-section.

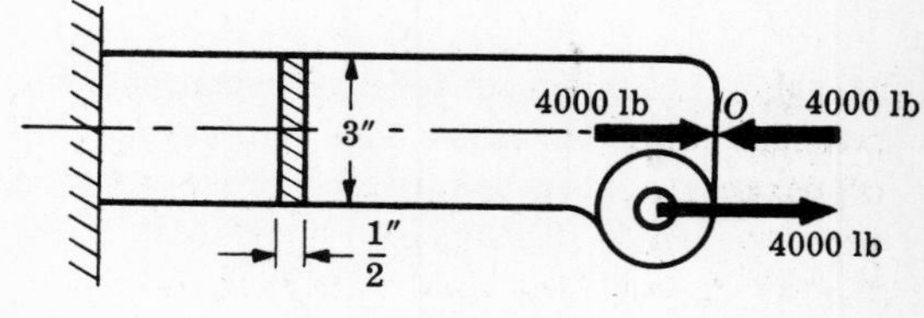

The applied force of 4000 lb may be replaced by a force acting through the centroid of the cross-section together with a couple. To do this two equal and opposite forces each equal to 4000 lb are introduced at the point O coinciding with the centroid of the cross-section. The one of these forces tha is directed to the right in the above diagram gives rise to a uniform tensile stress over the cross section. This stress is

$$s_1 = \frac{4000}{(1/2)(3)} = 2670 \text{ lb/in}^2$$

The force directed to the left together with the original force constitute a couple of magnitud 4000(1.5) = 6000 lb-in. At the extreme fibers this couple gives rise to bending stress given by

$$s_2 = \frac{Mc}{I} = \frac{6000(1.5)}{(1/2)(3^3)/12} = 8000 \text{ lb/in}^2$$

These bending stresses are tensile at the lower fibers and compressive at the upper fibers. Thus th resultant stresses obtained by superposition of these axial and bending stresses are as follows:

At the lower fibers: $s_3 = 2670 + 8000 = 10,670 \text{ lb/in}^2$

At the upper fibers: $s_4 = 2670 - 8000 = -5330 \text{ lb/in}^2$

The positive signs denote tension and the negative signs compression.

. Thin-walled cylindrical tubes are frequently subject to combined axial tension and internal pressure. If a shell of diameter 27 in. and wall thickness 0.125 in. is subject to a uniform internal pressure of 30 lb/in^2 together with an axial tension of 50,000 lb, determine the maximum tensile stress in the shell.

Due to the axial tension of 50,000 lb we have a uniformly distributed tensile stress given by

$$s_1 = \frac{P}{A} = \frac{50,000}{\pi(27)(0.125)} = 4700 \text{ lb/in}^2$$

acting over every cross-section. This stress acts on an element of the shell wall as shown in Fig. (a).

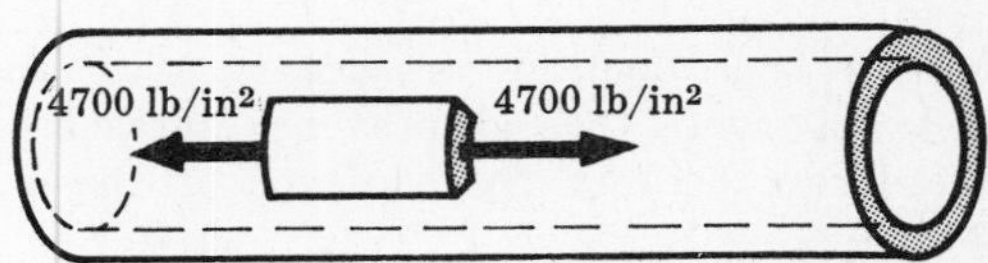

Fig. (a)

The stresses due to the uniform internal pressure of 30 lb/in^2 were determined in Problem 1, Chapter 3. There, the circumferential stress s_c was found to be $s_c = pr/h$, where r denotes the shell radius, h the shell thickness and p the internal pressure. Also, the longitudinal stress s_L was found to be $s_L = pr/2h$. For the values of p, r, and h given we have

$$s_c = \frac{pr}{h} = \frac{30(13.5)}{0.125} = 3240 \text{ lb/in}^2$$

$$s_L = \frac{pr}{2h} = \frac{30(13.5)}{2(0.125)} = 1620 \text{ lb/in}^2$$

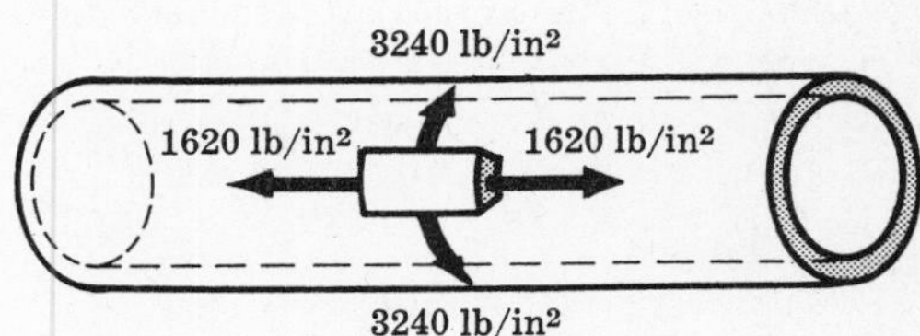

Fig. (b)

These stresses act on an element of the shell wall as shown in Fig. (b) above.

Since both loadings are acting simultaneously it is necessary to combine these stresses. Adding the stresses in the longitudinal direction we find a resultant longitudinal tension of $1620 + 4700 = 6320$ lb/in^2. The resultant circumferential stress is 3240 lb/in^2 since the axial loading does not give rise to any circumferential stress.

Thus the maximum tensile stress in the shell acts in the longitudinal direction and is of magnitude 6320 lb/in^2.

5. Consider a thin-walled cylindrical shell subject to combined axial tension and torsion. The shell is of diameter 16 in., and the wall thickness is 0.10 in. The shell is subject to an axial tension of 40,000 lb together with a torque of 400,000 lb-in. Determine the principal stresses in the shell. Also, find the maximum shearing stress.

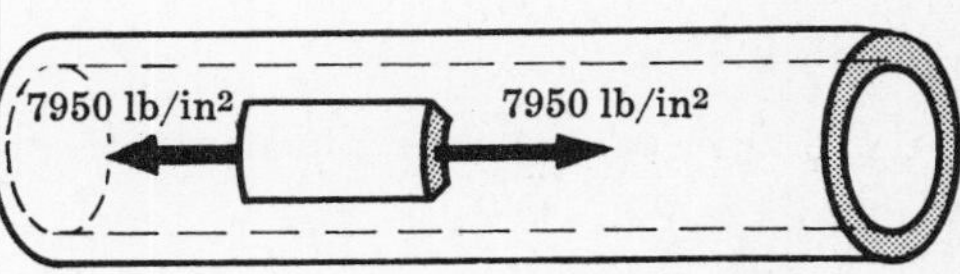

Fig. (a)

Due to the axial tension of 40,000 lb there exists a uniformly distributed tensile stress given by

$$s_x = \frac{P}{A} = \frac{40,000}{\pi(16)(0.10)} = 7950 \text{ lb/in}^2$$

acting over every cross-section. This stress appears as in Fig. (a).

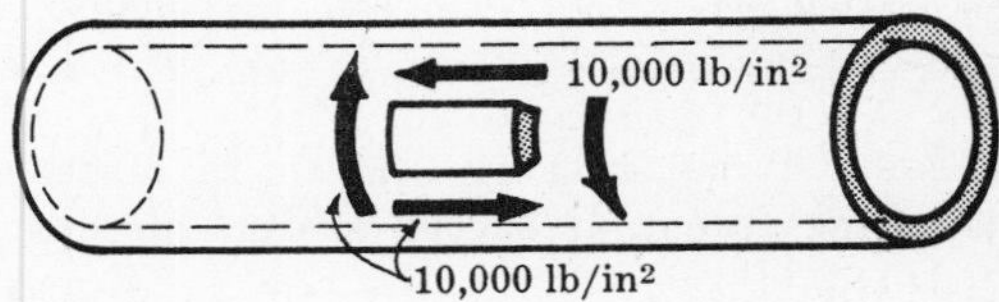

Fig. (b)

The shearing stresses due to the torque of 400,000 lb-in were determined in Problem 2, Chapter 5. The shearing stress in the wall of the shell was found to be $s_{xy} = T\rho/J$. For a thin-walled tube such as we have here, the polar moment of inertia is seen from Problem 9, Chapter 5 to be

$$J = 2\pi R^3 t = 2\pi(8^3)(0.10) = 321 \text{ in}^4$$

The shearing stress in the shell is thus

$$s_{xy} = \frac{T\rho}{J} = \frac{400,000(8)}{321} = 10,000 \text{ lb/in}^2$$

These stresses appear as in Fig. (*b*) above.

Since both loadings are acting simultaneously it is necessary to combine these stresses. In previous problems in this chapter the stresses to be combined due to various loadings have all acted in the same directions and the addition of these stress vectors reduced to simple algebraic addition.

In this problem the stress directions are different and the vector methods described in Chapter 15 must be employed. The case of one normal stress together with a shearing stress acting on an element was treated in Problem 7 of that chapter. Using the notation of that problem, we have here

$$s_x = 7950 \text{ lb/in}^2, \qquad s_{xy} = 10,000 \text{ lb/in}^2$$

From Problem 7 the principal stresses are

$$(s_n)_{max} = \tfrac{1}{2}s_x + \sqrt{(\tfrac{1}{2}s_x)^2 + (s_{xy})^2} = 7950/2 + \sqrt{(7950/2)^2 + (10,000)^2} = 14,800 \text{ lb/in}^2$$

$$(s_n)_{min} = \tfrac{1}{2}s_x - \sqrt{(\tfrac{1}{2}s_x)^2 + (s_{xy})^2} = 7950/2 - \sqrt{(7950/2)^2 + (10,000)^2} = -6800 \text{ lb/in}^2$$

These stresses occur on planes defined by Equation (*3*) of Problem 7:

$$\tan 2\theta_p = -\frac{s_{xy}}{\tfrac{1}{2}s_x} = -\frac{10,000}{7950/2} = -2.50 \qquad \text{or} \qquad \theta_p = 55°50', \ 145°50'$$

Substituting in Equation (*1*) of Problem 7, letting $\theta = 55°50'$ we have

$$s_n = 7950/2 - (7950/2)\cos 111°40' + 10,000 \sin 111°40' = 14,800 \text{ lb/in}^2$$

The maximum principal stress of 14,800 lb/in² thus occurs on a plane oriented 55°50' to the longitudinal axis of the shell.

From Equation (*8*) of Problem 7, the maximum shearing stresses are

$$s_s = \pm\sqrt{(\tfrac{1}{2}s_x)^2 + (s_{xy})^2} = \pm\sqrt{(7950/2)^2 + (10,000)^2} = \pm 10,800 \text{ lb/in}^2$$

These stresses occur on planes oriented at 45° to the planes on which the maximum normal stresses occur.

6. For the thin-walled tube discussed in Problem 5 determine the normal stress acting on a plane inclined at 30° to a generator.

In Problem 5 we found $s_x = 7950$ lb/in², $s_{xy} = 10,000$ lb/in².

From Problem 7, Chapter 15, Equation (*1*) we have the normal stress on any plane inclined at an angle θ with the direction of s_x to be given by

$$s_n = \tfrac{1}{2}s_x - \tfrac{1}{2}s_x \cos 2\theta + s_{xy}\sin 2\theta$$

Substituting $\theta = 30°$, $\quad s_n = 7950/2 - (7950/2)\cos 60° + 10,000 \sin 60° = 10,660 \text{ lb/in}^2$

7. A solid circular shaft 2.50 in. in diameter is subject to an axial tension of 50,000 lb as well as a twisting moment of 25,000 lb-in. Determine the maximum tensile stress in the shaft.

The axial tension gives rise to a uniform tensile stress given by

$$s_x = \frac{P}{A} = \frac{50,000}{\tfrac{1}{4}\pi(2.50)^2} = 10,200 \text{ lb/in}^2$$

The shearing stress caused by the twisting moment T was shown in Problem 2, Chapter 5 to be $s_{xy} = T\rho/J$ where ρ denotes the radial coordinate and J the polar moment of inertia of the cross-section. In Chapter 5 it was demonstrated, as is evident from the above equation, that the shearing stress is maximum at the outer fibers. The maximum shear is thus given by

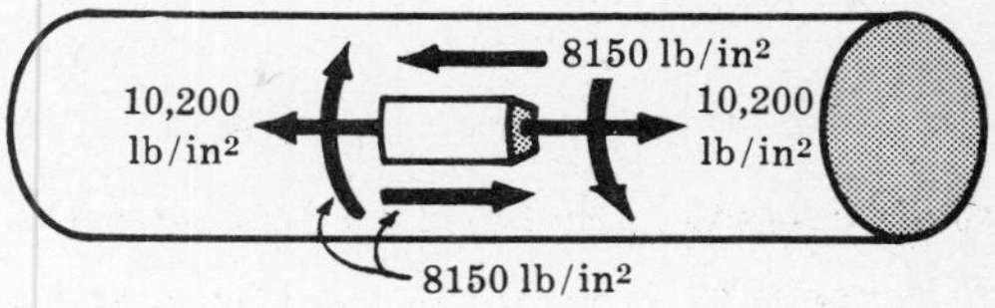

$$s_{xy} = \frac{T\rho}{J} = \frac{25,000(1.25)}{\pi(2.50)^4/32} = 8150 \text{ lb/in}^2$$

Thus an element at the outer surface of the shaft is subject to the stresses shown above.

The state of stress on an inclined plane of this element for this loading was studied in Problem 7, Chapter 15. The principal stresses were there found to be

$$(s_n)_{max} = \tfrac{1}{2}s_x + \sqrt{(\tfrac{1}{2}s_x)^2 + (s_{xy})^2} = 10,200/2 + \sqrt{(10,200/2)^2 + (8150)^2} = 14,700 \text{ lb/in}^2$$

$$(s_n)_{min} = \tfrac{1}{2}s_x - \sqrt{(\tfrac{1}{2}s_x)^2 + (s_{xy})^2} = 10,200/2 - \sqrt{(10,200/2)^2 + (8150)^2} = -4450 \text{ lb/in}^2$$

These stresses occur on planes defined by Equation (3) of Problem 7:

$$\tan 2\theta_p = -\frac{s_{xy}}{\tfrac{1}{2}s_x} = -\frac{8150}{10,200/2} = -1.60 \qquad \text{or} \qquad \theta_p = 61°0',\ 151°0'$$

Substituting in Equation (1) of Problem 7, letting $\theta = 61°0'$ we have

$$s_n = 10,200/2 - (10,200/2)\cos 122°0' + 8150 \sin 122°0' = 14,700 \text{ lb/in}^2$$

Thus the maximum tensile stress is 14,700 lb/in^2, occurring on a plane oriented 61°0' to the geometric axis of the shaft.

8. A shaft 2 in. in diameter is loaded by an axial compressive force of 50,000 lb together with a twisting moment of 30,000 lb-in. Determine the principal stresses and also the maximum shearing stress in the shaft.

The axial force gives rise to a uniform compressive stress given by

$$s_x = \frac{P}{A} = \frac{50,000}{\tfrac{1}{4}\pi(2)^2} = 15,900 \text{ lb/in}^2$$

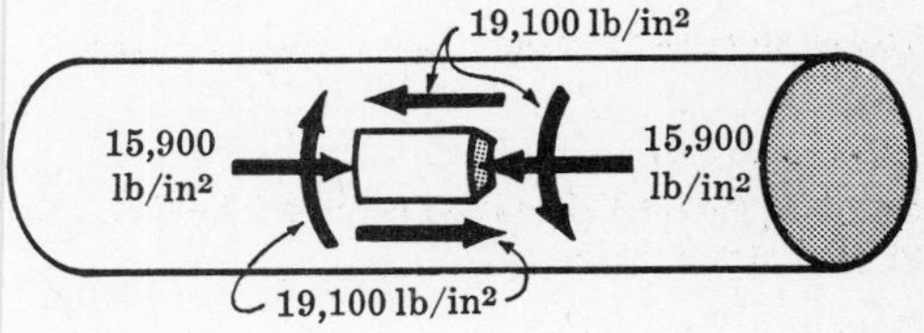

The shearing stress due to the applied twisting moment was shown in Problem 2, Chapter 5 to be $s_{xy} = T\rho/J$. This is maximum at the outer fibers of the shaft and becomes

$$s_{xy} = \frac{T\rho}{J} = \frac{30,000(1)}{\pi(2)^4/32} = 19,100 \text{ lb/in}^2$$

An element at the outer surface of the shaft is thus subject to the stresses shown above.

The principal stresses for such a loading on an element were derived in Problem 7, Chapter 15. They are

$$(s_n)_{max} = \tfrac{1}{2}s_x + \sqrt{(\tfrac{1}{2}s_x)^2 + (s_{xy})^2} = -15,900/2 + \sqrt{(-15,900/2)^2 + (19,100)^2} = 12,750 \text{ lb/in}^2$$

$$(s_n)_{min} = \tfrac{1}{2}s_x - \sqrt{(\tfrac{1}{2}s_x)^2 + (s_{xy})^2} = -15,900/2 - \sqrt{(-15,900/2)^2 + (19,100)^2} = -28,650 \text{ lb/in}^2$$

The maximum shearing stress is found from Equation 8, Problem 7, Chapter 15 to be

$$s_s = \pm\sqrt{(\tfrac{1}{2}s_x)^2 + (s_{xy})^2} = \pm\sqrt{(-15,900/2)^2 + (19,100)^2} = \pm 20,700 \text{ lb/in}^2$$

9. Consider a solid circular shaft subject to a constant twisting moment of 6000 lb-in as well as a constant bending moment of 2500 lb-in. The diameter of the shaft is 1.75 in. Determine the principal stresses in the body.

The twisting moment gives rise to shearing stresses that attain their peak values in the outer fibers of the shaft. From Problem 2, Chapter 5 these shearing stresses are given by $s_{xy} = T\rho/J$. At the outer fibers

$$s_{xy} = \frac{T\rho}{J} = \frac{6000(1.75/2)}{\pi(1.75)^4/32} = 5700 \text{ lb/in}^2$$

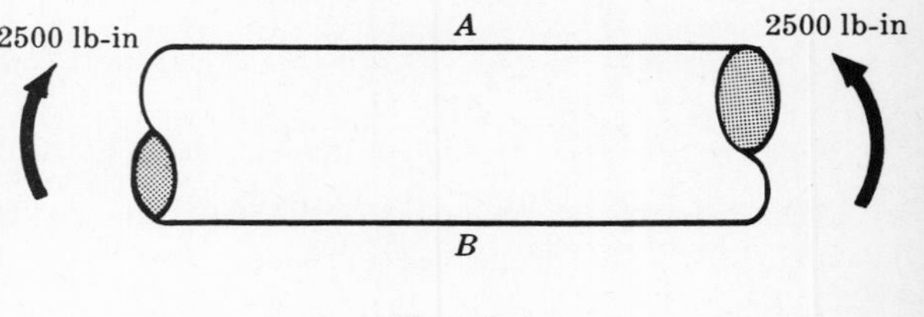

Fig. (a)

If the bending moments are assumed to lie in a vertical plane as shown in Fig. (a), then the extreme fibers designated as A and B will be subject to the peak bending stresses. From Problem 1, Chapter 8 the tensile stress at B is given by $s_x = My/I$. Here I denotes the moment of inertia of the cross-sectional area about a diameter. For a solid circular shaft this was found in Problem 11, Chapter 7 to be $I = \pi d^4/64$, where d represents the shaft diameter. Substituting,

$$s_x = \frac{My}{I} = \frac{2500(1.75/2)}{\pi(1.75)^4/64} = 4750 \text{ lb/in}^2$$

Consequently an element located at the lower extremity of the bar (anywhere along the lower fiber) is subject to the state of stress as shown in Fig. (b).

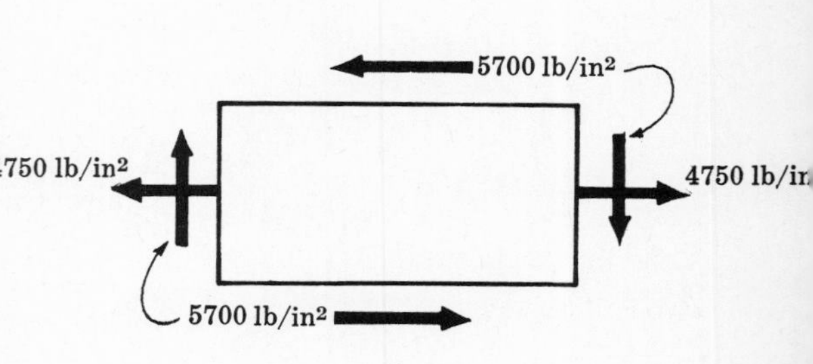

Fig. (b)

The principal stresses for an element stressed in this manner were obtained in Problem 7, Chapter 15. They are

$$(s_n)_{max} = \tfrac{1}{2}s_x + \sqrt{(\tfrac{1}{2}s_x)^2 + (s_{xy})^2} = 4750/2 + \sqrt{(4750/2)^2 + (5700)^2} = 8580 \text{ lb/in}^2$$

$$(s_n)_{min} = \tfrac{1}{2}s_x - \sqrt{(\tfrac{1}{2}s_x)^2 + (s_{xy})^2} = 4750/2 - \sqrt{(4750/2)^2 + (5700)^2} = -3820 \text{ lb/in}^2$$

These stresses occur on planes defined by Equation (3) of Problem 7:

$$\tan 2\theta_p = -\frac{s_{xy}}{\tfrac{1}{2}s_x} = -\frac{5700}{4750/2} = -2.40 \quad \text{or} \quad \theta_p = 56°15', \ 146°15'$$

Substituting in Equation (1) of Problem 7, letting $\theta = 56°15'$ we have

$$s_n = 4750/2 - (4750/2)\cos 112°30' + 5700 \sin 112°30' = 8580 \text{ lb/in}^2$$

Thus the maximum tensile stress is 8580 lb/in^2, occurring on a plane oriented 56°15′ to the geometric axis of the shaft.

10. Consider a hollow circular shaft whose outside diameter is 3 in. and having an inside diameter equal to one-half the outside diameter. The shaft is subject to a twisting moment of 20,000 lb-in as well as a bending moment of 30,000 lb-in. Determine the principal stresses in the body. Also, determine the maximum shearing stress.

The twisting moment gives rise to shearing stresses that attain their peak values in the outer fibers of the shaft. From Problem 2, Chapter 5 these shearing stresses are given by $s_{xy} = T\rho/J$. From Problem 1, Chapter 5 it is seen that for the hollow circular area

$$J = \frac{\pi}{32}(D_o^4 - D_i^4) = \frac{\pi}{32}(3^4 - 1.5^4) = 7.46 \text{ in}^4$$

where D_o denotes the outer diameter of the section and D_i represents the inner diameter. At the outer fibers the torsional shearing stresses are thus

$$s_{xy} = \frac{T\rho}{J} = \frac{20,000(1.5)}{7.46} = 4000 \text{ lb/in}^2$$

Let the bending moments lie in a vertical plane. Then the upper and lower fibers of the beam are subject to the peak bending stresses. These are found from the expression $s_x = My/I$. The moment of inertia I for the hollow circular cross-section may be seen from Problem 11, Chapter 7 to be

$$I = \frac{\pi}{64}(D_o^4 - D_i^4) = \frac{\pi}{64}(3^4 - 1.5^4) = 3.73 \text{ in}^4$$

Substituting,

$$s_x = \frac{My}{I} = \frac{30,000(1.5)}{3.73} = 12,000 \text{ lb/in}^2$$

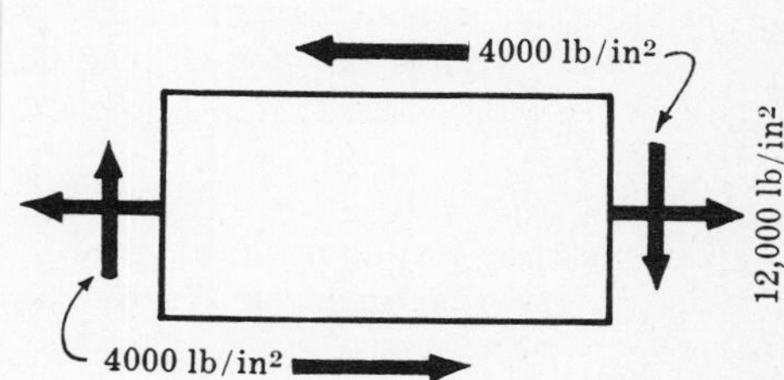

Thus an element located at the lower extremity of the shaft is subject to the stresses as shown in the adjacent figure.

From Problem 7, Chapter 15 the principal stresses for this element are

$$(s_n)_{max} = \tfrac{1}{2}s_x + \sqrt{(\tfrac{1}{2}s_x)^2 + (s_{xy})^2} = 12,000/2 + \sqrt{(12,000/2)^2 + (4000)^2} = 13,200 \text{ lb/in}^2$$

$$(s_n)_{min} = \tfrac{1}{2}s_x - \sqrt{(\tfrac{1}{2}s_x)^2 + (s_{xy})^2} = 12,000/2 - \sqrt{(12,000/2)^2 + (4000)^2} = -1200 \text{ lb/in}^2$$

These stresses occur on planes defined by Equation (*3*) of Problem 7, Chapter 15 :

$$\tan 2\theta_p = -\frac{s_{xy}}{\tfrac{1}{2}s_x} = -\frac{4000}{12,000/2} = -\frac{2}{3} \quad \text{or} \quad \theta_p = 73°10',\ 163°10'$$

Substituting in Equation (*1*) of Problem 7, Chapter 15, letting $\theta = 73°10'$ we have

$$s_n = 12,000/2 - (12,000/2)\cos 146°20' + 4000 \sin 146°20' = 13,200 \text{ lb/in}^2$$

Thus the maximum tensile stress is 13,200 lb/in², occurring on a plane oriented 73°10' to the geometric axis of the shaft. The other principal stress, $(s_n)_{min} = -1200 \text{ lb/in}^2$, occurs on a plane oriented 163°10' to the axis.

The maximum shearing stress is given by Equation 8 of Problem 7, Chapter 15. It is

$$s_s = \pm\sqrt{(\tfrac{1}{2}s_x)^2 + (s_{xy})^2} = \pm\sqrt{(12,000/2)^2 + (4000)^2} = \pm 7200 \text{ lb/in}^2$$

This stress occurs on planes oriented at 45° to the planes found above on which the principal stresses act.

1. The shaft shown in Fig. (*a*) below rotates with constant angular velocity. The belt pulls create a state of combined bending and torsion. Neglect the weights of the shaft and pulleys and assume that the bearings can exert only concentrated force reactions. The diameter of the shaft is 1.25 in. Determine the principal stresses in the shaft.

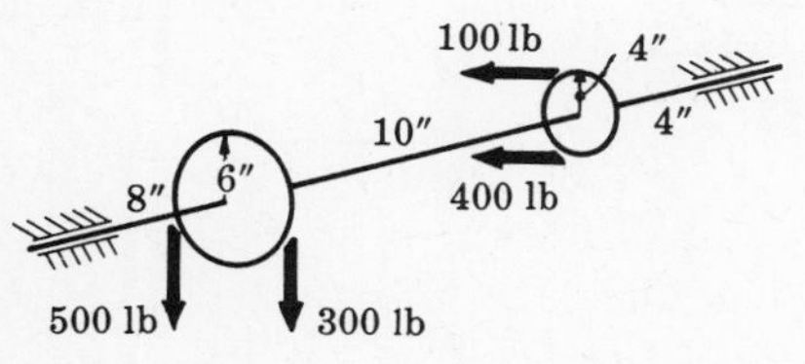

Fig. (*a*)

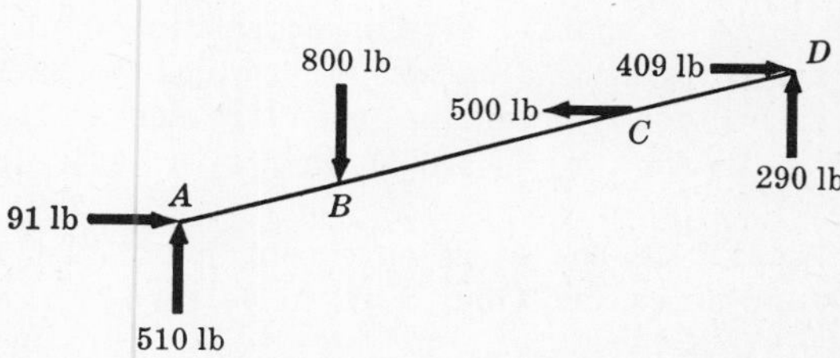

Fig. (*b*)

The transverse forces acting on the shaft are not parallel and the bending moments caused by the must be added vectorially to obtain the resultant bending moment. This vector addition need be carried out at only a few apparently critical points along the length of the shaft. The loads causi bending together with the reactions they produce are shown above in Fig. (*b*) and are considered a passing through the axis of the shaft.

The bending moment diagram for a vertical plane appears as in the upper shaded portion of the adjoining diagram.

The bending moment diagram for a horizontal plane appears as in the lower shaded portion of the adjoining diagram.

4080 lb-in
1160 lb-in
A B C
A B C
728 lb-in
1636 lb-in

The resultant bending moment at B is $M_B = \sqrt{(4080)^2 + (728)^2} = 4140$ lb-in.

The resultant bending moment at C is $M_C = \sqrt{(1160)^2 + (1636)^2} = 2000$ lb-in.

The twisting moment between the two pulleys is constant and equal to

$$T = (400-100)4 = 1200 \text{ lb-in}$$

Since the torque is the same at B and C, the critical element lies at the outer fibers of the shaf at point B. The maximum bending stress is given by

$$s_x = \frac{My}{I} = \frac{(4140)(1.25/2)}{\pi(1.25)^4/64} = 21,500 \text{ lb/in}^2$$

The maximum shearing stress, occurring at the outer fibers of the shaft, is given by

$$s_{xy} = \frac{T\rho}{J} = \frac{1200(1.25/2)}{\pi(1.25)^4/32} = 3100 \text{ lb/in}^2$$

The principal stresses were found in Problem 7, Chapter 15 to be

$$(s_n)_{max} = \tfrac{1}{2}s_x + \sqrt{(\tfrac{1}{2}s_x)^2 + (s_{xy})^2} = 21,500/2 + \sqrt{(21,500/2)^2 + (3100)^2} = 22,000 \text{ lb/in}^2$$

$$(s_n)_{min} = \tfrac{1}{2}s_x - \sqrt{(\tfrac{1}{2}s_x)^2 + (s_{xy})^2} = 21,500/2 - \sqrt{(21,500/2)^2 + (3100)^2} = -400 \text{ lb/in}^2$$

SUPPLEMENTARY PROBLEMS

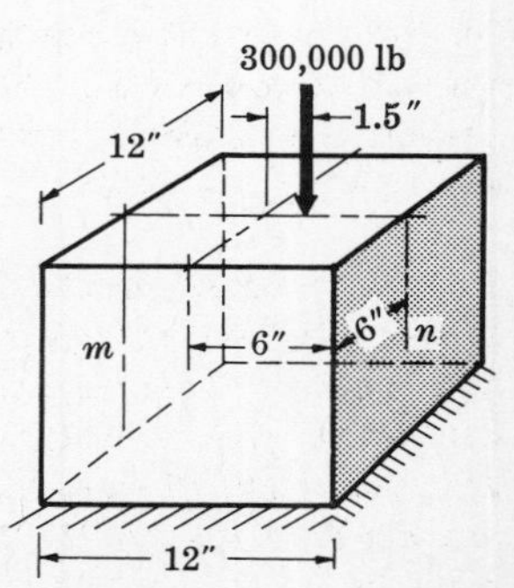

12. A short block is loaded by a compressive force of 300,000 lb. The force is applied with an eccentricity of 1.5 in. as shown in the adjoining diagram. The block is 12 in. by 12 in. in cross-section. Determine the stresses at the outer fibers m and n. *Ans.* $s_m = -520$ lb/in², $s_n = -3640$ lb/in²

13. In Problem 12 how large an eccentricity must exist if the resultant stress at fiber m is to be zero? *Ans.* 2.00 in.

14. The member shown in Fig. (*a*) below is subject to an eccentrically applied load of 5000 lb. If the permissible working

stress in the material is 10,000 lb/in^2, determine the width d of the member. The thickness is 1 in. *Ans.* 2.32 in.

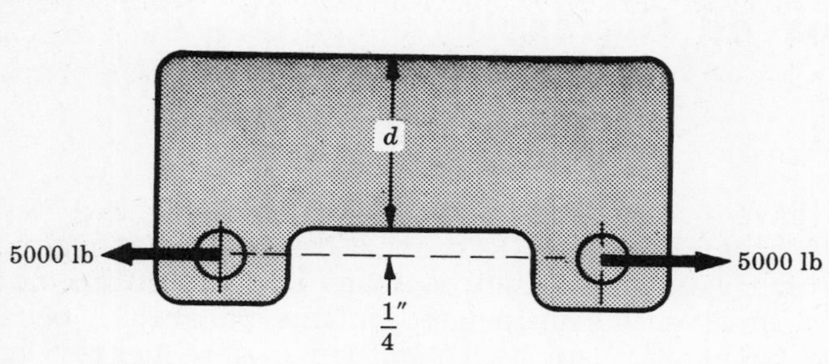

Fig. (*a*) Prob. 14

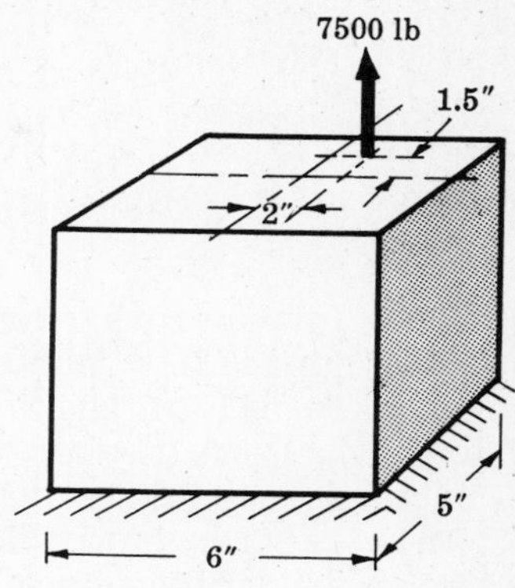

Fig. (*b*) Prob. 15

15. A block is loaded by the eccentric tensile force as shown in Fig. (*b*) above. Determine the maximum tensile stress. *Ans.* 1200 lb/in^2

16. A thin-walled cylinder is 10 in. in diameter and of wall thickness 0.10 in. The cylinder is subject to a uniform internal pressure of 100 lb/in^2. What additional axial tension may act simultaneously in order that the maximum tensile stress does not exceed 20,000 lb/in^2? *Ans.* 55,000 lb

17. For the thin-walled tube discussed in Problem 16, determine the normal stress acting on a plane inclined at 30° to a generator. Both the axial tension and the internal pressure are acting. *Ans.* 8750 lb/in^2

18. A thin-walled cylindrical shell is subject to an axial compression of 50,000 lb together with a torsional moment of 30,000 lb-in. The diameter of the cylinder is 12 in. and the wall-thickness 0.125 in. Determine the principal stresses in the shell. Also, determine the maximum shearing stress. Neglect the possibility of buckling of the shell.
Ans. $(s_n)_{max}$ = 120 lb/in^2, $(s_n)_{min}$ = −10,680 lb/in^2, s_s = 5400 lb/in^2

19. A shaft 2.50 in. in diameter is subject to an axial tension of 40,000 lb together with a twisting moment of 35,000 lb-in. Determine the principal stresses in the shaft. Also determine the maximum shearing stress. *Ans.* $(s_n)_{max}$ = 16,180 lb/in^2, $(s_n)_{min}$ = −8020 lb/in^2, s_s = 12,100 lb/in^2

20. A shaft 2.25 in. in diameter is subject to an axial tension of 25,000 lb as well as a twisting moment of 30,000 lb-in. Determine the principal stresses in the shaft.
Ans. $(s_n)_{max}$ = 16,900 lb/in^2, $(s_n)_{min}$ = −10,700 lb/in^2

21. A shaft 6.50 in. in diameter is subject to an axial compression of 150,000 lb as well as a twisting moment of 250,000 lb-in. Determine the principal stresses in the shaft as well as the maximum shearing stress. *Ans.* $(s_n)_{max}$ = 2890 lb/in^2, $(s_n)_{min}$ = −7410 lb/in^2, s_s = 5150 lb/in^2.

22. Consider a solid circular shaft subject to a twisting moment of 20,000 lb-in together with a bending moment of 30,000 lb-in. The diameter of the shaft is 3 in. Determine the principal stresses as well as the maximum shearing stress in the shaft.
Ans. $(s_n)_{max}$ = 12,450 lb/in^2, $(s_n)_{min}$ = −1150 lb/in^2, s_s = 6800 lb/in^2

23. The shaft shown rotates with constant angular velocity and is subject to combined bending and torsion due to the indicated belt pulls. The weights of the shaft and pulleys may be neglected and the bearings can exert only concentrated force reactions. The diameter of the shaft is 1.75 in. Determine the principal stresses in the shaft.
Ans. $(s_n)_{max}$ = 16,600 lb/in^2, $(s_n)_{min}$ = −750 lb/in^2

650 lb
12″
15″
15″
D
B
C
10″
10″
A
350 lb
560 lb
200 lb

CHAPTER 17

Reinforced Concrete

INTRODUCTION. In previous chapters we have treated the problem of stresses and deformations in a number of different types of structural elements subject to various loads. In all of the problems involving beams and columns in particular, it has been assumed that the member was built up from a single homogeneous material. Occasionally it is advantageous to combine two materials so as to obtain a more nearly optimum design. This is frequently done in the case of concrete reinforced by steel, and occasionally wooden sections reinforced by steel cover plates. Only the former will be discussed here, although the same basic principles apply to both cases.

NATURE OF REINFORCED CONCRETE SECTIONS. By definition, concrete is a mixture of portland cement, fine aggregate, coarse aggregate and water. Plain concrete is brittle and weak in tension and thus is suitable only for relatively heavy members subject to compression. The tensile strength of concrete is about 1/10 of its compressive strength, and because of this if a beam were made of concrete it would fail in tension at rather low values of load and stress. Such a beam can be strengthened by adding steel bars on the side of the beam subject to tension. The beam cross-section then has the appearance as shown in the adjacent figure. Because the concrete adheres to the steel very well there is no sliding of the steel bars with respect to the concrete during bending of the beam.

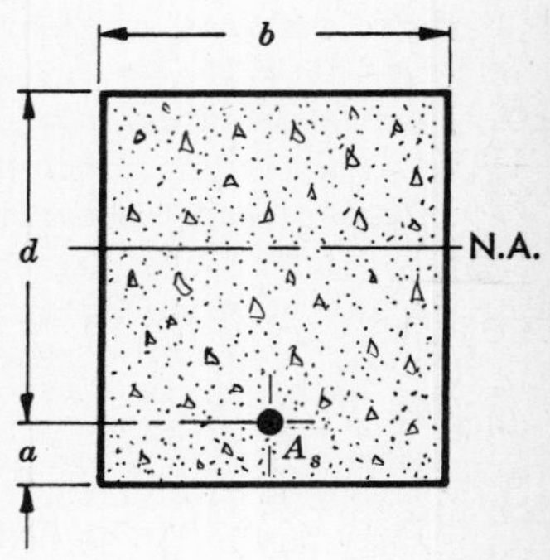

NATURE OF REINFORCEMENT. The reinforcement for concrete usually consists of steel rods, round or square. These range from 1/4 in. to 1 in. in diameter in the round variety and from 1/2 in. on a side to $1\frac{1}{4}$ in. on a side for square bars.

DISTRIBUTION OF LOAD BETWEEN STEEL AND CONCRETE. In the calculation of bending stresses in reinforced concrete beams it is customary to assume that all of the tension is resisted by the steel and all of the compression is taken by the concrete.

STRESS-STRAIN RELATIONS FOR CONCRETE. The stress-strain curve for concrete is illustrated in Figure 5*b*, Chapter 1. Evidently this is a nonlinear relation and hence the modulus of elasticity is not constant. However, in the interest of simplicity it is usually assumed that Hooke's law applies for concrete. From the stress-strain curve it is obvious that the modulus of elasticity decreases with an increase in stress. This variable modulus is partially accounted for by employing a lower value for the modulus than that obtained from compression tests at small compressive stresses.

VALUES OF THE MODULUS OF ELASTICITY. The compressive modulus of elasticity of concrete ranges from $2\cdot10^6$ lb/in^2 to $5\cdot10^6$ lb/in^2. The modulus of elasticity of the reinforcing steel is $30\cdot10^6$ lb/in^2. Consequently we may form the ratio of the modulus of the steel to that of the concrete and let n represent this ratio as follows:

$$n = E_s/E_c$$

where E_s represents the modulus of the steel and E_c that of the concrete. Evidently the ratio n varies from 15 to 6, with values between 10 and 15 being commonly employed.

TRANSFORMED CROSS-SECTION. The true cross-section of a steel-reinforced concrete beam has the appearance shown above. It is to be remembered that the moduli of elasticity of the steel and concrete are widely different. To facilitate analysis it is customary to transform the area of the steel A_s into an area of concrete nA_s which is equivalent with regard to elastic properties. The transformed cross-section then has the appearance shown in the adjacent figure, where the steel has been replaced by an area of concrete equal to nA_s which is represented by the narrow horizontal shaded rectangle. This concrete may be considered to have the same modulus as the concrete in the compression region but differing from it in its assumed ability to resist tension. Actually, in this diagram the shaded areas represent the effective portion of the cross-section and the neutral axis is denoted by N.A.

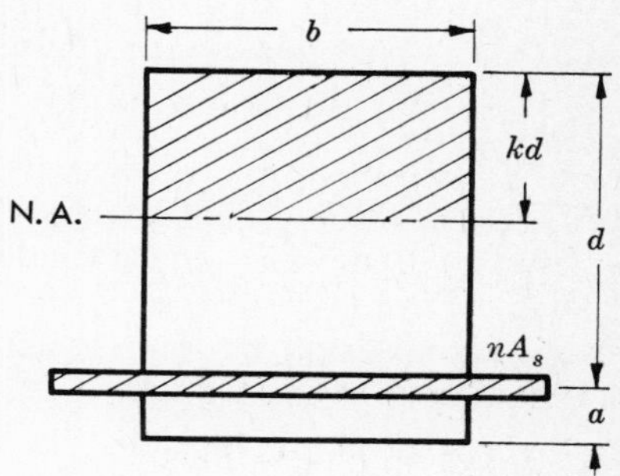

LOCATION OF THE NEUTRAL AXIS. The neutral axis, denoted by N.A. in the above diagram is located by equating the first moments of the areas above and below the axis about the axis. That is, the first moment of the compressive area above the axis is equal to the first moment of the transformed area withstanding tension below the axis. It is to be noted that the transformed area is used in this calculation. Examples of the determination of the location of the neutral axis are to be found in Problems 2, 3, 5-11.

PLACING OF REINFORCING STEEL. To protect the steel from fire damage the reinforcing bars should not be placed nearer the exposed surface than $1\frac{1}{2}$ in. This distance is illustrated in the first of the above diagrams, being denoted by the symbol a.

NOTATION. The symbols used in this chapter are somewhat different than those appearing elsewhere in the book. They are part of a notation adopted by the Joint Committee on Standard Specifications for Concrete and Reinforced Concrete and consequently have been widely accepted. Even though they do not conform with notations previously employed in this book they will be used here to acquaint the student with the accepted terminology of reinforced-concrete analysis and design. The symbols employed are defined as follows:

f_c = maximum compressive stress in the concrete

f_s = tensile stress in the steel (assumed constant over any one horizontal layer of reinforcing steel)

A_s = area of reinforcing steel

b = width of rectangular beam, or width of flange of T-beam

d = effective depth of beam, measured from the top of the beam to the center of the reinforcing steel

k = ratio of distance between extreme fibers and neutral axis to effective depth

j = ratio of distance between resultants of compressive and tensile stresses to effective depth.

BALANCED REINFORCEMENT. It is unlikely that the same bending moment when applied to the beam will simultaneously produce the allowable compressive stress in the concrete and the allowable tension in the steel. Frequently the maximum permissible stress in the concrete will be reached but the steel will not be stressed to its safe value, or visa versa. This is uneconomical because the full strength of both the steel and the concrete is not being utilized. If, however, the beam is designed so that both the steel and the concrete are simultaneously stressed to their allowable working stresses, then the beam is said to have *balanced reinforcement*. This may be accomplished by appropriately adjusting the size of the concrete and the area of the steel. For illustrations see Problems 9-11.

USE OF TABLES. The problems presented in this chapter are solved by direct application of the equation of equilibrium. Extensive tables exist to facilitate the solutions of such problems, and in design practice these tables are frequently used. For further discussions of their use the reader is referred to more advanced texts treating reinforced concrete design.

SOLVED PROBLEMS

1. A concrete beam of rectangular cross-section 10×18 in. is reinforced by three square steel bars, each one inch on a side. The bars are located as shown in Fig. (a) below. Determine the equivalent transformed cross-section. The moduli of elasticity of the steel and of the concrete are $30 \cdot 10^6$ lb/in^2 and $2 \cdot 10^6$ lb/in^2 respectively.

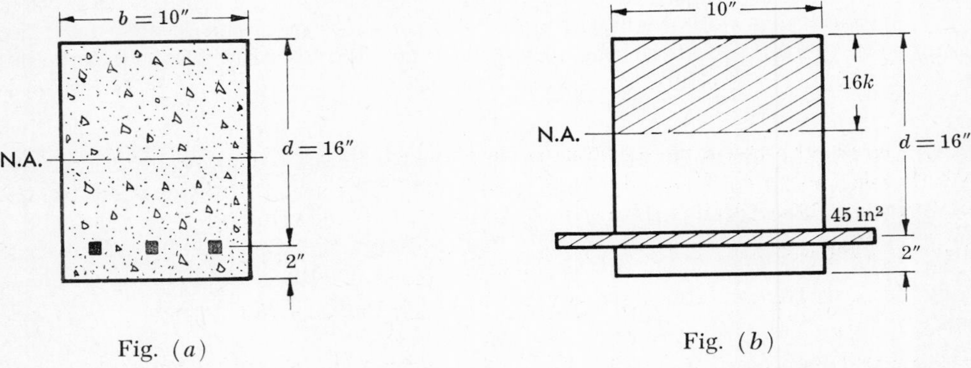

Fig. (a) Fig. (b)

The ratio of the modulus of elasticity of the steel to that of the concrete, which is usually denoted by n is

$$n = \frac{E_s}{E_c} = \frac{30 \cdot 10^6}{2 \cdot 10^6} = 15$$

Transforming the area of the steel A_s into an area of concrete nA_s equivalent as far as elastic properties are concerned, we find an equivalent concrete area of $3(15) = 45$ in^2. The transformed cross-section has the appearance as shown in Fig. (b) above.

This transformed section may be considered to consist entirely of concrete of modulus $2\cdot10^6$ lb/in^2. Note that only the shaded area is to be considered as effective; the upper area resists compression, the lower area resists tension. Thus, instead of the original steel-reinforcement we have now substituted a strip of concrete of area 45 in^2 which is assumed to be able to resist tension.

2. Locate the neutral axis of the steel-reinforced beam discussed in Problem 1.

The distribution of bending (or normal) stresses over the transformed cross-section follows the linear law, since plane cross-sections remain plane during bending and since Hooke's law is assumed to hold for the concrete. Consequently the neutral axis will coincide with the centroidal axis of the transformed section. This requires that the first moment about the neutral axis of the shaded area above the neutral axis be equal to that of the shaded area below. Thus, referring to Fig. (*b*) of Problem 1 and introducing the symbol k to designate the location of the neutral axis, we have

$$10(16k)(8k) = 45(16-16k) \qquad \text{or} \qquad k = 0.52$$

Thus the neutral axis lies $0.52(16) = 8.32$ in. below the top of the beam.

3. A concrete beam of T-section has the dimensions shown below. The beam is reinforced by three square steel bars, each one inch on a side. The moduli of elasticity of the steel and of the concrete are $30\cdot10^6$ lb/in^2 and $2\cdot10^6$ lb/in^2 respectively. Determine the location of the neutral axis.

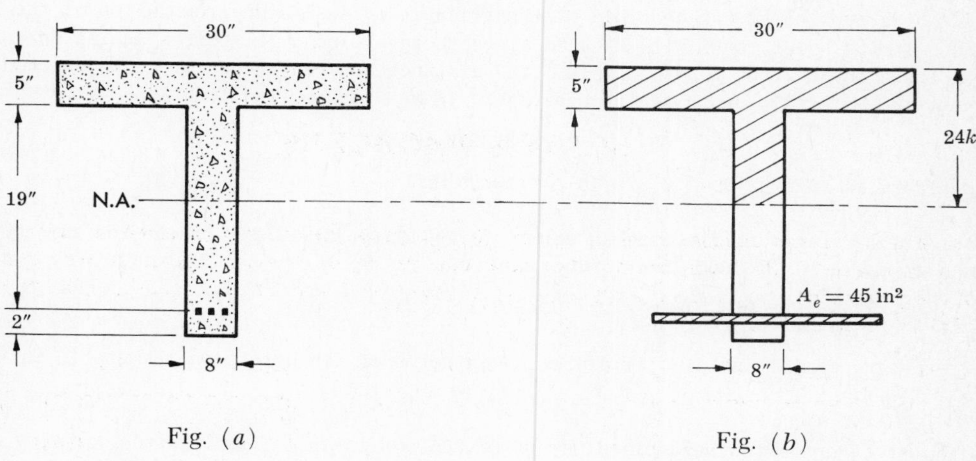

Fig. (*a*) Fig. (*b*)

First, the transformed cross-section is determined. The reinforcing steel is replaced by an area of concrete nA_s, where $n = 15$ and $A_s = 3$ in^2. The equivalent concrete area A_e is thus 45 in^2. The transformed section is shown at the right.

Equating first moments of these shaded areas about the neutral axis we have

$$30(5)(24k-2.5) + (24k-5)\left(\frac{24k-5}{2}\right)(8) = 45(24-24k) \qquad \text{or} \qquad k = 0.305$$

Thus the neutral axis lies $0.305(24) = 7.35$ in. below the top of the beam.

4. A reinforced concrete beam has the rectangular cross-section shown below. The reinforcement consists of three square steel bars, each one inch on a side. The beam is subject to a bending moment of 375,000 lb-in. Determine the maximum stress in the concrete and also the stress in the steel. Take $n = 15$.

We shall use the notation suggested by the Joint Committee on Standard Specifications for Concrete and Reinforced Concrete. In this notation f_c denotes the maximum compressive stress in the concrete and f_s represents the tensile stress in the steel.

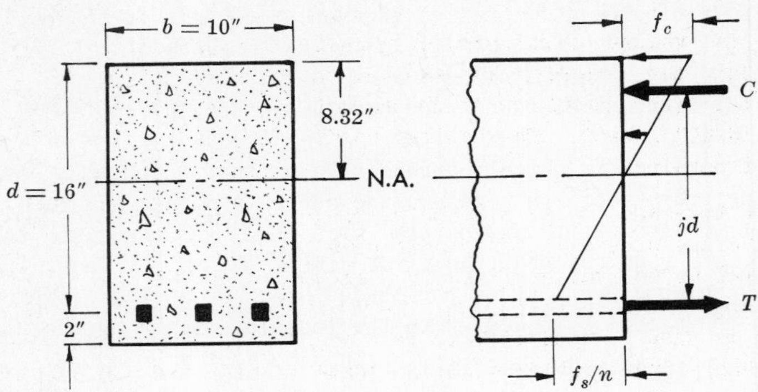

The neutral axis of this section has already been found in Problem 2 to lie 8.32 in. below the top of the beam. The average stress on the area above the neutral axis is $f_c/2$. Consequently the total compressive force in the concrete, represented by the vector C in the above diagram, is

$$C = (f_c/2)(10)(8.32)$$

The force C is applied at the centroid of the triangle or 8.32/3 in. from the top of the beam. The tensile force in the steel is $T = f_s \cdot A_s$, and since there are no axial forces acting on the beam we have $T = C$ and these forces thus form a couple. The distance between these forces is denoted by jd where d represents the effective depth of the beam, here 16 in., and j is a fraction. Thus

$$16j = 16 - 8.32/3 \qquad \text{or} \qquad j = 0.828$$

The moment arm is thus $jd = 0.828(16) = 13.25$ in.

The resisting moment of the beam is thus $(f_c/2)(10)(8.32)(13.25)$ lb-in and this must be equal to the applied moment of 375,000 lb-in. Thus

$$375{,}000 = (f_c/2)(10)(8.32)(13.25) \qquad \text{or} \qquad f_c = 680 \text{ lb/in}^2$$

The resisting moment may also be expressed in terms of the stress in the steel, in the form

$$T \cdot jd = (f_s \cdot A_s)jd$$

and this must be equal to the applied moment of 375,000 lb-in. Thus

$$375{,}000 = f_s(3)(13.25) \qquad \text{or} \qquad f_s = 9450 \text{ lb/in}^2$$

As a check we have

$$C = 340(10)(8.32) = 28{,}350 \text{ lb}, \qquad T = 9450(3) = 28{,}350 \text{ lb}$$

which is of course a necessary condition.

5. A reinforced concrete beam has the rectangular cross-section as shown in the adjacent diagram. The reinforcement consists of four square steel bars, each one inch on a side. If the allowable working stress in the concrete is 700 lb/in^2 and that in the steel is 16,000 lb/in^2, determine the maximum bending moment that may be applied to the beam. Take $n = 15$.

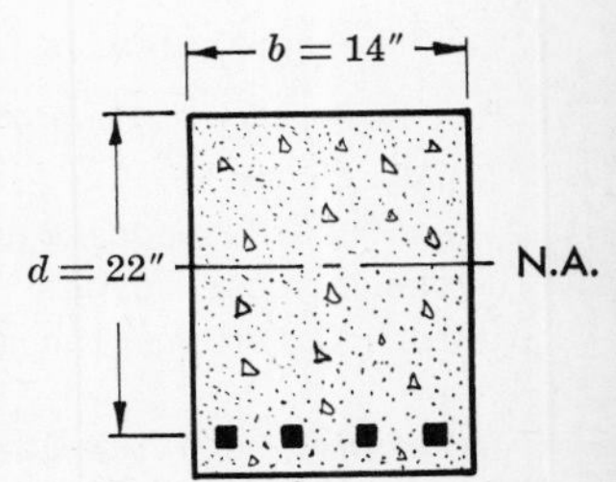

The transformed cross-section has the appearance shown in the diagram below. There, the reinforcing steel has been replaced by an equivalent area of concrete: $A_e = nA_s = 15(4) = 60$ in^2. For equality of first moments of areas about the neutral axis we have

$$14(22k)(11k) \quad = \quad 60(22 - 22k) \qquad \text{or} \qquad k = 0.46$$

The neutral axis thus lies 0.46(22) or 10.1 in. below the top of the beam.

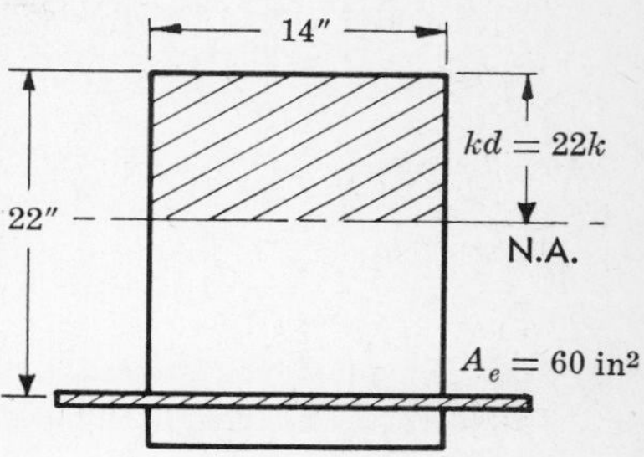

We shall determine two values of the bending moment, one based upon the assumption that the concrete is stressed to its maximum value of 700 lb/in^2, the other based upon the assumption that the steel is stressed to 16,000 lb/in^2. The desired bending moment is the minimum of these two values.

If the concrete is stressed to its allowable working stress of 700 lb/in^2 then the average stress in the concrete is 700/2 = 350 lb/in^2, since the normal stress varies linearly from 700 lb/in^2 at the upper fibers to zero at the neutral axis. The total compressive force in the concrete is then C = 350(14)(10.1) = 49,500 lb. Since this resultant force acts through the centroid of a triangle (see Problem 4) it is located a distance 10.1/3 = 3.36 in. below the top of the beam. The distance between the action line of C and the tension in the steel is (22 − 3.36) or 18.64 in. The resisting moment based upon the stress in the concrete is

$$M_c \quad = \quad 49{,}500(18.64) \quad = \quad 922{,}000 \text{ lb-in}$$

If the steel is stressed to its allowable working stress of 16,000 lb/in^2 the total tensile force in the steel is T = 16,000(4) = 64,000 lb. The resisting moment based upon the stress in the steel is

$$M_s \quad = \quad 64{,}000(18.64) \quad = \quad 1{,}192{,}000 \text{ lb-in}$$

The allowable bending moment is the minimum of these two values, or 922,000 lb-in.

6. A steel-reinforced concrete beam 15 ft long is supported at the ends and has a rectangular cross-section as shown below. The allowable stresses are 800 lb/in^2 in the concrete and 18,000 lb/in^2 in the steel. Take n = 12, and assume the dead weight of the concrete to be 150 lb/ft^3. Determine the maximum intensity of uniform load w that the beam may carry over its entire length.

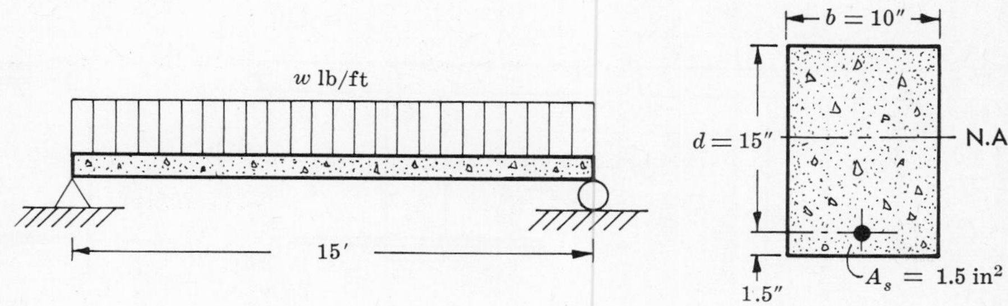

As in Problem 1, the reinforcing steel is replaced by an equivalent area of concrete A_e where

$$A_e \quad = \quad n \cdot A_s \quad = \quad 12(1.5) \quad = 18 \text{ in}^2$$

This equivalent area of concrete may be thought of as a thin horizontal strip located 15 in. below the top of the beam. We next locate the neutral axis from equality of first moments of areas about the axis, namely

$$10(15k)(7.5k) \quad = \quad 18(15 - 15k) \qquad \text{or} \qquad k = 0.385$$

The neutral axis thus lies 0.385(15) = 5.78 in. below the top of the beam.

We shall determine two values of the bending moment, one based upon the assumption that the concrete is stressed to 800 lb/in^2, the other based upon the assumption that the steel is stressed to 18,000 lb/in². The permissible bending moment is then the minimum of these two values.

If the concrete is stressed to 800 lb/in² the total compressive force in the concrete is

$$C = 400(10)(5.78) = 23{,}100 \text{ lb}$$

The distance between the action line of this force and the force in the steel is $15 - 5.78/3 = 13.07$ in. The resisting moment based upon the stress in the concrete is thus

$$M_c = 23{,}100(13.07) = 301{,}000 \text{ lb-in.}$$

If the steel is stressed to 18,000 lb/in² the total tensile force in the steel is $T = 18{,}000(1.5) = 27{,}000$ lb. The resisting moment based upon the stress in the steel is

$$M_s = 27{,}000(13.07) = 352{,}000 \text{ lb-in.}$$

The allowable moment is thus 301,000 lb-in.

The bending moment diagram for this loading has been determined in Problem 6, Chapter 6. From the results found there the bending moment is maximum at the center of the span and is given by

$$M_{max} = w_1 L^2/8$$

where w_1 represents the total load per unit length along the beam, and L denotes the length of the beam. The weight of the beam is appreciable and is included in w_1. The weight of the beam per foot of length is $\frac{10(16.5)}{144}(1)(150) = 172$ lb/ft. It is customary to neglect the difference in weight between the steel rods and the concrete, i.e. consider the beam as if it were entirely concrete.

Substituting in the above expression for M_{max},

$$\frac{301{,}000}{12} = \frac{(172 + w)(15)^2}{8} \qquad \text{or} \qquad w = 720 \text{ lb/ft}$$

7. A concrete beam of T-section has the dimensions shown below. The reinforcement consists of 2.5 in² of steel. A bending moment of 720,000 lb-in acts on the section. Determine the maximum stress in the concrete and also the stress in the steel. Take $n = 10$.

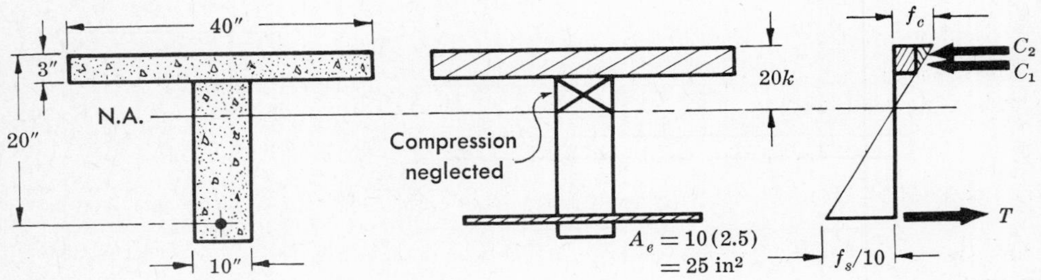

The transformed section is shown in the central diagram. The compression in the vertical stem below the flange is very small compared to that in the flange and will be neglected. We shall locate the neutral axis by considering the equality of first moments of the shaded areas:

$$40(3)(20k - 1.5) = 25(20 - 20k) \qquad \text{or} \qquad k = 0.235$$

The neutral axis thus lies $0.235(20) = 4.7$ in. below the top of the beam.

The variation of bending stresses is shown in the diagram at the right. The compressive stress on the lower edge of the horizontal flange is $(1.7/4.7)f_c = 0.362 f_c$. The total compressive force in the flange may be considered to be the sum of two forces: C_1 acting with a uniform intensity of $0.362 f_c$ over the flange, and C_2 acting with an intensity varying from 0 to $0.638 f_c$. Force C_1 acts at the center of the flange, 18.5 in. from the steel, and force C_2 acts at a distance of 19 in. from the steel.

The force T in the steel may be considered to consist of two parts, $T_1 = C_1$ and $T_2 = C_2$ where $T = T_1 + T_2$. Thus we have two couples acting. The resisting moment corresponding to C_1 is

$$(0.362\,f_c)(3)(40)(18.5) = 803\,f_c$$

The resisting moment corresponding to C_2 is

$$(0.319\,f_c)(3)(40)(19) = 730\,f_c$$

The total resisting moment is the sum of these, or $1533\,f_c$.

Since the resisting moment is equal to the bending moment, we have

$$1533\,f_c = 720{,}000 \qquad \text{or} \qquad f_c = 470 \text{ lb/in}^2$$

The total compressive force is given as the sum

$$[0.362(3)(40)(470) + 0.319(3)(40)(470)] = 38{,}400 \text{ lb.}$$

This is equal to the tensile force T acting in the steel. Consequently the tensile stress in the steel is $38{,}400/2.5 = 15{,}400 \text{ lb/in}^2$.

8. A concrete beam of T-section has the dimensions shown. The reinforcement consists of 3 in^2 of steel. Determine the maximum bending moment that can be carried by the beam. The limiting stresses are $f_s = 20{,}000 \text{ lb/in}^2$ and $f_c = 1350 \text{ lb/in}^2$. Take $n = 10$.

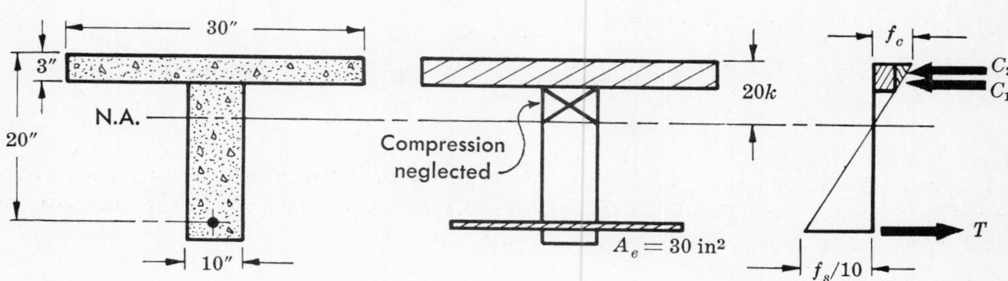

The transformed section is shown in the central diagram. Again, the compression in the stem will be neglected. The neutral axis is located from the equation

$$30(3)(20k - 1.5) = 30(20 - 20k) \qquad \text{or} \qquad k = 0.308$$

The neutral axis thus lies $0.308(20) = 6.16$ in. below the top of the beam.

The compressive stress in the outer fibers of the concrete is denoted by f_c and consequently the stress at the lower edge of the flange is $(3.16/6.16)f_c = 0.514\,f_c$. As in Problem 7 the total compressive force C in the flange is composed of a force C_1 acting with uniform intensity over the flange together with a force C_2 varying linearly over the depth of the flange. The force C_1 is given by

$$C_1 = (0.514\,f_c)(30)(3) = 46.3\,f_c$$

The force C_2 is given by

$$C_2 = (0.243\,f_c)(30)(3) = 21.9\,f_c$$

The total compression C in the concrete is the sum of these, or $68.2\,f_c$. The force C_1 acts at the center of the flange, 18.5 inches from the steel, and C_2 acts 19 inches from the steel. The resultant moment is thus given as the sum:

$$\begin{aligned}(46.3\,f_c)(18.5) &= 856\,f_c \\ (21.9\,f_c)(19) &= \underline{416\,f_c} \\ & \quad 1272\,f_c = \text{Resisting Moment}\end{aligned}$$

The moment arm of the resultant compression in the concrete is thus $(1272\,f_c)/(68.2\,f_c) = 18.7$ in.

If the concrete is stressed to the maximum allowable value of 1350 lb/in^2, then

$$C = 68.2(1350) = 92,000 \text{ lb}$$

If the steel is stressed to its maximum allowable value of 20,000 lb/in^2 then the tensile force T in the steel is given by

$$T = 20,000(3) = 60,000 \text{ lb}$$

This is the smaller of these two values and it determines the limiting bending moment. The maximum bending moment that can be carried is thus 60,000(18.7) = 1,120,000 lb-in.

9. Design a reinforced concrete beam of rectangular cross-section to withstand a bending moment of 700,000 lb-in. The allowable stresses are 20,000 lb/in^2 in the steel and 1350 lb/in^2 in the concrete. Take $n = 12$. The width of the beam is to be 12 in.

As may be seen from Problem 5, the bending moment that produces the maximum allowable stress in the concrete will not necessarily stress the steel to its allowable value. From the standpoint of economy it is desirable that both the concrete and the steel be simultaneously stressed to their maximum allowable values. The beam is then said to have "balanced reinforcement".

The beam cross-section as well as the transformed section are shown below. The problem is to determine d and A_s such that the allowable stresses will be realized simultaneously. That is, the ten-

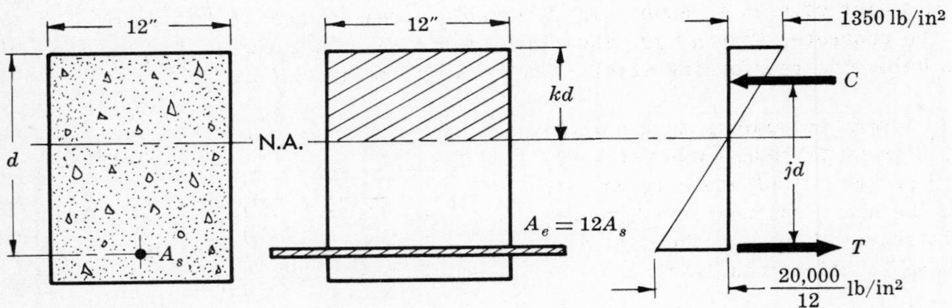

sile stress in the equivalent concrete, which has replaced the steel, is to be 20,000/12 lb/in^2 at the same time that the compressive stress in the concrete is 1350 lb/in^2. From the stress distribution shown above at the right, we have from similar triangles:

$$\frac{1350}{kd} = \frac{20,000/12}{d - kd} \qquad \text{or} \qquad k = 0.448$$

The distance between the forces C and T is thus $(d - 0.448d/3) = 0.850d$. The total compressive force in the concrete is

$$C = (1350/2)(0.448d)(12) = 3640d$$

The resisting moment is thus $(0.850d)(3640d) = 3100d^2$. This must equal the bending moment, hence

$$3100d^2 = 700,000 \qquad \text{or} \qquad d = 15.0 \text{ in.}$$

The compression in the concrete is thus $C = 3640(15) = 54,600$ lb. This is equal to the tensile force in the steel. The area of the steel is

$$A_s = 54,600/20,000 = 2.73 \text{ in}^2$$

10. Design a reinforced concrete beam of rectangular cross-section to withstand a bending moment of 400,000 lb-in. The allowable stresses are 18,000 lb/in^2 in the steel and 700 lb/in^2 in the concrete. Take $n = 10$. Make the depth of the beam equal to twice the width.

This is another problem involving "balanced reinforcement". The allowable stresses in the steel and in the concrete are to be realized simultaneously and thus we have the stress distribution shown in the adjoining figure. From similar triangles,

$$\frac{700}{kd} = \frac{18{,}000/10}{d-kd} \qquad \text{or} \qquad k = 0.280$$

The distance between the forces C and T is thus $(d - 0.280d/3) = 0.907d$. The total compressive force in the concrete is

$$C = (700/2)(b)(0.280d) = 98\,bd$$

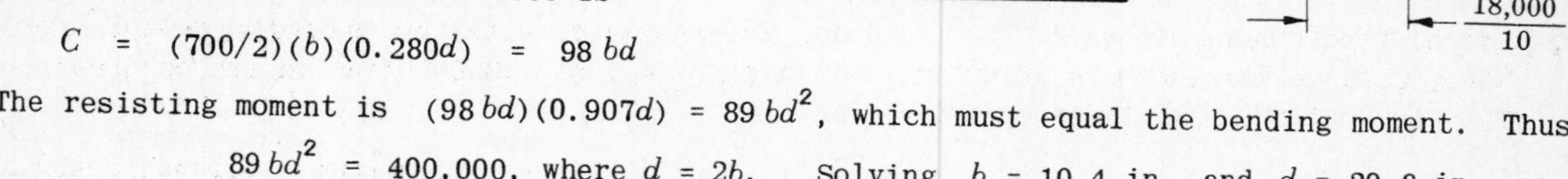

The resisting moment is $(98\,bd)(0.907d) = 89\,bd^2$, which must equal the bending moment. Thus $89\,bd^2 = 400{,}000$, where $d = 2b$. Solving, $b = 10.4$ in. and $d = 20.8$ in.

The compression in the concrete is thus $C = 98(10.4)(20.8) = 21{,}200$ lb. Since this is equal to the tensile force in the steel, the required steel area is

$$A_s = 21{,}200/18{,}000 = 1.18 \text{ in}^2$$

11. Design a reinforced concrete beam of rectangular cross-section to carry a concentrated load of 10,000 lb at the midpoint of a 15 ft span. The allowable stresses are 20,000 lb/in² in the steel and 1000 lb/in² in the concrete. Take $n = 10$. The width of the beam is to be 10 in. and the concrete is to extend 2 in. below the reinforcing steel. The concrete weighs 150 lb/ft³.

Again, it is most economical to use "balanced reinforcement". The allowable stresses are to be realized simultaneously at the midpoint of the span, and there we have the stress distribution shown in the adjoining diagram. From similar triangles,

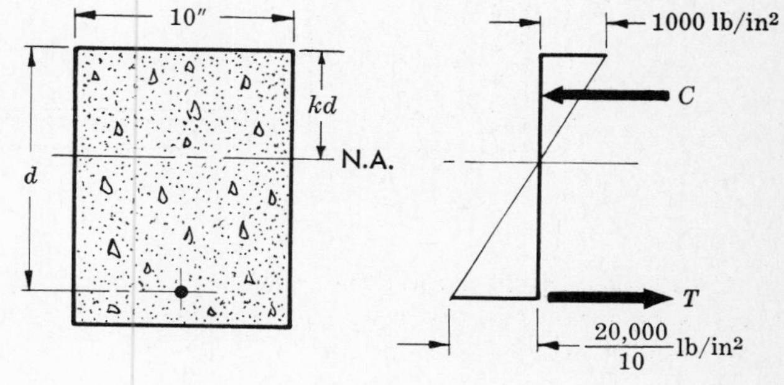

$$\frac{1000}{kd} = \frac{20{,}000/10}{d-kd} \qquad \text{or} \qquad k = 0.333$$

The distance between the forces C and T is thus $(d - 0.333d/3) = 0.889\,d$. The total compressive force in the concrete is

$$C = (1000/2)(10)(0.333d) = 1670\,d$$

The resisting moment is $(1670d)(0.889d) = 1480\,d^2$. From the symmetry of loading this maximum moment occurs at the center of the span.

The bending moment at the center of the span arises from the concentrated load of 10,000 lb as well as from weight of the beam. Due to the concentrated load, we have a bending moment of $10{,}000(15)/4 = 37{,}500$ lb-ft $= 450{,}000$ lb-in at the center of the span. Due to the weight we have a bending moment of

$$M_1 = wL^2/8$$

from Problem 6, Chapter 6. Here w represents the weight of the beam per unit length. Thus

$$w = \frac{10(d+2)}{144}(1)(150) = 10.4(d+2) \text{ lb/ft}$$

and

$$M_1 = \frac{10.4(d+2)(15)^2}{8} = 293(d+2) \text{ lb-ft} = 3520(d+2) \text{ lb-in}$$

Equating the resisting moment at the center of the span to the bending moment there, we have

$$1480\,d^2 = 450{,}000 + 3520(d+2) \qquad \text{or} \qquad d = 18.9 \text{ in.}$$

The compression in the concrete is thus $C = 1670(18.9) = 31{,}600$ lb. Since this is equal to the tensile force in the steel the required steel area is $A_s = 31{,}600/20{,}000 = 1.58 \text{ in}^2$.

12. A reinforced concrete beam is rectangular in cross-section and has a width denoted by b. The depth from the top of the beam to the reinforcing steel is d and the reinforcement consists of steel bars, of total area A_s. Determine the maximum shearing stress in the beam and also the shearing stress over the surface of the reinforcing bars.

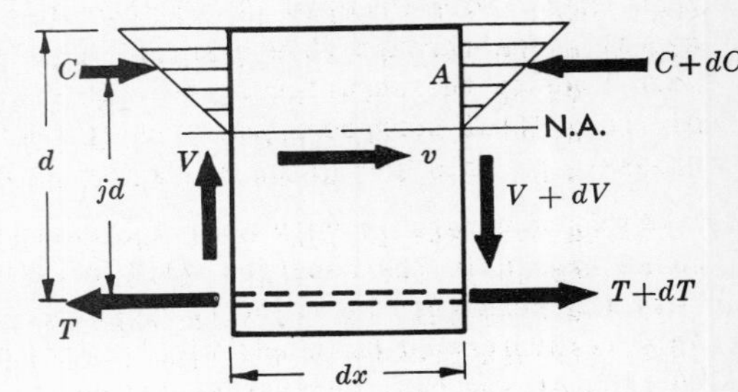

The shearing stress on a horizontal plane through the beam is determined by the same considerations as those given in Problem 21, Chapter 8. From the conclusion reached there, the maximum shearing stress will act over the neutral surface of the beam. This maximum shearing stress is denoted by the symbol v. The forces acting on two adjacent cross-sections appear as in the adjoining diagram. The increment in compression in the concrete between these two sections is denoted by dC. Consequently the shearing stress v over the neutral surface is found from the condition of horizontal equilibrium of the portion above the neutral axis:

$$v \cdot b \cdot dx = dC \qquad \text{or} \qquad v = \frac{1}{b} \cdot \frac{dC}{dx}$$

where b denotes the width of the beam. But the resisting moment is given by $M = C \cdot jd$ from which $\frac{1}{jd} \cdot \frac{dM}{dx} = \frac{dC}{dx}$. Substituting,

$$v = \frac{1}{jbd} \cdot \frac{dM}{dx} = \frac{V}{jbd}$$

where V denotes the transverse shearing force.

Next, summing moments about point A (on the line of action of the forces C and $C + dC$),

$$V \cdot dx + T \cdot jd - (T + dT) jd = 0 \qquad \text{or} \qquad dT = \frac{V\,dx}{jd}$$

But this increment in the tension in the steel is equal to the shearing forces distributed over the surfaces of the reinforcing bars. If Σ_o denotes the sum of the perimeters of the steel reinforcing bars, then the shearing stress over the surface of the bars (denoted by u) is given by

$$u = \frac{dT}{\Sigma_o\, dx} = \frac{V}{jd\,\Sigma_o}$$

This is usually referred to as the bond stress.

13. Consider a reinforced concrete beam of rectangular cross-section where $b = 10$ in., $d = 16$ in., and the reinforcement consists of three square steel bars, each one inch on a side. The maximum transverse shearing force is 7000 lb. Determine the maximum shearing stress in the beam and also the bond stress existing between the steel and the concrete. Take $n = 15$.

In Problem 12 the maximum shearing stress, occurring at the neutral axis, was found to be

$$v = \frac{V}{jbd}$$

This same cross-section was examined in Problem 2 and it was found that $k = 0.52$. Hence $j = 1 - k/3 = 0.828$. Substituting,

$$v = \frac{7000}{0.828(10)(16)} = 53 \text{ lb/in}^2$$

The bond stress acting over the surface of the steel bars was shown in Problem 12 to be

$$u = \frac{V}{jd\,\Sigma_o}$$

Here Σ_o denotes the sum of the perimeters of the reinforcing bars, which in this case is 12 in. Substituting,

$$u = \frac{7000}{0.828(16)(12)} = 44 \text{ lb/in}^2$$

14. Discuss diagonal tension in a reinforced concrete beam. What type of reinforcement should be used to prevent failure due to diagonal tension ?

As mentioned in Problem 12, the maximum horizontal shearing stress in such a beam occurs at the neutral surface. An element located at the neutral surface is subject to pure shear since the normal stresses are zero there. From Problem 13, Chapter 15, it is evident that on a 45° diagonal plane through this element there exists a tensile stress equal in magnitude to the shearing stress. An element cut at 45° to the direction of the neutral surface appears as in the adjoining diagram. The normal stresses on this 45° plane are known as diagonal tensile stresses.

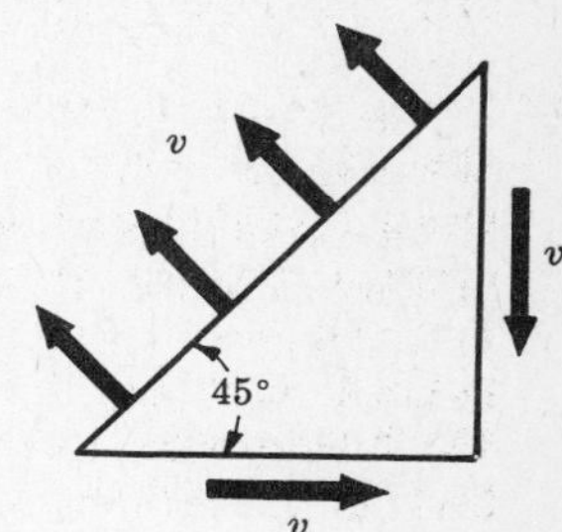

Since concrete is very weak in tension, these tensile stresses would cause cracks in the concrete with eventual failure of the beam. Therefore the beam must be reinforced against these diagonal tensile stresses. This reinforcement is in addition to that employed to strengthen the lower portion of the beam against longitudinal tension. Steel bars at right angles to the direction of cracking of the concrete are of course the most desirable type of reinforcement. However, vertical bars, called stirrups, are often used because of convenience in placing such bars, rather than inclined ones.

In calculating the stresses to be resisted by the diagonal tension reinforcement, it is to be noted that while on the neutral surface the diagonal tension makes an angle of 45° with the horizontal plane of the beam, this is not the case at other points in the beam. An element in the beam is usually subject to longitudinal normal as well as shearing stresses, and consequently the direction of the resultant tension is usually not at 45° with the horizontal. However, for simplicity it is usual to consider the diagonal tension to act at 45° at all points. The vertical stirrups will resist the vertical components of this 45° diagonal tension, and the horizontal components will be resisted by the horizontal reinforcement.

The concrete can resist a working shearing stress (and hence a diagonal tensile stress) equal to $0.03 f_c'$, where f_c' is the 28 day compressive ultimate strength. The diagonal tensile stress v' to be taken by the steel is thus

$$v' = v - 0.03 f_c'$$

Frequently it is assumed that $v' = 2v/3$. The vertical stirrups have the appearance shown in the adjoining figure. The spacing between stirrups is denoted by s and the total allowable force in one stirrup by P. From equilibrium we readily have

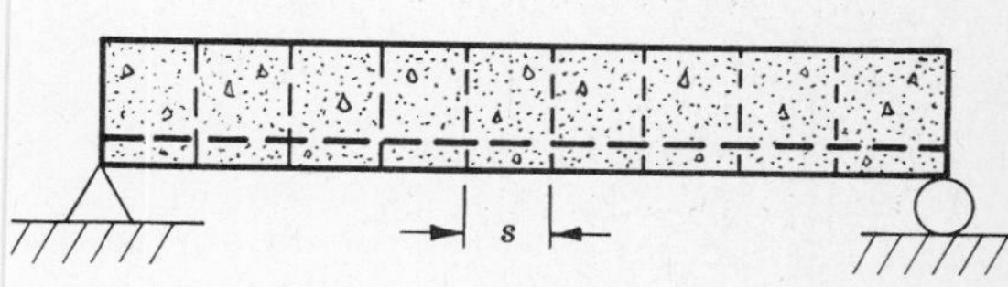

$$v' \cdot b \cdot s = P \qquad \text{or} \qquad s = \frac{P}{v'b}$$

This determines the spacing of the stirrups. The value of P is determined as the product of the cross-sectional area of the stirrup and the allowable tensile stress in the stirrup.

15. A reinforced concrete beam 10 in. wide and 14 in. in effective depth is subject to a transverse shear of 12,000 lb. What should be the spacing of 3/8 in. diameter vertical stirrups, allowing a unit shear of 40 lb/in^2 in the concrete and tensile stress of 16,000 lb/in^2 in the stirrups ?

From Problem 12 the maximum shearing stress, occuring at the neutral surface, is

$$v = \frac{V}{jbd} = \frac{12,000}{(7/8)(10)(14)} = 98 \text{ lb/in}^2$$

where an average value of $j = 7/8$ is satisfactory for all shearing computations.

The diagonal tension to be taken by the steel is thus $v' = 98 - 40 = 58$ lb/in^2

The allowable force in one U-shaped stirrup is $P = \frac{\pi}{4}(\frac{3}{8})^2(2)(16,000) = 3550$ lb.

From Problem 14 the spacing s of the stirrups is $s = \frac{P}{v'b} = \frac{3550}{58(10)} = 6.1$ in.

16. Determine the axial stresses in the steel and in the concrete of the tied column shown. The axial load is 200,000 lb and $n = 10$.

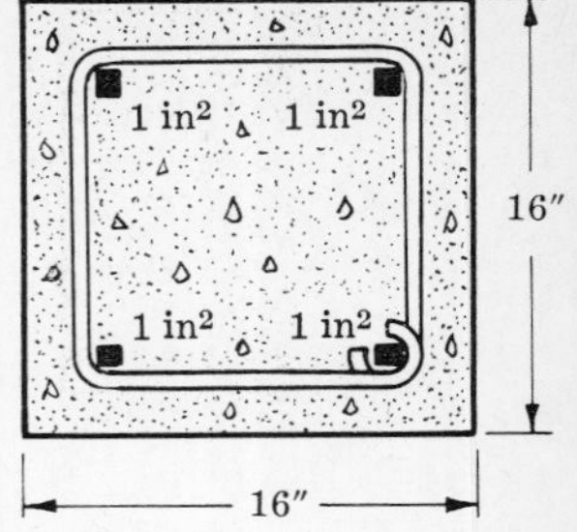

Again the transformed section technique is convenient. The 4 in² area of reinforcing steel is transformed into an equivalent concrete area of $10(4) = 40$ in², a net increase in area of 36 in². The effective area of concrete to be considered is now $(256+36) = 292$ in².

The axial stress in the concrete is $200,000/292 = 685$ lb/in² and that in the steel is $10(685) = 6850$ lb/in².

Lateral movement of the steel rods is prevented by ties in the form of round bars bent into a square contour and surrounding the reinforcing steel as shown. These ties are placed at regular intervals along the length of the column. For example, the Joint Committee on Standard Specifications for Concrete and Reinforced Concrete stipulates the maximum spacing of ties made from ¼ in. round bars to be 12 in.

17. Design a square reinforced concrete tied column to support a concentric load of 300,000 lb. The 28 day strength of the concrete is 3000 lb/in² and the working stress in the steel is 20,000 lb/in². Take $n = 8$ and let the cross-sectional area of the steel be four percent of the gross area of the concrete section.

According to the specifications of the Joint Committee on Standard Specifications for Concrete and Reinforced Concrete the allowable column load P is given by

$$P = 0.8(0.225 f_c' A_g + f_s A_s)$$

where f_c' = 28 day strength of the concrete

A_g = gross cross-sectional area of concrete

f_s = 40 percent of the elastic limit of the steel, usually taken to be 20,000 lb/in² for hard grade steel

A_s = cross-sectional area of reinforcing steel.

Substituting, $300,000 = 0.8[0.225(3000)A_g + 20,000(0.04 A_g)]$ or $A_g = 254 \text{ in}^2$

A suitable concrete section is thus $\sqrt{254}$ or 15.92 in., say 16 inches, on a side.

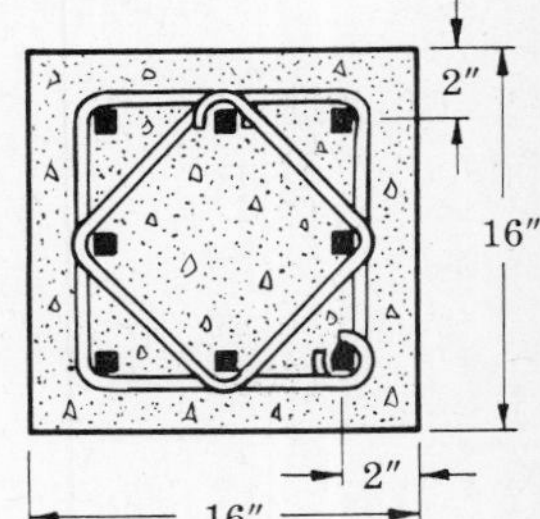

The area of the reinforcing steel is $A_s = 0.04(254) = 10.2$ in². This required area is provided by using 8 square steel bars, each 1-1/8 in. on a side. These may be arranged so that a cross-section has the appearance as shown in the adjoining diagram. Actually, these eight rods must be restrained against lateral movement and for this reason ties are added at intervals along the length of the column. These ties consist of small-diameter steel bars and each tie lies in a cross-section as shown.

SUPPLEMENTARY PROBLEMS

18. A rectangular concrete beam has the dimensions as shown in Fig. (*a*) below. The beam is reinforced by six circular steel bars, each 1-1/4 in. in diameter. Determine the location of the neutral axis. Take $n = 15$. *Ans.* 14.6 in. below the top of the beam.

19. A concrete beam of T-section has the dimensions as shown in Fig. (*b*) below. The beam is reinforced by three circular steel bars, each 1-1/4 in. in diameter. Determine the location of the neutral axis neglecting the compression in the vertical stem below the flange. Re-calculate taking these compressive stresses into account. Take $n = 15$.
Ans. 8.05 in. below the top of the beam. 7.96 in. below the top of the beam.

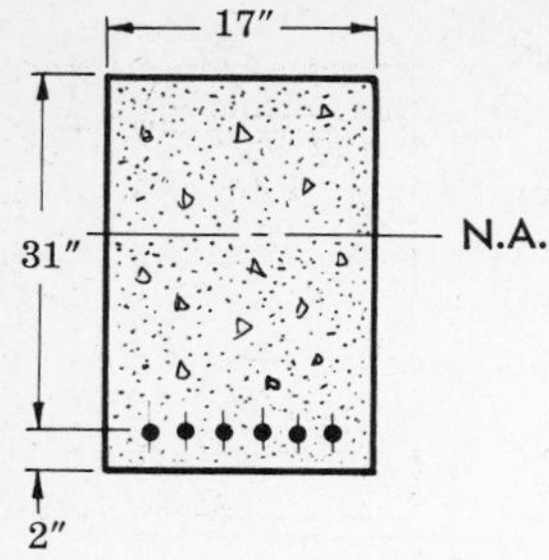

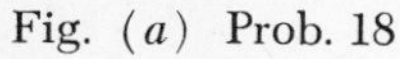

Fig. (a) Prob. 18

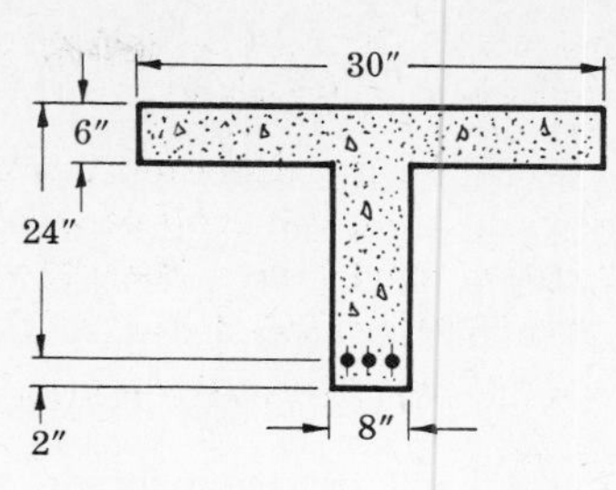

Fig. (b) Prob. 19

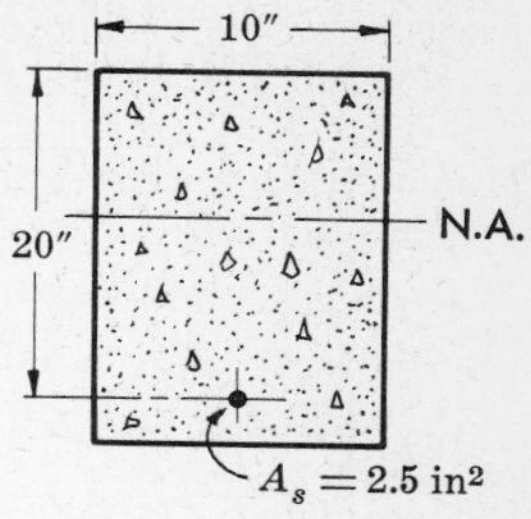

Fig. (c) Prob. 20

20. A rectangular concrete beam has the dimensions as shown in Fig. (c) above. The beam is reinforced by steel bars having a total cross-sectional area of 2.5 in^2. The beam is subject to a bending moment of 60,000 lb-ft. Determine the maximum stress in the concrete and also the stress in the steel. Take $n = 15$. *Ans.* $f_c = 940$ lb/in^2, $f_s = 17{,}000$ lb/in^2

21. A reinforced concrete beam of rectangular cross-section has a width of 7 in., an effective depth d of 11 in. and the total area of the reinforcing steel is 1.5 in^2. If the allowable working stress in the concrete is 600 lb/in^2 and that in the steel is 18,000 lb/in^2, determine the maximum bending moment that the beam can carry. Take $n = 15$. *Ans.* 110,000 lb-in

22. A steel-reinforced concrete beam 20 ft long is supported at the ends and has a rectangular cross-section. The width of the beam is 12 in., the effective depth from the top to the reinforcing steel is 18 in., the concrete extends 1-1/2 in. below the steel, and the area of the reinforcing steel is 2.5 in^2. The allowable stresses are 700 lb/in^2 in the concrete and 16,000 lb/in^2 in the steel. Consider $n = 15$ and take the dead weight of the concrete to be 150 lb/ft^3. Determine the maximum concentrated force that may be applied at the center of the beam. *Ans.* 6060 lb

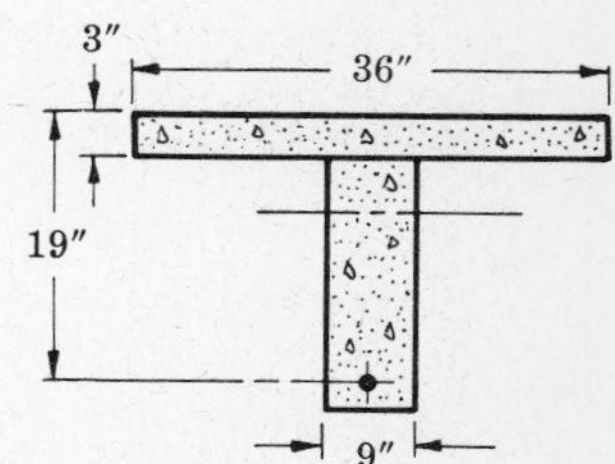

23. A concrete beam of T-section has the dimensions as shown in the adjacent figure. The reinforcement consists of 3 in^2 of steel. A bending moment of 850,000 lb-in acts on the section. Determine the maximum stress in the concrete and also the stress in the steel. Take $n = 10$.
Ans. $f_c = 620$ lb/in^2, $f_s = 16{,}000$ lb/in^2

24. Let us consider a T-section similar to that of Problem 23. The flange width is 27 in., the flange thickness 3.5 in., and the effective depth from the top of the beam to the reinforcing steel is 31 in. The reinforcement consists of 7 in^2 of steel. A bending moment of 3,300,000 lb-in acts on the section. Determine the maximum stress in the concrete and also the stress in the steel. Take $n = 10$.
Ans. $f_c = 1360$ lb/in^2, $f_s = 16{,}100$ lb/in^2

25. A concrete T-beam has a flange width of 40 in., a flange thickness of 3 in., and an effective depth from the top of the beam to the reinforcing steel of 24 in. The reinforcement consists of 2 in^2 of steel. Determine the maximum bending moment that can be carried by the beam. The allowable stresses are $f_s = 20{,}000$ lb/in^2 and $f_c = 1350$ lb/in^2. Take $n = 15$. *Ans.* 906,000 lb-in

26. Design a reinforced concrete beam of rectangular cross-section to withstand a bending moment of 350,000 lb-in. The allowable stresses are 18,000 lb/in^2 in the steel and 900 lb/in^2 in the concrete. The width of the beam is to be 8 in. Take $n = 15$. *Ans.* $d = 16.3$ in., $A_s = 1.40$ in^2

27. Design a reinforced concrete beam of rectangular cross-section to withstand a bending moment of 350,000 lb-in. The allowable stresses are 18,000 lb/in^2 in the steel and 900 lb/in^2 in the concrete. The depth of the beam is to be 1-1/2 times the width. Take $n = 15$.
Ans. $b = 9.8$ in., $d = 14.7$ in., $A_s = 1.54$ in^2

28. Design a reinforced concrete beam of rectangular cross-section to support a uniform load of 1000 lb/ft. The beam is 20 ft long and 12 in. in width. The allowable stresses are 18,000 lb/in^2 in the steel and

900 lb/in^2 in the concrete. Take $n = 12$. The concrete is to extend 2 in. below the reinforcing steel. The concrete weighs 150 lb/ft^3. *Ans.* d = 20.9 in., A_s = 2.34 in^2

29. A reinforced concrete beam of rectangular cross-section has a width b = 17 in. and a depth d = 31 in. from the top of the beam to the reinforcing steel. The reinforcement consists of six circular steel bars each 1-1/4 in. in diameter. Take $n = 15$. The maximum transverse shearing force is 25,000 lb. Determine the maximum shearing stress in the beam and also the bond stress existing between the steel and the concrete. *Ans.* v = 56.6 lb/in^2, u = 40.5 lb/in^2

30. Determine the axial stress in the steel and in the concrete of the tied column shown in the adjacent figure. The axial load is 125,000 lb and $n = 12$. *Ans.* f_c = 740 lb/in^2, f_s = 8900 lb/in^2

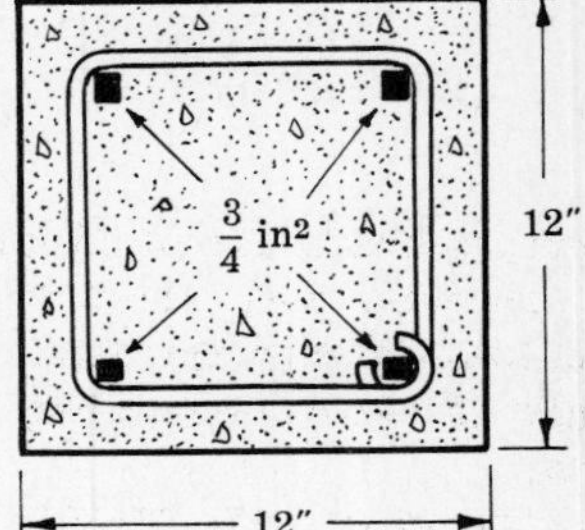

31. Design a square reinforced concrete tied column to withstand a concentric load of 500,000 lb. The 28 day strength of the concrete is 3000 lb/in^2 and the working stress in the steel is 20,000 lb/in^2. Take $n = 8$ and let the cross-sectional area of the steel be three percent of the gross area of the concrete section.
Ans. A 22-1/2 inch square section reinforced by 10 1-1/4 in. square steel bars.

INDEX

NOTES

NOTES

NOTES

NOTES

NOTES

NOTES

NOTES

NOTES

NOTES

NOTES

NOTES

NOTES